Pharmacology for Chemists

Drug Discovery in Context

Pharmacology for Chemists

Drug Discovery in Context

Edited by

Raymond Hill

Imperial College London, UK
Email: raymond.hill@imperial.ac.uk

Terrry Kenakin

University of North Carolina, USA
Email: kenakin@email.unc.edu

and

Tom Blackburn

TPBioventures LLC, USA
Email: tblackburn@tpbioventures.com

THE QUEEN'S AWARDS
FOR ENTERPRISE:
INTERNATIONAL TRADE
2013

Print ISBN: 978-1-78262-142-3
EPUB ISBN: 978-1-78801-245-4

A catalogue record for this book is available from the British Library

The Royal Society of Chemistry is a charity, registered in England and Wales, Number 207890, and a company incorporated in England by Royal Charter (Registered No. RC000524), registered office: Burlington House, Piccadilly, London W1J 0BA, UK, Telephone: +44 (0) 207 4378 6556.

Visit our website at www.rsc.org/books

Printed in the United Kingdom by CPI Group (UK) Ltd, Croydon, CR0 4YY, UK

Foreword

I am delighted to contribute this Introduction to *Pharmacology for Chemists,* in which experienced authors cover the drug discovery and development process from laboratory to clinic. I am sure this book will become essential reading for newcomers and experienced scientists across the drug-discovery continuum, as well as medicinal chemists. I have always believed that drug-discovery scientists should have an integrated understanding of the range of challenges involved in translating new ideas into clinical candidates and that they should appreciate the strengths and limitations of all the disciplines involved. *Pharmacology for Chemists* is a major step forward in meeting these ambitions.

Before speculating on the future it is worth analysing our past achievements and then laying out the future challenges before us. Dramatic progress has been achieved in overall healthcare quality over the past fifty years with life expectancy increasing significantly in both developed and developing nations, and major medical threats such as hypertension, elevated cholesterol and HIV AIDS have been brought under better control, for example. However, we cannot rest on our laurels as the World's population increases and ages, and serious conditions such as cardiovascular disease (CVD), cancer, diabetes and obesity are increasing on a global scale. Forty percent of Europeans will be affected by some sort of brain disorder and anti-microbial resistance continues unabated with the WHO predicting impending disasters, but only four new classes of antibiotics have been introduced over the past forty years. In the face of these formidable challenges, the relentless demand for new medicines will continue unabated, and new paradigms for research focus, collaboration and funding will be required.

Pharmacology for Chemists: Drug Discovery in Context
Edited by Raymond Hill, Terry Kenakin and Tom Blackburn
© The Royal Society of Chemistry 2018
Published by the Royal Society of Chemistry, www.rsc.org

The past decade has seen tremendous consolidation in the pharmaceutical sector where the negative impact of mergers and acquisitions on productivity has finally sunk in. Innovative research cannot be properly nurtured in massive organisations with multi-billion budgets spread over numerous locations and with ever-changing leadership. Technology can be expanded in a modular and global manner, but innovation simply does not scale. For the future, such large groups should be broken down into nimble multi-disciplinary teams, which should be largely autonomous but accountable, with a move away from consensus management and upward decision making. Drug discovery should become a personal and shared experience not a metrics-driven mechanical event, and organisation should be driven by critical mass not absolute scale.

Decentralisation will continue apace as Pharma strives to control fixed costs by externalising routine activities to CROs while working more closely with academic communities. Some disease areas will become virtual with Pharma scientists located in academic and bio-tech laboratories in order to provide early access to new biology, but translation to successful drug discovery projects still has to be realised and inevitable tensions between publications and IP resolved. However, such interactions may address the serious concern that over fifty percent of academic publications cannot be repeated as proper quality control will be demanded before committing significant industry resource. It will also be important to safeguard core expertise in Pharma as successful collaborations require complementary intellectual contributions from both partners, with coherence on objectives. A development-only industry would lose unique and valuable synergies.

Pre-competitive collaborations will become more important as cost constraints force reductions in budgets, risk and duplication. It seems reasonable to assume that most Pharma portfolios share high similarity with multiple parallel approaches to the same targets, often with similar scaffolds. For example, various companies took neuro-kinin and endothelin antagonists to the clinic with little to show for it, while the cumulative costs of parallel renin programmes was absolutely staggering. Such duplicative failures might be avoided through pre-competitive collaborations between industry and academia for target validation and identifying patient populations who would respond to new mechanisms of action. Such concerted efforts should have significant impact on candidate attrition, which is currently unsustainable and continually erodes the very foundations of drug discovery and development. If validated targets and patient subsets do enter the public domain, then robust IP will depend largely on innovative medicinal chemistry, which will become too valuable to contract out.

Some twelve years ago, I suggested to a sceptical audience that the future pharmaceutical industry would be located largely in the US with outposts in Europe and Japan, which appears to have come to pass. Household names have been consigned to the past, thousands of jobs have been lost, state-of-the-art research facilities closed and Pharma's capacity to meet the medical needs of the 21st century compromised. Taking refuge in the record number of FDA approvals in 2016 may be short sighted as half target rare diseases, and not the chronic conditions that afflict the majority of the population. Given the vagaries of VC funding, it seems unlikely that Biotech will make up all the shortfall, or that pharmaceutical R&D would make a major shift Eastwards in the foreseeable future.

In order to regain capacity, now would be an opportune time to strengthen drug discovery in the public sector by co-localising industry-experienced medicinal chemists alongside world-class biologists and clinicians with a real commitment to the discovery of new medicines. In many cases, a fundamental change in mind set will be required for medicinal chemists to be accepted as equal partners, rather than a service function, and critical mass will be required not just one or two chemists here or there. Of course, there are research institutes and academic groups focused on drug discovery, but not on the scale required and often focused on cancer or neglected disease. Integration of Pharma veterans within the wider community would take some time as there is little appreciation of the skills base required for medicinal chemistry but at steady state, barriers between "academic" and "industry" researchers may soften with increased permeability across previously defined disciplines and sectors without compromising quality control. Long-term investment in the most challenging disease areas such as antibacterials and neuroscience would be encouraged and there would also be important roles for Public–Private Partnerships. Overall, there is a strategic and pressing need to strengthen drug discovery initiatives outside of Pharma and Biotech and concerted efforts from interested parties will be required to ensure that research capabilities are commensurate with future medical needs.

Sir Simon Campbell

Contents

Pharmacology for Chemists: Drug Discovery in Context
Edited by Raymond Hill, Terry Kenakin and Tom Blackburn
© The Royal Society of Chemistry 2018
Published by the Royal Society of Chemistry, www.rsc.org

3 Structure-based Drug Discovery and Advances in Protein Receptor Crystallography   45
Miles Congreve and Fiona H. Marshall

4 Actions of Drugs on the Autonomic Nervous System   73
Thomas P. Blackburn

12   Predicting Dose and Selective Efficacy in Clinical Studies from Preclinical Experiments: Practical Pharmacodynamics   483
R. G. Hill

Subject Index   504

1 What is Pharmacology?

Humphrey Rang

British Pharmacological Society, London, UK
Email: humphrey.rang@bps.ac.uk

1.1 Introduction

Pharmacology is the study of the action of drugs on living systems –
neatly paraphrased as the chemical control of physiology and
pathology. It lies at the interface of chemistry and biology. *Drugs*, in
this context, are chemicals of known structure that are administered
as external agents – whether deliberately or accidentally – to the
organism, and produce an observable effect on its function. Living
organisms are, of course, complex chemical machines, which produce
and use many of their own chemicals as a means of controlling their
own functions. Not surprisingly, exposure to other chemicals (*i.e.*
drugs) from the outside world is liable to confuse and subvert the
internal signals, and that in essence is what pharmacology is all
about. An understanding of pharmacology plays an essential role in
the discovery and application of drugs as therapeutic agents, where
the aim is to provide benefit to individuals by the alleviation of
symptoms and disabilities, improved prognosis, prolongation of life,
or disease prevention. A drug that does none of these things, even
though it has been exquisitely engineered to interrupt what was
thought to be a key step in the pathogenesis of the disorder, is of no
use as therapy, though it may prove to be a valuable research tool.

Pharmacology comprises two main components, namely
pharmacodynamics, which is concerned with the effects that drugs

Pharmacology for Chemists: Drug Discovery in Context
Edited by Raymond Hill, Terry Kenakin and Tom Blackburn
© The Royal Society of Chemistry 2018
Published by the Royal Society of Chemistry, www.rsc.org

produce on living systems (*i.e.* what the drug does to the body), and *pharmacokinetics*, which describes the mechanisms by which the drug is absorbed, distributed, metabolised and excreted (*i.e.* what the body does to the drug). To explain fully the effects of a drug in an intact organism, both need to be understood.

Given the extreme chemical complexity of living organisms, and the delicately balanced regulatory mechanisms that have evolved over millions of years to allow organisms to survive environmental threats, it is not surprising that the intrusion of a foreign chemical is, in general, more likely to do harm than good.[†] The aim of drug discovery research is to find those few compounds that – against all odds – can deliver benefit to individuals affected by disease. Therapeutic benefit depends not only on choosing the right compound, but delivering it in the right dose, to the right patient, by the best route, at the right time and under the right circumstances. These important aspects are the concern of the subdiscipline of clinical pharmacology.

When developed as therapeutic agents, drugs are incorporated into *medicines*, which normally include other substances to enable them to be administered as pills, solutions for injection, skin patches, aerosols or other dosage forms (Box 1.1).

1.2   Origins and Antecedents

The word "pharmakon" in ancient Greek, could mean a medicine or a poison; in the ancient world there was little distinction. Attempts to heal the sick by the use of "medicines" – largely herbal and mineral in origin, and based on spiritualist dogma rather than science – was in the hands of spiritual healers, and the tradition survives to the present day, carried on by "medicine men" and "witch doctors", whose stock-in-trade is not only to heal the sick, but also to inflict harm on enemies. Compilations of such remedies go back thousands of years and it was the task of healers to produce them according to closely guarded recipes. Such traditional practices, based on dogma rather than science, remain popular to this day, despite the fact that evidence that they deliver benefit is generally weak or nonexistent. Prior to the 19th century, even though "medicines" based on traditional dogma had been catalogued and used for thousands of years,

---

[†]We do not expect to correct a fault in, say, the navigation system of an aircraft by spraying a chemical into the works, so it may seem remarkable that physiological malfunction can sometimes be put right by a circulating drug. What makes it possible is that living systems, unlike electronic ones, deploy chemical signalling to control their function, providing points of attack for chemical interventions.

Box 1.1 Some definitions

Pharmacology. The study of the action of drugs on living systems. **Clinical pharmacology** is a branch of pharmacology concerned with the action of drugs used clinically to treat patients.

Drug. A substance of known chemical structure that produces a functional effect when added exogenously to a living system. Many endogenous chemical mediators that regulate normal physiological functions in higher animals can also be administered as drugs, but most drugs are synthetic chemicals or natural products not found in higher animals.

Medicine. A preparation containing one or more drugs, designed for therapeutic use. Medicines usually contain additional materials to improve their suitability for clinical use, for example as an injectable solution, a pill to be swallowed, or an ointment for topical application.

Therapeutic efficacy. The disease-related benefit to humans that a medicine produces. The benefit may be:
- relief of symptoms or disabilities associated with disease;
- improved prognosis, *i.e.* slowing or reversal of the progress of the disease;
- prolongation of life;
- prevention of disease.

---

the understanding of drug action in terms of scientific principles – in other words, the emergence of pharmacology – was impossible. Chemistry had not yet advanced to the point of defining compounds in terms of their structure; physiology and pathology could not yet describe the functioning and malfunctioning of the body; traditional teaching emphasised the importance of esoteric procedures for concocting herbal preparations as a prerequisite for their clinical use. There were, however, a few earlier breaks with tradition – discoveries that found application in modern medicine. For example, Thomas Sydenham (1624–89) described in 1666 the use of opium (containing morphine) to control pain and "Jesuit's bark" (containing quinine) to control "intermittent fever" (*i.e.* malaria); William Withering (1741–99) described the use of foxglove (digitalis) to treat "dropsy" (heart failure) in 1785. But until the nineteenth century these preparations were generally assumed to owe their properties to the vital forces possessed by "organic" substances. Chemistry did not come into it, nor did any understanding of biological mechanism.

Materia medica – the description of natural products and their medicinal uses, based on beliefs passed from generation to generation down the ages – began to be commercialised in the mid-17th century by the apothecary's trade, the forerunner of the modern pharmaceutical industry, its aim being to satisfy the demand for medicines containing ingredients that were difficult to obtain, and prepared in an approved manner. At the same time, the "Age of Enlightenment", a gradual shift from dogma to science in the practice of medicine began to grow, when luminaries such as Robert Boyle (in chemistry) and William Harvey (in physiology) started to use evidence based on careful observation and experiment, as opposed to received wisdom, as a basis for understanding the natural world. The idea of living organisms as machines governed by the same physical laws as everything else in the material world, and of chemistry as the underlying basis of every substance and structure, slowly took root over the next two centuries, and dogmatic beliefs began to be challenged by discoveries based on empirical observations, many of which have stood the test of time.

1.3 The Emergence of Pharmacology as a Science

Pharmacology as a distinct biomedical discipline began in 1847, when Buchheim (1820–1879) established the first university department with that name in Dorpat. His was a bold vision, for at that time the medicines in use were mainly plant extracts of unknown composition and a few, mostly poisonous, inorganic compounds, such as salts of mercury and arsenic. Complicated mixtures were recommended, prepared and administered in accordance with elaborate rituals. Vomiting, sweating, diarrhoea and fever commonly resulted, and were regarded as evidence of the treatment's effectiveness in ridding the body of harmful "toxins". **A famous quotation from an eminent contemporary physician, Oliver Wendell Holmes in 1860 dismissed them thus: "If all the materia medica, as currently used, could be thrown into the sea, it would be the better for mankind, and the worse for the fishes".**[1] It is certainly true that the "remedies" in use at that time were not based on any understanding of how they produced their effects, or of the underlying pathological dysfunction that needed to be corrected (beyond the "toxins" notion mentioned above), and the idea of testing their therapeutic efficacy rarely surfaced. Nevertheless, Buchheim saw that the challenge, then as now, was to understand better the mechanisms by which they produced their effects in order to put their medicinal uses on a rational, and hopefully effective, basis. Even though organic chemistry had hardly come into being, he had the vision to see that physiology, pathology and chemistry were all advancing rapidly in the very fertile scientific environment of the mid-nineteenth century, to a point where a new interdisciplinary science could emerge.

1.3.1 Chemistry Makes its Entry

One of the essential foundations of pharmacology – the use of structural formulae to define chemical compounds – did not exist until the middle of the nineteenth century. A key figure was August Kekule, a German chemist who described both the tetravalent nature of carbon and the aromatic ring structure of benzene, two essential principles that allowed accurate structural formulae of organic molecules to be produced for the first time. The idea that the biological effects of plant extracts was likely to result from their chemical constituents rather than mysterious vital forces was implicit in the work of Buchheim and other pharmacological pioneers in the nineteenth century, but it was not until 1905 that a German pharmacist, Serturner, isolated crystals of morphine from opium poppies, tested it on himself and nearly died – the first irrefutable evidence that opium worked by chemistry, not by magic. Serturner's achievement was followed quickly by others who similarly extracted and purified chemical compounds from medicinal plants and showed them to possess distinctive pharmacological properties. Studying how such plant-derived substances as nicotine, atropine, curare, strychnine and ergot alkaloids, produce their effects, and relating them to the emerging knowledge of physiology and pathology, gave pharmacology the scientific foundations that it needed. But still, at this time, synthetic compounds, as opposed to natural products, played only a very minor part.

1.3.2 Pathology and Physiology Lay Important Foundations

The other essential foundations of pharmacology, namely physiology and pathology, also flourished in the nineteenth century. Some key milestones are worth noting. The cell theory, proposed by the German pathologist Rudolf Virchow (1821–1902), identified the cell as the fundamental unit of all living organisms, and proposed that cellular dysfunction – cells dying, dividing, migrating, or otherwise functioning incorrectly – was the basic cause of disease. Louis Pasteur (1822–1896), a French chemist, proposed the germ theory of infection in 1878, having famously demonstrated the role of air-borne micro-organisms in fermentation. He showed that cholera could be transferred from chicken to chicken by inoculation with fresh, but not "stale" material. In fact, inoculation with stale material actually protected against future infection, so Pasteur inadvertently discovered the phenomenon of immunisation.

The first physiological studies aimed at pinning down the site of action of poisons were performed by Francois Magendie (1783–1855) in Paris, who showed that the convulsant action of strychnine was due to its action on the spinal cord, rather than elsewhere in the brain, nerves or muscles. His pupil, Claude Bernard (1813–1878) used a similar anatomical approach to pinpoint the paralysing effect of the arrow-poison curare to the junction between motor nerve and muscle fibre.

1.3.3 The Receptor Concept is Established

The realisation that drugs act very specifically at precise anatomical sites led in later years to the search for cellular components and actual molecules involved. A particularly important stage in the development of modern pharmacology was the emergence of the receptor concept, pharmacology's Big Idea (reviewed by Rang, 2006[2]) of a "lock and key" mechanism by which drug molecules act on specific cellular molecules to produce their effects. This idea had been expressed, in a philosophical way, centuries earlier: **"Did we but know the mechanical affections of the particles of rhubarb, hemlock, opium and a man......we should be able to tell beforehand that rhubarb will purge, hemlock kill and opium make a man sleep......"[3] (John Locke, 1690, Essay concerning human understanding)**. These "mechanical affections", which we would today call chemical interactions, are what pharmacology is all about. The Cambridge physiologist, J. N. Langley (1852–1925), first used the term "receptive substance" in 1878 to describe the hypothetical endogenous substance in salivary glands with which pilocarpine (which causes salivary secretion) and atropine (which blocks pilocarpine's action) combine and compete with each other for binding. A. V. Hill (1886–1977), a student in Langley's laboratory, applied the Law of Mass Action to describe quantitatively the interaction of drug and receptor molecules, and this quantitative approach was further developed by later pharmacologists. We now know that specific receptors exist in all cells and tissues, and are key players in the numerous chemical signalling pathways used by all living organisms to control their physiological functions. The subversion of these signalling pathways by introducing alien chemicals is the basis is modern pharmacology. Understanding pharmacology in this mechanistic way, underpinned now through the application of molecular biological approaches and by detailed knowledge of the structure and function of receptor molecules, has become crucial for the discovery of new therapeutic drugs.

Receptors form one important class of targets for therapeutic drugs. Enzymes, transporter molecules, ion channels, *etc.* are other types of target, described elsewhere in this book. Identifying such drug targets, and explaining how drugs are able to act on them to influence the function of the cells and tissues that express them, is a central theme in modern pharmacology, and an important starting point for drug discovery and therapeutic innovation.

1.3.4 Many Chemical Mediators are Identified

The idea that internal secretions – chemical substances liberated into the bloodstream by organs such as the thyroid gland, testis and liver – play an important physiological role, emerged in the 17th century, and slowly gained ground over the next 200 years. The term "hormone" was coined in 1905 by Bayliss and Starling, who showed that the duodenum, in response to gastric acid production, produced a substance "secretin" that caused the pancreas to release digestive enzymes. Around the same time, many physiologists described the effects of removing individual glands – adrenal, thyroid, pituitary, pancreas, *etc.* – and of injecting various gland extracts, on different physiological functions, though chemical techniques for isolating and identifying the mediators involved were not yet available. The realisation that the release of chemical transmitters is the mechanism by which nerve cells communicate with each other, and with other cells and tissues came initially from work by Dale and Loewi, who identified acetylcholine as the transmitter released by parasympathetic nerve endings (winning the Nobel Prize in 1936), and the identification of other chemical mediators became – and remains – a major focus of pharmacology. Many large families of mediators are now recognised. These include:

- low molecular weight amines, such as acetylcholine, noradrenaline, histamine, dopamine, 5-hydroxytryptamine;
- peptides, such as insulin, oxytocin and angiotensin;
- protein mediators, such as growth hormone, interferon and a wide variety of cytokines;
- lipid mediators, such as prostaglandins and leukotrienes;

- steroids, such as oestrogens and adrenocortical hormones;
- amino acids, such as glutamate, glycine and γ-amino butyric acid (GABA);
- purines, such as adenosine, ADP and ATP;
- small molecules such as nitric oxide, carbon monoxide and hydrogen sulfide.

New members of each of these groups of mediators are still being discovered, particularly in the protein group, where modern techniques in molecular biology, cell biology and genomics have made a big impact in recent years.

1.4   Receptors and Drug Targets

From a pharmacological perspective, this profusion of mediators gives rise to many potential drug targets. Each mediator acts by binding to a specific recognition site located on a protein – the receptor – through which it produces its physiological effects. The four main types of receptor are:

- G-protein coupled receptors (GPCRs);
- ligand-gated ion channels;
- kinase-linked receptors;
- nuclear receptors.

Apart from nuclear receptors, most receptors are proteins that span cell membranes, accessible to mediators acting on the extracellular surface, and controlling events within the cell.

Most mediators act on more than one receptor, producing different effects on different cells. GPCRs are a particularly large and important type; approximately 350 GPCRs for endogenous mediators have been identified in the human genome, and roughly half the drugs in clinical use target GPCRs. Ligand-gated ion channels, activated by transmitters such as acetylcholine and glutamate, are important mainly in mediating fast synaptic neurotransmission. Kinase-linked receptors, located on the cell surface, respond to mediators such as growth factors, cytokines and insulin. Activation of the intracellular kinase moiety of the receptor protein initiates a cascade of protein phosphorylation reactions within the cell, culminating in the functional response. Nuclear receptors are intracellular proteins, responding to mediators such as steroids and thyroid hormones that are able to enter the cell; these receptors act by controlling gene expression in the nucleus.

Drug targets are the endogenous molecules (mostly proteins) to which drug molecules bind as the first step in producing their pharmacological effects. The various type of receptors for endogenous mediators, described above, constitute one important class of drug targets, but other types of functional protein, and also DNA, are also important.

1.5   Pharmacology in Drug Discovery

The pathologist, Paul Ehrlich (1854–1915), was impressed by the ability of chemical dyes to stain biological specimens in a very specific way, and argued that this selective binding to particular cell types might be used as a basis for finding drugs that would bind to and kill pathogenic organisms. Arsenic, in various forms, had been used as a poison and a medicine for thousands of years, so Ehrlich, working with an organic chemist, embarked on the first systematic attempt at drug discovery by chemical synthesis. They made and tested hundreds of organic arsenic compounds, based on aniline dyes (the constituents of many of the biological stains that had engaged Ehrlich's attention) as possible treatments for trypanosomiasis (sleeping sickness, a common and serious infectious tropical disease). From this came, in 1907, Compound 606, named Salvarsan, the first effective drug for this disorder, and the beginning of the era of anti-microbial chemotherapy – arguably the biggest therapeutic success story to date.

Synthetic chemistry had given rise to clinically useful drugs before this, though by serendipity rather than design. Diethyl ether was discovered in the 16th century (known then as "sweet oil of vitriol" because it was made from alcohol and sulfuric acid) and gained notoriety in the 19th century as a party drug ("ether frolics"). Nitrous oxide (laughing gas) had similar origins, and the ability of both of these agents to produce reversible insensibility led in the mid-19th century to their introduction as surgical anaesthetic agents – a vital breakthrough that allowed surgery to develop from agonising butchery to humane intervention. It was in the late 18th century that chemistry began to take over from alchemy, and the production and purification novel compounds of known structure became possible. But understanding and determination of chemical structure, and synthetic methods were still very limited, and it was not until the end of the 19th century that synthetic chemistry really took off, and some of the products were discovered to have medical uses. Among the earliest drugs that came from this were the local anaesthetic, procaine (1905), and the sedative, barbital (1907), both forerunners of important classes of clinically used drugs.

Chemistry-led drug discovery grew rapidly in the 20th century, and quickly became the leading source of new therapeutic agents, and, as a byproduct, new research tools that proved valuable in the study of physiological and pathological processes. Intervening in metabolic pathways, by synthesising "antimetabolites" – analogues of endogenous metabolites – was an approach followed for many years by the highly successful drug discovery team led by Hitchings (1905–98) and Elion (1918–99), working at Burroughs Wellcome in the USA. Earlier, Domagk (1895–1964) in Germany had developed sulfonamides, the first effective antibacterial drugs, which were later shown to work by inhibiting the synthesis of folic acid, a metabolite essential for bacterial growth. Hitchings and Elion sought other inhibitors by making and testing a range of purine and pyrimidine analogues, which acted as inhibitors of the enzyme dihydrofolate reductase. Their antimetabolite approach, begun in 1944, generated not only antibacterial drugs, but also a range of other chemotherapeutic agents, active against protozoa and human cancers. The same sulfonamide-based chemical lineage later gave rise to novel diuretics (acetazolamide, chlorothiazide), antidiabetic drugs (sulfonylureas) and antihypertensive drugs (diazoxide) – an extraordinary example of chemical inventiveness leading to important new therapeutic drug classes. Domagk was awarded the Nobel Prize in 1939, Hitchings and Elion in 1988.

Analytical chemistry later also played an important role in providing tools for identifying the signalling molecules – hormones, neurotransmitters, inflammatory mediators, *etc.* – that play such a major role in physiological regulation, and whose dysfunction commonly leads to disease. Chemists became very successful at inventing new drugs by synthesising analogues and derivatives of known structures. An intuitive sense – hard to pin down – possessed by successful medicinal chemists, guiding them to the kind of structures likely to yield clinically useful drugs, was an important driver of these inventions. The compounds would be handed over to biologists for testing on animals, and anything that looked interesting could be further tested and developed as a medicine. This compound-led strategy sustained a successful drug industry for many years. Nevertheless, natural products continue to be a fruitful source of new useful drugs, most notably in the discovery of penicillin, relying on the inventiveness of evolution rather than of human chemists.

From the mid-20th century "target-led" drug discovery began to rival the compound-led approach. The coming together of chemistry, physiology and pathology under the banner of pharmacology drew attention to the importance of "drug targets", namely the endogenous molecules – in most cases proteins – to which drugs bind in order to produce their effects. Such protein targets are of many kinds, including enzymes, receptors for endogenous mediators, transporter molecules, *etc.* and new ones are constantly being identified. James Black (1924–2010), a British pharmacologist working in industry, was a leader in this new target-led approach. Selecting the recently identified β-adrenoceptor as a promising target for treating cardiovascular disease, he and his team developed the first β-adrenoceptor antagonist, pronethalol, to be approved for clinical use (1965). Pronethalol was quickly withdrawn owing to adverse effects, to be followed by practolol (which had even more severe toxicity), and finally by propranolol (1973), which proved to be a valuable treatment for a range of cardiovascular and other disorders, and is still widely used. Black's team went on following this approach with another major success, the first H_2-histamine receptor antagonist, cimetidine (1975) used to treat gastric and duodenal ulcers. For this work he won the Nobel Prize in 1988. The example set by these early target-led drug discovery projects was quickly followed by pharmaceutical companies worldwide, and became the main source of new therapeutic drugs up to the late 1990s – a particularly fruitful period for drug discovery. Target-led drug discovery began with no knowledge of the molecular nature of the targets in question, the chemistry being led mainly by knowledge of the chemical nature of the relevant physiological mediators. From the 1980s, when receptors and other drug targets began to be isolated as proteins, sequenced and cloned, these new molecular approaches gave a big boost to drug discovery, both by identifying and characterising the many subtypes of receptors, transporters, enzymes and other targets, and also by providing a range of much faster and more powerful methods by which chemical leads could be screened and tested. The subdiscipline of molecular pharmacology, which emerged at this time, grew rapidly in importance, and gained a powerful boost when the human genome sequence was published in 2003. The use of genomic techniques to identify and characterise human drug targets is now a necessary part of most drug discovery projects (covered in later chapters).

1.6   Pharmacology Today

As we have seen, pharmacology arose through the convergence of medicine, chemistry, pathology and physiology, its purpose being to throw light on how medicines and poisons produce their effects. Biochemistry and molecular and cell biology joined the party as these newer disciplines emerged in the 20th century. One important spin-off from molecular biology was the emergence of *biopharmaceuticals* in the form of protein-based therapeutic agents, such as insulin, growth hormone and a variety of monoclonal antibodies, produced by genetically engineered bacteria or eukaryotic cells as an alternative to drugs made by synthetic chemistry. Biopharmaceuticals now constitute about one-third of newly approved therapeutic agents. Since the sequencing on the human genome in 2003 genomics has had a major impact on pharmacology and drug discovery, mainly by providing abundant new information about potential new disease-relevant human drug targets, and also paving the way for "personalised" therapeutics that aims to take into account an individual's genetic make-up as a guide to maximising the efficacy and reducing drug side effects. The multidisciplinary nature of pharmacology, present throughout its history, remains its abiding characteristic, and has grown in complexity as biomedical science has progressed. Pharmacology, you could say, is sustained by hybrid vigour, rather than intellectual purity.

As well as being a key driver of drug discovery and development, pharmacology figures in many other aspects of modern life. A few examples follow:

- Drugs prescribed by clinicians are often ineffective in a significant proportion of patients, and commonly cause adverse effects.[4,5] Selecting the right drug, the right dose, and where possible in the right patient, which can significantly diminish these problems, is the domain of clinical pharmacologists. The emergence of genomics-based personalised medicine is a likely to provide powerful new tools for clinical pharmacologists.
- Pharmacological knowledge is essential in the design and conduct of clinical trials of new medicines, which have to be conducted according to strict protocols governing standards of ethics, experimental design and statistical analysis in order to pass scrutiny by regulatory authorities as a condition of approval of the new drug for clinical use.

- Widely consumed "social" drugs, including alcohol, nicotine and caffeine have been subjected to extensive pharmacological research, the results of which provide the basis for official advice regarding their possible health risks.
- Drug abuse and addiction are serious problems in many countries, and present difficult challenges for prevention, remediation, legislation and policing; understanding the pharmacological properties of abused substances, including the mechanisms by which they produce psychological reward, as well as dependence and harm, is essential in planning rational control measures. The continuing emergence of new synthetic "street drugs" is a particular problem for pharmacologists and legislators.
- Drugs in sport present problems of a different kind, mainly concerned with detection, where understanding of the routes of metabolism and excretion of banned compounds, coupled with sensitive analytical methods, is the basis of most of the control measures that are used.

References

1. O. W. Holmes, Annual Address before the Massachusetts Medical Society, 1860.
2. H. P. Rang, *Br. J. Pharmacol.*, 2006, **147**(Suppl. 1), S9–S16.
3. J. Locke, *An Essay Concerning Human Understanding*, 1690, p. 332.
4. *Goodman and Gilman's The Pharmacological Basis of Therapeutics*, ed. L. Brunton, B. A. Chaber and B. Knollman, McGraw-Hill Education, 12th edn, 2011.
5. J. M. Ritter, R. J. Flower, G. Henderson and H. P. Rang, *Rang and Dale's Pharmacology*, Elsevier, 8th edn, 2016.

2 The Pharmacological Analysis of Drug Activity in Receptor Systems

Terry Kenakin

University of North Carolina School of Medicine, 120 Mason Farm Road, Room 4042 Genetic Medicine Building, CB# 7365, Chapel Hill, NC 27599-7365, USA
Email: kenakin@email.unc.edu

2.1 Introduction

This chapter discusses the application of pharmacodynamics to the quantification of functional drug effects in tissues. The drug discovery process strives to identify therapeutic molecules for the treatment of disease. Since drugs are rarely studied and developed in the therapeutic system (*i.e. in vivo* human disease), their activity must first be observed in test systems, and that activity converted to system-independent scales. From there, predictions of activity in other systems, including the therapeutic one, can be made. Thus, pharmacodynamics in drug discovery involves the conversion of descriptive data (what we see in the assay) to predictive data (what will happen in other systems). There are four properties of molecules that can be used to predict nearly all drug behaviors in all systems:[1]

1. Mode of interaction with the target; orthosteric *vs.* allosteric with respect to the endogenous ligand binding site.

Pharmacology for Chemists: Drug Discovery in Context
Edited by Raymond Hill, Terry Kenakin and Tom Blackburn
© The Royal Society of Chemistry 2018
Published by the Royal Society of Chemistry, www.rsc.org

2. Affinity of the molecule for the target (K_A = equilibrium dis-association constant of the target–receptor complex).
3. Efficacy (or efficacies) of the molecule: ability to change the target behavior toward the host cell.
4. Target dissociation rate constant: estimation of target coverage *in vivo*.

If these four parameters are known, then the tissue effects of the molecule often can be predicted in all tissues. The corollary to this is that when this information is known then a better selection can be made for ultimate drug candidates; it is worth considering each of these four parameters.

2.1.1 Orthosteric *vs.* Allosteric Mode of Action

This discussion will be centered on the most ubiquitous drug target in physiology, namely the seven transmembrane receptor (7TMR). There are several hundreds of these in the human genome controlling a huge number of physiological processes; they respond to a range of activators from small molecule neurotransmitters and hormones to large proteins and even light quanta. Synthetic therapeutic drugs can produce activation, inhibition or modification (increased or decreased natural signaling) of 7TMRs. There are two major modes of action open to synthetic drug molecules as they interact with 7TMRs. The first is they can bind to the natural agonist binding site, in which case there would not be a receptor species present in a tissue binding the natural and synthetic ligand bound simultaneously; this is referred to as orthosteric binding (mutually exclusive binding)- see Figure 2.1A. If the drug has efficacy, it will function as an agonist and may produce a signal similar to or, if biased, different from the natural agonist. If the molecule does not possess efficacy, it will not produce 7TMR-mediated response and by virtue of its occupation of the natural agonist binding site will produce antagonism of natural signaling. There are gradations of this scale as, for example, orthosteric partial agonists may produce a low level of agonism while producing antagonism of the natural agonist. For the purposes of this present discussion, the salient point is that the drug interacts with the receptor through the natural agonist binding site and this leads to a pre-emptive type of interaction with natural signaling, *i.e.* the repertoire of receptor signaling will change from that controlled by the natural agonist to one controlled by the synthetic drug molecule.

A. Orthosteric Interaction   B. Allosteric Interaction

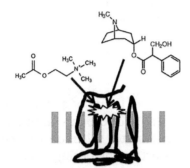   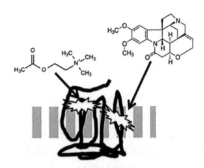

Figure 2.1   Two modes of interaction of ligands with receptors. Orthosteric ligands bind to the same site as the endogenous agonist, while allosteric ligands bind to a separate site on the receptor.

A second mode of action for synthetic drugs is to bind to a site on the receptor separate from that utilized by the natural agonist – see Figure 2.1B; this is an allosteric mode of action. In this case, the allosteric modulator affects natural signaling through a change in conformation of the receptor. This results in a permissive system in that the receptor species of interest has both the natural agonist and the allosteric modulator bound to the receptor simultaneously. This can lead to a much more complex and varied effect on natural signaling ranging from inhibition to potentiation. Moreover, if the natural agonist produces activation of multiple signaling pathways in cells, then allosteric modulators can change the relative stimulation of these pathways by the natural agonist, *i.e.* change the quality of natural signaling.

It is very important to determine how a new molecule interacts with the receptor as the therapeutic qualities and behaviors *in vivo* of orthosteric and allosteric molecules therapeutically can be very different. Secondly, it is important to know which mode of action is operative for a new molecule so that the correct pharmacodynamic model can be applied to data to generate the quantitative parameters that will predict activity within limited ranges of concentration. However, orthosteric and allosteric molecules can produce very similar types of effects and thus it may require special types of experiments to identify the mode of action of a given new molecule.

There are two principles operative for the identification of allosteric mechanisms: the first is that the allosteric effect may be saturable, *i.e.* the allosteric effect reaches a maximal asymptotic value when the allosteric site on the receptor is completely occupied.

Thus, if the allosteric effect results in antagonism of the natural signaling response, this antagonism will have a finite value. Therefore, to identify an allosteric mode of action, as wide a range of test ligand concentrations possible should be tested in order to possibly detect saturation of effect. Secondly, allosteric effects are notoriously probe dependent, *i.e.* an allosteric effect on the response to one receptor probe (*i.e.* agonist) may be different from the allosteric effect observed for another receptor probe. For instance, the allosteric ligand aplaviroc blocks the binding of the chemokine CCL3 to the CCR5 chemokine receptor but does not block the binding of the chemokine CCL5 to the receptor.[2] Therefore, as many receptor probes as possible need to be tested to possibly identify allosteric probe dependence.

There are numerous cases where both orthosteric and allosteric ligands produce similar effects on agonist concentration–response curves; in lieu of evidence to the contrary, it may not be possible to discern the mechanism of action of a given ligand on a receptor with a limited dataset. This is because the identification of allosteric mechanism is a "one way" experiment; if saturation of effect or probe dependence is observed, then evidence of allosterism is provided. If it is not, then it still may be possible that an allosteric effect is operative but that the appropriate range of concentrations of ligand or the appropriate receptor probes may not have been tested to detect allosterism. Therefore, the determination of orthosteric or allosteric mode of action would not be conclusive.

2.2 Data-driven Analysis

The experiments that would be carried out in the process of characterization of a new ligand obtained as a possible compound of interest in a pharmacological screen can be ordered in terms of two key pieces of information:

1. What direct target-mediated effects on cell function are produced? If direct response to addition of the molecule are observed then this indicates that one of the properties of the compound is agonism and the compound is an agonist.
2. What is the effect of the compound on the natural signaling system (*i.e.* naturally occurring hormones, neurotransmitters, autacoids)? If the molecule blocks natural signaling then the compound is an antagonist.

It should be noted that the labels "agonist" and "antagonist" are not mutually exclusive and it will be seen that a given molecule may be an agonist for some signaling pathways and an antagonist for others. For example, the angiotensin ligand TRV120 is a simple competitive antagonist of angiotensin responses when Gq signaling is used as the response readout but an agonist when beta-arrestin is used as response.[3] Additionally, if the compound is a weak agonist that does not produce the same maximal effect as the natural signaling agonist (*i.e.* is a partial agonist), then it will produce concomitant agonism and antagonism.

The first experiment is to determine if the molecule produces overt pharmacological response in a functional assay. If a response is obtained, and assuming it is selective for the biological target of interest, then a dose–response relationship for that response must be defined through testing of a range of concentrations. Usually, this yields a sigmoidal curve of response as a function of the logarithms of the concentration; this characterizes the activity of that molecule on that receptor mediating that particular cellular response. It should be noted that there may be more than one physiological function mediated by the target and that a molecule can be selective with respect to which signaling pathway is activated; this is referred to as "biased" signaling and will be dealt with later in this chapter. There are two practical functions of concentration–response curves:

1. They concisely summarize drug activity through measures of drug potency (through the effective concentration producing 50% of the maximal response to the drug, denoted EC_{50} and normally treated as a negative logarithm in the form of the pEC_{50} where $pEC_{50} = -\log(EC_{50})$) and also the maximal response produced by the drug,
2. They allow comparison of the effect of the drug to quantitative pharmacological models in order to determine possible mechanisms of action of the drug on a molecular level. As a first step, such curves can be fit to the Hill equation that relates concentration of the drug [A] to a response as a fraction of the maximal response max through a function of slope n with location parameter along the concentration axis of EC_{50n}:

$$\text{Response} = \frac{[A]^n E_{max}}{[A]^n + EC_{50^n}} \tag{2.1}$$

Data is fit to eqn (2.1) and values for E_{max} and EC_{50} are derived that characterize the activity of the molecule for that particular target mediating that particular signaling system.

If no observable response is seen upon addition of the molecule to a functional assay, then a rapid test to gauge the concentration range for possible antagonism is made with an IC_{50} curve. The IC_{50} is the inhibitor concentration producing 50% blockade of a defined submaximal preadded concentration of standard agonist for the target in the functional system. For instance, to assess muscarinic receptor blockade for a test compound, a concentration of the standard muscarinic receptor acetylcholine is added to the system to produce a steady-state effect of approximately 80% of the maximal response to acetylcholine. Then, in the continued presence of this agonist concentration, a concentration range of the test compound is added until blockade is produced. The IC_{50} value gives a convenient first test of antagonist potency.

At this point, it is useful to assess what information has been gained from the data-driven process. It can be seen that:

1. The molecule is either an agonist or antagonist for a given functional system controlled by the receptor of interest.
2. If an agonist, a measure of the potency is obtained from the pEC_{50} and a measure of the efficacy from the maximal response (*i.e.* if it is submaximal with respect to the natural agonist in the assay then it is a partial agonist with low efficacy).
3. The curve furnishes data for comparison to the standard model for pharmacological agonism, namely the Black/Leff operational model[4] (*vide infra*). This will determine the affinity and efficacy of the compound as an agonist, relative to a standard agonist, in that particular signaling system. This also will be a starting point for assessing possible signaling bias for other signaling systems mediated by the same receptor.
4. If an antagonist, a measure of the potency is gained from the IC_{50}. This also indicates the concentrations at which more detailed studies of the antagonism to gauge mechanism of action should be tested.

At this point two separate sets of experiments are done depending on whether agonism or antagonism is observed. The first to be considered is agonism.

2.3   Characterizing Agonist Activity

Observation of a sigmoid concentration–response curve for direct effect indicates that the molecule possess affinity and efficacy for the target receptor. The standard model for quantifying receptor agonism is the Black/Leff operational model.[4] To do this, data is fit to the operational model equation given as:

$$\text{Response} = \frac{[A]/K_A \tau E_{max}}{[A]/K_A(1+\tau)+1} \tag{2.2}$$

where E_{max} is the maximal response capability of the system, K_A is the equilibrium dissociation constant of the agonist–receptor complex (a surrogate of affinity, *vide infra*) and τ is the efficacy of the molecule for production of response in the system. Various levels of descriptive and predictive data can be obtained from this analysis. If the molecule is a partial agonist (*i.e.* the maximal response of the molecule is below E_{max}) then unique values for τ and K_A can be obtained and these can be very useful for predicting agonist response in other systems. If the molecule produces full agonism (maximal response of the molecule is equal to E_{max}) then an infinite array of combinations of τ and K_A can describe the curve and no unique values either for efficacy or affinity can be gained. However, the ratio τ/K_A will still be a unique identifier for the ability of the molecule to produce response. In fact, the parameter $\log(\tau/K_A)$, termed the transduction coefficient,[5] is a valuable number that indicates the "power" of the molecule to produce agonism for a given response pathway. Figure 2.2A shows data for a full and partial agonist fit to the operational model with the resulting values of τ, K_A and $\log(\tau/K_A)$. The ratio of $\log(\tau/K_A)$ values (in the form of $\Delta\log(\tau/K_A)$) is a meaningful ratio in that it is a system-independent ratio of the relative power of the two agonists to produce the defined response. In Figure 2.2A, the $\Delta\log(\tau/K_A)$ value for the two agonists is 1.57; Figure 2.2B shows the effect of the same two agonists in a different, more sensitive, assay. While the values of τ (the efficacy) change for the agonists (because the tissue-dependent component of τ is the receptor density in the tissue), the $\log(\tau/K_A)$ values also change in this more sensitive tissue. However, the $\Delta\log(\tau/K_A)$ value is still the same. This makes this index a valuable descriptor and predictor of the relative agonist effects of different agonists.

It is often observed that agonists are biased toward certain signaling pathways for receptors that are pleiotropically coupled to multiple cellular signaling pathways. This can actually be a therapeutically

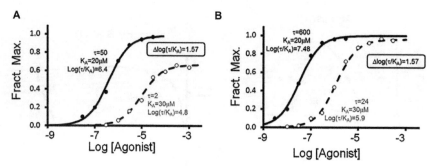

Figure 2.2   Predicting agonist response with the Black/Leff operational model. Panel A: Concentration–response curves for two agonists (filled circles solid line: $\tau = 50$, $K_A = 20$ μM and open circles broken line $\tau = 2$, $K_A = 30$ μM) in a relatively sensitive functional system. The $\Delta\log(\tau/K_A)$ value for the relative power of these agonists is 1.57. Panel B. In a more sensitive system, these same two agonists should have the same $\Delta\log(\tau/K_A)$ value thereby predicting the locations of their respective concentration–response curves as shown. The Value of $\Delta\log(\tau/K_A)$ links the relative locations and maxima of these two agonists.

favorable circumstance if certain pathways are linked to favorable therapeutic outcomes (and thus a bias to emphasize these can be obtained) or harmful drug effects (in which case a bias to de-emphasize these signals can be built into the molecule). For instance, an opioid agonist that signals through G proteins but does not cause the opiate receptor to associate with β-arrestin may yield analgesia with less propensity to produce respiratory depression.[6] The trans-ducer coefficient ($\log(\tau/K_A)$) can be used to quantify the amount of bias for any given agonist. An example of a biased opioid agonist referred to previously is TRV130 that has a 4-fold bias toward G protein signaling *vs.* β-arrestin interaction leading to reduced gas-trointestinal and respiratory dysfunction for given analgesia when compared to morphine.[7]

Figure 2.3A shows concentration–response curves for two agonists in two functional assays for signals emanating from a single receptor where each of these pathways yields a separate signal that may have differing cellular consequences. A visual representation of the relative propensity of these agonists to activate each of the signaling pathways can be obtained with a bias plot that simply expresses the response in one of the pathways as a function of the response seen in the other pathway. Figure 2.3B shows the bias plot for the two agonists shown in Figure 2.3A and it can be seen that there are different amounts of bias for the pathways found in these agonists. Specifically, agonist$_1$ is biased toward Function$_A$ and agonist$_2$ is biased toward Function$_B$. These differences can be quantified through calculation of

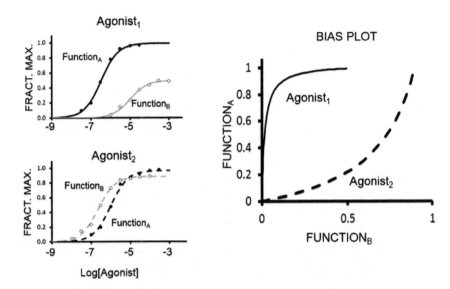

Figure 2.3   Biased agonism. Concentration–response curves for two agonists (Agonist$_1$ and Agonist$_2$) for two functional responses (*i.e.* G protein and β-arrestin activation) are shown on the left panels. Right panel (bias plot) shows the response of each agonist for Function$_A$ as a function of the amount of response produced in Function$_B$ for the two agonists. It can be seen that whereas Agonist$_1$ is biased toward Function$_A$ (*i.e.* there is relatively more response for Function$_A$ produced for a given generation of response in Function$_B$), the opposite is true for Agonist$_2$ (which is biased toward Function$_B$).

Table 2.1   Calculation of ligand-signaling bias.

Function 1					Function 2		
	$\log(\tau/K_A)$	$\Delta\log(\tau/K_A)$	$\Delta\Delta\log(\tau/K_A)$	Bias	$\Delta\log(\tau/K_A)$	$\log(\tau/K_A)$	
Agonist$_1$	6.48					6	Agonist$_1$
Agonist$_2$	4.7	1.78	2.38	239.9	−0.6	6.6	Agonist$_2$

$\Delta\Delta\log(\tau/K_A)$ values for the two functions. The differences in the $\log(\tau/K_A)$ values of each agonist for each function yield $\Delta\log(\tau/K_A)$ values for the two agonists for each function. For example, as shown in Table 2.1, the Function$_A$ $\log(\tau/K_A)$ value for agonist$_1$ is 6.48 and for agonist$_2$ is 4.7 resulting in a $\Delta\log(\tau/K_A)$ value for this function of 1.78 – see Table 2.1. This means that agonist$_1$ is $10^{1.78} = 60.2$-fold more powerful as an agonist for Function$_A$ than is agonist$_2$. The same is done for Function$_B$, where it can be seen that the $\log(\tau/K_A)$ for agonist$_1$ is 6.0 and for agonist$_2$ is 6.6. In contrast to what was seen for Function$_A$, in this case agonist$_2$ is $10^{0.6} = 4$-fold more powerful as an activator of Function$_B$ than is agonist$_1$. The relative bias for the two agonists is then calculated by subtracting the $\Delta\log(\tau/K_A)$ values to yield a $\Delta\Delta\log(\tau/K_A)$ value of 2.38 that, in this case, results in a final conclusion that agonist$_1$ is $10^{2.38} = 240$-fold biased toward Function$_A$ than is agonist$_2$. The bias calculation yields a vector that can be expressed either way therefore this also suggests that agonist$_2$ is 240-fold more biased toward Function$_B$ than is agonist$_1$.

In general, it can be seen that the full characterization of the agonism for a given new molecule involves quantification of agonist effect in a number of functional assays and that the efficacy of a given molecule therefore has a quality as well as quantity as various mixtures of activation of signaling pathways are produced. This quality of efficacy can be visualized by expressing $\Delta\log(\tau/K_A)$ values on multiple axes to yield a two-dimensional diagram quantifying unique patterns of activation. Figure 2.4A shows such a web diagram for an agonist with varying levels of activation of 3 signaling pathways in cells. Such diagrams are useful to characterize the mixtures of efficacies inherent in any one molecule (for experimental examples for β-adrenoceptors see ref. 8 and for κ-opioid receptors see ref. 9).

The same scale can be used to characterize receptor subtype selectivity. Figure 2.4B shows $\Delta\Delta\log(\tau/K_A)$ values for a muscarinic

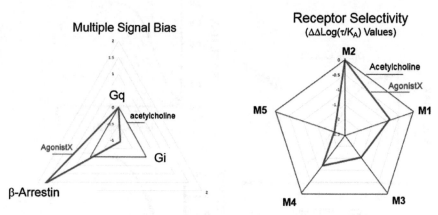

Figure 2.4   Graphical depictions of signaling bias and receptor selectivity through $\Delta\Delta\log(\tau/K_A)$ values. Left panel shows $\Delta\Delta\log(\tau/K_A)$ values for the relative activation of G_q, G_i and β-arrestin signaling pathways by test agonists compared to the standard, acetylcholine. This graph readily shows that this agonist is biased heavily toward β-arrest and somewhat away from G_i relative to acetylcholine. Right panel shows $\Delta\Delta\log(\tau/K_A)$ values for the relative agonism of a test agonist (compared to acetylcholine) on five muscarinic receptor subtypes. This graph shows that, for comparable activation of m2 receptors, the test agonist is selective with less activity toward m1, m3, m4 and m5 receptors.

agonist acting on five muscarinic subtype receptors (m1 to m5); the $\Delta\Delta\log(\tau/K_A)$ values represent the relative power of this agonist to activate the receptor compared to a standard agonist which in this case is acetylcholine. The comparison to the standard agonist cancels any effects of assay sensitivity and receptor expression, making these ratios system independent and thus predictive for all systems.

2.4   Antagonism

If no agonist response to the test compound is observed, the next test is to determine if the molecule blocks the effects of standard agonists for the biological target receptor. A rapid approach to doing this is to determine a pIC_{50} for the molecule as an antagonist of a natural agonist. The pIC_{50} is the negative logarithm of the molar concentration of antagonist that reduces the response to a pre-equilibrated concentration of agonist by 50%. Figure 2.5A shows the effects of a test antagonist on the steady-state response to an agonist producing 80% maximal response in the functional assay. It can be seen that the

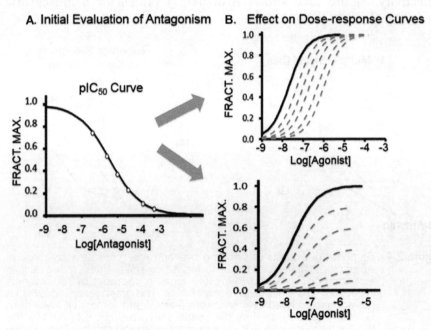

Figure 2.5   A rapid assessment of antagonism through an IC_{50} curve shown in the left panel could result from two types of antagonist; surmountable and insurmountable as shown in the right panels.

pIC_{50} for this test antagonist is 6.0, indicating that 50% inhibition is obtained when 1 μM of the test compound is added to the assay. This simple experiment yields valuable information mostly in the form of identifying the concentrations of the test antagonist that produce antagonism; this is valuable data for the next step in the characterization of the antagonism in which the effects of various concentrations of the test antagonist on full agonist concentration–response curves are observed; this next step is instrumental in determining the mechanism of action of the antagonism (in the form of the effects of a range of concentrations of the antagonist on a full agonist concentration–response curve – see Figure 2.5B). The pattern of curves observed in this procedure will indicate the pharmacological model that should be used to estimate the system-independent measure of antagonism ($pK_B = -$logarithm of the K_B, which is the equilibrium dissociation constant of the antagonist–receptor complex). At this point it is again useful to assess the information gained from these simple experiments in terms of describing the pharmacological effects of a new test molecule:

1. A measure of the efficacy and affinity of the agonist is obtained from characterization of the concentration–response curve to the test compound. In addition, a signaling pathway for which the agonist has activity may be identified.
2. If the compound is a partial agonist, a measure of the affinity is obtained from the pEC_{50}. Similarly, a measure of the relative power for agonism of the molecule with respect to the natural agonist may be determined in the form of $\Delta\log(\tau/K_A)$ values.
3. Observance of possible agonism in a number of pathways allows an assessment of the range of efficacies possessed by the molecule (quality of efficacy).
4. Observance of $\log(\tau/K_A)$ values for other signaling pathways enable assessment of possible biased signaling for the molecule.

If agonism is not observed but rather an IC_{50} curve emerges from a test of antagonism:

1. The concentration range where the molecule interacts with the receptor is identified.
2. The magnitude of the minimal ordinate value for the pIC_{50} curve may indicate partial or inverse agonism or allosteric antagonism (*vide infra*)

2.4.1   Pharmacological Analysis of Orthosteric Antagonism

Once antagonism has been identified in a test molecule, the mechanism of action of that molecule can be identified. This can be done by observing the pattern of concentration–response curves to a full agonist produced by a range of concentrations of the test ligand. The first mechanism to be considered is the orthosteric binding of the test molecule to the same site utilized by the natural agonist for the receptor. The procedure to quantify this effect is comprised of determining a control concentration–response curve to the natural agonist, pre-equilibration of the assay system with a fixed concentration of the test antagonist and then reassessment of the sensitivity of the assay system to the agonist in the presence of the test antagonist. The outcome of such an approach for orthosteric antagonists depends on the rate of dissociation of the test antagonist from the receptor relative to that of the agonist. If the antagonist has rapid offset kinetics (associate and dissociate from the receptor rapidly), then the concentration–response curves to the full agonist will be shifted to the right by the test antagonist with no diminution of the maximal response to the full agonist (an effect referred to as "surmountable antagonism" – see Figure 2.6A). Alternatively, if the test antagonist has slow offset kinetics, whereby it binds to the receptor upon equilibration and demonstrates very slow (even pseudoirreversible) offset kinetics, then the maximal response to the full agonist may be depressed (a condition referred to as "insurmountable

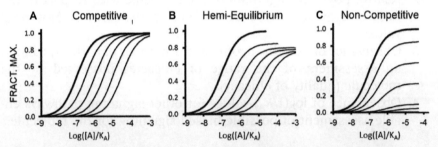

Figure 2.6   Three orthosteric antagonists with differing rates of offset from the receptor. Panel A shows a rapid offset antagonist that would produce surmountable competitive antagonism. Panel B shows an antagonist with a slower rate of offset in a functional assay system that captures only rapid agonist response (*i.e.* calcium transient). This produces a hemiequilibrium condition whereby high concentrations of agonist are unable to produce the control maximal response. A new somewhat depressed maximum is produced. Panel C shows a slow offset antagonist that is essentially irreversible with respect to re-equilibration with the agonist during the production of agonist response; insurmountable antagonism results.

antagonism" – see Figure 2.6C). Antagonists may have midrange offset kinetics to create a "hemiequilibrium" state consisting of behaviors consistent with rapid kinetics (shifts to the right of the full agonist concentration–response curves) but also of a slow offset antagonism resulting in a partial and limited depression of the maximal response to the full agonist – see Figure 2.6B. Once the pattern of antagonism is identified, then the concentration–response curves can be fit to models of simple competitive kinetics or noncompetitive kinetics. As a preface to the discussion of orthosteric antagonism, it is useful to discuss another mode of action of ligands with receptors, namely allosteric.

2.5 Allosteric Modulation

If it is known that the interaction of the antagonist with the receptor is allosteric, then data is fit to the functional model for allosteric receptor interaction. This model describes the binding of an allosteric molecule B to a receptor R that can also bind the agonist A; this can occur because molecule B binds to a separate site on the receptor protein surface.[10,11] The ligand-bound protein species, either bound to the agonist A, or the allosteric molecule B or both molecules are then treated as possible response-producing species according to the Black/Leff operational model for agonism. The total scheme for the system is shown in Figure 2.7; the equation for this model is given as[12–14]

$$\text{Response} = \frac{(\tau_A[A]/K_A(1 + \alpha\beta[B]/K_B) + \tau_B[B]/K_B)E_m}{([A]/K_A(1 + \alpha[B]/K_B) + \tau_A(1 + \alpha\beta[B]/K_B)) + [B]/K_B(1 + \tau_B) + 1}$$

$$(2.3)$$

The important allosteric parameters in this model are α, defined as the effect of the allosteric molecule on the affinity of the agonist (ratio of the affinity of A for the receptor in the presence and absence of allosteric molecule) and β, the effect of the allosteric molecule on the efficacy of the agonist (ratio of efficacy of the agonist in the presence and absence of the allosteric molecule). This model also allows the allosteric molecule to have direct efficacy for excitation of the receptor through the efficacy term τ_B. Allosteric antagonists can produce surmountable or insurmountable antagonism depending on the values of β and the sensitivity of the functional system. This latter factor is characterized by a receptor reserve that is related to the fraction of activated receptors required to produce the system maximal response.

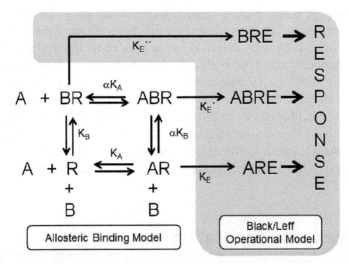

Figure 2.7 Schematic diagram of the functional allosteric receptor model based on the allosteric binding model[10,11] and the Black/Leff operation model[4] for receptor function.

Changes in α cause only shifts in the location parameter of the agonist concentration–response curve and no change in the maximal response, while changes in β cause changes in maximal response to partial agonists and mixtures of changes in maximal response and concentration–response curve location parameters for full agonists. The parameters K_B, α and β are characteristic of modulator-agonist pairs and serve to quantify modulator activity. In general, allosteric modulators can produce completely different patterns of agonism for the receptor that can be categorized in terms of three phenotypes with two variants:

1. Negative allosteric modulators (NAMs): The product $\alpha\beta$ is <1 and the modulator causes surmountable and/or insurmountable blockade of agonist response. When $\tau_B > 0$, then a variant of this phenotype is a NAM-agonist – see Figure 2.8A.
2. Positive allosteric modulators (PAMs): The produce $\alpha\beta$ is >1 and the modulator potentiates agonist response through a mixture of sinistral displacement of curves and possible increases in the maximal response. A variant when $\tau_B > 0$ is a PAM-agonist – see Figure 2.8B.
3. PAM-Antagonists: These modulators increase the affinity of the receptor for the agonist but decrease the efficacy – see Figure 2.8C.

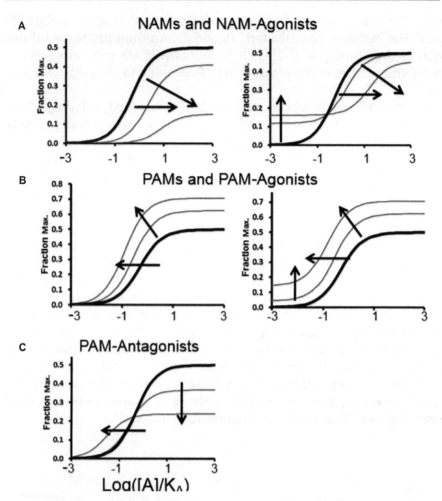

Figure 2.8   Three distinct phenotypes (with two variants) of allosteric ligands. NAMs (negative allosteric modulators) produce antagonism through an $\alpha\beta$ product <1; this can translate to surmountable and/or insurmountable effects on agonist concentration–response curves. A variant of this phenotype has the NAM producing direct agonism as well. Panel B shows the PAM (positive allosteric modulator) phenotype where an $\alpha\beta$ product >1 leads to potentiation of agonist response; a variant shows direct agonism for the PAM as well. Panel C shows a unique phenotype whereby the modulator increases the affinity of the agonist ($\alpha > 1$) but decreases its efficacy ($\beta < 1$) to cause sinistral shifts of the concentration–response curves with depression of maximal response.

As shown in Figure 2.8, reduction in sensitivity to the agonist coupled with positive agonism indicates partial agonism that could be orthosteric or allosteric in nature. The principles of saturation of effect and/or probe dependence then need to be applied to differentiate these mechanisms. It should be noted that there are no general rules for values of α and β any modulator–agonist (or radioligand) pair. For instance, the allosteric ligand alcuronium produces inhibition of the binding of [³H] QNB to muscarinic M2 receptors $(\alpha\beta<1)$ but a stimulation of the binding of [³H] atropine to the same receptor $(\alpha\beta>1)$.[15]

Once an allosteric mechanism has been identified, it is important to characterize that activity in terms of the system-independent parameters K_B, α and β as the impact of these different parameters on agonist concentration–response curves can be different in functional systems of different sensitivities. For example, Figure 2.9 shows the effects of a positive allosteric modulator $(\alpha=20,\ \beta=5)$ with weak direct agonist activity $(\tau_B=3\times10^{-5}\ \tau_A)$ in three tissues with varying receptor density (leading to three different sensitivities to the agonist). In a tissue of low sensitivity $(\tau_A=1)$, the agonist maximal response is increased. In a tissue of greater intrinsic sensitivity $(\tau_A=30)$ where the agonist is already a full agonist, then the PAM produces only a sinistral shift of the concentration–response curve. Finally, in a highly sensitive tissue $(\tau_A=150)$, the weak direct agonism of the PAM produces an agonist response that accompanies the sinistral shift of the agonist concentration–response curve. While these phenotypes in the various tissues are quite different, they are consistently described by the molecular parameters determining the allosteric effect; these parameters then offer a system-independent scale for medicinal chemists to optimize activity.

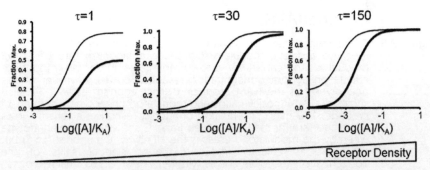

Figure 2.9 The effects of a PAM-agonist $(\tau_B=3\times10^{-5}\ \tau_A;\ \alpha=20;\ \beta=5)$ in three functional systems with differing sensitivity to the agonist. In a poorly sensitive system $(\tau_A=1)$, the PAM-agonist produces an increased maximal response. In a more sensitive system $(\tau_A=30)$, there is no further increase in the maximal response since the agonist is already a full agonist but only a sinistral displacement of the concentration–response curves. In a system of yet higher sensitivity $(\tau_A=150)$, the direct agonism of the PAM-agonist is expressed.

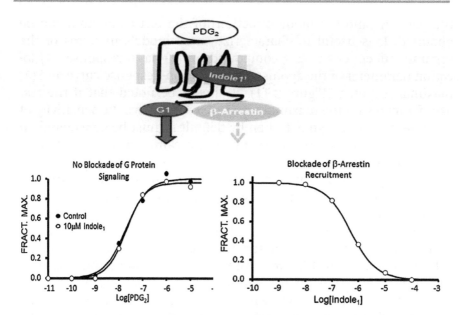

Figure 2.10 Induction of signaling bias for the CRTH2 receptor activated by PDG2. After binding of Indole1, PDG2 selectively activates only β-arrestin and not Gq protein. Figure made from information in ref. 16.

Another important feature of allosteric ligands comes from their permissive nature, *i.e.* they cobind to the receptor with other ligands. This gives allosteric ligands the ability to modify the bias of the natural endogenous agonist signaling. For instance, under normal circumstances, the agonist prostaglandin D2 binds to the CRTH2 receptor to activate Gq protein and cause association of the receptor with β-arrestin, However, the allosteric modulator indole1 (1-(4-ethoxyphenyl)-5-methoxy-2-methylindole-3-carboxylic acid) changes the quality of that natural signal by selectively blocking G protein effects and allowing the β-arrestin signal to remain unchanged[16] (see Figure 2.10). Similar induction of signaling bias has been reported for other NAMs (NK2 receptors;[17] calcium-sensing receptors[18,19]) and PAMs (GLP-1 receptors;[20] mGlutamic acid 5 receptors[21]).

2.6 Assignment of Receptor Models to Characterize Activity

Once the effects of a range of concentrations of a test compound on agonist response is tested in a functional assay, the correct mathematical model can be applied and system-independent measures of the potency (and the mode of action) of the test compound can be obtained. It is useful to characterize these models in terms of the separate effects of the test compound on (1) basal response, (2) location parameter of the agonist concentration–response curve and (3) maximal response (Figure 2.11). It should be noted that if the test agonist increases the maximal response or increases the sensitivity of the receptor to the agonist, then by default it must be considered an allosteric ligand since this type of behavior cannot be produced by an orthosteric ligand.

A starting point for the discussion of the application of different pharmacological models to assess test ligand potency is to discuss a test ligand that does not alter the basal response or the maximal response to the agonist (surmountable antagonist – see Figure 2.11). Orthosteric surmountable antagonism can be quantified through Schild analysis.[22] In this procedure, the antagonism is measured as the shift to the right of the agonist concentration–response curve (denoted as a dose ratio DR where DR = EC_{50} of the agonist in the presence of the antagonist divided by the EC_{50} of the agonist in the absence of the antagonist). DR values are measured for a range of

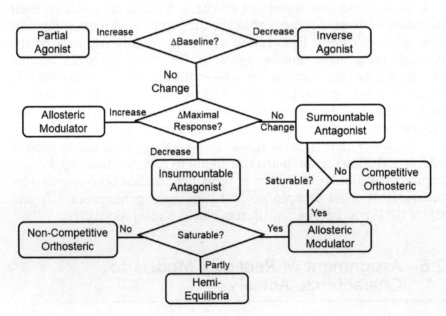

Figure 2.11 Logistical scheme for the determination of the mechanism of action of antagonists, partial agonists, inverse agonists and allosteric modulators. The basic flow centers on the effects on basal response, maximal response and saturability of effect.

antagonist concentrations and used to construct a Schild plot according to the Schild equation:[22]

$$\log(DR - 1) = \log([B]) - \log K_B \qquad (2.4)$$

This is an equation for a straight line, thus a regression of a set of $\log(DR - 1)$ values on $\log([B])$ values should produce a straight line with an intercept of $-\log(K_B)$ which is the pK_B; this is referred to as a Schild regression – see Figure 2.12. The pK_B is a characteristic estimate of the equilibrium dissociation constant of the antagonist–receptor complex unique to the antagonist–receptor pair and is independent of the tissue in which it is measured. Orthosteric

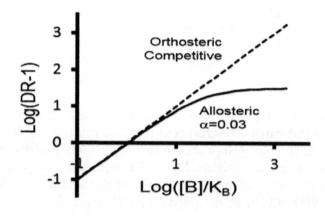

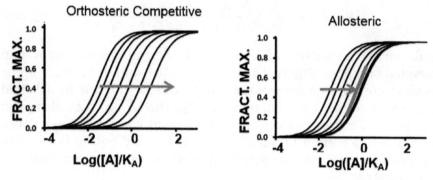

Figure 2.12   Quantification of surmountable antagonism. Orthosteric competitive antagonism yields limitless dextral displacements of the agonist concentration–response curves to furnish a linear Schild regression. In contrast, an allosteric surmountable antagonist will produce dextral displacement until the allosteric site is saturated and then the effect will not continue. Once the limiting maximal DR value is approached, the Schild regression will be nonlinear with an asymptote determined by the magnitude of the allosteric co-operativity factor α. Nonlinearity of the Schild regression for a test surmountable antagonist thus becomes a test for allostery.

competitive antagonism can explicitly be modeled with the following equation:[23]

$$\text{Response} = \frac{([A]/EC_{50})E_m}{([A]/EC_{50}) + [B]/K_B + 1} \tag{2.5}$$

where the agonist and antagonist are A and B, respectively, and K_B is the equilibrium dissociation constant for the antagonist; EC_{50} refers to the midpoint agonist concentration for the control concentration–response curve (absence of antagonist). Thus, the K_B can be obtained through the construction of a Schild plot or through direct fitting of agonist concentration–response curves according to eqn (2.5). However, an advantage to constructing a Schild plot is determining linearity. This is because a surmountable allosteric antagonist might also be identified by observing nonlinearity in the Schild regression; this occurs because of allosteric saturation of effect, *i.e.* a limiting DR value is obtained upon saturation of the allosteric binding site-see Figure 2.12.

If the test antagonist changes the baseline response in the assay, then this indicates other forms of receptor interaction. If the baseline is elevated this suggests that the test ligand has positive efficacy and is a partial agonist (Figure 2.13). The equation for partial agonism incorporates the efficacy and affinity of the agonist (τ_A, K_A) and the efficacy and affinity of the partial agonist (τ_B, K_B):

$$\text{Response} = \frac{(\tau_A[A]/K_A + \tau_B[B]/K_B)E_m}{[A]/K_A(1 + \tau_A) + [B]/K_B(1 + \tau_B) + 1} \tag{2.6}$$

This yields a pattern of curves with dextral displacement and elevated baselines (Figure 2.14A). The parallel shift to the right of the agonist concentration–response curves can be utilized in a Schild regression to yield an estimate of pK_B. Partial agonists also yield unique IC_{50} curves characterized by elevated basal responses (basal response to a concentration of agonist is reduced only to the level of agonism produced by the partial agonist); these effects are shown in Figure 2.14A.

If the basal response is depressed below the control level, then this indicates that the functional system is constitutively active and the basal response is elevated due to the presence of receptors spontaneously and the test ligand is an inverse agonist (see Figure 2.13). As with partial agonists, the parallel dextral displacement of the agonist concentration–response curves can be used to construct a Schild regression to calculate the pK_B – Figure 2.14B. The IC_{50} curve for inverse

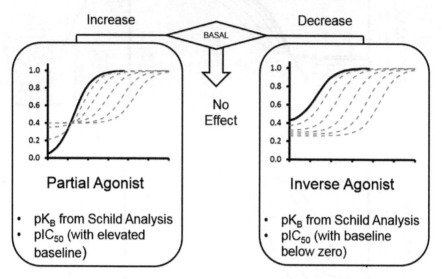

Figure 2.13 Options for effects of test ligands on basal responses. If basal response is elevated, this indicates that the ligand has positive efficacy and is a partial agonist. If the basal response is reduced, this indicates that the functional system is constitutively active with an elevated baseline due to receptors spontaneously in an active state producing response. Reduction of this elevated basal indicates that the ligand has negative efficacy and is an inverse agonist. In both cases antagonist potency can be estimated through an IC_{50} and/or a pK_B through Schild analysis.

agonists in constitutively active systems will show a characteristic depression of the maximal response below the control maximum (Figure 2.14B).

The next criterion for evaluation of test antagonism is the effect on the maximal response to the agonist (Figure 2.11). As noted earlier, if maximal response is increased, then an allosteric effect is indicated. If it is depressed, then there are a number of possibilities – see Figure 2.15. The next question to be asked is whether the depression of the maximal response reaches a saturable level above zero (where the maximal response to the agonist is depressed by the test antagonist to a new level below control and then assumes this new level for all subsequent concentrations of the test agonist). It is important to note the concomitant position of the agonist concentration–response curves when this is observed. If higher concentrations of the test antagonist continue shifting the agonist concentration–response curves to the right and they retain this new depressed maximum then

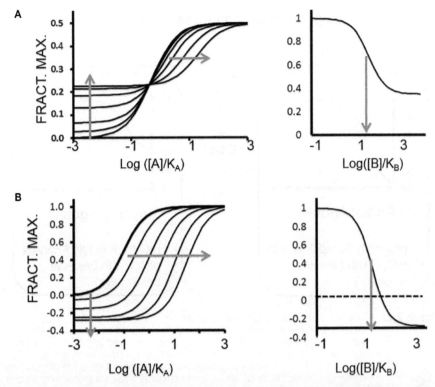

Figure 2.14 Measurement of antagonist potency of partial agonists (A) or inverse agonists (B). In both cases, the parallel displacement of the agonist concentration–response curves can be used to furnish dose ratios for Schild analysis; the resulting Schild plot can then be used to obtain a pK_B. The IC_{50} curve for a partial agonist will have a maximum determined by the agonist response of the partial agonism, *i.e.* it will be elevated above zero. For inverse agonists, the IC_{50} curve will have a maximum below the starting basal response if the system is constitutively active.

this indicates that a hemiequilibrium condition is operable – see Figure 2.15. This is seen when the orthosteric antagonist has too slow a rate of offset to allow full receptor occupancy by the agonist (see Figure 2.6B); under these circumstances the pK_B value can be estimated with a Schild regression obtained from DR values in the parallel range of the concentration–response curves. However, if the new depressed maximal response is accompanied by a saturation in the position of the concentration–response curve, *i.e.* there is no further shift to the right with higher concentrations of test agonist, then this indicates that allosteric saturation of effect has taken place and the test agonist should be analyzed with the functional allosteric equation (eqn (2.3)).

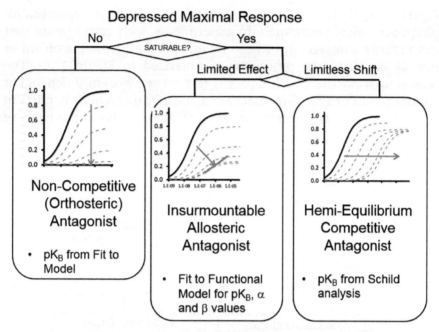

Figure 2.15 Logistic diagram for determining mechanism of action of antagonists that depress the maximal agonist response. If the depression of the maximum is not obviously saturable (*i.e.* response can be blocked to basal levels) then this is consistent with orthosteric noncompetitive antagonism. Alternatively, it is possible that this still could be a NAM effect with a very small value for β. If the depression of the maximum is saturable (*i.e.* converges to an asymptotic value above zero), then it must be determined if the dextral displacement of the concentration–response curves follow suit (saturate to a maximal value) or continue in a dose-dependent manner. In the latter circumstance, this is consistent with a hemiequilibrium condition whereby competitive antagonism is operative with a depressed maximum due to the hemiequilibrium. If the shifts of the curve and the depression of maxima saturate to a level above zero, this is consistent with insurmountable allosteric antagonism (NAM effect).

If a true depression of the maximal response curve to the full agonist is produced (*i.e.* if high concentrations of antagonist can completely depress the agonist response to baseline), then the data may be fit to the model of orthosteric noncompetitive antagonism (Figure 2.14); the responses are modeled with the equation for orthosteric noncompetitive antagonism as shown below:

$$\text{Response} = \frac{(\tau[\text{A}]/K_\text{A})E_m}{[\text{A}]/K_\text{A}(1 + \tau + [\text{B}]/K_\text{B}) + [\text{B}]/K_\text{B} + 1} \qquad (2.7)$$

Thus, experimental data can be fit to eqn (2.7) to yield estimates of the pK_B for the noncompetitive antagonist. As can be seen from

Figure 2.11, there is a logical progression for applying quantitative pharmacological techniques to experiments with test ligands that can identify allosteric or orthosteric antagonism, partial agonism or inverse agonism (the process is summarized in Figure 2.16). The reason to apply these techniques is to measure system independent parameters of drug activity that can be used by medicinal chemists to change the activity through synthesis. Table 2.2 shows the types of

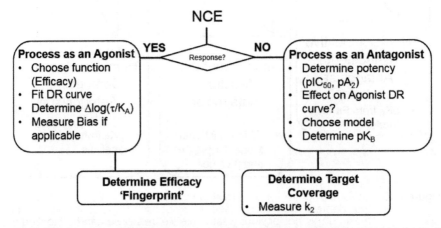

Figure 2.16 Summary of pharmacological procedures to quantify the agonist and antagonist activity of test ligands and determine their mechanism of action. A major determinant of how the functional assay can be used is to see if a direct response to the test agonist is observed. Whether or not a response is observed also may be determined by the type of functional assay used, *i.e.* signaling pathway being monitored; the total pattern of agonist constitutes the efficacy fingerprint of the ligand.

Table 2.2 System-independent parameters of drug activity.

Ligand Type	Parameter	Description
Competitive Antagonist	K_B	Equilibrium dissociation constant of antagonist–receptor complex
Partial Agonist	K_B	Equilibrium dissociation constant of antagonist–receptor complex
	τ_B	Efficacy of Partial Agonist
Inverse Agonist	K_B	Equilibrium dissociation constant of antagonist–receptor complex
	τ_B	Negative Efficacy of Inverse Agonist
Allosteric NAMs/PAMs	K_B	Equilibrium dissociation constant of antagonist–receptor complex
	τ_B	Efficacy of allosteric modulator
	α	Effect on agonist affinity
	β	Effect on agonist efficacy

ligands that can be characterized and the parameters that are relevant.

The important pharmacological activity for a test molecule is the effect it will have on endogenous tissue function either directly or through endogenous agonists (*i.e.* hormones, neurotransmitters) and the foregoing discussion centers on functional tests to obtain these data. However, biochemical binding studies also can be useful in some cases to elucidate the mode of action of test ligands.

2.7 Biochemical Binding Studies

In general, there are two major approaches to the determination of ligand association with a receptor; function and binding. By far the more relevant data is obtained through functional experimentation as this represents what will be observed pharmacologically *in vivo* for therapy. However, there are biochemical binding techniques that can be applied to measure the physical interaction between a ligand and a protein that can be useful. It should be noted that these techniques often are deceptively simple in that they are based on simple models such as the Langmuir adsorption isotherm[24] but in reality involve much more complex combinations of the mass action equation.[25] The Langmuir adsorption isotherm (based on a previous model developed by Hill,[26] was developed to quantify the adsorption of chemicals to metal surfaces. If a traceable version of the test molecule is available (*i.e.* radioactive analog) then binding can yield an estimate of the amount of test molecule bound to the biological target at any concentration through the estimation of the binding K_d (k_2/k_1, where k_2 and k_1 refer to the respective rates of dissociation and association of the molecule with the target). Alternatively, the test ligand can be used to displace (or in the case of allosteric ligands, modify the affinity of) radioligands prebound to the receptor to yield a K_I which represents the equilibrium dissociation constant of the ligand–receptor complex. A detailed discourse of binding technology is beyond the scope of this present chapter but suffice to say that this technology can identify the concentrations of a ligand that occupy 50% of a given population of receptor in a tissue preparation that subsequently can be used to predict a concentration range where the ligand would be predicted to produce a drug effect. While these approaches can be fruitful for orthosteric association of ligands with receptors, they may fall short for describing allosteric interactions. This is because the affinity of ligands in allosteric systems can be dependent upon the presence of the cobinding ligand. Therefore, unless the binding system is identical to the physiologically relevant system, differences emanating from allosteric co-operative effects are observed and the resulting parameters are inaccurate.

There are instances where binding experiments can yield uniquely illuminating data. Figure 2.17 shows the functional and binding effects of the PAM-antagonist Org27569. This molecule produces noncompetitive functional antagonism of responses of the cannabinoid receptor to the agonist CP55940.[27] However, paradoxically it also increases the affinity of the receptor toward CP55940, a fact confirmed in binding. Specifically, Org27569 increases the formation of the agonist–receptor complex but this complex cannot signal, thus antagonism is the result. The unique increase in agonist binding but decrease in agonist function is made evident with the dual technologies of binding and function.

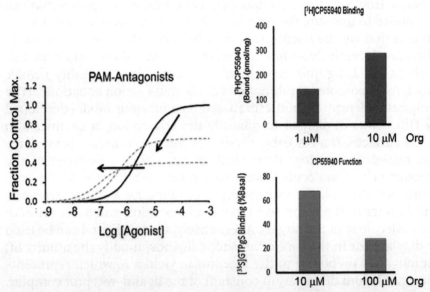

Figure 2.17 Unique function and binding profile of the PAM-Antagonist Org27569. Concentration–response curves show the effects of a PAM-antagonist on agonist response; for the cannabinoid PAM-antagonist Org27569, this would be the functional effects produced on curves for the cannabinoid agonist CP55940. Panels on the right show the effects of Org27569 on the binding of [³H]-CP55940 and maximal functional response to CP55940; it can be seen that while the response is inhibited, the binding is stimulated. This compound stimulates the formation of the receptor–agonist complex into a protein species that cannot signal.
Data redrawn from ref. 27.

2.8   Kinetic Studies of Drug Function

The foregoing analyses are basically snapshots of steady-states or equilibrium conditions that may not be totally accurate *in vivo* since these therapeutic systems are not ever in equilibrium and therefore real time is an important variable to be considered. This introduces the concept of "target coverage" that basically quantifies receptor occupancy by the therapeutic ligand.[28] To describe this, the rate of offset of the ligand from the receptor, once bound, is required. The equilibrium dissociation constant of the ligand–receptor complex denotes potency and formally is the ratio of the rate of offset of the ligand form the receptor divided by the rate of onset of the ligand to the receptor. Specifically, ligands bind to the target with a rate constant k_1 and also dissociate from that target once bound with a rate constant k_2. These are first-order reactions, thus the rate of association of the ligand with the protein is $[A]k_1\theta_0$, where the concentration is $[A]$ and θ_0 refers to the fraction of binding sites for the ligand that are free to bind ligand. Similarly, the rate of dissociation is given by $k_2\theta_A$, where θ_A is the fraction of sites already occupied by the ligand ($\theta_A = 1 - \theta_0$). At equilibrium, the rate of association equals the rate of dissociation and $[A]k_1(1 - \rho_A) = k_2\rho_A$. This yields $\theta_A = [A]/([A] + k_2/k_1)$, which is the Langmuir adsorption isotherm as used in pharmacology with the naming modification K_A (k_2/k_1); this represents the concentration of ligand that binds to 50% of the available sites and functions as an inverse correlate of affinity (affinity $= K_A^{-1}$). Therefore, it can be seen that two ligands could have identical potencies but quite different rates of action. For instance, a ligand can have $K_A = 50$ nM through a ratio of $k_1 = 10^6$ s^{-1} and $k_2 = 0.05$ s^{-1} M (fast acting) or $k_1 = 10^5$ s^{-1} and $k_2 = 0.005$ s^{-1} M (slow acting). The therapeutically relevant parameter for antagonist activity is target coverage that is defined as the length of time the biological target is under the influence of the drug *in vivo*. This parameter requires description in real time since the drug concentration is never constant but rather is dependent on the rate of drug absorption and clearance.

For instance, Figure 2.18 shows two theoretical antagonists both with K_B values of 50 nM.

One, however, has a rapid rate of onset and offset and the other much slower kinetics. As seen in Figure 2.18, the target coverage of the latter antagonist (as measured by the area under the curve of the offset curve) is much greater than the fast antagonist and if target coverage is a requirement for drug therapy, the slower antagonist

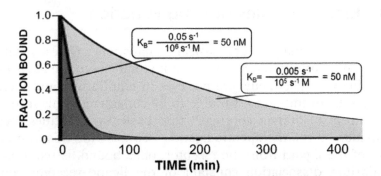

Figure 2.18 Differential target coverage for two equipotent antagonists with differ-ent rates of receptor dissociation. For equal concentrations of an-tagonist, it can be seen that the receptor occupancy of the fast offset ligand ($k_2 = 0.05$ s^{-1}) essentially wanes to zero by 70 min, whereas the occupancy of the slow offset antagonist ($k_2 = 0.005$ s^{-1}) is still 20% at 400 min. The *in vivo* coverage for this latter antagonist will be much higher than the rapid offset compound.

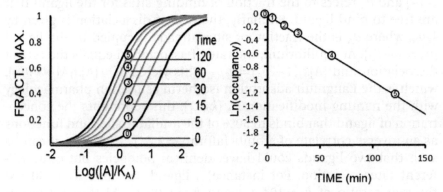

Figure 2.19 *In vitro* measurement of antagonist rate of dissociation. After determin-ation of a control concentration–response curve to the agonist, the system is equilibrated with a concentration of antagonist. After equili-bration, a concentration–response curve to the agonist is obtained in the presence of the antagonist and then the preparation washed with drug-free medium. A defined concentration of agonist is added peri-odically over the wash period (in this case 120 min) and the magnitude of the response to that concentration used to estimate a full concentration–response curve to the agonist at the time of measure-ment. The calculated concentration–response curves are used to estimate a receptor occupancy of the antagonist that is then plotted as a function of time on a natural logarithmic scale. The slope of the linear offset curve yields k_2, the rate of dissociation of the antagonist from the receptor.

would be of more value. Therefore, a measurement of the rate of offset, in addition to antagonist potency, is an essential piece of data for complete characterization of a drug candidate. This can be done through specially designed kinetic experiments. Figure 2.19 shows the effect of a given concentration of antagonist on an agonist concentration–response curve; the dextral displacement is used to determine the equilibrium receptor occupancy in the presence of the antagonist (through the Schild equation). The preparation is then washed with drug-free medium (to remove ambient antagonist in the receptor compartment) and the response to a defined concentration of agonist tested periodically to measure response. The response to this agonist concentration is used to estimate a full concentration–response curve to the agonist at the time of measurement and these estimated curves are used to estimate the residual receptor occupancy by the antagonist. The relationship between the receptor occupancy and wash time is used to generate a value for k_2, the rate of offset of the antagonist from the receptor according to:

$$\rho_t = \rho_0 e^{-k_2 t} \tag{2.8}$$

where ρ_t is the receptor occupancy at time t, t is time and ρ_0 the receptor occupancy in the presence of the antagonist. A semi-logarithmic plot of ρ_t vs. t yields a linear relationship the slope of which is k_2 (see Figure 2.19).

2.9 Conclusions

This chapter summarizes the techniques available to apply theoretical pharmacology to experimental data to provide system independent measures of drug effect. These can, in turn, be used to make predictions for drug effects in therapeutic systems from data obtained in test experimental systems.

References

1. T. Kenakin, *ACS Chem. Biol.*, 2009, **4**, 249.
2. K. Maeda, H. Nakata, Y. Koh, T. Miyakawa, H. Ogata, Y. Takaoka, S. Shibayama, K. Sagawa, D. Fukushima and J. Moravek, *J. Virol.*, 2004, **78**, 8654.
3. J. D. Violin, S. M. DeWire, D. Yamashita, D. H. Rominger, L. Nguyen, K. Schiller, E. J. Whalen, M. Gowen and M. W. Lark, *J. Pharmacol. Exp. Ther.*, 2010, **335**, 572.
4. J. W. Black and P. Leff, *Proc. R. Soc. London, Ser. B*, 1983, **220**, 141.
5. T. Kenakin, C. Watson, V. Muniz-Medina, A. Christopoulos and S. Novick, *ACS Chem. Neurosci.*, 2012, **3**, 193.
6. K. M. Raehal, J. K. L. Walker and L. M. Bohn, *J. Pharmacol. Exp. Ther.*, 2005, **314**, 1195.
7. S. M. DeWire, D. S. Yamashita, D. Rominger, G. Liu, C. L. Cowan, T. M. Graczyk, X.-T. Chen, P. M. Pitis, D. Gotchev, C. Yuan, M. Koblish, M. W. Lark and J. D. Violin, *J. Pharmacol. Exp. Ther.*, 2013, **344**, 708.
8. B. A. Evans, M. Sato, M. Sarwar, D. S. Hutchinson and R. J. Summers, *Br. J. Pharmacol.*, 2010, **159**, 1022.

9. L. Zhou, K. M. Lovell, K. J. Frankowski, S. R. Slauson, A. M. Phillips, J. M. Streicher, E. Stahl, C. L. Schmid, P. Hodder, F. Madoux, M. D. Cameron, T. E. Prisinzano, J. Aubé and L. M. Bohn, *J. Biol. Chem.*, 2013, **288**, 36703.
10. J. M. Stockton, N. J. Birdsall, A. S. Burgen and E. C. Hulme, *Pharmacology*, 1983, **23**, 551.
11. F. J. Ehlert, *Mol. Pharmacol.*, 1988, **33**, 187.
12. T. Kenakin, *Nat. Rev. Drug Discovery*, 2005, **4**, 919.
13. M. R. Price, G. L. Baillie, A. Thomas, L. A. Stevenson, M. Easson, R. Goodwin, A. McLean, L. McIntosh, G. Goodwin and G. Walker, *Mol. Pharmacol.*, 2005, **68**, 1484.
14. F. J. Ehlert, *J. Pharmacol. Exp. Ther.*, 2005, **315**, 740–754.
15. L. Hejnova, S. Tucek and E. E. el-Fakahany, *Eur. J. Pharmacol.*, 1995, **291**, 427–430.
16. J. M. Mathiesen, T. Ulven, L. Martini, L. O. Gerlach, A. Heinemann and E. Kostenis, *Mol. Pharmacol.*, 2005, **68**, 393–402.
17. E. L. Maillet, N. Pellegrini, C. Valant, B. Bucher, M. Hibert, J. J. Bourguignon and J. L. Galzi, *FASEB J.*, 2007, **21**, 2124.
18. A. E. Davey, K. Leach, C. Valant, A. D. Conigrave, P. M. Sexton and A. Christopoulos, *Endocrinology*, 2012, **153**, 1232.
19. A. E. Cook, S. N. Mistry, K. J. Gregory, S. G. Furness, P. M. Sexton, P. J. Scammells, A. D. Conigrave, A. Christopoulos and K. Leach, *Br. J. Pharmacol.*, 2015, **72**, 185.
20. C. Koole, D. Wootten, J. Simms, C. Valant, R. Sridhar, O. L. Woodman, L. J. Miller, R. J. Summers, A. Christopoulos and P. M. Sexton, *Mol. Pharmacol.*, 2010, **78**, 456.
21. S. J. Bradley, C. J. Langmead, J. M. Watson and R. A. Challiss, *Mol. Pharmacol.*, 2011, **79**, 874.
22. O. Arunkakshana and H. O. Schild, *Br. J. Pharmacol.*, 1959, **14**, 48.
23. T. Kenakin and M. Williams, *Biochem. Pharmacol.*, 2014, **87**, 40.
24. I. Langmuir, *J. Am. Chem. Soc.*, 1916, **38**, 2221.
25. T. P. Kenakin, *Br. J. Clin. Pharmacol.*, 2016, **81**, 41.
26. A. V. Hill, *J. Physiol.*, 1910, **40**(Suppl.), iv–vii.
27. G. L. Baillie, J. G. I. Horswil, S. Anavi-Goffer, P. H. Reggio, D. Bolognini, M. E. Abood, S. McAllister, P. G. Strange, G. J. Stephens, R. G. Pertwee and R. A. Ross, *Mol. Pharmacol.*, 2013, **83**, 322.
28. R. A. Copeland, D. L. Pompliano and T. D. Meek, *Nat. Rev. Drug Discovery*, 2006, **5**, 730.

3 Structure-based Drug Discovery and Advances in Protein Receptor Crystallography

Miles Congreve and Fiona H. Marshall*

Heptares Therapeutics Ltd, Biopark, Welwyn Garden City, Hertfordshire AL7 3AX, UK
*Email: fiona.marshall@heptares.com

3.1 Introduction

The cost of taking a compound through clinical drug development to the market place is now considered to be around $2 billion.[1] This ever-increasing figure puts pressure on the pharmaceutical industry, in the light of demanding regulatory requirements and restrictions on healthcare costs, to reduce R and D expenditure, reduce attrition and increase productivity. The genomic era of drug discovery gave the promise of new leads from high-throughput screening (HTS) against single protein targets in recombinant biochemical and cell-based assays. However, such methods have not proved generally effective and have not resulted in increased productivity. HTS frequently generates hits that are less than ideal chemical starting points having high molecular weight (MW) and lipophilicity that then prove difficult to optimise, in particular with regard to their pharmacokinetic properties. Lead optimisation frequently involves addition of further functional groups that inflate MW and lipophilicity increasing overall

Pharmacology for Chemists: Drug Discovery in Context
Edited by Raymond Hill, Terry Kenakin and Tom Blackburn
© The Royal Society of Chemistry 2018
Published by the Royal Society of Chemistry, www.rsc.org

toxicological liabilities.[2,3] The size of HTS libraries has continued to grow as assays have become miniaturised, potentially increasing success rates. However, in practice such libraries will frequently include large numbers of historical compounds reflecting organisations prior targets of interest and only cover a very small proportion of 3-dimensional chemical space.[4]

The popularity of cell-based assays in screening has meant that the precise mechanism of action of the compound at the target is often not known, at least initially. This hampers lead optimisation, which must proceed by an empirical trial and error approach, reducing the scope of targets for which rational design approaches can be employed and making the overall process more challenging. In the case of G protein-coupled receptors (GPCRs) an analysis of the 10 first in class novel chemical entities (NCEs) approved during the decade 2000–2009 showed that many of these drugs are at the upper limits of the Lipinski criteria for drug-like molecules in terms of both MW and lipophilicity.[5]

One strategy used to address these problems has been the development of fragment-based drug discovery (FBDD).[6,7] Fragments are low MW compounds (usually 100–250 Da) that readily fit into a diverse range of binding pockets in proteins. A library of 1000 fragments can sample a much greater proportion of the 3-dimensional chemical space available than that of a large traditional HTS library.[8] Although fragments usually bind with a much lower affinity and must be screened at a higher concentration, they are then selected based on the efficiency of binding,[9] and can offer an excellent starting point for hit generation, generating smaller and more polar leads.

Structure-based drug design (SBDD) is an approach in which an X-ray (or less frequently NMR) structure of the target protein is obtained in complex with ligands of interest, allowing rational optimisation of hits and leads to a drug candidate. SBDD is often combined with FBDD or virtual ligand screening (VLS) methods for hit identification. VLS is an approach in which liganded X-ray structures are used as the template for *in silico* docking of libraries of compounds, followed by subsequent compound screening in order to identify hits at the binding site of interest.[10] SBDD provides a detailed description of the protein–ligand interactions that can be exploited by medicinal and computational chemists to optimise ligand properties. Soluble enzyme targets that can be readily purified and crystallised were the first set of proteins to benefit from SBDD and FBDD methods and many drugs derived from these methods have now reached the market.[11] Arguably, the first medicines designed using SBDD were the HIV protease inhibitors for the treatment of HIV/AIDS such as Saquinavir, Ritonavir and Indinavir in the mid-1990s. Another success story is the development of kinase inhibitors where there are now close to 30 drugs approved, the first of which were Imatinib, Gefitinib and Erlotinib launched around 10 years ago for treatment of various cancers.

SBDD methods are increasingly being applied to target specific protein–protein interactions that were previously considered undruggable, such as Bcl-x$_L$, and MDM2-p53.[12] These very difficult targets would probably not be feasible without the insights provided using structural biology. In addition, resolution of a protein–ligand complex of a molecule to its target gives profound confidence in the molecular mode of action of the compound and encouragement to continue the challenging optimisation process.

In the last 7 years a number of technological developments have resulted in progress in obtaining X-ray structures of membrane proteins such as GPCRs, opening up the potential to apply SBDD methods to this target class.[13,14] Unlike the relatively simple way in which drugs that target enzymes disrupt the reaction catalysed by their target protein, GPCR ligands have a much more complex effect on signalling and activation of cellular processes governed by these receptors. A molecular understanding of how small molecule compounds can alter the conformation and thereby activity of proteins is changing the way we define the pharmacology of compounds. For example, definitions of orthosteric or allosteric modulators must now include reference to the binding site on proteins as well as their behaviour in assays. SBDD for GPCRs promises to revolutionise our understanding of this important class of drug targets and is a particular focus of this review.

In this chapter we describe recent developments in the practice of SBDD and the computational methods that can be applied across a diversity of target classes. New technological developments will be discussed that have driven breakthroughs in membrane crystallography and led to the structures of over 30 GPCRs to be solved. SBDD applied to four specific target classes, kinases, proteases, protein–protein interactions and GPCRs, are then briefly reviewed. This is a large and rapidly moving field which is significantly impacting on drug-development pipelines and the future success of the pharmaceutical industry. As such, this review will not look to be comprehensive but instead focus on some recent examples that illustrate the benefits of rational design approaches.

3.2   Principles and Practice of Structure-based Design

Medicinal chemistry is a heavily data driven discipline in which each iteration of synthetic targets is designed using the previous generation of biological data that has been generated. The quality of the medicinal chemistry campaign is critically determined by the quality of the information on those compounds that have already been tested and derive the structure–activity relationship (SAR). Typically, as well as the binding and/or functional effects of compounds against the target of interest and their selectivity against related targets, some or all of the samples will be tested in a range of further assays. These experiments help understand compound solubility and physicochemical properties, metabolic fate in an *in vitro* setting and their activity against key antitargets such as the hERG channel and cytochrome p450 metabolising enzymes. The medicinal chemists will then decide on a further round of analogues to synthesise based on the available data to produce better derivatives in terms of both their biological profile and also their drug-like properties (termed multi-parameter optimisation or MPO).[15]

One of the problems with this approach is that there are so many options for synthetic targets from any given lead compound. Filtering down these options can be very difficult and if only empirical SAR is available the ease of synthesis becomes a significant bias in the selection process, leaving many opportunities for modifications untried. Another issue is so-called "flat SAR" in which the data being generated is largely uninformative for the next round of targets. However, when structural biology has been enabled and there is an understanding of the precise mode of binding of the compound series to the target protein, SBDD allows computational and medicinal chemists to choose targets that make sense within the SAR and MPO trajectory but also to carefully optimise the fit of the ligand within the binding pocket. Intriguingly, targets suggested by the SBDD approach would often not be made empirically as they are more complex and challenging to synthesise. Often, these less straightforward targets can lead to breakthroughs in activity that move the project forward and build confidence that a drug candidate can be identified in a reasonable time frame.

SBDD also encourages the "atom by atom" optimisation of a ligand, especially if the starting point is a fragment hit (FBDD).[16] The concept of ligand efficiency (LE) introduced in 2004 normalises binding affinity for molecules of different size, allowing the quality of the ligand to be tracked during optimisation and furthermore encourages a rational understanding of the contribution of each atom to the potency of a compound.[9] Hydrogen bonding contacts between the ligand and the protein are easily visualised in X-ray crystal structures and can be optimised or new contacts can be introduced. Because of this, compounds derived from SBDD are often relatively polar, a property now understood to help reduce attrition due to toxicological findings.[2] The understanding of molecular interactions has improved significantly over the last 10 years and is the subject of a recent and excellent review.[17] As well as direct interactions to a given binding site the role of water molecules in forming indirect contacts (through-water H-bonding networks) and of solvation and desolvation of both ligands and binding sites are starting to be rationalised and increasingly considered in SBDD campaigns.[18] The role of water may be particularly important in hydrophobic binding pockets such as those often seen in GPCRs as opposed to solvent-exposed clefts more common in enzymes.[19]

Despite the tremendous progress and the broad application of SBDD methods, there is still a great deal to learn about the factors that govern the energetics of binding and therefore the affinity to the target for a given ligand, making it very difficult to predict ligand affinity *in silico*.[20] The flexibility of small molecules and of protein targets is important and hard to incorporate into the design process; there is much research ongoing in the field of molecular dynamics to address this.[21] The thermodynamics of binding of ligands can much more easily be measured today with modern instruments, but the relative contributions of entropy and enthalpy to binding is difficult to understand, although some gross trends related to molecular complexity are emerging.[22] Finally, the kinetics of binding of compounds is another parameter that can be measured and in best cases optimised (structure kinetics relationship, SKR), for example when it is desirable to have a slow off-rate of binding to drive a pharmacodynamic effect *in vivo*.[23] Today, the factors that determine on- and off-rates are poorly understood and SKR is often an empirical process where drug kinetics are rationalised rather than truly designed. Importantly, however, SBDD and structural biology tend to go hand in hand with biophysical methods because it is only when good quantities of pure protein are available that both disciplines are enabled. This has the effect that SBDD programs are often in the fortunate position to be able to both understand how molecules bind to their protein target but also to characterise the biophysical parameters associated with the interaction. Overall, the quality of data on targets being synthesised by medicinal chemists is extremely high, further increasing the probability of success of the project.

The impact of SBDD on the way that medicinal chemists work and on best practice to design drugs is illustrated by the plot shown in Figure 3.1. Using the RCSB PDB databank of deposited X-ray crystal structures of biological macromolecules (http://www.rcsb.org/pdb/home/home.do) and limiting citations to the Journal of Medicinal Chemistry, the figure plots the number of X-ray structures deposited from these journal publications (PDB files) by year from when the first structure was published (4TIM, 1992) in a cumulative manner. It is reasonable to assume that (for this journal) the structures described are being used for drug design. The figure clearly highlights how the medicinal chemistry community is using X-ray data of ligands bound to protein targets for SBDD in a rapidly expanding fashion over time and that SBDD is now a major driver of medicinal chemistry projects whenever it is available.

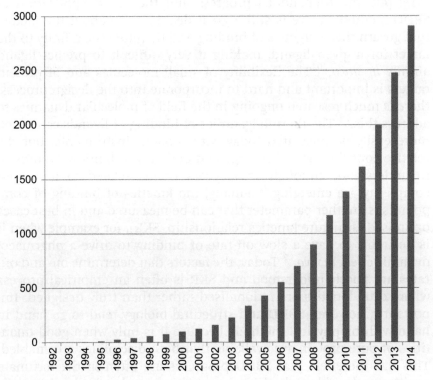

Figure 3.1 Cumulative structures deposited in RCSB from Journal of Medicinal Chemistry citations by year.

3.2.1 Assessment of Druggability

Structural biology studies detailing which parts of a ligand form energetically favourable interactions with its target protein (including electrostatic, charge–charge, hydrogen bonding, van der Waals and hydrophobic contacts) and knowledge of the shape and orientation of the ligand are tremendously helpful for drug design. In addition, the effect on binding-site water energetics and solvation can be considered. As well as useful for SBDD, X-ray crystallographic information describing the features of a binding site, especially when a ligand is bound, gives an assessment of the tractability of the target to small molecule drug discovery, termed druggability. There is an extensive literature on this topic, with discussions of binding-site topography, the balance of polar and lipophilicity required, induced fit by ligands and a number of other considerations.[24,25] A comparison of the properties of GPCR bindings sites compared with some well-studied enzyme targets has been reported.[19] There are also a number of studies using FBDD as an approach to assessing druggability; the success or failure of fragment hit identification *versus* a protein target can be used as an indication of its tractability and the determination of fragment cocomplexes by X-ray crystallography will describe any binding site "hot spots".[26] Overall, information from structural biology on a given target is a useful guide to how much effort it is likely to take to discover a small molecule drug, or indeed may suggest that a biological agent such as a peptide or monoclonal antibody approach might be more appropriate.

3.3 New Techniques in Membrane Protein Crystallography

X-ray structure determination of membrane proteins is highly challenging due to their poor natural abundance, high intrinsic flexibility, difficulty in extraction from the membrane, instability in detergents used for crystallisation and problems in the formation of well-ordered 3D crystals.[27,28] In recent years there has been a concerted effort to overcome these diverse problems, resulting in an exponential growth of membrane protein structures solved (currently over 1200 in the Protein Database). This section will focus primarily on the technological developments that have led to the structures of G protein-coupled receptors, although many of the techniques described can also be used for other classes of membrane proteins.

The first challenge for membrane proteins is that they are naturally expressed at low levels. Even in recombinant expression systems, only small quantities of purified protein can be generated compared to soluble proteins. Although bacterial cells such as *E. coli* provide a simple and readily scalable system for expression, many eukaryotic membrane proteins do not express well or fold properly in bacterial cells. However, the system has proved useful for the purification of a number of GPCRs and resulted in the structure of a stabilised neurotensin receptor.[29] Yeast cells can also be grown easily and can be modified to express some GPCRs. However, the cell wall can hinder the isolation of membranes and differences in membrane lipid composition may affect the functionality of some proteins. The structures of the histamine H_1 receptor[30] and adenosine A_{2A} receptor[31] have been obtained following expression in *Pichia pastoris*.

Insect cells have been the most useful system for membrane protein expression with the majority of GPCR structures solved using protein derived from insect cells infected with the baculovirus expression system.[27] Several ion-channel structures have also used insect cell-expression systems including the acid-sensing ion channel ASIC1[32] and the P2X4 ligand-gated channel.[33]

Mammalian cell expression represents the most physiologically relevant system for expression of membrane proteins for drug discovery purposes; however, the level of expression is lower than other systems and the scale up of cells in culture is more difficult. Nevertheless, a number of groups have used stable or transient expression in mammalian cells to generate protein for structural studies. Baculovirus transduction of mammalian cells (BacMam) is an emerging strategy for increasing expression levels in mammalian cells.[34]

Protein instability and conformational flexibility is another impediment to successful crystallisation. The addition of high affinity ligands can increase protein stability and drive a specific conformational state based on the pharmacology of the specific ligand. Engineering proteins to be more stable through mutagenesis has provided particularly successful for GPCRs. Conformational thermostabilisation can be used to obtain structures in antagonist conformation – for example the β_1 adrenergic receptor[35] and adenosine A_{2A} receptor[36] as well as in agonist conformation – for example the neurotensin receptor.[37] Thermostabilising mutations have also been used for ion channels such as the NMDA[38] and AMPA receptors[39] and the serotonin transporter.[40] A major advantage of using the thermostabilisation method is that costructures can be obtained with very weak ligands including fragments.[36,41] This is of significant importance when using costructures to inform lead optimisation during drug discovery since early chemistry is likely to be focused on weaker starting points.

A number of additional protein-engineering strategies are utilised to facilitate crystallisation. Constructs are routinely truncated to reduce or remove flexible regions such as N- and C-termini as well as loops. The use of fusion proteins is now routinely used for GPCR structures. In this method, soluble proteins that readily crystallise, such as T4 lysozyme, are fused to the GPCR, most commonly in an intracellular loop or at the N-terminus. These act to reduce the flexibility of these regions and also behave as crystallisation chaperones increasing the surface for crystal contacts within the protein lattice.[42]

Solubilisation of the membrane protein from the cell membrane and subsequent crystallisation is critically dependent on the choice of detergent. Detergents effectively replace and mimic the membrane lipid environment surrounding the protein in a spherical micelle. Shorter-chain detergents leave sufficient regions of the protein exposed to allow vapour diffusion crystallisation conditions, which are usually used for soluble proteins. However, the majority of GPCRs cannot survive in such harsh environments unless they have been thermostabilised. The stabilised β_1 receptor[35] and A_{2A} receptors[36] were crystallised using vapour diffusion. New detergents are being developed as tools for membrane protein work,[28] such as the neopentyl glycol class,[43] which provide improved stability at low concentrations. An important development in this area is the use of LCP (lipidic cubic phases) crystallisation. The cubic phase is obtained by mixing lipid, buffer and protein at a certain ratio to allow the formation of lipid bilayers separated by water channels. This has the advantage of providing a less harsh environment that more closely mimics the plasma membrane. More than 250 membrane protein structures have been solved to date using this approach including 62 GPCR and 28 transporter structures (http://cherezov.scripps.edu/structures.htm). LCP is particularly effective for GPCR structures when combined with fusion proteins.

The advances described above have increased our ability to generate crystals of membrane proteins. However, collection of diffraction data from resulting crystals remains a challenge due to their small size, fragility, poor order and sensitivity to radiation damage.[28] Advances in X-ray sources, in particular the use of dedicated microfocus beamlines at a number of synchrotrons including the Diamond Light source in Oxford, UK (http://www.diamond.ac.uk/Home.html) have greatly facilitated data collection from such crystals.

X-ray free electron lasers (XFEL) is a recent development that offers an exciting new approach to membrane protein structures.[44] XFEL generate pulses of light that are a million times brighter than those from standard synchrotrons. The femtosecond pulses allow rapid collection of structural information before disintegration of the sample. Femtosecond nanocrystallography is a technique whereby a continuous jet stream of microcrystals crosses the X-ray beam. This method combined with LCP has resulted in several membrane protein structures, including the frizzled family receptor smoothened[45] and the serotonin 5HT$_{2B}$ receptor.[46]

Developments across all aspects of membrane protein structural biology have led to a dramatic increase in the numbers of structures solved and have enabled a new era of structure-based drug design for many therapeutically important targets.

3.4 Structural Biology of G Protein-coupled Receptors

Here, the progress in our understanding of the 3-dimentional structures of GPCRs and their ligand binding sites that have emerged over the last 9 years are briefly outlined.

3.4.1 Orthosteric Binding Sites

The opsin visual pigment receptor family is one of the oldest and largest groups of GPCR with opsins present in almost all vertebrate species as well as some invertebrates such as squid. The opsin ligand 11-*cis*-retinal is covalently bound to the protein and undergoes 11-*cis* to all-*trans* isomerisation in response to light, triggering conformational changes in the protein that lead to activation of downstream signalling cascades.[47,48] The 11-*cis*-retinal binding site is entirely enclosed and lies between TM5 and TM6 with the covalent Schiff's base linkage to Lys296 (7.43) (numbers in parentheses refer to the Ballesteros–Weinstein numbering system that enables comparison across different GPCRs and has recently been updated).[49] The retinal binding site can be considered as the prototypical orthosteric binding site for the wider Rhodopsin family known as Class A GPCRs. Within this subclass a similar binding site is found for the aminergic receptors including the adrenergic receptors,[50] dopamine receptors[51] and histamine receptors.[30] The natural ligands of these receptors sit in approximately the same region of the receptor as retinal does in rhodopsin, making interactions with TM3, 5 and 7. In the aminergic family there is a highly conserved acidic residue at positon 3.32. Since these freely diffusing ligands need to readily pass in and out of the receptor, the binding sites are more open than found in rhodopsin.

The first structure of a peptide receptor solved was that of the chemokine receptor CXCR4 in complex with a cyclic peptide antagonist (CVX15) and a small molecule antagonist (IT1t).[52] These ligands both bind higher up in the receptor than ligands to the aminergic receptor. Indeed, the large peptide ligand protrudes out of the top of the receptor. It is likely that the endogenous peptide agonist also binds and interacts with extracellular regions of the receptor in addition to accessing the outer parts of the transmembrane domain. The small-molecule antagonist binds at the deeper end of the same pocket as the peptide ligand. Interestingly, in this structure the second extracellular loop folds over the top of the receptor to form a cap, which likely contributes to the slow off-rate and high affinity of this antagonist. Subsequently, the related chemokine receptor CCR5 structure was solved in complex with maraviroc used to treat HIV. Maraviroc sits in a more horizontal orientation deeper in the pocket than the CXCR4 ligands and is considered to be allosteric in its mode of action since it does not directly compete with the natural ligand CCL10 nor the HIV envelope protein gp120.[53] Binding of maraviroc to the receptor is considered to modify the receptor conformation in such a way that the receptor is no longer accessible to orthosteric ligands or the virus itself.

X-ray structures of all four opioid receptors have been solved in complex with a variety of antagonist ligands. A morphine derivative β-funaltrexamine was bound to the μ-opioid receptor,[54] natrindole to the δ-opioid receptor,[55] JDTic to the κ receptor[56] and the peptide mimetic C-24 to the nociception receptor.[57] A common feature of the opioid receptors is the binding site is very open compared to those in the aminergic family. This is in part due to the conformation of ECL2 that forms a β-strand. These open binding sites have presumably evolved to bind the natural peptide ligands and this is particularly noticeable in the nociception structure (NOP) where the binding site is further enlarged compared to the other members of the family through a shift in the extracellular end of TM5, thus enabling the binding of the larger (17 amino acid) endogenous ligand. The deeper part of the binding pocket is considered to be key for driving activation of the receptor. It is in this region that small-molecule morphine-related agonists and antagonists bind within the receptor.

3.4.2 Allosteric Binding Sites

The majority of GPCR drugs currently on the market are competitive with the native ligands; however, it is clear that many alternative binding sites exist on GPCRs that can stabilise the antagonist conformation to which the agonist has low affinity. Allosteric modulators (both positive or negative) have a number of potential advantages over orthosteric drugs including improved selectivity, temporal modulation of agonist activation, altered signalling and an improved side-effect profile.[58,59]

Using unbiased molecular dynamic (MD) simulations[60] to study the process of ligand binding to β-adrenergic receptors it was found that many ligands, including both agonists and antagonist, took the same pathway of entry to both the β_1 and β_2 receptors, initially making contact with a region close to the extracellular surface, which could be considered as an "outer vestibule" at a distance of 15 Å from the orthosteric binding site. This could be considered an outer binding site for the ligand, allowing formation of an early encounter complex on the way to the orthosteric site. These transient binding sites may be exploited by allosteric modulators. For example, the muscarinic M_2 positive allosteric modulator (PAM) LY2119620 was cocrystallised with the orthosteric agonist iperoxo[61] and found to bind to the extracellular vestibule directly above the orthosteric binding site. Binding of the allosteric modulator to this site results in closure of the entrance to the orthosteric site, and presumably the PAM effect is caused by a resultant slowing of the dissociation of the agonist. Binding and closure of the vestibule by the allosteric modulator may also directly stabilise the agonist state of the receptor allowing the PAM to show direct agonist activity. In contrast, negative allosteric modulators may act to restrict access of the orthosteric ligand to its binding pocket and/or by stabilising the antagonist conformation.

Structural studies are also identifying allosteric binding sites on other GPCRs. Of particular interest is the binding site of the ago-allosteric modulator TAK-875 (fasiglifam) on the free fatty acid receptor FFAR1 (previously known as GPR40), an important therapeutic target for the treatment of diabetes. TAK-875 binds to a site close to the extracellular region between TM3–TM5 and ECL2 and is positioned such that the ligand protrudes into the lipid bilayer. This suggests that entry of the ligand is likely *via* the lipid membrane through the side of the receptor. Further examination of FFAR1 indicated the existence of two additional binding pockets in the receptor; one of which would allow a ligand to bind between TM4 and TM5 and one that may represent the orthosteric pocket, closer to the canonical binding site found in other Class A receptors.[62,63] The identification of three potential allosteric binding sites was previously identified on the receptor using mutagenesis studies, and furthermore, ligands binding at these sites all demonstrated strong positive cooperativity with the endogenous fatty acid.[64] FFAR1 has been shown to couple both to Gs and to Gq and activation of both are required to illicit maximal stimulation of incretins. Whilst endogenous fatty acids stimulate only Gq signalling, some synthetic agonists have been identified that can activate both Gq and Gs signalling.[65] This has very important consequences for the design of drugs directed at this receptor.

Class B and Class C GPCRs both have large extracellular protein domains at their N termini that are involved in ligand binding, although drugs targeted at these receptors are primarily allosteric modulators that bind within the 7-transmembrane domain. The CRF_1 receptor is a member of the secretin Class B subgroup of GPCRs that bind large peptide hormone ligands. CRF_1 is involved in regulation of the hypothalamic pituitary axis and is a drug target for stress-related disorders or disorders of adrenal function. The X-ray structure of CRF_1 showed that the antagonist CP-376395, which was identified using high-throughput screening, bound to a novel and unexpected location deep within the 7-TM region, 13–23 Å lower than the classical Class A binding site.[66] Druggability analysis of this binding pocket compared to the open orthosteric site at the top of the receptor suggests that future drug discovery efforts at Class B receptors should focus on this and other allosteric binding sites.[67] X-ray structures of the Class C metabotropic glutamate receptors $mGlu_1$[68] and $mGlu_5$[69] also revealed allosteric ligands bound within the 7-TM domain. In contrast to the Class B receptor there is no interaction of the orthosteric ligand glutamate with the 7-TM domain. Instead, glutamate binds exclusively to the so-called venus fly trap domain triggering a conformational change between a pair of receptors within a dimer to activate the receptor. Unlike other GPCRs, which have a ligand that binds to the TM domain, the top of the receptor only has a very narrow entrance to the allosteric binding pocket. The highly constrained shape of the binding pocket in $mGlu_5$ observed in the crystal structure helped explain the predominance of the alkyne linker found in many $mGlu_5$ negative allosteric modulators.[70]

3.5 Structure-based Drug Discovery Applied to Protein Classes

The rest of this review will outline a number of examples of SBDD for some key target classes. Recent studies are included and in most cases have been selected to illustrate a particular SBDD approach.

3.5.1 Kinases

Protein kinases are enzymes that catalyse the transfer of a phosphate group to a specific serine, threonine, tyrosine, or histidine of a protein substrate. Kinases are used to transmit signals inside the cell; phosphorylation of molecules can enhance or inhibit their activity and modulate their ability to interact with other molecules. The addition (and removal) of phosphoryl groups provides a way to regulate cellular processes because various kinases can respond to different conditions or signals in a complex network of signalling cascades. Mutations in kinases that lead to a loss-of-function or gain-of-function can cause disease in humans, particularly various types of cancers.[71] Initially, there was scepticism that kinases would be useful drug targets because the family contains over 500 members all containing an ATP binding domain (the site that inhibitors will usually bind) that would have significant structural similarity. However, kinases have proven quite tractable for medicinal chemists and the selectivity challenge can usually be overcome using SBDD or by developing compounds that hit multiple kinases for which the overall spectrum of activity is acceptable or perhaps beneficial. Many kinases have proven to be good drug targets and multiple drugs are now approved, mainly for the treatment of cancer.[72] An extensive database of protein–ligand structures is now available as a resource for research in the kinase field such that the binding mode of ligands can now potentially be predicted *in silico* for the design and optimisation of compounds. As well as the ATP binding site, allosteric sites have started to emerge and are a source of new intellectual property in this crowded area; the first allosteric kinase inhibitor, Trametinib (MEK kinase), has now been launched.[73] Kinases show significant flexibility in the ATP binding domain giving rise to different classes of inhibitors depending on which conformation of the protein they bind to (competitive inhibitors are type I, noncompetitive type II and III).[74]

FBDD has been extensively applied to kinase targets, where structural biology has been very successful at producing X-ray structures for many proteins and with multiple ligands including very low affinity fragments.[75] Indeed, the first launched drug developed using FBDD is the B-RAF inhibitor Vemurafenib, marketed in 2011 for the treatment of melanoma.[76] The drug was developed from a small fragment 1 (Scheme 3.1) by a process called "fragment evolution", where the fragment is substituted around the scaffold to occupy additional regions of the binding site in a productive fashion.[77] Because the initial fragment is so small and polar, SBDD is greatly facilitated allowing much scope to investigate multiple modifications suggested by *in silico* docking to optimise the fit to the binding site. In most cases, including this study, the starting fragment maintains its initial position and interactions with the protein, which tend to be of high quality; however, there are reports of fragments switching binding modes during optimisation, confounding the process.[78]

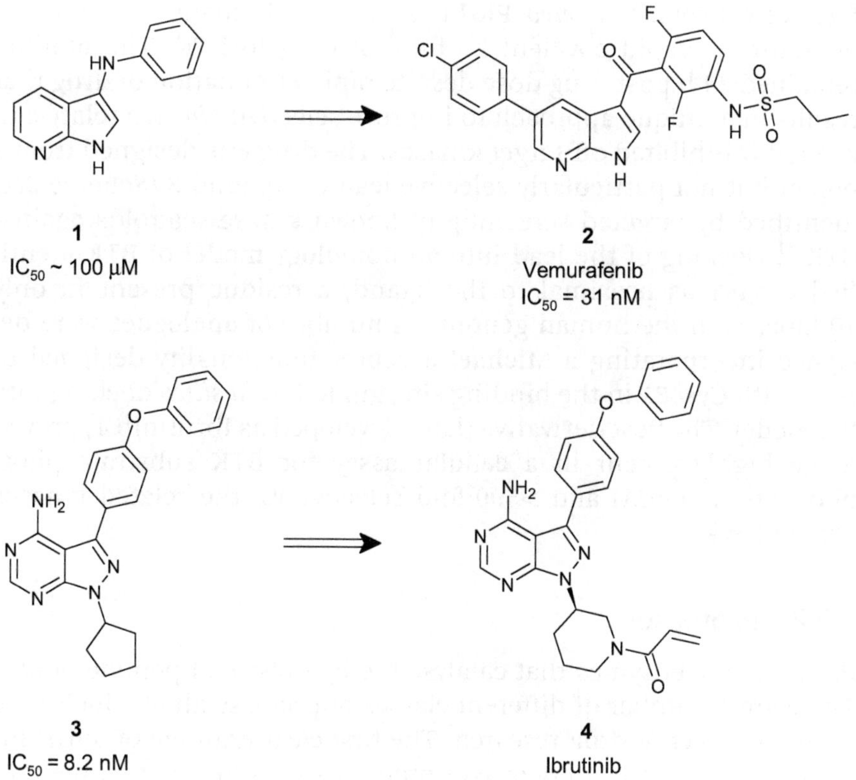

1

$IC_{50} \sim 100\ \mu M$

2

Vemurafenib

$IC_{50} = 31\ nM$

3

$IC_{50} = 8.2\ nM$

4

Ibrutinib

$IC_{50} = 0.72\ nM$

Scheme 3.1 Examples of SBDD for kinase targets.

Vemurafenib 2 was designed to bind to both the wild-type B-RAF (100 nM) and preferentially to the V600E mutant (31 nM) that is very common as a genetic driver of melanoma. It is otherwise a quite selective agent; in a 200-member kinase panel the most relevant additional activity was inhibition of c-Raf-1 (48 nM).

Using SBDD to inform introduction of a reactive functional group into an inhibitor is a strategy gaining popularity. A covalently binding ligand will potentially show very potent irreversible inhibition of the target and impart very high selectivity, provided that the reacting group does not show a high degree of reactivity with general cellular nucleophiles.[79] Covalent inhibitors have long been used to target proteases in which the enzyme has an activated nucleophilic group within the binding site to carry out the proteolysis (see Section 3.5.2). However, it has also been possible to target residues inside a binding pocket that are not usually involved in any chemistry, directed by SBDD. In 2013, the Bruton's tyrosine kinase (BTK) inhibitor Ibrutinib was launched for the treatment of B-cell lymphoproliferative cancers.[80] *In vivo* PK/PD studies in humans and animals demonstrate rapid covalent binding of drug to BTK. The binding remained 24 h post drug dose despite rapid elimination of drug that results in a unique approach to improve selectivity *in vivo* relative to reversibly inhibited off-target kinases. The drug was designed from a potent but not particularly selective lead compound 3 (Scheme 3.1) identified by targeted screening of known kinase scaffolds against BTK.[81] Docking of the lead into an homology model of BTK identified Cys481 as proximal to the ligand, a residue present in only 10 kinases in the human genome. A number of analogues were designed incorporating a Michael acceptor functionality designed to react with Cys481 in the binding site, guided by *in silico* docking into the model. The best derivative (later developed as Ibrutinib 4) proved to be highly potent in a cellular assay for BTK substrate phosphorylation (1 nM) and >500-fold selective *vs.* the related kinases Lyn and Syk.

3.5.2 Proteases

Proteases are enzymes that catalyse the hydrolysis of peptide bonds. There are a number of different classes of protease all of which have been the target of drug research. The first clear example of SBDD for proteases is the discovery of HIV protease drugs for the treatment of HIV/AIDs.[11] HIV protease is an aspartyl protease that exists as a symmetrical homodimer. The enzyme plays an essential part in the life cycle of HIV and there are now 10 FDA approved drugs that inhibit the wild type and a broad range of mutated forms that arise to give drug resistance. Other drugs discovered using SBDD that inhibit serine and aspartyl proteases include Aliskiren (Renin), Rivaroxaban, Edoxaban and Apixaban (Factor Xa), Dabigatran (thrombin), Boceprevir, Telaprevir and Simeprevir (HCV NS3/4A protease).

The most recent success story for SBDD within the protease arena is the discovery of drugs targeting the HCV protease NS3/4A. In a similar way to HIV protease this enzyme activity is essential for HCV to replicate and for many years was recognised as an excellent drug target. However, the binding site revealed by X-ray crystallography was not compatible with the binding of conventional drug-like small molecules.[82] A huge effort shared by many pharmaceutical and biotechnology companies very slowly established how heavily modified tetrapeptides could be identified with potent activity, often involving incorporation of an electrophilic group that bound co-valently with the nucleophilic serine residue that carries out the enzymes' proteolytic reaction. This eventually led to the discovery of Boceprevir and Telaprevir; it is unlikely that without the insight and understanding of how molecules bind to this enzyme from multiple X-ray protein–ligand complexes that these drugs would have been discovered.[83]

One of the latest NS3/4A inhibitor to be launched as a drug is Simeprevir (TMC435).[84] Notably, this is the first drug in which a covalent linkage to the enzyme is not formed, so it is a competitive reversible inhibitor. Also, Simeprevir is the first drug that does not contain a central scaffold that is derived from proline, instead containing a cyclopentane moiety, such that it does not contain an amino acid core. Building on the body of research available the drug was designed as an analogue of BILN2061 **5** (Scheme 3.2), the first inhibitor to enter clinical trials and show efficacy in patients.[85] Analogues were designed based on *in silico* docking studies *vs.* the protease binding site using the novel cyclopentane scaffold and incorporating heterocyclic groups to bind to the S2 pocket in the site. A strategy often used in protease research, where ligands are based on peptide substrates, is to incorporate a macrocyclic bridge between two of the amino acid substituents. Macrocyclic compounds, in which the flexibility of the compound has been significantly reduced potentially reinforcing the bioactive conformation, can have a significant impact on the pharmaceutical properties of high MW compounds that do not obey Lipinski's rules for oral drugs.[86] This strategy has been used extensively for HCV protease inhibitors and here macrocycles were

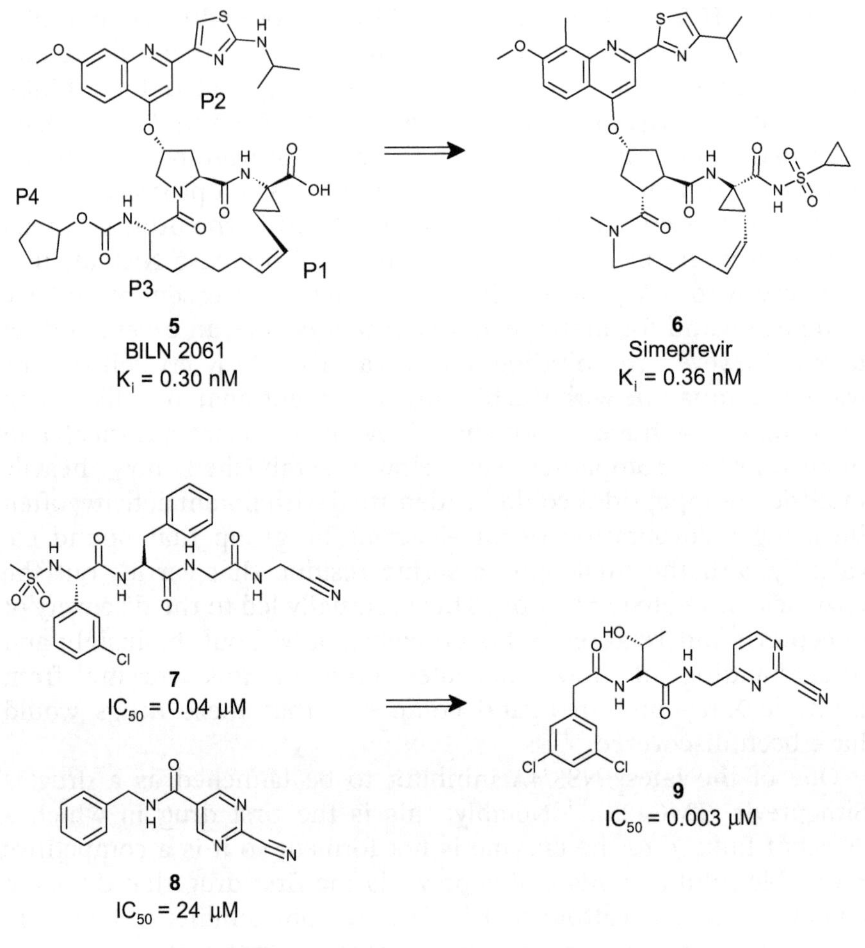

Scheme 3.2   Examples of SBDD for protease targets.

designed using SBDD joining the P1 and P3 positions (P1, P2, P3 numbering based on the substrate peptide). Incorporation of the macrocycle allowed simplification of the molecule without loss of potency and this coupled with modifications at P2 yielded Simeprevir **6**. Macrocycles can be seen as an example of the medicinal chemistry strategy of reducing the numbers of rotatable bonds to optimise potency and improve drug-like properties, here informed using SBDD.[87]

Another SBDD strategy is to use the X-ray binding data of molecules from different chemical series to design new hybrid compounds that contain groups from separate ligands. In FBDD, one approach is termed "fragment linking" where two (essentially) nonoverlapping compounds are joined together.[77] In a recent report, SBDD efforts focused on the protease target Adenain used a linking strategy to join a small peptide **7** and a fragment hit **8** together to derive a hybrid lead compound **9** (Scheme 3.2).[88] In this elegant study the linked molecules showed potent inhibition of the target and the predicted binding pose was validated by X-ray crystallography with lead compounds in complex with the protein. Linking of two separate hits to form a new molecule is very challenging in the absence of X-ray co-complex information and this work shows the power of this approach where the structural biology informs the design process.

3.5.3 Protein–Protein Interactions

The targeting of protein–protein interfaces (PPIs) of regulatory multi-protein complexes has become a significant focus in drug discovery due to the biological data supporting many as good targets for intervention for the treatment of human diseases. However, from structural biology studies most interfaces would be classed as "undruggable" by conventional analyses, being large, flat and featureless. Over the past decade, there has been an increasing understanding that some PPIs have druggable "hotspots" that contribute much of the free energy of interaction.[12] In addition, there has been the recognition that many PPIs use a continuous epitope from one protein and a well-defined groove on the other.[89] This has encouraged the development of stapled α-helical peptides and other proteomimetic approaches and stimulated FBDD to identify small low-affinity "footholds" that might subsequently be elaborated to useful inhibitors.

An excellent example of the potential and challenges of developing inhibitors of PPIs is the discovery of ABT-737 and subsequently Navitoclax (ABT-263) and ABT-199 at Abbott Laboratories.[90] These compounds mimic the key α-helix that binds Bcl-x_L, a protein critical in programmed cell death that is overexpressed in many tumours, inhibiting apoptosis. In this case FBDD and fragment linking were used; two different fragments **10** and **11** (Scheme 3.3) that were found to bind simultaneously to Bcl-x_L at quite distinct sites were evolved and then linked into a single molecule ABT-737 **12**.[91] Having established that potent nonpeptidic molecules could be designed, a large effort then eventually led to further agents with progressible pharmaceutical properties that support clinical study for the treatment of cancer.

An obvious challenge of PPIs is identifying inhibitors with properties that lend themselves to finding oral drugs. As noted earlier,

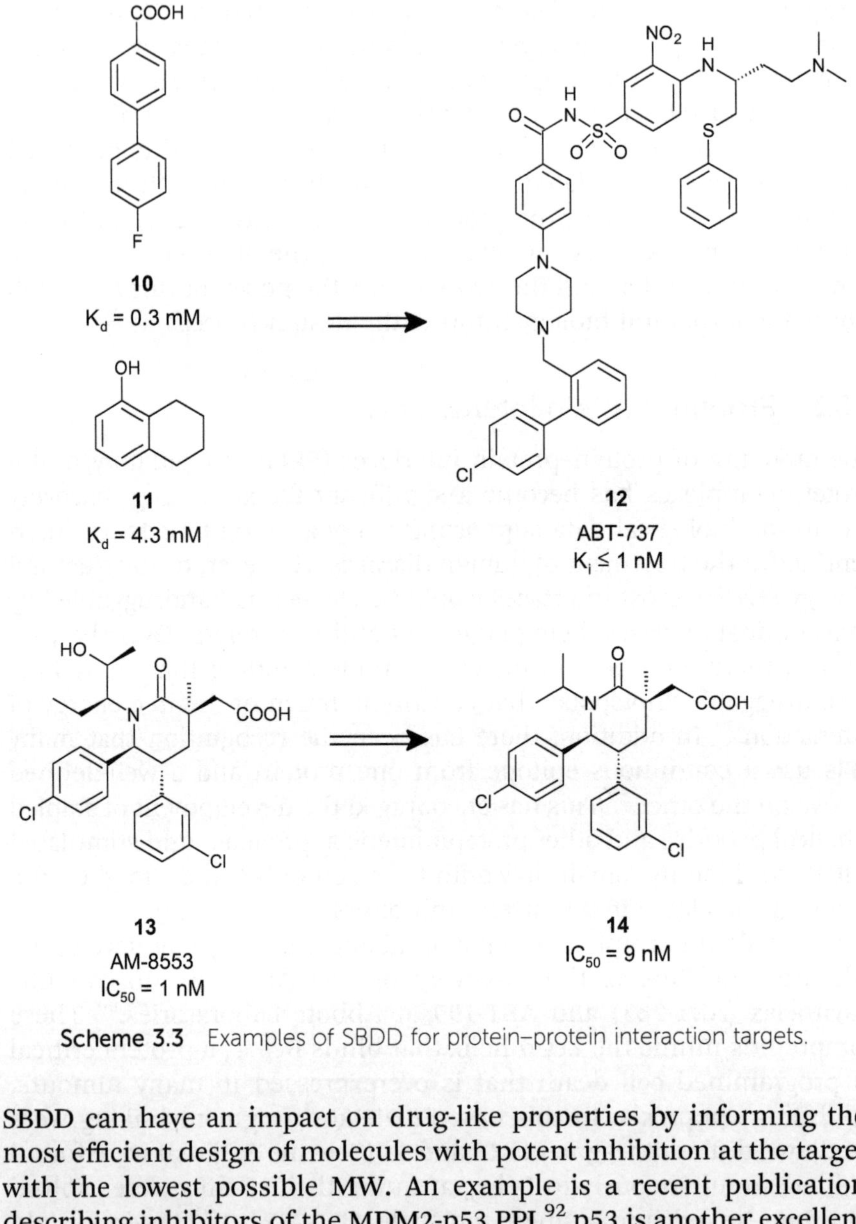

Scheme 3.3 Examples of SBDD for protein–protein interaction targets.

SBDD can have an impact on drug-like properties by informing the most efficient design of molecules with potent inhibition at the target with the lowest possible MW. An example is a recent publication describing inhibitors of the MDM2-p53 PPI.[92] p53 is another excellent cancer target, inducing cell death when the cell is under stress and is mutated in ∼50% of human cancers.[93] MDM2 binds to p53 and serves to regulate it; inhibitors of this PPI have been progressed into clinical studies for the treatment of cancers (including RG7112, RG7388, SAR299155, MK-8242, AMG 232, CGM-097 and DS-3032b) and the development of many of these agents has been supported by multiple liganded X-ray structures and SBDD.[94] Building on earlier work leading to the discovery of AM-8553 **13** (Scheme 3.3), Yu and coworkers used X-ray data to guide careful optimisation of derivatives with a key objective of improved PK properties. This led to a simplified derivative **14** with a reduced MW of 434 Da but comparable potency at the target. This compound had significantly improved PK properties in rat and mouse and good *in vivo* efficacy. The strategy of using SBDD to reduce MW (and also lipophilicity) without loss of potency is particularly valuable for less-druggable targets where molecular properties are challenging and do not initially support identification of compounds with good PK. SBDD in the same series has also led to the discovery of AMG 232 (now in clinical trials), in which a sulfone substituent was introduced that improved affinity by forming a new hydrophobic interaction within the protein cleft and had excellent PK and efficacy preclinically.[95]

3.5.4 G Protein-coupled Receptors

To date, there are limited numbers of examples published where SBDD has been carried out for GPCRs in which a crystal structure of a ligand has been solved for the chemical series being optimised. One example from our laboratories is the discovery and optimisation of β_1-Adrenergic Receptor (β_1AR) fragments using SBDD.[41] Using an SPR screen with thermostabilised β_1AR, fragment hits such as **15** (Scheme 3.4) were identified and their activity was confirmed in an orthogonal radioligand binding assay. Initially guided by docking of molecules into a crystal structure of β_1AR (with the X-ray ligand removed) a number of analogues of the initial fragments were purchased and screened. This gave hit molecules such as **16** with improved affinity and ligand efficiency. With these results in hand, it was possible to prospectively design compounds with the anticipation that further interactions, known to be important for this receptor from other X-ray structures, could be introduced. This approach allowed more polar molecules to be identified that reduced the lipophilicity of the series. This strategy yielded lead compounds such as indole **17**, with high affinity, good ligand efficiency and solubility. Crystal structures of the optimised fragments in complex with β_1AR were subsequently solved giving a start point for further structure-based optimisation.

A major area of GPCR research is discovery of agonists and antagonists of adenosine receptors A_1, A_{2A}, A_{2B} and A_3. Multiple crystal structures of ligands are now available for Adenosine A_{2A} in both agonist and antagonist conformations.[96] In our own laboratories we

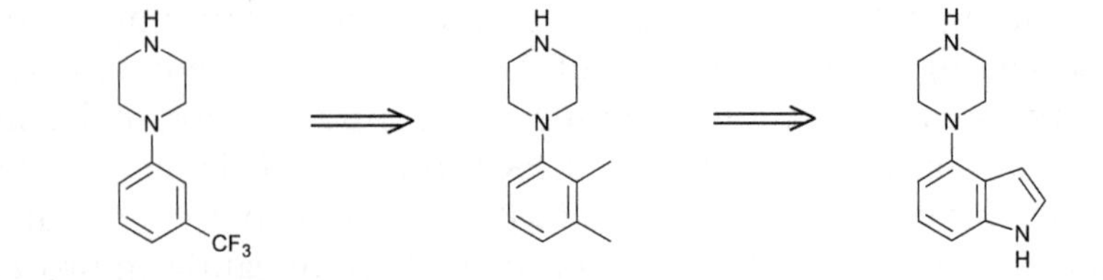

15
β_1AR pK_D 4.8 (LE 0.41)

16
β_1AR pK_i 6.3 (LE 0.61)

17
β_1AR pK_i7.2 (LE 0.65)

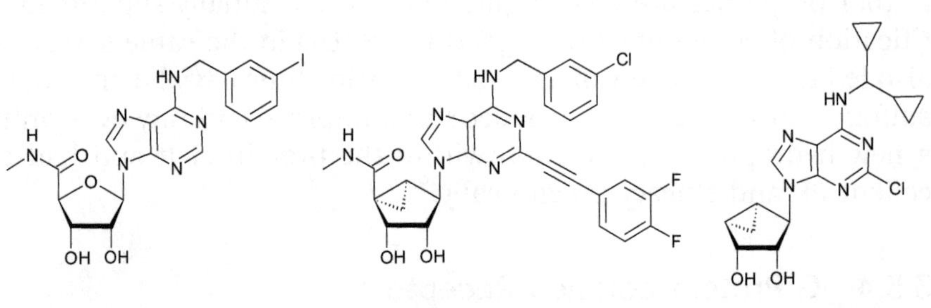

18
IB-MECA (CF101)
A_3AR pK_i 8.7

19
MRS5698
A_3AR pK_i8.5

20
MRS5474
A_1AR pK_i 7.3

Scheme 3.4 Examples of SBDD for GPCR targets.

have used SBDD to design a clinical candidate for use as an Adenosine A_{2A} antagonist and this has been reviewed elsewhere.[97] Adenosine analogues have been studied for many years as agonists of adenosine receptors. Jacobson and coworkers have worked extensively on adenosine receptor agonists and more recently have used the agonist-bound X-ray structures of adenosine A_{2A} for SBDD. Previous work had led to identification of the A_3 agonist IB-MECA **18**. The X-ray ligand binding information can now be used to design derivatives that probe both the ribose pocket (between TM3, 5 and 7) and the extracellular loop regions of the binding site. One compound discovered was MRS5698 **19**, containing a carbocyclic replacement for the ribose ring, having very high A_3 selectivity. This selectivity is derived from the phenylacetylene substituent and can be rationalised by a predicted outward movement of the portion of TM2 near the EL region of the A_3 receptor, in order to preserve hydrogen-bonding interactions with conserved residues in TMs 3, 5, and 7 that lock the adenosine moiety in its binding pocket.[98] This outward movement of TM2, was proposed to be less likely to occur in A_{2A} because the EL region of this AR subtype is conformationally constrained by four disulfide bridges. Further structure-based exploration of this series yielded A_1 receptor agonists including MRS5474 **20** in which the bis-cyclopropylmethyl substituent is combined with truncation of the carboxamide group switching the selectivity profile, which again can be rationalised using the available X-ray data.[99]

3.6 Conclusions

The practice of SBDD is now a critical part of many drug discovery campaigns and enablement of structural biology for protein classes outside of soluble enzyme targets is highly desirable. The many examples in the literature of successful application of SBDD to design therapeutic agents, exemplified by key examples in this chapter, serve to highlight this point. Currently, pharmaceutical companies and academic institutions are seeking to extend their capabilities for structural biology in an attempt to access SBDD for a broader range of therapeutic targets and it is clear that the use of FBDD and computational medicinal chemistry will be positively impacted by these investments. It is likely that the discovery of a new generation of important drugs will be enabled by these developments over the next 10–15 years.

Acknowledgements

The authors would like to thank Rebecca Nonoo for assistance in preparation of this manuscript.

References

1. S. M. Paul, D. S. Mytelka, C. T. Dunwiddie, C. C. Persinger, B. H. Munos, S. R. Lindborg and A. L. Schacht, *Nat. Rev. Drug Discovery*, 2010, **9**, 203.
2. P. D. Leeson and J. R. Empfield, *Annu. Rep. Med. Chem.*, 2010, **45**, 393.
3. M. H. Hann, *Med. Chem. Commun.*, 2011, **2**, 349.
4. P. D. Leeson, A. M. Davis and J. Steele, *Drug Discovery Today: Technol.*, 2004, **1**, 189.
5. M. Congreve, C. J. Langmead, J. S. Mason and F. H. Marshall, *J. Med. Chem.*, 2011, **54**, 4283.
6. D. A. Erlanson, *Top. Curr. Chem.*, 2012, **317**, 1.
7. M. Congreve, G. Chessari, D. Tisi and A. J. Woodhead, *J. Med. Chem.*, 2008, **51**, 3661.
8. T. Fink, H. Bruggesser and J. Reymond, *Angew. Chem., Int. Ed.*, 2005, **44**, 1504.
9. A. L. Hopkins, C. R. Groom and A. Alex, *Drug Discovery Today*, 2004, **9**, 430.

10. T. Zhu, S. Cao, P. Su, R. Patel, D. Shah, H. B. Chokshi, R. Szukala, M. E. Johnson and K. E. Hevener, *J. Med. Chem.*, 2013, **56**, 6560.
11. A. A. Alex and D. S. Millan, *Drug Design Strategies: Quantitative Approaches*, ed. D. J. Livingstone and A. M. Davis, Royal Society of Chemistry, Cambridge, 2012, vol. 13, ch. 5, pp. 108–163.
12. H. Jubb, A. P. Higueruelo, A. Winter and T. L. Blundell, *Trends Pharmacol. Sci.*, 2012, **33**, 241.
13. V. Katritch, V. Cherezov and R. C. Stevens, *Trends Pharmacol. Sci.*, 2012, **33**, 17.
14. M. Congreve, J. M. Dias and F. H. Marshall, *Prog. Med. Chem.*, 2013, **53**, 1.
15. I. Yusof, F. Shah, T. Hashimoto, M. D. Segall and N. Greene, *Drug Discovery Today*, 2014, **19**, 680.
16. M. L. Verdonk and D. C. Rees, *ChemMedChem*, 2008, **3**, 1179.
17. C. Bissantz, B. Kuhn and M. Stahl, *J. Med. Chem.*, 2010, **53**, 5061.
18. D. Alvarez-Garcia and X. Barril, *J. Med. Chem.*, 2014, **57**, 8530.
19. J. S. Mason, A. Bortolato, M. Congreve and F. H. Marshall, *Trends Pharmacol. Sci.*, 2012, **33**, 249.
20. A. R. Leach, B. K. Shoichet and C. E. Peishoff, *J. Med. Chem.*, 2006, **49**, 5851.
21. P. Cozzini, G. E. Kellogg, F. Spyrakis, D. J. Abraham, G. Costantino, A. Emerson, F. Fanelli, H. Gohlke, L. A. Kuhn, G. M. Morris, M. Orozco, T. A. Pertinhez, M. Rizzi and C. A. Sotriffer, *J. Med. Chem.*, 2008, **51**, 6237.
22. G. G. Ferenczy and G. M. Keserű, *J. Chem. Inf. Model.*, 2012, **52**, 1039.
23. M. Vilums, A. J. M. Zweemer, Z. Yu, H. de Vries, J. M. Hillger, H. Wapenaar, I. A. E. Bollen, F. Barmare, R. Gross, J. Clemens, P. Krenitsky, J. Brussee, D. Stamos, J. Saunders, L. H. Heitman and A. P. IJzerman, *J. Med. Chem.*, 2013, **56**, 7706.
24. T. H. Keller, A. Pichota and Z. Yin, *Curr. Opin. Chem. Biol.*, 2006, **10**, 357.
25. P. J. Hajduk, J. R. Huth and C. Tse, *Drug Discovery Today*, 2005, **10**, 1675.
26. P. J. Hajduk, J. R. Huth and S. W. Fesik, *J. Med. Chem.*, 2005, **48**, 2518.
27. S. Maeda and G. F. X. Schertler, *Curr. Opin. Struct. Biol.*, 2013, **23**, 381.
28. I. Moraes, G. Evans, J. Sanchez-Weatherby, S. Newstead and P. D. Shaw Stewart, *Biochim. Biophys. Acta*, 2014, **1838**, 78.
29. P. Egloff, M. Hillenbrand, C. Klenk, A. Batyuk, P. Heine, S. Balada, K. M. Schlinkmann, D. J. Scott, M. Schütz and A. Plückthun, *Proc. Natl. Acad. Sci. U. S. A.*, 2014, **111**, E655.
30. T. Shimamura, M. Shiroishi, S. Weyand, H. Tsujimoto, G. Winter, V. Katritch, R. Abagyan, V. Cherezov, W. Liu, G. W. Han, T. Kobayashi, R. C. Stevens and S. Iwata, *Nature*, 2011, **475**, 65.
31. T. Hino, T. Arakawa, H. Iwanari, T. Yurugi-Kobayashi, C. Ikeda-Suno, Y. Nakada-Nakura, O. Kusano-Arai, S. Weyand, T. Shimamura, N. Nomura, A. D. Cameron, T. Kobayashi, T. Hamakubo, S. Iwata and T. Murata, *Nature*, 2012, **482**, 237.
32. E. B. Gonzales, T. Kawate and E. Gouaux, *Nature*, 2009, **460**, 599.
33. T. Kawate, J. C. Michel, W. T. Birdsong and E. Gouaux, *Nature*, 2009, **460**, 592.
34. A. Goehring, C. H. Lee, K. H. Wang, J. C. Michel, D. P. Claxton, I. Baconguis, T. Althoff, S. Fischer, K. C. Garcia and E. Gouaux, *Nat. Protoc.*, 2014, **11**, 2574.
35. T. Warne, M. J. Serrano-Vega, J. G. Baker, R. Moukhametzianov, P. C. Edwards, R. Henderson, A. G. Leslie, C. G. Tate and G. F. Schertler, *Nature*, 2008, **454**, 486.
36. A. S. Doré, N. Robertson, J. C. Errey, I. Ng, K. Hollenstein, B. Tehan, E. Hurrell, K. Bennett, M. Congreve, F. Magnani, C. G. Tate, M. Weir and F. H. Marshall, *Structure*, 2011, **19**, 1283.

37. J. F. White, N. Noinaj, Y. Shibata, J. Love, B. Kloss, F. Xu, J. Gvozdenovic-Jeremic, P. Shah, J. Shiloach, C. G. Tate and R. Grisshammer, *Nature*, 2012, **490**, 508.
38. C. H. Lee, W. Lü, J. C. Michel, A. Goehring, J. Du, X. Song and E. Gouaux, *Nature*, 2014, **511**, 191.
39. L. Chen, K. L. Dürr and E. Gouaux, *Science*, 2014, **345**, 1021.
40. S. Abdul-Hussein, J. Andréll and C. G. Tate, *J. Mol. Biol.*, 2013, **425**, 2198.
41. J. A. Christopher, J. Brown, A. S. Doré, J. C. Errey, M. Koglin, F. H. Marshall, D. G. Myszka, R. L. Rich, C. G. Tate, B. Tehan, T. Warne and M. Congreve, *J. Med. Chem.*, 2013, **56**, 3446.
42. E. Chun, A. A. Thompson, W. Liu, C. B. Roth, M. T. Griffith, V. Katritch, J. Kunken, F. Xu, V. Cherezov, M. A. Hanson and R. C. Stevens, *Structure*, 2012, **20**, 967.
43. P. S. Chae, S. G. Rasmussen, R. R. Rana, K. Gotfryd, R. Chandra, M. A. Goren, A. C. Kruse, S. Nurva, C. J. Loland, Y. Pierre, D. Drew, J. L. Popot, D. Picot, B. G. Fox, L. Guan, U. Gether, B. Byrne, B. Kobilka and S. H. Gellman, *Nat. Methods*, 2010, **12**, 1003.
44. G. K. Feld and M. Frank, *Curr. Opin. Struct. Biol.*, 2014, **27**, 69.
45. U. Weierstall, D. James, C. Wang, T. A. White, D. Wang, W. Liu, J. C. Spence, R. Bruce Doak, G. Nelson, P. Fromme, R. Fromme, I. Grotjohann, C. Kupitz, N. A. Zatsepin, H. Liu, S. Basu, D. Wacker, G. W. Han, V. Katritch, S. Boutet, M. Messerschmidt, G. J. Williams, J. E. Koglin, M. Marvin Seibert, M. Klinker, C. Gati, R. L. Shoeman, A. Barty, H. N. Chapman, R. A. Kirian, K. R. Beyerlein, R. C. Stevens, D. Li, S. T. Shah, N. Howe, M. Caffrey and V. Cherezov, *Nat. Commun.*, 2014, **5**, 3309.
46. W. Liu, D. Wacker, C. Gati, G. W. Han, D. James, D. Wang, G. Nelson, U. Weierstall, V. Katritch, A. Barty, N. A. Zatsepin, D. Li, M. Messerschmidt, S. Boutet, G. J. Williams, J. E. Koglin, M. M. Seibert, C. Wang, S. T. Shah, S. Basu, R. Fromme, C. Kupitz, K. N. Rendek, I. Grotjohann, P. Fromme, R. A. Kirian, K. R. Beyerlein, T. A. White, H. N. Chapman, M. Caffrey, J. C. Spence, R. C. Stevens and V. Cherezov, *Science*, 2013, **342**, 1521.
47. J. H. Park, P. Scheerer, K. P. Hofmann, H. W. Choe and O. P. Ernst, *Nature*, 2008, **454**, 183.
48. P. Scheerer, J. H. Park, P. W. Hildebrand, Y. J. Kim, N. Krauß, H. W. Choe, K. P. Hofmann and O. P. Ernst, *Nature*, 2008, **455**, 497.
49. V. Isberg, C. de Graaf, A. Bortolato, V. Cherezov, V. Katritch, F. H. Marshall, S. Mordalski, J. P. Pin, R. C. Stevens, G. Vriend and D. E. Gloriam, *Trends Pharmacol. Sci.*, 2015, **36**, 22.
50. V. Cherezov, D. M. Rosenbaum, M. A. Hanson, S. G. F. Rasmussen, F. S. Thian, T. S. Kobilka, H. J. Choi, P. Kuhn, W. I. Weis, B. K. Kobilka and R. C. Stevens, *Science*, 2007, **318**, 1258.
51. E. Y. T. Chien, W. Liu, Q. Zhao, V. Katritch, G. W. Han, M. A. Hanson, L. Shi, A. H. Newman, J. A. Javitch, V. Cherezov and R. C. Stevens, *Science*, 2010, **330**, 1091.
52. B. Wu, E. Y. T. Chien, C. D. Mol, G. Fenalti, W. Liu, V. Katritch, R. Abagyan, A. Brooun, P. Wells, F. C. Bi, D. J. Hamel, P. Kuhn, T. M. Handel, V. Cherezov and R. C. Stevens, *Science*, 2010, **330**, 1066.
53. J. Garcia-Perez, P. Rueda, J. Alcami, D. Rognan, F. Arenzana-Seisdedos, B. Lagane and E. Kellenberger, *J. Biol. Chem.*, 2011, **286**, 33409.
54. A. Manglik, A. C. Kruse, T. S. Kobilka, F. S. Thian, J. M. Mathiesen, R. K. Sunahara, L. Pardo, W. I. Weis, B. K. Kobilka and S. Granier, *Nature*, 2012, **485**, 321.
55. S. Granier, A. Manglik, A. C. Kruse, T. S. Kobilka, F. S. Thian, W. I. Weis and B. K. Kobilka, *Nature*, 2012, **485**, 400.

56. H. Wu, D. Wacker, M. Mileni, V. Katritch, G. W. Han, E. Vardy, W. Liu, A. A. Thompson, X. Huang, F. I. Carroll, S. W. Mascarella, R. B. Westkaemper, P. D. Mosier, B. L. Roth, V. Cherezov and R. C. Stevens, *Nature*, 2012, **485**, 327.
57. A. A. Thompson, W. Liu, E. Chun, V. Katritch, H. Wu, E. Vardy, X. Huang, C. Trapella, R. Guerrini, G. Calo, B. L. Roth, V. Cherezov and R. C. Stevens, *Nature*, 2012, **485**, 395.
58. P. J. Conn, C. W. Lindsley, J. Meiler and C. M. Niswender, *Nat. Rev. Drug Discovery*, 2014, **13**, 692.
59. D. Wootten, A. Christopoulos and P. M. Sexton, *Nat. Rev. Drug Discovery*, 2013, **12**, 630.
60. R. O. Dror, A. C. Pan, D. H. Arlow, D. W. Borhani, P. Maragakis, Y. Shan, H. Xu and D. E. Shaw, *Proc. Natl. Acad. Sci. U. S. A.*, 2011, **108**, 13118.
61. A. C. Kruse, A. M. Ring, A. Manglik, J. Hu, K. Hu, K. Eitel, H. Hübner, E. Pardon, C. Valant, P. M. Sexton, A. Christopoulos, C. C. Felder, P. Gmeiner, J. Steyaert, W. I. Weis, K. C. Garcia, J. Wess and B. K. Kobilka, *Nature*, 2013, **504**, 101.
62. A. Srivastava, J. Yano, Y. Hirozane, G. Kefala, F. Gruswitz, G. Snell, W. Lane, A. Ivetac, K. Aertgeerts, J. Nguyen, A. Jennings and K. Okada, *Nature*, 2014, **513**, 124.
63. J. Shonberg, R. C. Kling, P. Gmeiner and S. Löber, *Bioorg. Med. Chem.*, 2014, DOI: 10.1016/j.bmc.2014.12.034.
64. D. C. Lin, Q. Guo, J. Luo, J. Zhang, K. Nguyen, M. Chen, T. Tran, P. J. Dransfield, S. P. Brown, J. Houze, M. Vimolratana, X. Y. Jiao, Y. Wang, N. J. Birdsall and G. Swaminath, *Mol. Pharmacol.*, 2012, **82**, 843.
65. M. Hauge, M. A. Vestmar, A. S. Husted, J. P. Ekberg, M. J. Wright, J. Di Salvo, A. B. Weinglass, M. S. Engelstoft, A. N. Madsen, M. Lückmann, M. W. Miller, M. E. Trujillo, T. M. Frimurer, B. Holst, A. D. Howard and T. W. Schwartz, *Mol. Metab.*, 2014, **4**, 3.
66. K. Hollenstein, J. Kean, A. Bortolato, R. K. Y. Cheng, A. S. Doré, A. Jazayeri, R. M. Cooke, M. Weir and F. H. Marshall, *Nature*, 2013, **499**, 438.
67. A. Bortolato, A. S. Doré, K. Hollenstein, B. G. Tehan, J. S. Mason and F. H. Marshall, *Br. J. Pharmacol.*, 2014, **171**, 3132.
68. H. Wu, C. Wang, K. J. Gregory, G. Won Han, H. P. Cho, Y. Xia, C. M. Niswender, V. Katritch, J. Meiler, V. Cherezov, P. Jeffrey Conn and R. C. Stevens, *Science*, 2014, **344**, 58.
69. A. S. Doré, K. Okrasa, J. C. Patel, M. Serrano-Vega, K. Bennett, R. M. Cooke, J. C. Errey, A. Jazayeri, S. Khan, B. Tehan, M. Weir, G. R. Wiggin and F. H. Marshall, *Nature*, 2014, **511**, 557.
70. K. A. Bennett, A. S. Doré, J. A. Christopher, D. R. Weiss and F. H. Marshall, *Curr. Opin. Pharmacol.*, 2015, **20**, 1.
71. M. Huang, A. Shen, J. Ding and M. Geng, *Trends Pharmacol. Sci.*, 2014, **35**, 41.
72. J. G. Moffat, J. Rudolph and D. Bailey, *Nat. Rev. Drug Discovery*, 2014, **13**, 588.
73. T. Yoshida, J. Kakegawa, T. Yamaguchi, Y. Hantani, N. Okajima, T. Sakai, Y. Watanabe and M. Nakamura, *Oncotarget*, 2012, **3**, 1533.
74. L. Garuti, M. Roberti and G. Bottegoni, *Curr. Med. Chem.*, 2010, **17**, 2804.
75. C. W. Murray, M. L. Verdonk and D. C. Rees, *Trends Pharmacol. Sci.*, 2012, **33**, 224.
76. G. Bollag, J. Tsai, J. Zhang, C. Zhang, P. Ibrahim, K. Nolop and P. Hirth, *Nat. Rev. Drug Discovery*, 2012, **11**, 873.
77. D. C. Rees, M. Congreve, C. W. Murray and R. Carr, *Nat. Rev. Drug Discovery*, 2004, **3**, 660.
78. C. P. Mpamhanga, D. Spinks, L. B. Tulloch, E. J. Shanks, D. A. Robinson, I. T. Collie, A. H. Fairlamb, P. G. Wyatt, J. A. Frearson, W. N. Hunter, I. H. Gilbert and R. Brenk, *J. Med. Chem.*, 2009, **52**, 4454.

79. M. E. Bunnage, E. L. Piatnitski Chekler and L. H. Jones, *Nat. Chem. Biol.*, 2013, **9**, 195.

80. L. A. Honigberg, A. M. Smith, M. Sirisawad, E. Verner, D. Loury, B. Chang, S. Li, Z. Pan, D. H. Thamm, R. A. Miller and J. J. Buggy, *Proc. Natl. Acad. Sci. U. S. A.*, 2010, **107**, 13075.

81. Z. Pan, H. Scheerens, S. Li, B. E. Schultz, P. A. Sprengeler, L. C. Burrill, R. V. Mendonca, M. D. Sweeney, K. C. K. Scott, P. G. Grothaus, D. A. Jeffery, J. M. Spoerke, L. A. Honigberg, P. R. Young, S. A. Dalrymple and J. T. Palmer, *ChemMedChem*, 2007, **2**, 58.

82. S. R. LaPlante and M. Llinàs-Brunet, *Curr. Med. Chem.: Anti-Infect. Agents*, 2005, **4**, 111.

83. K. M. Marks and I. M. Jacobson, *Antiviral Ther.*, 2012, **17**, 1119.

84. Å. Rosenquist, B. Samuelsson, P. Johansson, M. D. Cummings, O. Lenz, P. Raboisson, K. Simmen, S. Vendeville, H. de Kock, M. Nilsson, A. Horvath, R. Kalmeijer, G. de la Rosa and M. Beumont-Mauviel, *J. Med. Chem.*, 2014, **57**, 1673.

85. D. Lamarre, P. C. Anderson, M. Bailey, P. Beaulieu, G. Bolger, P. Bonneau, M. Bös, D. R. Cameron, M. Cartier, M. G. Cordingley, A. Faucher, N. Goudreau, S. H. Kawai, G. Kukolj, L. Lagacé, S. R. LaPlante, H. Narjes, M. Poupart, J. Rancourt, R. E. Sentjens, R. St George, B. Simoneau, G. Steinmann, D. Thibeault, Y. S. Tsantrizos, S. M. Weldon, C. Yong and M. Llinàs-Brunet, *Nature*, 2003, **426**, 186.

86. S. Avolio and V. Summa, *Curr. Top. Med. Chem.*, 2010, **10**, 1403.

87. D. F. Veber, S. R. Johnson, H. Y. Cheng, B. R. Smith, K. W. Ward and K. D. Kopple, *J. Med. Chem.*, 2002, **45**, 2615.

88. A. Mac Sweeney, P. Grosche, D. Ellis, K. Combrink, P. Erbel, N. Hughes, F. Sirockin, S. Melkko, A. Bernardi, P. Ramage, N. Jarousse and E. Altmann, *Med. Chem. Lett.*, 2014, **5**, 937.

89. A. M. Watkins and P. S. Arora, *Eur. J. Med. Chem.*, 2014, DOI: 10.1016/j.ejmech.2014.09.047.

90. T. Oltersdorf, S. W. Elmore, A. R. Shoemaker, R. C. Armstrong, D. J. Augeri, B. A. Belli, M. Bruncko, T. L. Deckwerth, J. Dinges, P. J. Hajduk, M. K. Joseph, S. Kitada, S. J. Korsmeyer, A. R. Kunzer, A. Letai, C. Li, M. J. Mitten, D. G. Nettesheim, S. Ng, P. M. Nimmer, J. M. O'Connor, A. Oleksijew, A. M. Petros, J. C. Reed, W. Shen, S. K. Tahir, C. B. Thompson, K. J. Tomaselli, B. Wang, M. D. Wendt, H. Zhang, S. W. Fesik and S. H. Rosenberg, *Nature*, 2005, **435**, 677.

91. M. Bruncko, T. K. Oost, B. A. Belli, H. Ding, M. K. Joseph, A. Kunzer, D. Martineau, W. J. McClellan, M. Mitten, S. Ng, P. M. Nimmer, T. Oltersdorf, C. Park, A. M. Petros, A. R. Shoemaker, X. Song, X. Wang, M. D. Wendt, H. Zhang, S. W. Fesik, S. H. Rosenberg and S. W. Elmore, *J. Med. Chem.*, 2007, **50**, 641.

92. M. Yu, Y. Wang, J. Zhu, M. D. Bartberger, J. Canon, A. Chen, D. Chow, J. Eksterowicz, B. Fox, J. Fu, M. Gribble, X. Huang, Z. Li, J. J. Liu, M. C. Lo, D. McMinn, J. D. Oliner, T. Osgood, Y. Rew, A. Y. Saiki, P. Shaffer, X. Yan, Q. Ye, D. Yu, X. Zhao, J. J. Zhou, S. H. Olson, J. C. Medina and D. Sun, *ACS Med. Chem. Lett.*, 2014, **5**, 894.

93. J. Chen, X. Wu, J. Lin and A. J. Levine, *Mol. Cell. Biol.*, 1996, **16**, 2445.

94. Y. Zhao, D. Bernard and S. Wang, *BioDiscovery*, 2013, **8**, 4.

95. D. Sun, Z. Li, Y. Rew, M. Gribble, M. D. Bartberger, H. P. Beck, J. Canon, A. Chen, X. Chen, D. Chow, J. Deignan, J. Duquette, J. Eksterowicz, B. Fisher, B. M. Fox, J. Fu, A. Z. Gonzalez, F. Gonzalez-Lopez De Turiso, J. B. Houze, X. Huang, M. Jiang, L. Jin, F. Kayser, J. Liu, M. Lo, A. M. Long, B. Lucas, L. R. McGee, J. McIntosh, J. Mihalic, J. D. Oliner, T. Osgood, M. L. Peterson, P. Roveto, A. Y. Saiki, P. Shaffer, M. Toteva, Y. Wang, Y. C. Wang, S. Wortman, P. Yakowec,

X. Yan, Q. Ye, D. Yu, M. Yu, X. Zhao, J. Zhou, J. Zhu, S. H. Olson and J. C. Medina, *J. Med. Chem.*, 2014, **57**, 1454.

96. K. A. Jacobson, *In Silico Pharmacol.*, 2013, **1**, 22.
97. S. Andrews and B. Tehan, *Med. Chem. Commun.*, 2013, **4**, 52.
98. D. K. Tosh, F. Deflorian, K. Phan, Z. G. Gao, T. C. Wan, E. Gizewski, J. A. Auchampach and K. A. Jacobson, *J. Med. Chem.*, 2012, **55**, 4847.
99. D. K. Tosh, S. Paoletta, F. Deflorian, K. Phan, S. M. Moss, Z. G. Gao, X. Jiang and K. A. Jacobson, *J. Med. Chem.*, 2012, **55**, 8075.

4 Actions of Drugs on the Autonomic Nervous System

Translational Pharmacology BioVentures LLC, PO Box 3126, 56, 14th St., Hoboken, New Jersey, NJ 03070, USA
Email: tblackburn@tpbioventures.com

4.1 Introduction

In this chapter we will review the basic anatomical, physiological and pharmacological organization, nomenclature and function of the autonomic nervous system and discuss the rational basis for the development of therapeutic agents that target specific aspects of autonomic function on organ systems.

Three main anatomical divisions control the coordinated activity of the autonomic nervous system (ANS): *sympathetic*, *parasympathetic* and *enteric* nervous system. The organs of the body are innervated by branches of both *sympathetic* and *parasympathetic* (Figure 4.1) and provide a link between the central nervous system (CNS) and peripheral organs.[1,5] The enteric nervous system comprises the intrinsic nerve plexus of the gastrointestinal tract, which is closely integrated with the *sympathetic* and *parasympathetic* nervous system.[1,2,4] However, unlike the sympathetic and *parasympathetic* systems the enteric nervous system is able to act independently of the CNS.[4] Stimulation of the body's sympathetic system generally depresses all physiological functions, except under extreme stress when the body is severely challenged in an emergency situation, *e.g.* the so-called *"fight-or-flight"* response.[6] In

Pharmacology for Chemists: Drug Discovery in Context
Edited by Raymond Hill, Terry Kenakin and Tom Blackburn
© The Royal Society of Chemistry 2018
Published by the Royal Society of Chemistry, www.rsc.org

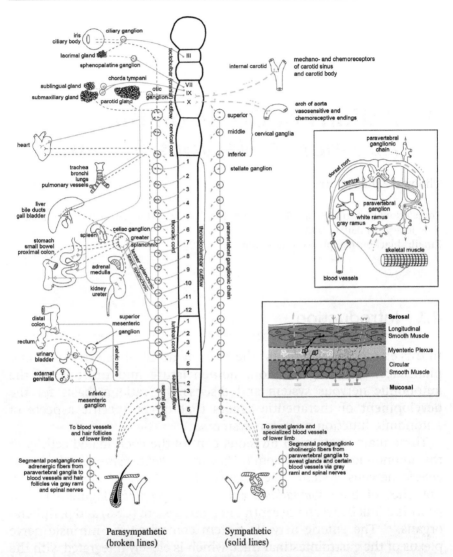

Figure 4.1 Basic organization of the mammalian autonomic nervous system. C, cervical; GI, gastrointestinal; L, lumbar; M, medullary; S, sacaral; T thoracic. Adapted with permission from ref. 7. Brunton, L.L., Lazo, J.S. Parker, K.L., (2005). Goodman and Gilman's, The Pharmacological Basis of Therapeutics 11th Edition, Pub: McGraw Hill, New York. Copyright © 2006 by The McGraw-Hill Companies, Inc.

contrast, the *parasympathetic* controls those physiological functions that are necessary to cope with the energy demands of the body. With a few exceptions, the effects of *parasympathetic* stimulation on a given gland or organ is to oppose the actions of *sympathetic* nerve stimulation (See Table 4.1).[3] It is important to note that the ANS is predominately

Table 4.1 Responses of effector organs to autonomic nerve impulses. Reproduced with permission from ref. 7. L.L. Brunton, J.S. Lazo and K.L. Parker, (2005). Goodman and Gilman's, The Pharmacological Basis of Therapeutics 11th Edition, Pub: McGraw Hill, New York. Copyright © 2006 by The McGraw-Hill Companies, Inc.

Organ system	Sympathetic effect[a]	Adrenergic receptor subtype[b]	Parasympathetic effect[a]	Cholinergic receptor subtype[b]
Intestine				
Motility and tone	Decrease[a]+	$\alpha_1, \alpha_2, \beta_1, \beta_2$	Increase+++	M_3, M_2
Sphincters	Contraction+	α_1	Relaxation (usually)+	M_3, M_2
Secretion	Inhibition	α_2	Stimulation++	M_3, M_2
Gallbladder and ducts	Relaxation+	β_2	Contraction+	M
Kidney				
Renin secretion	Decrease+; increase++	$\alpha_1; \beta_1$	No innervation	—
Urinary bladder				
Detrusor	Relaxation+	β_2	Contraction+++	$M_3 > M_2$
Trigone and sphincter	Contraction++	α_1	Relaxation++	$M_3 > M_2$
Ureter				
Motility and tone	Increase	α_1	Increase (?)	M
Uterus	Pregnant contraction:	α_1	Variable	M
	Relaxation	β_2		
	Nonpregnant relaxation	β_2		
Sex organs, male	Ejaculation+++	α_1	Erection+++	M_3

Table 4.1 (*Continued*)

Organ system	Sympathetic effect[a]	Adrenergic receptor subtype[b]	Parasympathetic effect[a]	Cholinergic receptor subtype[b]
Skin				
Pilomotor muscles	Contraction++	α_1		
Sweat glands	Localized secretion++	α_1		M_3, M_2
Spleen capsule	Generalized secretion+++	α_1	—	—
	Contraction+++	β_2	—	
Adrenal medulla	Relaxation+			
	—			
	Secretion of epinephrine and norepinephrine			$N(\alpha_3)_2(\beta_4)_3$:M (secondarily)
Skeletal muscle	Increased contractility; glycogenolysis: K^+ uptake	β_2	—	—
Liver	Glycogenolysis and gluconeogenesis+++	α_1	—	—
Pancreas				
Acini		β_2	Secretion++	M_3, M_2
		α		
Islets (βcells)	Decreased secretion+	α_2		M_3, M_2
	Decreased secretion+++	β_2	—	
Fat cells	Increased secretion+	α_1, β_1, β_2, β_3		—
	Lipolysis+++; (thermogenesis)	α_2		

Organ	Sympathetic effect	Receptor	Parasympathetic effect	Receptor
Salivary glands	Inhibition of lipolysis; K$^+$ and water secretion+	α_1	K$^+$ and water Secretion+++	M$_3$, M$_2$
Nasopharyngeal glands	—	β	Secretion++	M$_3$, M$_2$
Pineal glands	Melanton synthesis	β_1	—	
Posterior pituitary	Antidiruetic secretion		—	
Eye				
Radial muscle, iris	Contraction (mydriasis)++	α_1		
Sphincter muscle, iris			Contraction (miosis)+++	M$_3$, M$_2$
Pupil	Dilation	α_1	Contraction	M$_3$
Ciliary muscle	Relaxation for far vision+	β_2	Contraction for near vision++	M$_3$, M$_2$
Lacrimal glands	Secretion+	α	Secretion+++	M$_3$, M$_2$
Heart				
Sinoatrial node	Increase in heart rate++	$\beta_1 > \beta_2$	Decrease in heart rate+++	M$_2 \gg$ M$_3$
Atria	Increase in contractility and conduction velocity++	$\beta_1 > \beta_2$	Decrease in contractility++ and shortened AP duration	M$_2 \gg$ M$_3$
Atrioventricular node	Increase in automaticity and conduction velocity++	$\beta_1 > \beta_2$	Decrease in conduction velocity; AV block+++	M$_2 \gg$ M$_3$
His-Purkinje system	Increase in automaticity and conduction velocity	$\beta_1 > \beta_2$	Little effect	M$_2 \gg$ M$_3$
Ventricle	Increase in contractility, conduction velocity, automaticity and rate of idioventricular pacemakers+++	$\beta_1 > \beta_2$	Slight decrease in contractility	M$_2 \gg$ M$_3$

Table 4.1 *(Continued)*

Organ system	Sympathetic effect[a]	Adrenergic receptor subtype[b]	Parasympathetic effect[a]	Cholinergic receptor subtype[b]
Blood vessels				
(Arteries and arterioles)		α_1, α_2, β_2		
Coronary	Constriction+; dilation++	α_1, α_2	No innervation[a]	—
Skin and mucosa	Constriction+++	α_1, β_2		—
Skeletal muscle	Constriction; dilation++	α_1	No innervation[a]	—
Cerebral	Constriction (slight)	α_1, β_2	Dilation (?)	—
Pulmonary	Constriction+; dilation	α_1, β_2	No innervation[a]	—
Abdominal viscera	Constriction+++; dilation+	α_1, α_2, β_1, β_2	No innervation[a]	
Salivary glands	Constriction+++	α_1, α_2, β_2	Dilation++	M_3
Renal	Constriction++; dilation++		No innervation[a]	
(Veins)				
Endothelium	Constriction; dilation		Activation of NO synthase	M_3
Lung				
Tracheal and bronchial smooth muscle	Relaxation	β_2	Contraction	$M_2 = M_3$
Bronchial glands	Decreased secretion, increased secretion	α_1 β_2	Stimulation	M_3, M_2
Stomach				
Motility and tone	Decrease (usually)+	α_1, α_2, β_1, β_2	Increase+++	$M_2 = M_3$
Sphincters	Contraction (usually)+	α_1	Relaxation (usually)+	M_3, M_2
Secretion	Inhibition	α_2	Stimulation++	M_3, M_2

Autonomic nerve endings

Sympathetic terminals			
Autoreceptor	Inhibition of NE release	—	$\alpha_{2A} > \alpha_{2C}(\alpha_{2B})$
Heteroreceptor		Inhibition of NE release	$M_2 + M_4$
Parasympathetic terminal		—	
Autoreceptor		Inhibition of Ach release	$M_2 + M_4$
Heteroreceptor	Inhibition ACh release		$\alpha_{2A} > \alpha_{2C}$

[a]Responses are designated + to +++ to provide an approximate indication of the importance of sympathetic and parasympathetic nerve activity in the control of the various organs and function listed.

[b]Adrenergic receptors: $\alpha_1 + \alpha_2$ and subtypes thereof: β_1-β_2-β_3-Cholinergic receptors: nicotinic (N): muscarinic (M), with subtypes 1–4. When a designation of subtype is not provided, the nature of the subtype has not been determined, unequivocally. Only the principal receptor subtypes are shown. Transmitters other than acetylcholine and norepinephrine contribute to many of the responses. In the human heart, the ratio of β_1 to β_2 is about 3:2 in atria and 4:1 in ventricles. M_2 receptors predominate in the heart but M_3 receptors are also present. The predominant α_1 receptor subtype in most blood vessels (both arteries and veins) is α_{1A}, although other α_1 subtypes are present in specific blood vessels. The α_{1D} is the predominant subtype in the aorta. Dilation predominates *in situ* owing to metabolic autoregulatory mechanisms. Over the usual concentration range of physiologically released circulating epinephrine, the β-receptor response (vasodilation) predominates in blood vessels of skeletal muscle and liver; α-receptor response (vasoconstriction) in blood vessels of other abdominal viscera. The renal and mesenteric vessels also contain specific dopaminergic receptors whose activation causes dilation (see review Goldberg *et al.*, 1978). Sympathetic cholinergic neurons cause vasodilation in skeletal muscle beds, but this is not involved in most physiological responses. The endothelium of most blood vessels NO, which causes vasodilation in response to muscarinic stimuli. However, unlike the receptors innervated by sympathetic cholinergic fibers in skeletal muscle blood vessels, these muscarinic receptors are not innervated and respond only to exogenously added muscarinic agonists in the circulation. While adrenergic fibers terminate at inhibitory β-receptors on smooth muscle fibers and at inhibitory α-receptors on parasympathetic (cholinergic) excitatory ganglion cells of the myenteric plexus, the primary inhibitory response is mediated *via* enteric neurons through NO, P2Y receptors and peptide receptors. Uterine responses depend on stages of menstrual cycle, amount of circulating estrogen and progesterone and other factors. Palms of hands and some other sites ("adrenergic seating"). There is significant variation among species in the receptor types that mediate certain metabolic responses. All three β adrenergic receptors have been found in human fat cells. Activation of β_3 adrenergic receptors produces a vigorous thermogenic response as well as lipolysis. The significance is unclear. Activation of β adrenergic receptors also inhibits leptin release from adipose tissue.

outside the influence of voluntary control.[1] For example, if nerve impulse transmission in a voluntary nerve to a striated muscle is disrupted for a period of time, the muscle becomes paralyzed and atrophies.

In contrast, smooth muscles and glands, innervated by autonomic nerves, show a degree of spontaneous activity, independent of intact innervation. The main processes that the ANS regulates are:

- contraction and relaxation of vascular and visceral smooth muscle;
- all exocrine and certain endocrine secretions;
- heart beat;
- energy metabolism.

Figure 4.2 shows a simple example of two end-to-end neurones that comprise the autonomic connection between the spinal cord and an effector organ. It is important to note that one presynaptic fiber may synapse with many postsynaptic fibers, thus multiplying the effects produced, affecting many other systems, including for example, the heart, kidney and immune system. However, in *ganglia* a large number of individual synapses between component neurones form the *autonomic ganglion* (pl. *ganglia*), a discrete mass of neurones. The cell bodies of the first neurone originating in the spinal cord or brain stem are termed preganglionic fibers and the second neurone that leads to and terminates at an effector organ are postganglionic fibers (Figure 4.2).

Sympathetic and *parasympathetic* synapses within the ganglia release neurotransmitter into the synaptic gap to facilitate synaptic

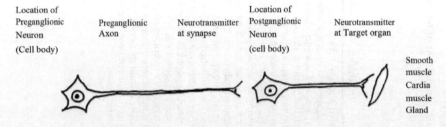

| Location of Preganglionic Neuron (Cell body) | Preganglionic Axon | Neurotransmitter at synapse | Location of Postganglionic Neuron (cell body) | Neurotransmitter at Target organ |

Smooth muscle
Cardia muscle
Gland

Figure 4.2   Basic organization of the autonomic nervous system into preganlionic and postganglionic neurones. In the sympathetic division, ganglia constitute the sympathetic chain, Preganglionic neurones, therefore, have relatively short axons, while postganglionic fibers are relatively long. In the parasympathetic division, ganglia are located close to the target organ. Consequently, preganglionic axons are long and post-ganglionic axons are short.

neurotransmission between the spinal cord and effector organ and are shown in Figure 4.1. The *parasympathetic* postganglionic neurones secrete acetylcholine, hence are designated cholinergic neurones, whereas, *sympathetic* postganglionic fibers release noradrenaline (norepinephrine), and are called noradrenergic or adrenergic (epinephrine) neurones. The IUPHAR-recognized nomenclature and terminology for the cholinergic nervous system is used synonymously with the term *parasympathetic* nervous system,[1,8,9] whereas, the *adrenergic* or *noradrenergic* system are known as the sympathetic nervous system, with one important exception, the preganglionic fibers found within ganglia.[10] The *sympathetic* and *parasympathetic* fibers in ganglia both secrete and utilize acetylcholine and are therefore designated cholinergic fibers.[1,2,10]

4.2 Post- and Presynaptic Receptors

All nerve synapses have no anatomical connections between the nerve terminal and the effector organ. Each nerve terminal contains a store of a neurochemical transmitter that is released by nervNO e impulses that travel across the synaptic gap/cleft to act on postsynaptic receptors located on the surface of the effector organ.[1,3] Thus, activating or inhibiting a cascade of biochemical pathways that results in a physiological event in the effector organ (see Figure 4.1 and Table 4.1 (*e.g.* pupil constriction or dilatation)). However, those receptors located on the surface of postganglionic nerve terminals (Figure 4.1) are designated as presynaptic receptors (or autoreceptors) and are activated/inhibited by the endogenous neurotransmitter present in the nerve terminal.[3,6] The release of neurotransmitter into the synaptic cleft is independent of nerve-impulse traffic at the postganglionic nerve terminal. In general, postsynaptic receptors are more sensitive to neurotransmitter release than presynaptic receptors. At times of excessive neurotransmitter release a self-limiting, negative feedback/reuptake mechanism (autoreceptor) switches off further release as a self-regulating mechanism. However, some postganglionic nerve terminals are promiscuous in that they can store more than one neurotransmitter in times of excessive release (*e.g.* acetylcholine, dopamine and 5-hydroxtypamine) in the synaptic cleft.[3] The therapeutic implications are yet to be fully understood. The pharmacology and physiology of the cholinergic parasympathetic nervous system will be discussed next.

4.3 Anatomy, Physiology and Pharmacology of Cholinergic Neurotransmission

The focus of this section is on the physiology of cholinergic neurotransmission and the different types of acetylcholine (ACh) receptors, their function, synthesis and release. The therapeutic properties of cholinergic agonists and antagonists are also discussed.

A more detailed description of the physiology of cholinergic neurotransmission can be found in the reference list (Rang and Dale, 2016).[3] The site of action of ACh synthesis and neurotransmitter release and drug site of action are shown in Figure 4.3. Further information on the release and metabolism of ACh is well documented in several excellent reviews featured in the reference list at the end of this chapter. For the sake of brevity, ACh is synthesized within the nerve terminal from choline, which is taken up into nerve terminals by a specific choline transporter.[3] Contrary to other neurotransmitters transporters; it specifically transports choline, not ACh. Free endogenous choline within the nerve terminal is acetylated by a cystol enzyme, cholineacetyltransferase (CAT), which is a rate-limiting step, transferring the acetyl group from acetyl coenzyme A (Figure 4.3).

ACh is continually being hydrolyzed and resynthesized by cholinesterase (see below) in presynaptic nerve terminals. Inhibition of cholinesterase causes the accumulation of ACh in the cytosol and leaks out *via* a choline carrier. ACh accumulates in cholinergic vesicles by means of a specific neurotransmitter transporter, one of a family of specific amine transporters, which is discussed later in the section on adrenergic transmission.

CH3-COO +

4.1

ACh accumulation and release is coupled to the large electrochemical gradient for protons that exists between intracellular organelles and the cytosol and can be specifically blocked by the experimental drug vesamicol (see Chapter 6). ACh release activates receptors on the postsynaptic cell and excess ACh is hydrolyzed by acetylcholinesterase (AChE). Hydrolysis of ACh by AChE between neuromuscular and ganglionic synapses is rapid (within 1 ms), whereas in smooth muscle, gland cells, and the heart it is much slower.[3]

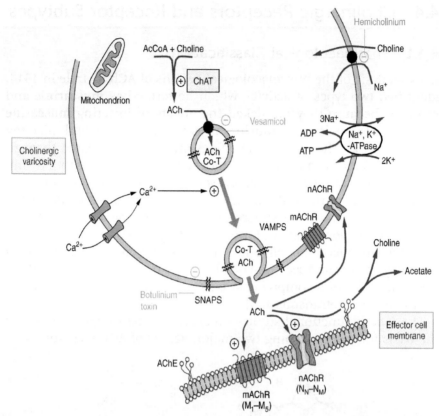

Figure 4.3 Schematic representations of a cholinergic neuroeffector junction showing features of the synthesis, storage, and release of acetylcholine (ACh) and receptors on which ACh acts. The synthesis of ACh in the varicosity depends on the uptake of choline *via* a sodium-dependent carrier. This uptake can be blocked by *hemicholinium*. Choline and the acetyl moiety of acetyl coenzyme A, derived from mitochondria, form ACh, a process catalyzed by the enzyme choline acetyltransferase (ChAT). ACh is transported into the storage vesicle by another carrier that can be inhibited by *vesamicol*. ACh is stored in vesicles along with other potential cotransmitters (Co-T) such as ATP and VIP at certain neuroeffector junctions. Release of ACh and the Co-T occurs on depolarization of the varicosity, which allows the entry of Ca^{2+} through voltage-dependent Ca^{2+} channels. Elevated $[Ca^{2+}]_{in}$ promotes fusion of the vesicular membrane with the cell membrane, and exocytosis of the transmitters occurs. This fusion process involves the interaction of specialized proteins associated with the vesicular membrane (VAMPs, vesicle-associated membrane proteins) and the membrane of the varicosity (SNAPs, synaptosome-associated proteins). The exocytotic release of ACh can be blocked by *botulinum toxin*. Once released, ACh can interact with the muscarinic receptors (mAChR), which are GPCRs, or nicotinic receptors (nAChR), which are ligand-gated ion channels, to produce the characteristic response of the effector. ACh also can act on presynaptic mAChRs or nAChRs to modify its own release. The action of ACh is terminated by metabolism to choline and acetate by acetylcholinesterase (AChE), which is associated with synaptic membranes.

4.4   Cholinergic Receptors and Receptor Subtypes

4.4.1   Pharmacological Classification

Early analysis of the pharmacological actions of ACh by Dale in 1914, identified two types of activity, which he termed as **muscarinic** and **nicotinic** because they mimicked the effects of injecting **muscarine** (**4.2**), from the poisonous mushroom *Amanita muscaria* and injecting **nicotine** (**4.3**).[2] The muscarinic effects are largely mediated through *parasympathetic* stimulation and blocked by atropine, with two important exceptions; (i) an indirect vasodilation on vascular endothelial cells to release nitric oxide inducing vasodilation and secretion from sweat glands, which are innervated by cholinergic fibers from the *sympathetic* nervous system. However, the larger doses of ACh stimulate nicotine-like effects on all autonomic ganglia, motor endplates of voluntary muscle and secretion of adrenaline from the adrenal medulla (see Table 4.1, for a comprehensive list of the main effects of ACh on cholinergic neurotransmission).

These early functional studies on acetylcholine receptors provided the basis for distinguishing two major classes of ACh receptor.

4.2   **4.3**

4.4.2   Nicotinic Receptors

There are three main classes of nicotinic ACh receptors (nAChRs), muscle, ganglionic and CNS, which comprises of a number of subunits, summarized in Table 4.1.[3,13] Nicotinic receptors found on muscle neuromuscular junctions and ganglia are responsible for transmission at sympathetic and *parasympathetic* ganglia. In the CNS, nAChRs are ubiquitous and heterogeneous in their molecular composition and location (Figure 4.4). The location and function of these receptors subtypes open the possibility of selective actions of nAChRs agonists and antagonist on neuromuscular, ganglionic and brain synapses as presented in Table 4.2.

  (i) **Ganglion Stimulants**
      Apart from ACh, nearly all nAChR agonists act on either ganglionic and CNS receptors or at motor endplates (Table 4.2).

Nicotine (which is a tertiary amine found in tobacco leaves and is part of pharmacology folklore (**4.3**)), like other nonsynthetic (lobeline, epibatidine) and synthetic agonists, *e.g.* varenicline (**4.4**), causes a complex peripheral response associated with generalized stimulation of autonomic ganglia. Nicotine readily penetrates the BBB and the average cigarette contains 6–11 mg. of which approximately 1–3 mg stimulates the brain and is highly addictive on chronic use. However, large doses lead to profound depression and death due to respiratory failure.[3]

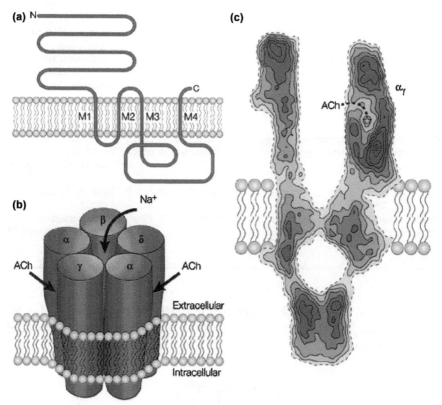

Figure 4.4 Structure of the nicotinic acetylcholine receptors. (a) The threading pattern of receptor subunits through the membrane. (b) A schematic representation of the quaternary structure, showing the arrangement of the subunits in the muscle-type receptor, the location of the two acetylcholine (ACh)-binding sites (between and α- and a γ-subunit, and an α- and a δ-subunit), and the axial cation-conducting channel. (c) A cross-section through the 4.6-Å structure of the receptor determined by electron microscopy of the tubular crystals of *Torpedo* membrane embedded in ice. The dashed line indicates the proposed path to binding site.[11]
Reprinted by permission from Macmillan Publishers Ltd: *Nature Reviews Neuroscience* (ref. 12), Copyright 2002.

Table 4.2 Nicotinic receptor agonists and antagonists. (Reproduced with permission from ref. 3.)

	Site of action	Response	Clinical effects
Agonist			
Nicotine	Autonomic ganglia	Stimulation then block	Drug of abuse, resulting in dependence
	CNS	Stimulation	
Lobeline	Autonomic ganglia	Stimulation	Used as a smoking cessation aid and drug addiction
	Sensory nerve terminals	Stimulation	
Epibatidine	Autonomic ganglia, CNS	Stimulation	Isolated from frog skin
			Highly potent
			No clinical use
Varenicline	CNS, autonomic ganglia	Stimulation	Used for nicotine addiction
Suxamethonium	Neuromuscular junction	Depolarization block	Used clinically as muscle relaxant
Decamethonium	Neuromuscular junction	Depolarization block	No clinical use
Antagonists			
Hexamethonium	Autonomic ganglia	Transmission block	No clinical use
Trimetaphan	Autonomic ganglia	Transmission block	Blood pressure lowering in surgery (rarely used)
Tubocurarine	Neuromuscular junction	Transmission block	Now rarely used
Pancuronium Atracurium Vecuronium Curonium	Neuromuscular junction	Transmission block	Widely used as muscle relaxants in anaesthesia

At lower doses tachyphylaxis due to desensitization is a characteristic of these receptors (where the nicotinic ion channel remains closed).

4.4

(ii) **Ganglion-blocking Drugs**

The pharmacology of ganglion-blocking drugs is complex with both divisions of the autonomic system blocked indiscriminately. They can act at the neuromuscular junction blocking the effects of ACh; produce a prolonged depolarization after initial ACh stimulation and by interference with the post-synaptic actions of ACh (see Figure 4.3). Along with nicotine, ganglion-blocking drugs include **hexamethonium (4.5)**, **trime-taphan (4.6)** and **D-tubocuraine (4.7)**. Paton and Zamis investigating a series of linear bisquaternary compounds discovered hexamethonium over 60 years ago.[13] Compounds with five or six carbon atoms (hexamethonium) in the methylene chain-linking two quaternary groups produced ganglion block, whereas compounds with nine or ten carbon atoms (decamethonium) produce neuromuscular block. Hexamethonium is of little clinical significance today, however, it is still recognized as the first antihypertensive agent.[3] These agents block all autonomic ganglia and their main effects are: hypotension, loss of cardiovascular reflexes, inhibition of secretions, gastrointestinal paralysis and impaired micturition. All these agents are of little clinical use, with only trimetaphan used to produce controlled hypotension in anesthesia during some emergency procedures.[3]

4.5   4.6   4.7

(iii) **Neuromuscular-blocking Drugs**
Drugs can block neurotransmission either presynaptically by blocking inhibiting synthesis or release, or by acting postsynaptically, the site of action of most clinically acting drugs used in anesthesia (Rang and Dale 2016).[3] Suxamethonium is the only drug that does not act as a noncompetitive antagonist.

(iv) **Nondepolarizing Neuromuscular-blocking Drugs**
The work of Claude Bernard in 1856 on curare (a mixture of naturally occurring alkaloids) showed that it caused paralysis by blocking neuromuscular neurotransmission independent of blocking nerve conduction or muscle contractility.[3] The most

important chemical moiety is D-tubocurarine (**4.7**) that has been superseded by several synthetic drugs with improved duration of action (pancuronium, vecuronium and atracurium). Galamine was the first clinically useful synthetic drug following curare but was replaced by drugs with fewer side effects. All the synthetic drugs are quaternary ammonium compounds, thus are poorly absorbed and generally rapidly excreted. The low oral absorption and inability to cross the placenta are important in their use in obstetric anaesthesia.[14] All nondepolarizing agents act as noncompetitive antagonists at ACh receptors on muscle endplates. They also block the facilitatory presynaptic autoreceptor and thereby inhibit the release of ACh during repetitive stimulation of the motor nerve, resulting in the phenomenon of "tetanic fade," often used by anesthetists to monitor postoperative recovery of neurotransmission.[14] Nondepolarizing blocking agents are mainly metabolized by the liver or excreted unchanged in the urine (pancuronium, vecuronium), the exception being, atracurium and mivacurium (in plasma). Atracurium (tetrahydroisoquinolinium class of neuromuscular blocking drugs) was designed to be chemically unstable at physiological pH and splits into two inactive moieties by cleavage at one of the quaternary nitrogen atoms. The rapid pH degradation and short half-life becomes critical during respiratory alkalosis caused by hyperventilation. Thus, rapid postoperative recovery of muscle strength is needed to reduce the risk of complications. Atracurium (**4.8**) was later replaced by cisatracurium (**4.9**), the *R-cis* isomer component of the ten stereoisomers that comprise atracurium, based on its superior clinical profile. Clinically, the cholinesterase inhibitor, neostigmine is often used to reverse the action of nondepolarizing drugs postoperatively. Atropine is also coadministered to prevent unwanted parasympathomimetic effects.[14]

4.8

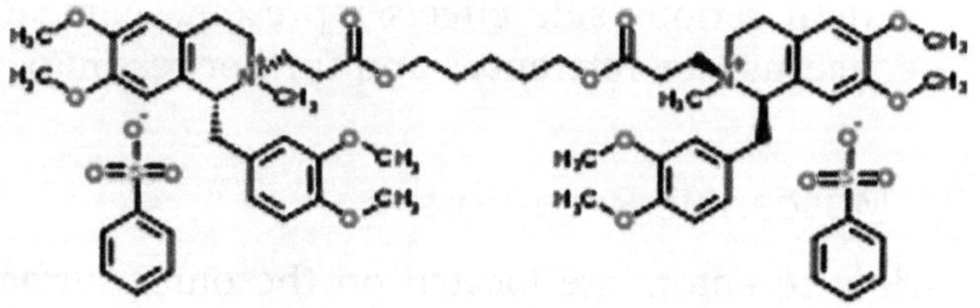

4.9

(v) Depolarizing Neuromuscular-blocking Drugs

Following the pioneering work of Paton and Zamis (1949)[13] on the bisquaternary ammonium compound decamethonium the later work of Burns and Paton (1951)[15] showed it induced a depolarizing blockade like ACh at the endplate region of the muscle fiber. More recently, suxamethonium (succinyldicholine), was found to be a depolarizing blocking agent that acts on nicotinic receptors like acetylcholine but unlike acetylcholine its action was more prolonged at the receptor (longer on-time) with a persistent depolarization, resulting in a flaccid paralysis due to prolonged relaxation of the muscle.

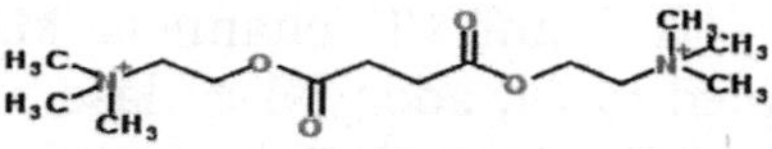

4.10

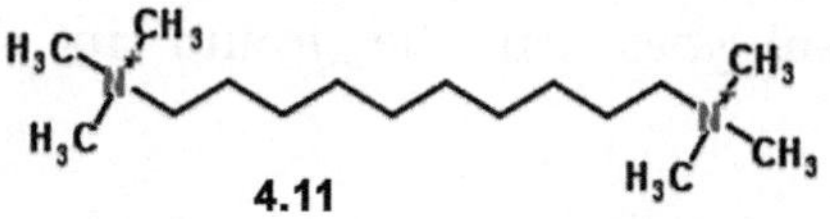

4.11

The structure of suxamethonium (**4.10**) is closely related to both decamethonium (**4.11**) and ACh (consisting of two ACh molecules linked to their acetyl groups). However, it has a short duration of action being rapidly metabolized by plasma cholinesterase. Drugs that inhibit acetylcholinesterase like physostigmine are ineffective in reversing the paralysis induced by the depolarizing neuromuscular blockers. The resulting accumulation of ACh at the endplate region has no effect, as a result of the prolonged depolarization block of the postsynaptic nicotinic receptor. Although it has limited clinical use (*e.g.* as a surgical adjuvant to relax spastic voluntary muscles in the abdominal wall), the fact that suxamethonium has several serious side effects represents clinical problems and concerns (see references and further reading).

4.4.3 Muscarinic Receptors

Muscarinic receptors are located on the outer surface of cell membranes and are G-protein-coupled.[8] However; some muscarinic receptors have alternative second messenger transduction mediated. On the basis of classical experimental pharmacological data and cloning studies, five subtypes of the muscarinic receptor have been identified: M1–M5:

- M1: in autonomic ganglia and various regions of the brain;
- M2: cardiac, ileal portion of the small intestine, and cerebellum;
- M3: in secretory glands and smooth muscle;
- M4: in brain;
- M5: in brain.

M1, M3, and M5 receptors are positively coupled to Gq to stimulate inositol triphosphate production and these receptors indirectly modulate some K^+, Ca^{2+}, and Cl^- channels. M2 and M4 receptors are negatively coupled to Gi, adenylyl cyclase and inhibit intracellular synthesis of cyclic AMP.[16] This response indirectly affect K^+ and Ca^{2+} channels. Both groups inhibit MAP kinase and their anatomical location, function and pharmacology and further reading on the complexity of the various transduction mechanisms of muscarinic subtypes can be found in the recommended reading list.

4.4.4 Drugs Affecting Muscarinic Receptors

(i) **Muscarinic Agonists and Partial Agonists**
Acetylcholine is the endogenous agonist acting at Class A, rhodopsin-like G-protein receptors. Numerous studies have shown that structural analogs (AC-42, AC-26058, 77-LH-28-1, N-desmethylclozapine, TBPB and LuAE51090) are functionally selective agonists for the M1 receptor at a binding site distinct from nonselective agonists. Also, there are two pharmacologically characterized allosteric sites on muscarinic receptors, one defined by it binding gallamine (M2), strychnine (M2) and brucine (M1), and the other by the binding of KT 5720 (M1),

WIN 62,577, WIN51, 508, LU2033298 (M4), BQCA (M1) and staurosporine. (See Alexander *et al.*, "The Concise Guide to Pharmacology," 2015[8] for more detailed information.)

4.12   4.13   4.14

Muscarinic agonists (or parasympathomimetics) in general stimulate parasympathetic receptors on whole-body organs. Acetylcholine and related choline esters act at both mAChRs and nAChRs, although more potently on mAChRs. **Bethanechol (4.12), pilocarpine (4.13)** and **cevimeline (4.14)** are the only parasympathomimetics currently prescribed in clinical practice. In determining SAR of the ACh molecule, the quaternary ammonium and ester groups bear the positive and negative charge, respectively, with the latter the key site for hydrolysis by cholinesterase and altering the relative activity on mAChRs and nAChRs. Older muscarinic agonists, carbachol and methachloline are used mainly in experimental pharmacological bioassays.

(a) **Therapeutic Effects of Muscarinic Agonists**

The responses of effector organs to muscarinic/parasympathomimetic excitation are shown in Table 4.1 for individual organs and the cholinergic receptor type involved. The effects on each organ are briefly reviewed below; more detailed and comprehensive reviews are listed in the bibliography.

Cardiovascular effects

- The main effects are a decrease in cardiac muscle activity and output due to a decrease in the force of contraction of the atria, because the ventricles have little parasympathetic innervation and low sensitivity to muscarinic agonists.
- The overall response is a general vasodilation due in part to nitric oxide-mediated effects resulting in a sharp fall in arterial blood pressure.

Smooth muscle effects

- Apart from the heart, muscarinic agonists contract gastrointestinal smooth muscle, increasing peristaltic activity, which can cause colicky pain.
- Bladder and bronchial smooth muscle contraction.

Sweating, lacrimation, salivation and bronchial secretion.

- Increased exocrine activity resulting in bronchial secretion, constriction and impairment of the airways.

Effects on the eye

The effects of parasympathetic stimulation on the eye can be broken down into the following effects on the eye muscles and secretion briefly described below (more detail reviews on the effects of drugs on the eye are listed in the references).

- Pupillae muscle: runs circumferentially in the iris and adjusts the curvature of the lens with the cillary muscle, but also helps in regulating intraocular pressure
- Cillary muscle: contraction pulls the cillary body forward and inward, relaxing the tension on the suspensory ligament on the lens, allowing the lens to bulge more, reducing its focal length.
- Aqueous humor: an increases in intraocular pressure associated with glaucoma impedes drainage of aqueous humor secretion, which is mitigated by muscarinic agonists acting on constrictor pupillae muscles to lower intraocular pressure.

Effects on the central nervous system.

- Muscarinic drugs that penetrate the blood–brain barrier induce tremor, hypothermia and increase in locomotion activity and cognitive enhancing properties.
- A number of M_1-selective agonists have recently been investigated for the possible use in treating cognitive disorders (*e.g.* xanomeline).

(b) **Therapeutic Effects of Muscarinic Antagonists**

Relatively few selective muscarinic acetylcholine antagonists have been identified, based on their rank order of affinity, but limited selectivity (*e.g.* 4-DAMO, MT7, darifencin (M3), pirenzepine (M1)) in order to identify the involvement of particular subtypes (see Table 4.3 and

Table 4.3 Muscarinic receptor agonists and antagonists.

Atropine	Nonselective antagonist	Belladonna alkaloid
	Well absorbed orally	Main side effects: urinary retention, dry mouth, blurred vision
	CNS stimulant	Dicycloverine (dicyclomine) is similar and used mainly as an antispasmodic agent
Hyoscine	Similar to atropine	Belladonna alkaloid (also known as scopolamine)
	CNS depressant	Causes sedation; other side effects as atropine
Hyoscine butylbromide	Similar to atropine but poorly absorbed and lacks CNS effects	Quaternary ammonium derivative
	Significant ganglion-blocking activity	Similar drugs include atropine methonitrate, propantheline
Ciotropium	Similar to atropine methonitrate	Quaternary ammonium compound
	Does not inhibit mucociliary clearance from bronchi	Ipratropium similar
Tropicamide	Similar to atropine	—
	May raise intraocular pressure	
Cyclopentolate	Similar to tropicamide	—
Darifenacin	Selective for M_3 receptors	Used for urinary incontinence with few side effects (increases salivary and lacrimal secretion)
Ipratropium	Nonselective muscarinic antagonist	Indicated for the treatment of reversible bronchospasm associated with chronic obstructive pulmonary disease (COPD).
Pirenzapine	Selective for M_1 receptors and promotes oligomerization of the receptor.	Used in the treatment of peptic ulcers, as it reduces gastric acid secretion by an action on ganglion cells and reduces muscle spasm. Also used in the treatment of myopia.

Alexander *et al.*, "The Concise Guide to Pharmacology," 2013/14, p. 1474 for more detailed information). Several nonselective antagonists are currently used in clinical practice; (atropine, dicycloverine, tolterodine, oxybutynin, ipratropin – M1 type), and the selective antagonists gallamine, darifencin and pirenzepine (Table 4.3).

Muscarinic receptor antagonists (parasympatholytics or anticholinergic drugs) are competitive inhibitors of ACh.

That is, one molecule of antimuscarinic agents blocks one ACh receptor.[17] A large number of antimuscarinic agents show greater affinity for the muscarinic receptor than the ACh receptor but noncompetitive antagonism is observed in the presence of large concentrations of ACh in the synaptic cleft as a result of acetylcholinesterase inhibition (AChI).

At the nerve terminal on smooth muscle, accumulation of ACh can result in spasticity due to constant stimulation, and antispasmodics are a standard treatment in conditions that are due to excessive secretions of ACh: *e.g.*, gastrointestinal hypermotility (diarrhea), bronchi, biliary, arterial and urinary duct spasms.[17]

Typical antimuscarinic drugs belong to the same botanical *solanaceous alkaloid* family, *e.g. hyoscyamine (atropine)* (**4.15**) and *hyoscine* (scopolamine) (**4.16**), pirenzepine (**4.17**)

4.15   **4.16**

These natural products like many of the lipophilic synthetic agents (containing carboxylic acids, heterocyclic or acyclic amino alcohols) are all competitive antagonists at the ACh muscarinic receptors. The therapeutic activity and the many pronounced side effects (dry mouth, flushing, mydriasis, cycloplegria of the eye) of these agents are not always specific to one muscarinic receptor subtype and result in poor patient compliance (see recommended reading for further information). However, some antimuscarinic agents *e.g. pipenzepine* (**4.17**) *and ipratropium* (**4.18**) are selective for M1 receptors and produce therapeutic gastrointestinal effects with limited blurred vision and dry mouth.[3] Like many compounds with low lipid solubility it exhibits limited ability to cross the blood–brain barrier (BBB) and result in adverse central effects.

4.17   **4.18**

(c) **Central Effects of Muscarinic Receptor Agonists**

The central effects of muscarinic agents are well documented (see the reviews in the recommended reading[3,16]), most are lipophilic and thus penetrate the BBB and produce an array of CNS effects. Depending on dose they can produce euphoria, hallucinations, delirium (high doses of atropine and hyoscyamine) and central depression, drowsiness, fatigue and dreamless sleep (low doses of hyoscine) depending on which population of muscarinic receptor subtypes are blocked by the antagonist.[3,16] Anticholinesterase drugs such as physostigmine oppose these effects, and are effective antidotes to atropine poisoning. Muscarinic receptor antagonists are also used for treating motion sickness, as antiemetics and reducing involuntary movement in patients with Parkinson's disease.[16]

(ii) **Inhibition of Acetylcholine Synthesis**

ACh is synthesized is presynaptic nerve terminals (Figure 4.3). The rate-limiting step in the process is the transport of choline into the nerve terminal. Hemicholinium, also known as hemicholine, is a drug that blocks the reuptake of choline by the high-affinity choline transporter (ChT), thereby inhibiting ACh synthesis. Vesamicol acts in a similar mechanism of action by slowly depleting ACh stores at the nerve terminal. Both drugs have no clinical application.[3]

(iii) **Inhibition of Acetylcholine Release**

ACh release is mediated by a nerve impulse resulting in entry of Ca^{2+} into the nerve terminal, which stimulates exocytosis and increases the quantal release of ACh. Drugs that inhibit Ca^{2+} entry include various amino glycoside antibiotics (*e.g.* streptomycin and neomycin) and $Mg.^{2+}$

Other agents include the neurotoxins botulinum toxin and β-bungarotoxin. The former is a protein produced by the anaerobic bacillus *Clostridium botulinum,* which specifically inhibits ACh release and causes an extremely serious type of contaminated food poisoning. Botulinum poisoning causes a progressive and irreversible parasympathetic and motor paralysis often leading to death.

Clinical uses of locally injected botulinum toxin (Botox) include for blepharospasm (persistent eyelid spasm), squint, siahorroea (excessive salivation) and fashionable cosmetic use to reduce wrinkles. However, botulinium toxin is antigenic, resulting in a loss of efficacy and immunogenicity, after repeated injections. β-bungarotoxin action is similar to botulinum toxin, although its active principle is a phospholipase rather than a peptidase. The snake venom from cobras also contains α-bungarotoxin that blocks postsynaptic receptors, thus causing a complete paralysis at both pre- and postsynaptic sites.

(iv) **Inhibition of Cholinesterase**

Drugs that inhibit peripheral cholinesterase form three main categories depending on their interaction with the active site, duration of action and ability to inhibit AChE and butyrylcholinesterase (BuChE). (i) Short acting (edrophonium) **(49)**; (ii) medium acting (neostigmine, physostigmine **(4.20)**); irreversible (organophosphates, dyflos **(4.21)**, ecothiophate).

edrophonium **(4.19)**,   physostigmine **(4.20)**   dyflos **(4.21)**

Both AChE and BuChE are serine hydrolases, which include many proteases such as trypsin. The site of action of AChE resides in two distinct regions: an anionic site (glutamate moiety), which binds the basic choline moiety of ACh; and an esteric (catalytic) site (histidine + serine). The acidic (acetyl) group of the substrate is transferred to the serine hydroxyl group, leaving (transiently) an acetylated enzyme molecule and a molecule of free choline. This results in a spontaneous hydrolysis of the serine acetyl group and a rapid and large turnover of ACh molecules (estimated to be in the region of over 5000 molecules of ACh hydrolyzed per second by a single active site).[18,19]

The effects of anticholinesterase are mainly due to an augmentation of cholinergic transmission at cholinergic autonomic synapses and at the neuromuscular junction. The actions of neostigmine and pyridostigmine are mainly on neuromuscular transmission and less so on the autonomic system, whereas physostigmine and organophosphates show the opposite action. Why there are two different types of action is largely unknown (see the reference section for further reading on anticholinesterase drugs). Anticholinesterase acting on autonomic synapses results in severe bradycardia, hypotension and difficulty in breathing and may be fatal in the case of organophosphate poisoning due to exposure to insecticides and nerve gases. However, the effects of AChEs on the neuromuscular junction cause muscle fasciculation and increased twitch tension, owing to a repetitive firing of the muscle fibers associated with prolongation of the end-plate potential (epp).[3,19]

The clinical uses of AChE drugs is mainly postoperative, to reverse the action of nondepolarizing drugs. Neostigmine is the drug of choice in combination with atropine to reduce the parasympathetic effects. Neostigmine or pyridostigmine are also used to treat myasthenia gravis, a rare disease of fatigue/failure of the neuromuscular transmission resulting from repetitive stimulation.[3,19]

Irreversible inhibitors of AChE may lead to muscular paralysis, convulsions, bronchial constriction, and death by asphyxiation. Organophosphates (OP), esters of phosphoric acid, are a class of irreversible AChE inhibitors. Cleavage of OP by AChE leaves a phosphoryl group in the esteratic site, which is slow to be hydrolyzed (on the order of days) and can become covalently bound. Irreversible AChE inhibitors have been used in insecticides (*e.g.*, malathion) and nerve gases for chemical warfare (*e.g.*, sarin and soman). Carbamates, esters of *N*-methyl carbamic acid, are AChE inhibitors that hydrolyze in hours and have been used for medical purposes (*e.g.*, physostigmine for the treatment of glaucoma). Reversible inhibitors occupy the esteratic site for short periods of time (seconds to minutes) and are used to treat of a range of central nervous system diseases. Tetrahydroaminoacridine (THA) and donepezil are FDA-approved to improve cognitive function in Alzheimer's disease. Rivastigmine is also used to treat Alzheimer's and Lewy body dementia, and pyridostigmine bromide is used to treat myasthenia gravis (see references and further reading at the end of the chapter). The Alzheimer disease drugs donepezil, galantamine, and rivastigmine are inhibitors of acetylcholinesterase as well.

An endogenous inhibitor of AChE in neurones is Mir-132 microRNA, which may limit inflammation in the brain by silencing the expression of this protein and allowing ACh to act in an anti-inflammatory capacity.[20]

(v) Cognitive Dysfunction: Alzheimer's Disease

It is well documented that a deficit in central cholinergic function is a major contributor to cognitive decline, dementia and Alzheimer's disease. Postmortem studies from Alzheimer patient brain tissue show shrinkage of brain tissue and a selective loss of cholinergic neurones in the forebrain nuclei. Degeneration of presynaptic M2 receptor, along with a significant reduction in nicotinic receptors underlies the pathophysiology of the disease, with no loss of postsynaptic M1 receptors. Despite extensive research and clinical studies the effects of direct- or indirect-acting cholinergic agents show limited symptomatic relief and peripheral and central intolerable side effects.

Current therapy has focused on novel inhibitors of acetylcholinesterase, tacrine (THA; 9-aminotetrahydroacridine) and bis-(7)-tacrine; the latter has been reported to have superior memory-enhancing effects compared to tacrine (**4.22**) and is a more potent competitive antagonist of GABA$_A$ receptors than tacrine, which may contribute to its memory-enhancing properties. The GABA$_A$ activity suggests that Alzheimer's disease may not solely be due to a defect in cholinergic transmission.

Donepezil (**4.23**) is one of a number of purported selective inhibitors of acetylcholinesterase that are reported to show less propensity to produce hepatotoxicity and cholinergic-related side effects than tacrine and are used to treat Alzheimer's disease and Parkinsonian-like side effects.

4.22   **4.23**

Finally, it has also been shown that the main active ingredient in cannabis, tetrahydrocannabinol, is a competitive inhibitor of acetylcholinesterase, an area of current therapeutic interest.[21]

4.5 Noradrenergic Receptors and Receptor Subtypes

In this section, the physiology and function of noradrenergic neurones and the properties of adrenoceptors (the receptors on which noradrenaline and adrenaline act), is discussed along with the various classes of drugs that affect them. For the sake of brevity, most of the pharmacological information is summarized in figures and tables later in this section and more detail can be found in the reference list for recommended reading.

4.5.1 Pharmacological Classification

The discovery of the hypertensive effects of adrenaline and its subsequent isolation from adrenal gland extracts in 1896 by Oliver and Schafer, led to the pioneering work of Sir Henry Dale in 1913.[3] He showed that adrenaline causes two distinct kinds of effect, namely vasoconstriction in certain vascular beds (which normally predominates and, together with its actions on the heart, results in a rise in arterial pressure) and vasodilatation in others. Further experiments at the Wellcome laboratories showed that the pressor vasoconstrictor component disappeared if the animal was first injected with an ergot derivative, and it was noticed that adrenaline then caused a fall, instead of a rise, in arterial pressure. These landmark findings *paralleled* Dale's demonstration of the separate muscarinic and nicotinic components of the action of acetylcholine. It was not until the work of Ahlquist in 1948 that the existence of several subclasses of adrenoceptor with distinct tissue distributions and actions was clearly shown (see Table 4.1).[3]

He determined the rank order of potency of several catecholamines, depending on what was being measured and based on these postulated two kinds of receptors, α and β, based on their agonist potencies as follows:

- α: noradrenaline > adrenaline > isoprenaline;
- β: isoprenaline > adrenaline > noradrenaline.

4.5.1.1 *Noradrenergic Receptors*

From these classical pharmacology studies, several α-adrenoceptors have now been identified ($\alpha_{1A}, \alpha_{1B}, \alpha_{1D}$) along with three β-adrenoceptors ($\beta_1, \beta_2, \beta_3$), all distinct from each other based on their physiological/pharmacological response and their association with a specific second-messenger system. The major physiological/pharmacological effects that are produced by these adrenoceptor subtypes and the main drugs that act on them are shown in Tables 4.4 and 4.5. More detailed descriptions of the agonists and antagonists acting at these receptors can be found in the recommended reading list and references.[1,3,6,10] It is important to note that both α and β receptor subtypes are expressed in smooth muscle cells, nerve terminals and endothelial cells, and their role in physiological regulation and pharmacological responses is still a matter of debate as many of the available drugs are not completely selective, which can caused unwanted side effects such as bronchoconstriction, as is the case with β_1-antagonists with some β_2 selectivity, *e.g.* propranolol.[3]

All noradrenergic neurones in the periphery are postganglionic sympathetic and their cell bodies are situated in sympathetic ganglia. The neurones generally have long axons that contain a series of varicosities all along the branching terminal network. These varicosities contain numerous synaptic vesicles, which are the sites of synthesis and release of noradrenaline and of coreleased mediators such as ATP and neuropeptide Y, which are stored in vesicles and released by exocytosis. The tissue content of noradrenaline closely parallels the density of the sympathetic innervation in most cases. With the exception of the adrenal medulla, sympathetic nerve terminals account for all the noradrenaline content of peripheral tissues. The highest concentration of noradrenaline is found in the heart, spleen, vas deferens and some blood vessels are particularly rich in noradrenaline ($5–50 \ \mathrm{nmol \, g^{-1}}$ of tissue). These tissues are widely used in pharmacological studies of noradrenergic transmission.[3,6,14] For further information on noradrenergic neurones, see the bibliography and recommended reading list.

4.5.2 Noradrenaline Synthesis

The biosynthetic pathway for noradrenaline synthesis is shown in Figure 4.6 and drugs that affect noradrenaline synthesis are summarized, release and uptake in Table 4.4(a–d). The metabolic precursor for noradrenaline is *L-tyrosine*, an aromatic amino acid that is

Table 4.4 Drugs that affect noradrenaline synthesis, release or uptake.

Drug	Main action	Uses/function	Unwanted effects	Pharmacokinetic profile	Comments
(a) Drugs affecting NA synthesis					
α-Methyl-*p*-tyrosine (AMPT)	Inhibits tyrosine hydroxylase	Occasionally used in phaeochromocytoma	Hypertension and sedation	High bioavailability Plasma $t_{1/2} \sim 3$–4 h	No clinical use
Carbidopa	Inhibits dopa decarboxylase	Used as adjunct to levodopa to prevent peripheral effects	—	Absorbed orally Does not enter brain	Used in combination with L-Dopa for Parkinson's Disease (see Chapter 5)
Methyldopa	False transmitter precursor for methylnoradrenaline, a potent α_2-agonist	Hypertension in pregnancy	Hypotension, drowsiness, diarrhoea, impotence, hypersensitivity reactions	Absorbed slowly by mouth Excreted unchanged or as conjugate Plasma $t_{1/2} \sim 6$ h	Central effects (see Chapter 5)
(b) Drugs that release NA (indirectly acting sympathomimetic amines)					
Tyramine	NA release	No clinical uses Present in various foods	As norepinephrine	Normally destroyed by MAO in gut Does not enter brain	
Amphetamine	NA release, MAO inhibitor, NET inhibitor, CNS stimulant	Used as CNS stimulant in narcolepsy, also (paradoxically) in hyperactive children Appetite suppressant Drug of abuse	Hypertension, tachycardia, insomnia Acute psychosis with overdose Dependence	Well absorbed orally Penetrates freely into brain Excreted unchanged in urine Plasma $t_{1/2} \sim 12$ h, depending on urine flow and pH	Methylphenidate and atomoxetine are similar (used in the treatment of ADHD.

Table 4.4 (Continued)

Drug	Main action	Uses/function	Unwanted effects	Pharmacokinetic profile	Comments
Ephedrine	NA release, β agonist, weak CNS stimulant action	Nasal decongestion	As amphetamine but less pronounced	Similar to amphetamine aspects	Contraindicated if MAO inhibitors are given
(c) Drugs that inhibit NA release					
Reserpine	Depletes NA stores by inhibiting VMAT	Hypertension (no clinical use today)	As methyldopa Also depression, parkinsonism, gynaecomastia	Poorly absorbed orally Slowly metabolized Plasma $t_{1/2} \sim 100$ h Excreted in milk	Long-acting antihypertensive effect
(d) Drugs affecting NA uptake					
Imipramine	Blocks neuronal transporter (NET) Also has atropine-like action	Antidepressant	Atropine-like side effects Cardiac dysrhythmias in overdose	Well absorbed orally 95% bound to plasma protein Converted to active metabolite (desmethyl-imipramine) Plasma $t_{1/2} \sim 4$ h	Desipramine and amitriptyline are similar See Chapter 5
Cocaine	Local anaesthetic; blocks NET CNS stimulant	Rarely used local anaesthetic Major drug of abuse	Hypertension- CNS stimulant, proconvulsive, and drug dependence	Well absorbed orally or intranasally	See Chapter 5

Table 4.5 Mixed (α- and β-) adrenoceptor agonists.[a]

Drug	Main action	Uses/function	Unwanted effects	Pharmacokinetic profile	Comments
Noradrenaline (Norepinephrine)	α/β agonist	Sometimes used for hypotension in intensive care Transmitter at postganglionic sympathetic neurons and in CNS	Hypertension, vasoconstriction, tachycardia (or reflex bradycardia), ventricular dysrhythmias	Poorly absorbed by mouth Rapid removal by tissues Metabolized by MAO and COMT Plasma $t_{1/2} \sim 2$ min As norepinephrine Given i.m. or s.c. (i.v. infusion in intensive-care settings)	
Adrenaline (Epinephrine)	α/β agonist	Asthma (emergency treatment), anaphylactic shock, cardiac arrest Added to local anaesthetic solutions Main hormone of adrenal medulla	As norepinephrine		
Isoprenaline	β agonist (nonselective)	Asthma (obsolete)	Tachycardia, dysrhythmias	Some tissue uptake, followed by inactivation (COMT) Plasma $t_{1/2} \sim 2$ h	Now replaced by salbutamol in treatment of asthma (see Chapter 5)
Dobutamine	β_1 agonist (nonselective)	Cardiogenic shock	Dysrhythmias	Plasma $t_{1/2} \sim 2$ min Given i.v.	

Table 4.5 (*Continued*)

Drug	Main action	Uses/function	Unwanted effects	Pharmacokinetic profile	Comments
Salbutamol	β_2 agonist	Asthma, premature labor	Tachycardia, dysrhythmias, tremor, peripheral vasodilation	Given orally or by aerosol. Mainly excreted unchanged. Plasma $t_{1/2} \sim 4$ h	
Clenbuterol	β_2 agonist	"Anabolic" action to increase muscle strength	As salbutamol	Active orally. Long acting	Illicit use in sport
Mirabegron	β_3 agonist	Symptoms of overactive bladder	Tachycardia	Active orally, given once daily	
Phenylephrine	α_1 agonist	Nasal decongestion	Hypertension, reflex bradycardia	Given intranasally. Metabolized by MAO. Short plasma $t_{1/2}$	—
Methoxamine	α partial agonist	Nasal decongestion	As phenylephrine	Given intranasally. Plasma $t_{1/2} \sim 1$ h	—
Clonidine	α_2 partial agonist	Hypertension, Migraine	Drowsiness, orthostatic hypotension, oedema and weight gain, rebound hypertension	Well absorbed orally. Excreted unchanged and as conjugate. Plasma $t_{1/2} \sim 12$ h	

[a]Catechol-O-methyltransferase (COMT); monoamine oxidase (MAO); NA, noradrenaline (NA); vesicular monoamine transporter (VMAT).

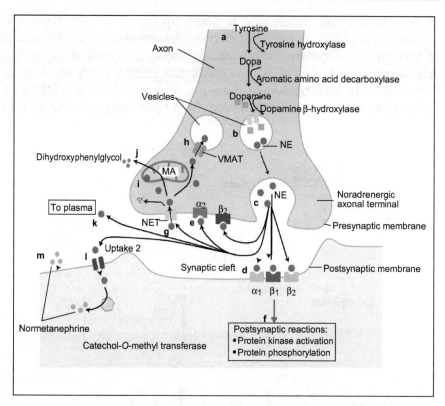

Figure 4.5 Diagram of a noradrenergic axonal terminal showing the release and reuptake of norepinephrine (a) Norepinephrine (NE) is synthesized from tyrosine – *via* hydroxylation to form dihydroxyphenylalanine (dopa), decarboxylation to form dopamine, and hydroxylation to form NE – and (b) stored in vesicles. (c) As a result of an appropriate stimulus (not shown), NE is released into the synaptic cleft. (d) Released NE activates the adrenergic receptors located on the postsynaptic membrane (α_1, β_1 and β_2) and also (e) presynaptic membrane (α_2, β_2), and causes (f) postsynaptic reactions such as protein kinase activation and protein phosphorylation. (g) The norepinephrine transporter (NET) is responsible for reuptake of the NE in the synaptic cleft, and terminates its action. (h) After reuptake by NET, a small portion of the NE is restored in vesicles [following uptake by vesicular amine transporter 2 (VMAT)]; (i) the rest is metabolized in the mitochondria by the enzyme monoamine oxidase (MOA), and (j) the end product dihydroxyphenylglycol (DHPG) is released into the circulation. (k) A small portion of the synaptic NE leaks into the circulation, or (l) is taken up by another system (uptake 2) and (m) metabolized to form normetanephrine (NMN). Because 70–90% of the synaptic NE is taken up by NET, a blockade of NET is likely to produce a shift towards the NMN pathway and away from the DHPG pathway.
T. Tellioglu, D. Robertson, Genetic or acquired deficits in the norepinephrine transporter: current understanding of clinical implications, *Expert Reviews in Molecular Medicine*, Volume 3, Issue 29, pp. 1–10, 2001, reproduced with permission.

present in the body fluids and is taken up by adrenergic neurones. *Tyrosine hydroxylase*, a cytosolic enzyme that catalyzes the conversion of tyrosine to *dihydroxyphenylalanine* (dopa), is found only in catecholamine-containing cells. It is a selective enzyme, only involved in catecholamine metabolism; it does not accept indole derivatives as substrates, and so is not involved in 5-hydroxytryptamine (5-HT)

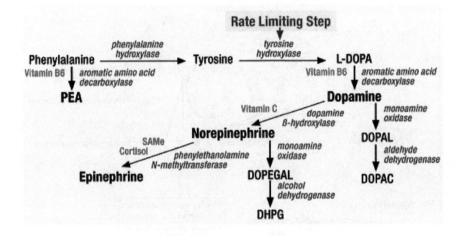

metabolism (see Chapter 5 for discussion on 5-HT-serotonin). This first hydroxylation step is the main control point for noradrenaline synthesis. Tyrosine hydroxylase is therefore the rate-limiting step in the biosynthetic pathway of noradrenaline synthesis and is inhibited by the tyrosine analog **α-methyltyrosine**, which is used experimentally to block noradrenaline synthesis.

The next step in catecholamine biosynthesis is the conversion of dopa to dopamine (see Chapter 5 for a discussion on dopamine), which is catalyzed by the nonselective cytosolic enzyme *dopa decarboxylase*. It also catalyzes the decarboxylation of various other L-aromatic amino acids, such as *L-histidine* and *L-tryptophan*, which are precursors in the synthesis of histamine and 5-HT, respectively. Unlike, tyrosine hydroxylase, dopa decarboxylase activity is not rate limiting for noradrenaline synthesis, Figure 4.6.

Dopamine-β-hydroxylase (DBH) is another relatively nonspecific enzyme. It is restricted to catecholamine-synthesizing cells and located in synaptic vesicles, in a membrane-bound form. It is coreleased from adrenergic nerve terminals with noradrenaline, as a small proportion in a soluble form within the vesicle. The released DBH is not subject to rapid degradation or uptake and its concentration in plasma and body fluids is a good index of overall sympathetic nerve activity. Noradrenaline is subject to rapid degradation and reuptake, Figure 4.6.

Numerous drugs inhibit DBH, including copper-chelating agents and **disulfiram (4.24), nepicostat (4.25) and tropolone (4.26).**

4.24   **4.25**   **4.26**

Figure 4.6   Norepinephrine biosynthesis pathway. Catecholamines biosynthesis: Both norepinephrine and epinephrine are first oxidatively deaminated by monoamine oxidase (MAO) to 3,4-dihydroxyphenylglycoaldehyde (DOPGAL) and then either reduced to 3,4-dihydroxyphenylethylene glycol (DOPEG) or oxidized to 3,4-dihydroxymandelic acid (DOMA). Alternatively, they can be initially methylated by catechol-O-methyltransferase (COMT) to normetanephrine and metanephrine, respectively. Most of the products of either type of reaction then are metabolized by the other enzyme to form the major excretory products in blood and urine, 3-methoxy-4-hydroxyphenylethylene glycol (MOPEG or MHPG) and 3-methoxy-4-hydroxymandelic acid (VMA). Free MOPEG is largely converted to VMA. The glycol and, to some extent, the O-methylated amines and the catecholamines may be conjugated to the corresponding sulfates or glucuronides.

These drugs can cause a partial depletion of noradrenaline stores and interference with sympathetic transmission. A rare genetic disorder, DBH deficiency, causes failure of noradrenaline synthesis resulting in severe orthostatic hypotension (see the recommended reading list).[22]

Phenylethanolamine N-*methyl transferase* (PNMT) catalyzes the *N*-methylation of noradrenaline to adrenaline. The enzyme is mainly located in the adrenal medulla, which contains populations of adrenaline-releasing and noradrenaline-releasing cells. The production of PNMT is induced by an action of the steroid hormones secreted by the adrenal cortex, after birth. PNMT is also found in certain parts of the brain, where adrenaline may function as a transmitter, but little is known about its role in the central nervous system (CNS).

Noradrenaline turnover can be measured under steady-state conditions by measuring the rate at which labeled noradrenaline accumulates when a labeled precursor (*e.g.* tyrosine or dopa), is administered. Noradrenaline turnover is defined as the time taken for an amount of noradrenaline equal to the total tissue content to be degraded and resynthesized. In peripheral tissues, it is generally about 5–15 h, but it becomes much shorter if sympathetic nerve activity is increased. Under normal physiological conditions, the rate of synthesis closely matches the rate of release, so that the noradrenaline content of tissues is constant regardless of how fast it is being released.

4.5.3   Drugs that Inhibit Noradrenaline Synthesis

Several drugs that inhibit noradrenaline synthesis are listed in Table 4.4, other agents are known for their neurotoxic/neurodegenerative effects and are mainly used in experimental studies as listed below;

- **6-Hydroxydopamine** (identical with dopamine except for an extra hydroxyl group) is selectively neurotoxic for noradrenergic neurones, because it is taken up and converted to a toxic metabolite. Used experimentally to eliminate noradrenergic neurones, not used clinically.
- **MPTP (1-methyl-4-phenyl-1, 2,3,5-tetrahydropyridine)**; is a similar selective neurotoxin acting on dopaminergic neurones.

4.5.4   Noradrenaline Storage

Nearly all of the noradrenaline in nerve terminals or chromaffin cells is contained in vesicles; only a little is free in the cytoplasm under normal circumstances. The concentration in the vesicles is very high $(0.3-1.0 \text{ mol} \, l^{-1})$ and is maintained by the *vesicular monoamine transporter* (VMAT), which is similar to the monoamine transporter responsible for noradrenaline uptake into the nerve terminal, but depends on a transvesicular proton gradient driving its uptake. Amine depleting drugs, such as **reserpine** (see Table 4.4(c)) block this transport and cause nerve terminals to lose their vesicular noradrenaline stores. The vesicles also contain two major constituents besides noradrenaline, namely ATP (about four molecules per molecule of noradrenaline) and a protein called *chromogranin A*. These substances are coreleased along with noradrenaline, and it is generally assumed that a reversible complex, depending partly on the opposite charges on the molecules of noradrenaline and ATP, is formed within the vesicle. This would serve both to reduce the osmolality of the vesicle contents and also to reduce the tendency of noradrenaline to leak out of the vesicles within the nerve terminal.[3,22]

4.5.5 Noradrenaline Release

Drugs that affect noradrenaline release on adrenergic receptor subtypes are summarized in Figure 4.5.

A unique feature of noradrenaline release is that a single neuron possesses many thousands of varicosities, so one impulse leads to the discharge of a few hundred vesicles, having a wide range of activities (see Table 4.1). This is in marked contrast to the highly localized release of acetylcholine from a single nerve terminal at the neuromuscular junction. Noradrenaline release is affected by a variety of substances that act on presynaptic receptors (see Rang and Dale[3]). Many different types of nerve terminal (cholinergic, noradrenergic, dopaminergic, 5-HT-ergic, peptidergic) are subject to this type of control, and many different mediators (acetylcholine acting through muscarinic receptors, catecholamines acting through α and β receptors, angiotensin II, prostaglandins, purine nucleotides, neuropeptides, *etc.*) can act on presynaptic terminals.[7] Presynaptic modulation represents an important physiological control mechanism throughout the nervous system.

Also, noradrenaline acting on presynaptic β_2-receptors, can regulate its own release, and that of coreleased ATP (see Rang and Dale[3]). The released noradrenaline exerts a local inhibitory effect on the terminals from which it came – the so-called *autoinhibitory feedback* mechanism (see Figure 4.5). Drugs acting at these presynaptic receptors can have large effects on sympathetic transmission. The physiological significance of presynaptic action is still somewhat contentious, as in most tissues; it is less influential than biochemical measurements of transmitter overflow would seem to imply. Thus, blocking autoreceptors causes large changes in noradrenaline *overflow* and the amount of noradrenaline released into the bathing solution or the bloodstream when sympathetic nerves are stimulated is not always correlated with changes in the tissue responses. In the sympathetic nervous system, the α_2 receptor operates an inhibitory feedback mechanism, which inhibits adenylyl cyclase and prevents the opening of calcium channels. Sympathetic nerve terminals also possess β_2 receptors; coupled to activation of adenylyl cyclases, which increase noradrenaline release and is also modulated by β_2 receptors at the sympathetic nerve terminal *via* action of adenylyl cyclase. The physiological significance is not yet clear.[3,7]

In summary, drugs can affect noradrenaline release in four main ways: (i) by directly blocking release (noradrenergic neuron-blocking drugs), (ii) by evoking noradrenaline release in the absence of nerve terminal depolarization (indirectly acting sympathomimetic drugs), (iii) by acting on presynaptic receptors that indirectly inhibit or enhance depolarization-evoked release; examples include α_2 agonists, angiotensin II, dopamine and prostaglandins, (iv) by increasing or decreasing available stores of noradrenaline (*e.g.* reserpine, MAO inhibitors).[3,7,22]

4.5.6 Drugs that Affect Release or Indirectly Act as Sympathomimetic Amines

This is a drug that does not itself interact with adrenoceptors. The most important drugs in this category are **tyramine, amphetamine,** and **ephedrine,** all structurally related to noradrenaline. Drugs used for their central effects (see Chapter 5) *e.g.*, **methylphenidate,** are used in clinical practice to treat attention deficit-hyperactivity disorder (ADHD).

These categories of drugs are similar to noradrenaline and are transported into nerve terminals by noradrenaline transporter (NET). They are taken up into the vesicles by VMAT, in exchange for noradrenaline, which escapes into the cytosol. MAO, Figures 4.5 and 4.6), degrades the remaining cytosolic noradrenaline while the rest escapes *via* NET, in exchange for these monoamines, to act on postsynaptic receptors (Figure 4.5.). Exocytosis is not involved in the release process, so their actions do not require the presence of Ca^{2+}. They are relatively nonspecific in their actions, and act partly by a direct effect on adrenoceptors, partly by inhibiting NET and enhancing the effect of the endogenous released noradrenaline and partly by inhibiting MAO.[3,23]

These drugs are strongly influenced by other drugs that modify noradrenergic transmission. Thus, reserpine and 6-hydroxydopamine abolish their effects by depleting the terminals of noradrenaline, whereas MAO inhibitors strongly potentiate their effects by preventing inactivation processes. However, because tyramine is a substrate for MAO and MAO metabolizes dietary tyramine before reaching the systemic circulation, ingestion of tyramine amine-rich foods such as fermented cheese can then provoke a sudden and dangerous rise in blood pressure. Inhibitors of NET, such as tricyclic antidepressants (*e.g.* imipramine and amitriptyline), interfere with the effects of indirectly acting sympathomimetic amines and other endogenous trace amines (methyltryptamine and N,N,-dimethyltryptamine) by preventing their uptake into the nerve terminals.[3,7]

This class of drugs and releasing agents such as amphetamine have important effects on the central nervous system (see Chapter 5) that depend on their ability to release not only noradrenaline, but also 5-HT and dopamine from nerve terminals in the brain. However, marked tolerance rapidly develops to these indirect sympathomimetics. For example, tyramine and amphetamine produce progressively smaller pressor responses, due to a depletion of the releasable store of noradrenaline. Tolerance to the central effects also develops on repeated administration, contributing to the liability of amphetamine and related drugs to cause dependence and significant abuse potential.

The peripheral actions of these agents include bronchodilatation, raised arterial pressure, peripheral vasoconstriction, increased heart rate and force of myocardial contraction, and inhibition of gut motility. Only ephedrine is used for its peripheral sympathomimetic effects, as a nasal decongestant.[3,7]

4.5.7 Uptake and Degradation of Catecholamines

Almost all released noradrenaline is terminated by reuptake of the transmitter into noradrenergic nerve terminals, Figure 4.6. Other cells in the vicinity also sequester a small amount. All circulating adrenaline and noradrenaline are degraded enzymically, but much more slowly than acetylcholine, where synaptically located acetylcholinesterase inactivates the transmitter in milliseconds.[3,22] The two main catecholamine-metabolizing enzymes (monoamine oxidase – MAO and catechol-o-methyltransferase – COMT) are located intracellular, so uptake into cells necessarily precedes metabolic degradation (Figure 4.6.).

About 75% of the noradrenaline released by sympathetic neurones is recaptured and repackaged into vesicles. This serves to cut short the action of the released noradrenaline, as well as recycling it. Non-neuronal cells in the vicinity, limiting its local spread, capture the remaining 25%. These two uptake mechanisms depend on distinct transporter molecules. Neuronal uptake is performed by the plasma membrane noradrenaline transporter (generally known as NET, the *norepinephrine transporter*), which belongs to the family of neuro-transmitter transporter proteins (NET, DAT, SERT, *etc.*) specific for different amine transmitters, these act as cotransporters of Na^+, Cl^- and the amine in question, using the electrochemical gradient for Na^+ as a driving force.[22] Packaging into vesicles occurs through the *vesicular monoamine transporter* (VMAT), driven by the proton gradient between the cytosol and the vesicle contents. Extraneuronal uptake is performed by the *extraneuronal monoamine transporter* (EMT), which belongs to a large and widely distributed family of organic cation transporters (OCTs). NET is relatively selective for noradrenaline, with high affinity and a low maximum rate of uptake, and it is important in maintaining releasable stores of noradrenaline (Figure 4.6.). EMT has lower affinity and higher transport capacity than NET, and transports adrenaline and isoprenaline as well as noradrenaline. The effects of several important drugs that act on noradrenergic neurones depend on their ability either to inhibit NET or to enter the nerve terminal with its help. The properties of neuronal and extraneuronal uptake are reviewed extensively in the recommended reading list.[22,23]

4.5.8 Inhibitors of Noradrenaline Uptake

Reuptake of released noradrenaline by NET is the main process by which its action is negated. Many drugs inhibit NET, and thereby enhance the effects of both sympathetic nerve activity and circulating noradrenaline. NET is not responsible for clearing circulating ad-renaline, so these drugs do not affect responses to this amine. There are several chemical classes of drugs that selectively inhibit the action of NET, earlier compounds include the *tricyclic antidepressants*, ex-amples of which include imipramine (4.27), reboxetine (4.28) and the natural product cocaine (4.29) a benzoylmethylecgonine. These drugs have their major effect on the central nervous system but also cause tachycardia and cardiac dysrhythmias, reflecting their peripheral effect on sympathetic transmission. Cocaine, known mainly for its abuse liability and local anesthetic activity, enhances sympathetic transmission, causing tachycardia and increased arterial pressure. The euphoria and excitement are a manifestation of the same mechanism acting in the brain.[3,7]

4.27   **4.28**   **4.29**

The overall therapeutic activity of noradrenaline reuptake inhibitors (NRIs) is governed by their affinity, efficacy and selectivity with respect to different types of adrenoceptor, with the right properties for specific clinical indications. The rich families of adrenoceptor drugs are used in several clinical indications but largely to relax smooth muscle in different organs of the body and those that block the cardiac stimulant effects of the sympathetic nervous system.[24]

4.5.9 Metabolic Degradation of Catecholamines

Two main intracellular enzymes metabolize endogenous and exogenous catecholamines: MAO and COMT. MAO (of which there are two distinct isoforms, MAO-A and MAO-B; (see the recommended reading list[3,14,22]) is bound to the surface membrane of mitochondria. It is abundant in noradrenergic nerve terminals but is also present in liver, intestinal epithelium and other tissues. MAO converts catecholamines to their corresponding aldehydes, which, in the periphery, are rapidly metabolized by *aldehyde dehydrogenase* to the corresponding carboxylic acid (3,4-dihydroxyphenylglycol being formed from noradrenaline; (see Figure 4.6). MAO can also oxidize other monoamines, including dopamine and 5-HT. It is inhibited by various drugs, which are used mainly for their effects on the central nervous system, where these three amines all have their own unique transmitter functions (see Chapter 5). These drugs have important side effects that are related to disturbances of peripheral noradrenergic transmission. Within sympathetic neurones, MAO controls the content of dopamine and noradrenaline, and the releasable store of noradrenaline increases if the enzyme is inhibited. MAO and its inhibitors are discussed in more detail in the recommended reading list.

The second major pathway for catecholamine metabolism involves methylation of one of the catechol hydroxyl groups by COMT to give a methoxy derivative. COMT is absent from noradrenergic neurones but present in the adrenal medulla and many other cells and tissues. The final product formed by the sequential action of MAO and COMT is *3-methoxy-4-hydroxyphenylglycol* (MHPG; see Figure 4.6). This is partly conjugated to sulfate or glucuronide derivatives, which are excreted in the urine, but most of it is converted to *vanillylmandelic acid* (VMA; Figure 4.6) and excreted in the urine in this form. In patients with tumors of chromaffin tissue that secrete these amines (a rare cause of high blood pressure), the urinary excretion of VMA is markedly increased, this being used as a diagnostic test for this condition.[22]

In the periphery, neither MAO nor COMT is primarily responsible for the termination of transmitter action, most of the released noradrenaline being quickly recaptured by NET. Circulating catecholamines are sequestered and inactivated by a combination of NET, EMT and COMT, the relative importance of these processes varying according to the agent concerned. Thus, mainly NET removes circulating noradrenaline, whereas adrenaline is more dependent on EMT. Isoprenaline, on the other hand, is not a substrate for NET, and is removed by a combination of EMT and COMT.

In the central nervous system (see Chapter 5), MAO is more important as a means of terminating transmitter action than it is in the periphery, and MAO knockout mice show a greater enhancement of noradrenergic transmission in the brain than do NET knockouts, in which neuronal stores of noradrenaline are almost depleted.[23,25,26] The main excretory product of noradrenaline released in the brain is MHPG.[3,23]

4.5.10   Drugs Acting on Noradrenergic Transmission

Many clinically important drugs, particularly those used to treat cardiovascular, respiratory and psychiatric disorders (see the CNS section), act by affecting noradrenergic neuron function, acting on adrenoceptors, transporters or catecholamine-metabolizing enzymes.[22,24] The properties of the most important drugs in this category are summarized below and in Table 4.6.

4.5.11   Noradrenergic Neuron-blocking Drugs

Noradrenergic neuron-blocking drugs antagonize the effects of noradrenaline, adrenaline, isoprenaline and other direct- and

Table 4.6 Adrenoceptor antagonists.

Drug	Main action	Uses/function	Unwanted effects	Pharmacokinetic profile	Comments
(a) α-Adrenoceptor receptor antagonists					
Phenoxybenzamine	α antagonist (nonselective, irreversible) Uptake 1 inhibitor	Phaeochromocytoma	Postural hypotension, tachycardia, nasal congestion, impotence	Absorbed orally Plasma $t_{1/2} \sim 12$ h	Action outlasts presence of drug in plasma, because of covalent binding to receptor
Prazosin	α_{1a} antagonist/ inverse agonist	Hypertension	As for phenoxybenzamine	Absorbed orally Metabolized by liver Plasma $t_{1/2} \sim 4$ h	Doxazosin and terazosin are similar but longer acting
Tamsulosin	α_{1A} antagonist (uroselective) α_{1B} inverse agonist	Prostatic hyperplasia	Failure of ejaculation	Absorbed orally Plasma $t_{1/2} \sim 5$ h	Selective for α_{1A} adrenoceptor
Yohimbine	α_2 antagonists	Not used clinically Claimed to be aphrodisiac	Excitement, hypertension	Absorbed orally Metabolized by liver Plasma $t_{1/2} \sim 4$ h	
(b) β-Adrenoceptor receptor antagonists					
Propranolol	$\beta_{1\&2}$ antagonist	Angina, hypertension, cardiac dysrhythmias, anxiety, tremor, glaucoma	Bronchoconstriction, cardiac failure, cold extremities, fatigue and depression, hypoglycaemia	Absorbed orally Extensive first-pass metabolism About 90% bound to plasma protein Plasma $t_{1/2} \sim 4$ h	Timolol is similar and used mainly to treat glaucoma

Table 4.6 (*Continued*)

Drug	Main action	Uses/function	Unwanted effects	Pharmacokinetic profile	Comments
Metoprolol	β_1 antagonist	Angina, hypertension, dysrhythmias	As propranolol, less risk of bronchoconstriction and sleep disturbances	Absorbed orally Mainly metabolized in liver Plasma $t_{1/2} \sim 3$ h	Atenolol is similar, with a longer half-life
Nebivolol	β_1 antagonist Enhances nitric oxide synthesis	Hypertension	Fatigue, headache	Absorbed orally $t_{1/2} \sim 10$ h	—
(c) mixed (α-/β-) receptor antagonists					
Labetalol	α/β antagonist	Hypertension in pregnancy	Postural hypotension, bronchoconstriction	Absorbed orally Conjugated in liver Plasma $t_{1/2} \sim 4$ h	See Chapter 6
Carvedilol	β/α_1 antagonist	Heart failure	As for other β blockers Initial exacerbation of heart failure Renal failure	Absorbed orally $t_{1/2} \sim 10$ h	Additional actions may contribute to clinical benefit See Chapter 6

indirect-acting stimulants of noradrenergic receptors. **Guanethidine** was the first antagonist discovered in the mid-1950s, along with **phenoxybenzamine** and **tolazoline** when alternatives to ganglion-blocking drugs were used in the treatment of hypertension. The main effect of these drugs is to inhibit the release of noradrenaline from sympathetic nerve terminals. This has little effect on the adrenal medulla, and none on nerve terminals that release transmitters other than noradrenaline. Later drugs very similar to these compounds include bretylium (**4.30**), bethanidine (**4.31**) and debrisoquin (**4.32**).

4.30   **4.31**   **4.32**

4.5.12 Adrenoceptor Agonists

Examples of adrenoceptor agonists (also known as *directly acting sympathomimetic* drugs) are given in Table 4.5, and the characteristics of individual drugs are summarized in Table 4.4. The major physiological effects mediated by different types of adrenoceptors are summarized in Table 4.1 and briefly described below.

(i) **Smooth Muscle**
- α_1-adrenoceptor receptor mediated contraction by intracellular Ca^{2+} release.
- Vascular smooth muscle possesses both α_1 and α_2 receptors.
- May not be β_1 or β_2 (see the section on the GI system).
- α_1 receptors mainly responsible for neuronally mediated vasoconstriction.
- α_2 receptors present on the muscle fiber surface and are activated by circulating catecholamines.
- Stimulation of β receptors causes relaxation of smooth muscle by increasing cAMP formation (see the CVS chapter).
- β-receptor activation enhances Ca^{2+} extrusion and intracellular Ca^{2+} binding, both effects acting to reduce intracellular Ca^{2+} concentration.
- β_2 receptor appears to be responsible for relaxation.
- The β-receptor that is responsible for this effect in gastrointestinal smooth muscle is unclear.

- In the vascular system, β_2-mediated vasodilatation is (particularly in humans) is mainly endothelium dependent and mediated by nitric oxide release (see the CVS chapter).
- β_2-mediated vasodilatation occurs in many vascular beds and is especially marked in skeletal muscle.
- Bronchial smooth muscle is relaxed by activation of β_2-adrenoceptors.
- Selective β_2 agonists are important in the treatment of asthma.
- Uterine smooth muscle responds similarly, and these drugs are also used to delay premature labor.
- Bladder detrusor muscle is relaxed by activation of β_3-adrenoceptors, and selective β_3-adrenoceptor agonists have recently been introduced to treat symptoms of overactive bladder (see the recommended reading list).
- α_1-adrenoceptors mediate a long-lasting trophic response, stimulating smooth muscle proliferation in various tissues, for example in blood vessels and in the prostate gland.
- *Benign prostatic hyperplasia* is commonly treated with α-adrenoceptor antagonists.

(ii) **Nerve Terminals**
- Presynaptic adrenoceptors are present on both cholinergic and noradrenergic nerve terminals.
- The main effect (α_2-mediated) is inhibitory.
- A weaker facilitatory action of β receptors on noradrenergic nerve terminals is also reported.

(iii) **Heart**
- Catecholamines, acting on β_1 receptors, exert a powerful stimulant effect on the heart (see the CVS chapter).
- Heart rate (*chronotropic effect*) and the force of contraction (*inotropic effect*) are increased with a marked increased cardiac output and cardiac oxygen consumption.
- The cardiac efficiency (see CVS chapter) is reduced.
- Catecholamines can also cause disturbance of the cardiac rhythm, culminating in ventricular fibrillation.
- Paradoxically, adrenaline is used therapeutically to treat ventricular fibrillation arrest as well as other forms of cardiac arrest.
- Hypertrophy of cardiac muscle occurs in response to activation of both β_1 and α_1 receptors, probably related to the hypertrophy seen in vascular and prostatic smooth muscle.
- The pathophysiology of hypertension and of cardiac failure and their associated with sympathetic overactivity is

discussed in more detail in Rang and Dale[3] and the later CVS section.

(iv) **Metabolism**

- Catecholamines are important in the conversion of energy stores (glycogen and fat) to freely available fuels (glucose and free fatty acids[27]).
- They cause an increase in the plasma concentration of the latter substance (for the detailed biochemical mechanisms see the recommended reading[27]).
- These effects vary from species to species, but in most cases the effects on carbohydrate metabolism of liver and muscle (Table 4.1) are mediated through β_1 receptors and the stimulation of lipolysis and thermogenesis is produced by β_3 receptors (see Table 4.1).
- Activation of α_2 receptors inhibits insulin secretion, an effect that further contributes to the hyperglycaemia.
- *Leptin* produced by adipose tissue (see Rang and Dale[3] for further reading) is also inhibited.
- Importantly, adrenaline-induced hyperglycaemia in humans is blocked completely by a combination of α and β antagonists but not by either on its own.[27]

(v) **Other Effects**

- Skeletal muscle is affected by adrenaline, acting on β_2 receptors, although the effect is far less dramatic than that on the heart.
- The twitch tension of Type II fast-contracting fibers (white muscle) is also increased by adrenaline, particularly if the muscle is fatigued.
- The Type 1 slow twitch (red muscle) is reduced by a still unclear action on contractile proteins.
- Adrenaline and other β_2 agonists cause a marked tremor, the shakiness that accompanies fear, excitement or the excessive use of β_2 agonists (*e.g.* **salbutamol**) in the treatment of asthma.
- Likely the result of an increase in muscle spindle discharge, coupled with an effect on the contraction kinetics of the fibers.
- Such effects combining to produce instability in the reflex control of muscle length. β-receptor antagonists are sometimes used to control pathological tremor.
- Cardiac dysrhythmias are associated with β_2 agonists and thought to be partly due to hypokalemia, caused by an increase in K^+ uptake by skeletal muscle.
- β_2 agonists also cause long-term changes in the expression of sarcoplasmic reticulum proteins that control contraction

kinetics, and thereby increase the rate and force of contraction of skeletal muscle.

- Clenbuterol, an "anabolic" drug used illicitly by athletes to improve performance (see the CVS chapter and Rang and Dale[3]), is a β_2 agonist that acts in this way.
- Catecholamines, acting on β_2 receptors inhibit histamine release by human and guinea pig lung tissue in response to anaphylactic challenge.
- Lymphocytes and other immune cells express adrenoceptors (mainly β-adrenoceptor).[28]
- Lymphocyte proliferation, lymphocyte-mediated cell killing, and production of many cytokines are inhibited by β-adrenoceptor agonists.[28]

The physiological and clinical importance of these effects has not yet been established. For a review of the effects of the sympathetic nervous system on immune function, see the recommended reading list.[28]

(vi) **Therapeutic Use**
- The main therapeutic uses of adrenoceptor agonists are summarized below and in Table 4.5.
- The most important use of β-adrenoceptor agonists for the treatment of asthma in Table 4.5 (see the recommended references for more detailed information).

4.5.13 Adrenoceptor Antagonists

The main drugs are listed in Table 4.4a–c, and further information is given in the recommended reviews list. Most are selective for α or β receptors, and many are also subtype-selective;

(i) **α-Adrenoceptor Antagonists**
The main groups of α-adrenoceptor antagonists are:
- nonselective between subtypes (*e.g.* **phenoxybenzamine (4.33), phentolamine**)
- α_1-selective (*e.g.* **prazosin (4.24), doxazosin, terazosin**)
- α_2-selective (*e.g.* **yohimbine, idazoxan (4.35)**)

4.33   **4.34**   **4.35**

The *ergot derivatives* (*e.g.* **ergotamine, dihydroergotamine**) also block α-receptors as well as having many other actions, notably on 5-HT receptors and were used in the treatment of migraine. They are described fully in (Rang and Dale[3]).

(ii) **Nonselective α-Adrenoceptor Antagonists**
- **Phenoxybenzamine:** nonspecific α receptors antagonist, active at acetylcholine, histamine and 5-HT receptors, with a prolonged action is due to covalently binding to the receptor.
- **Phentolamine** relatively more selective, but it binds reversibly and has a short duration of action.
- Both drugs cause a fall in arterial pressure (because of a block of α-receptor-mediated vasoconstriction) and postural hypotension in human studies.
- Cardiac output and heart rate are increased due to a reflex response to the fall in arterial pressure that is mediated by β receptors.
- Blocking of α_2 receptors tends to increase noradrenaline release, which has the effect of enhancing the reflex tachycardia that occurs with any blood pressure-lowering agent.
- Phenoxybenzamine is still used clinically in preparing patients with *phaeochromocytoma* for surgery.
- It produces irreversible antagonism with a resultant depression in the maximum of the agonist dose–response curve, which is desirable in a situation where surgical manipulation of the tumor may release a large bolus of pressor amines into the circulation.

Labetalol and **carvedilol** are mixed α_1- and β-receptor-blocking drugs, although clinically they act predominantly on β receptors. The combined activities of these drugs in one molecule may account for their efficacy over existing single receptor target agents. Carvedilol is used mainly to treat hypertension and heart failure, whereas labetalol is used to treat hypertension in pregnancy.

4.36   **4.37**

(iii) **Selective α_1-antagonists**
- **Prazosin** was the first selective α_1 adrenoceptor antagonist
- Drugs with longer half-lives (*e.g.* **doxazosin, terazosin, alfuzosin**), have the advantage of allowing once-daily dosing and are now preferred. All have inverse agonist effects on α_1 subtypes
- These highly selective α_1 adrenoceptor antagonists cause vasodilatation and fall in arterial pressure, but less tachycardia than occurs with nonselective α-receptor antagonists, as they do not increase noradrenaline release from sympathetic nerve terminals
- The α_1-receptor antagonists cause relaxation of the smooth muscle of the bladder neck and prostate capsule, and inhibit hypertrophy of these tissues, and are therefore useful in treating urinary retention associated with *benign prostatic hypertrophy*
- **Tamsulosin**, a α_{1A}-receptor selective antagonist that shows some selectivity for the bladder, although an inverse agonist at α_{1B} and α_{1D}. It causes less hypotension than drugs such as prazosin, which act on α_{1B} receptors to control vascular tone
- Consensus suggests that α_{1A} receptors may play a part in the pathological hypertrophy not only of prostatic and vascular smooth muscle, but also in the cardiac hypertrophy that occurs in hypertension and heart failure

(iv) **Selective α_2-Adrenoceptor Antagonists**
- **Yohimbine** is a naturally occurring indole alkaloid, which has a rich pharmacology
- Various other synthetic analogs include **idazoxan, tamsulosin and SNAP 5089**

4.38   **4.39**

(v) **Therapeutic Uses and Unwanted Effects of α-Adrenoceptor Antagonists**
- The main uses of α-adrenoceptor antagonists are related to their cardiovascular actions, and are summarized below.

- All α-adrenoceptor antagonists have only limited therapeutic applications.
- In hypertension, the nonselective agents have a tendency to produce tachycardia and cardiac dysrhythmias, and gastrointestinal symptoms.
- The selective α₁-receptor antagonists (especially the longer-acting compounds **doxazosin** and **terazosin**), do not affect cardiac function appreciably, and postural hypotension is less troublesome than with prazosin or nonselective α-receptor antagonists.
- Their main use is in treating severe hypertension, but are not used as first-line agents and are used along with other antihypertensives (see Chapter 5).
- Unlike other agents, they cause a modest decrease in low-density lipoprotein, and an increase in high-density lipoprotein cholesterol (see the reviews in the recommended reading), although the beneficial effects is uncertain.
- They are also used to control urinary retention in patients with benign prostatic hypertrophy.
- Overall, the therapeutic uses of α₁-receptor antagonists have been disappointing.
- Apart from *phaeochromocytoma* – a catecholamine-secreting tumor of chromaffin tissue, which causes severe and initially episodic hypertension.
- A combination of α- and β-receptor antagonists is the most effective way of controlling the blood pressure in this condition due to an excessive release of CAs in this condition.
- A combination of phenoxybenzamine and atenolol is an effective therapy for this purpose.

4.5.14 β-Adrenoceptor Receptors

β-adrenoceptor agonists and antagonists were developed through modifications of the noradrenaline molecule to allow selective interaction with the β_1 and β_2 receptors (*e.g.*, isoproterenol), rather than the combined α- and β-effects of adrenaline. Further modification allows selectivity for β_2 receptors on bronchial smooth muscle to achieve bronchodilation without the tachycardia associated with activation of β_1 receptors on cardiac muscle.[3,7] Agents available by inhalation that are selective for the β_2-adrenergic receptor (*e.g.*, albuterol, levalbuterol, salmeterol, formoterol) are preferred for asthma therapy compared with non-β_2 selective agents (*e.g.*, metaproterenol), because they provide equivalent bronchodilation with less cardiac stimulation. Isoproterenol, a nonselective β-agonists, is no longer used to treat asthma exacerbations, due to the better safety and duration of action profiles of other available medications.[3,7]

The $β_2$-receptor is a G protein-coupled transmembrane receptor that activates the enzyme adenylyl cyclase. Activation of adenylyl cyclase produces cyclic adenosine monophosphate (cAMP). The exact mechanism by which cAMP causes smooth muscle relaxation is not fully understood, but likely involves activation of protein kinase A and changes in intracellular calcium concentrations.[30,31] Activation of the $β_2$-receptor also affects potassium channels through a separate mechanism. The function of the $β_2$-adrenergic receptor and the role of polymorphisms of the receptor in individual responses to $β_2$-agonists are discussed in the CVS section.

Variations in the molecular structure of beta agonists affect the onset and duration of bronchodilation. As an example, prolongation of the bronchodilator effect (relative to isoproterenol) is achieved by modifications that reduce susceptibility to degradation by COMT and MAO. In addition, the long, lipophilic side chains of formoterol and salmeterol attach to the plasma membrane and increase the duration of binding of the drugs to the adrenergic receptor. The lipophilic side chain of salmeterol leads to incorporation of the drug into the cell membrane and activation of the $β_2$-adrenergic receptor through an alternate binding site, rather than the usual site in the aqueous surface of the cell membrane.[3,7] It is thought that accessing the alternate binding site deeper in the cell membrane slows the onset of action of salmeterol. In contrast, formoterol has a different lipophilic side chain and its onset of action is comparable to that of albuterol (also known as salbutamol).[3]

4.5.15 β-Adrenoceptor Antagonists

Sir James Black and colleagues at ICI Pharmaceuticals PLC, 10 years after Ahlquist[32] had postulated the existence of β-adrenoceptors, discovered the first of a series of β-adrenoceptor antagonists in 1960. The first compound synthesized was pronethalol, by replacing the bulky chlorine atoms of **dichloroisoprenaline (40)** with a second benzene ring to form pronethalol. However, this drug had fairly low potency, was short acting and was a partial agonist and was quickly superseded by **propranolol (41)**.[33] with greater potency and blocked both $β_1$ and $β_2$ receptors equally. Following this work, the potential clinical advantages of drugs with some partial agonist (intrinsic sympathomimetic) activity, and/or with selectivity for $β_1$ receptors, led

to the development of a number of other drugs *e.g.* practolol, oxprenolol and alprenolol for their antianginal, antiarrhthmic and antihypertensive properties first observed with pronethalol.[33] With an
improved safety profile, **atenolol (42)** a β_1-cardioselective analog of
practolol with no partial agonist activity and low lipophilicity later
became the drug of choice for these indications. Several alternatives
to atenolol were subsequently developed, some with dual mechanism
of action, **carvedilol** (a nonselective β-adrenoceptor antagonist with
additional α_1-blocking activity) and **nebivolol (43)** (a β_1-selective antagonist that also causes vasodilatation by inducing endothelial nitric
oxide production).[3] The therapeutic advantages of these drugs over
conventional β-adrenoceptor antagonists were in treating heart failure (see also Chapter 6).[34] The most important compounds in this
category are set out in Table 4.6.

4.40   **4.41**   **4.42**   **4.43**

The pharmacological effects of β-receptor antagonists are highlighted in Table 4.6. The most important effects are on the cardiovascular system and on bronchial smooth muscle but these effects
depend on the degree of sympathetic tone (see the recommended
reading list for reviews).

Only modest changes in heart rate, cardiac output or arterial
pressure are observed with propranolol at rest but it markedly reduces
the effect of exercise or excitement on these parameters in normal
subjects. β-receptor antagonists with partial agonist activity, such as
oxprenolol, increase the heart rate at rest but reduce it during exercise
(Rang and Dale[3]). An expected finding in patients with hypertension
on β-blockers (although not normotensive subjects) is a gradual fall in
arterial pressure that takes several days to develop fully. The mechanism is complex and involves the following: (i) reduction in cardiac
output, (ii) reduction of renin release from the juxtaglomerular cells
of the kidney, (iii) a central action, reducing sympathetic activity.

Drugs with additional vasodilator properties (*e.g.* carvedilol and
nebivolol), are more effective antihypertensives. Blockade of the

facilitatory effect of presynaptic β-receptors on noradrenaline release may also contribute to the antihypertensive effect. The antihypertensive effect of β-receptor antagonists is clinically very useful, as reflex vasoconstriction is preserved, postural and exercise-induced hypotension is less troublesome than with many other antihypertensive drugs (Rang and Dale[3]).

Other known therapeutic effects of β-receptor antagonists include an antidysrhythmic effect on the heart and cardiac failure (see the CVS section). However, it is well documented that in asthmatic subjects, nonselective β-receptor antagonists (such as propranolol) can cause severe bronchoconstriction, which does not respond to the usual doses of drugs such as salbutamol or adrenaline. The β_1-selective antagonists are the drugs of choice in these patients, although selectivity is still a real concern at higher doses. Similarly, the involvement of β-receptors in the hyperglycaemic actions of adrenaline is a concern in diabetic patients; the use of β-receptor antagonists increases the likelihood of exercise-induced hypoglycemia, because the normal adrenaline-induced release of glucose from the liver is reduced.[3]

4.5.16 Therapeutic Uses and Unwanted Adverse Effects

The main uses of β-receptor antagonists are connected with their effects on the cardiovascular system, and are discussed more fully discussed in Chapter 6: The Heart and the Cardiovascular System (CVS). They are briefly summarized below and further reading can be found in the reference list at the end of the chapter.

(i) **Therapeutic Uses of β_1-Adrenergic Agonists**
- Mainly used to treat acute but potentially reversible heart failure (*e.g.* dobutamine in cardiac surgery and septic shock) See the cardiovascular section for further reading.
- Act to increase heart rate, contractility and automaticity.
- Reduce cardiac efficiency (due to reduce oxygen consumption).

(ii) **Unwanted Side Effects of β_1-Receptor Agonists**
- β_1-agonists: tachycardia, arrthymias.

(iii) **Therapeutic Uses of β_2-Adrenergic Antagonists**
- cardiovascular;
- angina pectoris;
- myocardial infarction, and following infarction;
- prevention of recurrent dysrhythmias (especially if triggered by sympathetic activation);

- heart failure (in well-compensated patients);
- hypertension (no longer first choice);
- Other uses:
  - glaucoma (*e.g.* **timolol** eye drops);
  - thyrotoxicosis as adjunct to definitive treatment (*e.g.* preoperatively);
  - anxiety to control somatic symptoms (*e.g.* palpitations, tremor);
  - migraine prophylaxis;
  - benign essential tremor (a familial disorder).

It is clear that β-adrenoceptor-blocking drugs have a complex pharmacology and physiology depending on their characteristics as partial agonists (see Table 4.6), selective and nonselective β-adrenoceptor-blocking activity and other pharmacological properties.[29,31] Examples of which include; **alprenolol, oxprenolol** that under resting conditions increase heart rate, while at the same time opposing the tachycardia produced by sympathetic stimulation (see also Chapter 6). This has been interpreted as a partial agonist effect, although there is evidence that mechanisms other than β-receptor activation may contribute to the tachycardia. Furthermore, the translational pharmacology and receptor specificity found for some compounds in laboratory animals is seldom found in humans. Moreover, in normal hearts cardiac stimulation is mediated through β_1 receptors, but in heart failure β_2 receptors contribute significantly (Rang and Dale[3]).

Furthermore, a complex signal transduction is also evident as β-adrenoceptor agonists and partial agonists may act not only through cAMP formation, but also through other signal transduction pathways (*e.g.* the mitogen-activated protein [MAP] kinase pathway), and that the relative contribution of these signals differs for different drugs.[30,31] These complex pathways show different levels of constitutive activation, which is reduced by ligands that function as inverse agonists. β-Adrenoceptor antagonists in clinical use differ in respect of these properties, and drugs classified as partial agonists may actually activate one pathway while blocking another (see Rang and Dale[3]). Finally, genetic variants of both β_1 and β_2 receptors occur in humans, and the influence the effects of agonists and antagonists is still unclear (see the recommended reviews and Chapter 6 for more detailed information).[29]

References

1. D. W. Robertson, 2004, *Primer on the Autonomic Nervous System*, Academic Press, New York.

2. G. Burnstock, Autonomic Transmission: 60 years since Henry Dale, *Ann.Rev.-Pharamacol.*, 2009, **49**, 1–30.

3. *Rang & Dale's Pharmacology*, ed. H. P. Rang, J. M. Ritter, R. J. Flower and G. Henderson, Elsevier Churchill Livingstone, 8th edn, 2016.

4. R. K. Goyal and I. Hirano, The enteric nervous system, *N. Engl. J. Med.*, 1996, **334**, 1106–1115.

5. J. R. Cooper, F. E. Bloom and R. H. Roth, *The Biochemical Basis of Neuropharmacology*, Oxford University Press, New York, 2002, 8th edn.

6. W. Jänig and E. M. Lachlan, Characteristics of function-specific pathways in the sympathetic nervous system, *Trends Neurosci.*, 1992, **15**, 475–481.

7. L. L. Brunton, J. S. Lazo and K. L. Parker, *Goodman and Gilman's, The Pharmacological Basis of Therapeutics*, McGraw Hill, New York, 11th edn, 2005.

8. S. P. H. Alexander, H. E. Benson, E. Faccenda, *et al.*, Concise Guide to Pharmacology [Acetylcholine receptors (muscarinic), p. 1474. Nicotinic acetylcholine receptors p. 1597], *Br. J. Pharmacol.*, 2013, **170**, 1449–1896.

9. S. Neil, *et al.*, Introduction IUPHAR/BPS Guide to PHARMACOLOGY, 2014.

10. D. B. Bylund, *et al.*, International Union of Pharmacology nomenclature of adrenoceptors, *Pharmacol. Rev.*, 1994, **46**, 121–136.

11. J. P. Changeux and S. Edelstein, *Nicotinic Acetylcholine Receptors*, Odile Jacob, New York, 2005.

12. A. Karlin, *Nat. Rev. Neurosci.*, 2002, **3**, 102–114.

13. W. D. M. Paton and E. J. Zamis, Clinical potentialities of certain bisquaternary salts causing neuromuscular and ganglionic block, *Nature.*, 1948, **162**, 810.

14. B. G. Katzung, *Basic and Clinical Pharmacology: Introduction to Autonomic Pharmacology*, The McGraw Hill Companies, 8th edn, 2001, pp. 75–91.

15. B. D. Burns and W. D. M. Paton, Depolarization of the motor end-plate by decamethonium and acetylcholine, *J. Physiol.*, 1951, **115**, 41–73.

16. K. Leach *et al.*, Structure-function studies of muscarinic acetylcholine receptors, *Handb. Exp. Pharmacol.*, 2012, 29–48.

17. B. G. Katzung, *Basic and Clinical Pharmacology: Introduction to Autonomic Pharmacology*, The McGraw Hill Companies, 8th edn, 2001, pp. 75–91.

18. D. Blumenthal, L. Brunton, L. S. Goodman, K. Parker, A. Gilman, J. S. Lazo and I. Buxton, 5: Autonomic Pharmacology: Cholinergic Drugs, *Goodman & Gilman's The Pharmacological Basis of Therapeutics*, McGraw-Hill, New York, 1996, p. 1634.

19. C. Giuliano *et al.*, Increases in cholinergic neurotransmission measured by using choline-sensitive microelectrodes: enhanced detection by hydrolysis of acetylcholine on recording sites, *Neurochem. Int.*, 2008, **52**(7), 1343–1350.

20. G. Shaltiel *et al.*, Hippocampal microRNA-132 mediates stress-inducible cognitive deficits through its acetylcholinesterase target, *Brain Struct. Funct.*, 2013, **218**(1), 59–72.

21. L. M. Eubanks *et al.*, A Molecular Link Between the Active Component of Marijuana and Alzheimer's Disease Pathology, *Mol. Pharm.*, 2006, **3**(6), 773–777.

22. G. Eisenhofer, I. J. Kopin and D. S. Goldstein, Catecholamine metabolism: a contemporary view with implications for physiology and medicine, *Pharmacol. Rev.*, 2004, **56**, 331–349.

23. R. R. Gainetdinov and M. G. Caron, Monoamine transporters: from genes to behaviour, *Annu. Rev. Pharmacol. Toxicol.*, 2003, **43**, 261–284.

24. B. K. Kobilka, Structural insights into adrenergic receptor function and pharmacology, *Trends Pharmacol. Sci.*, 2011, **32**, 213–218.

25. J. C. Shih, Cloning, after cloning, knock-out mice, and physiological functions of MAO A and B, *Neurotoxicology*, 2004, **25**(1–2), 21–30.

26. M. Philipp and L. Hein, Adrenergic receptor knockout mice: distinct functions of 9 receptor subtypes, *Pharmacol. Ther.*, 2004, **101**, 65–74.

27. K. Nonogaki, New insights into sympathetic regulation of glucose and fat metabolism, *Diabetologia*, 2000, **43**, 533–549.
28. I. J. Elenkov, R. L. Wilder, G. P. Chrousos, *et al.*, The sympathetic nerve – an integrative interface between two supersystems: the brain and the immune system, *Pharmacol. Rev.*, 2000, **52**, 595–638.
29. O. E. Brodde, β1 - and β2 -Adrenoceptor polymorphisms and cardiovascular diseases, *Fund. Clin. Pharmacol.*, 2008, **22**, 107–125.
30. J. G. Baker, *et al.*, Agonist and inverse agonist actions of β-blockers at the human β2 -adrenoceptor provide evidence for agonist-directed signaling, *Mol. Pharmacol.*, 2003, **64**, 1357–1369.
31. J. G. Baker, *et al.*, Evolution of β-blockers: from anti-anginal drugs to ligand-directed signaling, *Trends Pharmacol. Sci.*, 2011, **32**, 227–234.
32. R. P. Ahlquist, A study of the adrenotropic receptors, *Am. J. Physiol.*, 1948, **153**(3), 586–600.
33. J. W. Black, Comparison of some properities of pronethalol and propranolol, *Br. J. Pharmacol.*, 1997, (Suppl 1), 285–299.
34. M. R. Bristow, Treatment of chronic heart failure with beta-adrenergic receptor antagonists: a convergence of receptor pharmacology and clinical cardiology, *Circ. Res.*, 2011, **109**, 1176–1194.

5 Actions of Drugs on The Brain and CNS Disorders

Thomas P. Blackburn

Translational Pharmacology BioVentures LLC, PO Box 3126, 56, 14th St., Hoboken, New Jersey, NJ 07030, USA
Email: tblackburn@tpbioventures.com

5.1 Function of Various Brain Regions

It is beyond the scope of this chapter to discuss the function of all the various regions of the human brain. However, the structure and function are shown in Table 5.1 and the structure and components of the limbic system is shown in Figure 5.1. It is the limbic and extra-pyramidal system that is the focus of this chapter and its role in the control of mood, emotion, memory, and motivational activities associated with normal and aberrant pathophysiology and the mechanism of action of CNS agents.[1]

5.2 The Development of Psychotropic Drugs (1950–2010)

Figure 5.2 shows an outline of the discovery and development of psychotropic drugs over the last several decades. Their therapeutic

Pharmacology for Chemists: Drug Discovery in Context
Edited by Raymond Hill, Terry Kenakin and Tom Blackburn
© The Royal Society of Chemistry 2018
Published by the Royal Society of Chemistry, www.rsc.org

Table 5.1 The structure and functions of various brain regions.

Brain structure	Main function
Forebrain: Telencephalon	
Cerebellum	Information processing
Basal ganglia	Movement, motor control
Amygdala[a]	Reward and emotions
Hippocampus[a]	Memory
Olfactory bulb[a]	Sense of smell
Forebrain: Diencephalon	
Thalamus[a]	Integrates sensory information
Hypothalamus[a]/Pituitary	Regulates body temperature, feeding, reproduction and circadian rhythms
Midbrain	
Optic lobes	Processes visual, auditory and touch sensory information
Tegmentum	Reflex responses to sensory information
Hindbrain	
Medulla oblongata	Generates rhythmic breathing
	Regulates heart rate and blood pressure
Pons	Regulates several autonomic functions, *e.g.* biting, chewing and swallowing.
Cerebellum	Maintains body posture and coordinates movement
	Integrates information from proprioceptors

[a]Limbic system brain structures.

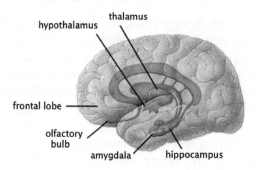

Figure 5.1 A sagittal section of the human brain showing components of the limbic system and surrounding structures.

use in psychiatric and neurological disorders will be discussed along with the diagnosis, classification, and pathogenesis of the various mental and neurological disorders, along with more recent advances in drug discovery technology and drug development.

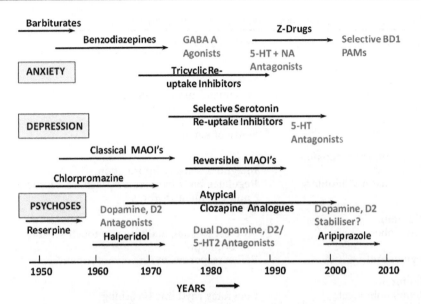

Figure 5.2 A summary of classes of psychotropic medicines and their principal modes of action.
Reprinted from Neurobiology of Disease, Volume 61, D. J. Nutt and J. Attridge, CNS drug development in Europe – Past progress and future challenges, pp. 15, Copyright 2014, with permission from Elsevier.

5.3 CNS Disorders: Anxiety, Depression, Dipolar Disorders and Schizophrenia

Anxiety and depressive disorders involve all major bodily functions, mood, and thoughts, affecting the ways in which an individual eats and sleeps, feels about themselves, and thinks. Without treatment, the symptoms of mood disorders can last for weeks, months or years. Fortunately, however, appropriate treatment can help most individuals suffering from anxiety or depression, with an increasing number of treatment options now available for individuals, *e.g.*, those with major depression disorder (MDD), accompanied by a growing body of evidence-based medicine describing the effectiveness, efficacy, and safety that provides clinicians with multiple options to determine the most appropriate treatment antidepressant treatment for each patient based increasingly on personalized medicine studies and cognitive behavioral therapy.

5.3.1 Anxiety Disorders

Anxiety is a fear/stress reaction and is a normal part of life. You feel anxious when faced with challenges at work, before taking an exam, or making an important life decision. Pathological anxiety is fear that is sufficiently severe as to be physically and mentally disabling. Such a reaction may be a response to a threatening situation (*e.g.* public speaking) or to a nonthreatening event (*e.g.* social event). Episodes of paroxysmal severe anxiety associated with severe autonomic symptoms (*e.g.* chest pain, dyspnoea and palpitations) are termed panic attacks and often accompany a generalized anxiety disorder. Other anxiety disorders that are recognized clinically include: social anxiety, phobias, post-traumatic stress disorder and obsessive-compulsive disorder. The major types of anxiety disorders and their signs and symptoms are briefly listed below; more information can be found in the recommended reading list.

5.3.2 Generalized Anxiety Disorder

People with generalized anxiety disorder display excessive anxiety or worry for months and face several anxiety-related symptoms.

Generalized anxiety disorder symptoms include:

- restlessness or feeling wound-up or on edge;
- being easily fatigued;
- difficulty concentrating or having their minds go blank;
- irritability;
- muscle tension;
- difficulty controlling the worry;
- sleep problems (difficulty falling or staying asleep or restless, unsatisfying sleep).

5.3.3 Panic Disorder

People with panic disorder have recurrent unexpected panic attacks, which are sudden periods of intense fear that may include palpitations, pounding heart, or accelerated heart rate; sweating; trembling or shaking; sensations of shortness of breath, smothering, or choking; and feeling of impending doom.

Panic disorder symptoms include:

- sudden and repeated attacks of intense fear;
- feelings of being out of control during a panic attack;
- intense worries about when the next attack will happen;
- fear or avoidance of places where panic attacks have occurred in the past.

5.3.4 Social Anxiety Disorder

People with social anxiety disorder (sometimes called "social phobia") have a marked fear of social or performance situations in which
they expect to feel embarrassed, judged, rejected, or fearful of offending others.

Social anxiety disorder symptoms include:

- Feeling highly anxious about being with other people and having
  a hard time talking to them.
- Feeling very self-conscious in front of other people and worried
  about feeling humiliated, embarrassed, or rejected, or fearful of
  offending others.
- Being very afraid that other people will judge them.
- Worrying for days or weeks before an event where other people
  will be present.
- Staying away from places where there are other people present.
- Having a hard time making friends and keeping friends.
- Blushing, sweating, or trembling around other people.
- Feeling nauseous or sick to your stomach when other people are
  around.

Evaluation for an anxiety disorder often begins with a visit to a
primary care provider. Some physical health conditions, such as an
overactive thyroid or low blood sugar, as well as taking certain
medications, can imitate or worsen an anxiety disorder. A thorough
mental health evaluation is also helpful, because anxiety disorders
often coexist with other related conditions, such as depression or
obsessive-compulsive disorder.

5.3.5 Bipolar Affective Disorder

Bipolar affective disorder (BPAD or manic-depressive illness) is a
common, recurrent, and severe psychiatric disorder characterized by
episodes of mania, depression, or mixed states (simultaneously occurring manic a depressive symptoms). BPAD is frequently unrecognized
and goes untreated for many years without clinical vigilance. Newer
screening tools assist physicians in making the diagnosis and several
drugs are now available to treat the acute mood episodes of BPAD and
to prevent further episodes with maintenance treatment. Two consensus reviews on depression and one on BPAD have been published by
the American College Neuropsychopharmacology (ACNP),[2] the other by

the British Association of Psychopharmacology (BAP) depression.[3] Consensus reviews on BPAD have also been published by BAP[4] and the American Psychiatric Association (APA).[5] Together, these encompass the major findings and lessons learned from antidepressant and BPAD research over the last two decades, providing treatment guidelines for these conditions (see below). They also bring some consensus disparities that exist in the clinical diagnosis of depression and mania.

5.4 Types of Depression

Depressive disorders exhibit different phenotypes with variations in the number of symptoms, their severity, and persistence according to DSM-IV[6] and ICD-10[7] classifications:

Major depression disorder: This disorder manifests as a combination of symptoms (see below).

Dysthymia: This is a less severe form of depression involving long-term, chronic symptoms that do not disable but keep individuals from functioning well or from feeling good. Many people with dysthymia experience major depressive episodes at some time in their lives.

Bipolar affective disorder (BPAD), also called manic-depressive illness, less prevalent other forms of depressive disorder. BPAD is characterized by cycling mood changes: severe highs (mania, BPAD I) and low (depression, BPAD II).

(For comprehensive review on CNS disorders, diagnosis and treatment see the review by Blackburn and Wasley,[8] (2007)).

5.4.1 Disease Prevalence

In 1987, the selective serotonin (5-HT) reuptake inhibitors (SSRis; *e.g.*, fluoxetine[1]) were introduced into clinical practice. Their improved safety and side-effect profile resulted in a move away from older antidepressant agents, the antidepressants (TCAs) and monoamine oxidase inhibitors (MAOIs). Drugs for depression are now prescribed for 1 in every 3 patients in clinical practice. Unipolar major depression is ranked fourth as a disease burden measured in disability adjusted for life years. Despite available antidepressant medications, unipolar major depression is ranked second behind ischemic disease as a potential disease burden by 2020.[7] The risk of unipolar major depression, especially for developed countries, is 1 in 10. There is considerable evidence-based medicine that depression is associated with increased cardiovascular and infectious diseases as well as immunological and endocrine changes. The WHO has predicted that depression will become the leading cause of human disability by 2020.[1]

5.4.2 Unmet Medical Need

Depression is a global phenomenon, poorly understood and treated.[9] While existing antidepressants are far from perfect, several key clinical improvements are urgently needed to improve patient compliance and safety: (i) improvements in efficacy; (ii) speed of onset; (iii) safety/tolerability of NCEs; and (iv) a reduction in remission rates and relapse/recurrence. A National Institutes of Mental Health (NIMH) National Comorbidity Survey of more than 9000 US adult using DSM-IV-TR Text Revision 2001 criteria found that 6% of those studied had a debilitating mental illness for which treatment was difficult to obtain, with only one-third of those in care receiving minimally adequate treatment, appropriate drugs, or a few hours of therapy over a period of several months.[8] In general, sadly, little has changed over the past two decades. Depressed individuals incur almost twice the medical cost burden as nondepressed patients, the main part (80%) being for medical care rather psychiatric or psychological services, with the bulk of antidepressant prescriptions (>80% worldwide) being written by primary-care physicians and prescribed along the same lines as the unnecessary misuse of antibiotics.

The ideal effective treatment for moderate-to-severe depression includes a combination of somatic therapies (pharmacotherapy or electroconvulsive therapy (ECT)). ECT has been rejuvenated for the treatment of the most severe melancholic depressions, particularly in the elderly (who are more prone to adverse effects of drugs) and in patients who do not respond to antidepressants.

5.4.3 Current Treatments

The fact that SSRIs and other second-generation antidepressants are superior to the older antidepressants, such as TCAs, raised physicians' clinical expectations of the newer agents. Similar expectations exist for the treatment for BPAD, where the newer atypical antipsychotic agents ($5\text{-HT}_{2A}/\text{DA}_2$ antagonists), anticonvulsants and gabapentin (Figure 5.2) in establishing themselves as first-line therapy over the gold standard, lithium.[7]

5.4.4 Classification and Subclassification of Major Depression Disorder and Bipolar (Affective) Disorder

(i) Major Depressive Disorder: The symptom criteria for major depression in DSM-IV-TR and ICD- guidelines are very similar (see the recommended reading list[6,7] for more details), although the coding systems are somewhat different. One difference is that ICD-10 has a separate, optional subdiagnosis for depression with and without somatic symptoms. The latter is not present in the DSM-IV system. Both sets of guidelines have depressive disorder subdiagnoses for the following:

(ii) Bipolar Disorder: This describes a spectrum of disorders in which episodes of depression occur, interspersed with periods of normal mood. It is also known as bipolar depression or manic depression, characterized by cycles of mania and depression, which cause a person with bipolar disorder to experience swings. Biomarkers are currently being investigated in psychiatry and neurology through a wide variety of procedures and the scientific community and industry are slowly integrating the knowledge gained in the use of biomarkers into databases for use.

5.4.5 Biomarkers in Depression and Bipolar Disorder

There are several advantages, disadvantages, and limitations of selected biotechnologies for assessing the access of a drug to the brain.[8] Brain imaging technology is a rapidly improving methodology for studying the interaction of a new compound entity (NCE) with its target, making this a preferred technique. However, there are a limited number of targets for which validated positron emission tomography (PET) or single photon emission computed tomography (SPECT) ligands are available.[67] Furthermore, recent PET studies with a neurokinin-1 (NK1) receptor antagonist showed that the NCE had greater than 90% occupancy of central NK1 receptors and was active in an early Phase II efficacy study of depression but no subsequent larger Phase II studies, making this a striking example of a novel target with an excellent PET ligand that fails to be supported by clinical data and thereby calls into question the entire clinical hypothesis and the NCE (see Chapter 10). According to Bieck and Potter, a single approach may not provide the answer to addressing the question of brain penetration and drug efficacy.[10] Instead, a multimodal approach to biomarkers in CNS disorders may well be the answer, using a combination of imaging technology (where PET ligands are feasible) and CSF studies.[8,10,67]

5.4.6 Pathogenesis of Depression

The etiology of depression and BPAD is unknown. Depression is polygenic in nature with both genetic and epigenetic components, making the use of genetically engineered rodents as models for drug discovery precarious.[11,12] Moreover, emerging understanding of the biochemical mechanisms is compromised by the fact that most of the drugs used to treat depression and bipolar disorders (*e.g.*, lithium and antidepressants in general) have complex and ill-defined pleiotypic actions.

5.4.7 The Monoamine Theory of Depression

The monoamine theory of depression and drugs acting on monoamine neurotransmission has dominated treatment of depression for over 30 years. Indeed, the monoamine reuptake inhibitors and the MAOIs (see Figure 5.3) were shown to have antidepressant activity by chance, and the discoveries of their modes of action were instrumental in developing the monoamine theory. In the days when the monoamine theory of depression was evolving, the focus was more on norepinephrine (NE) than 5-HT (5-HT) or dopamine (DA). The theory developed from observations that reserpine depletes monoamines and caused depression, whereas the MAOIs and monoamine reuptake inhibitors enhanced monoamine function and thereby relieved depression. More recent studies indicated that it may be the inhibition of glial sites by normetanephrine (NMN) or other inhibitors of Uptake 2 that results in an enhanced accumulation of NMN in the synapse. The hypothesis proposed by Schildkraut and Mooney suggests that drugs or other agents that increase levels of NMN or otherwise inhibit the extraneuronal monoamine transporter, Uptake 2, in the brain will have the same clinical effects as NE reuptake inhibitor antidepressant drugs.[12] This hypothesis, as well as others below, continue to influence pharmaceutical research for the "Holy Grail," that is, a fast-acting antidepressant. The neurogenesis hypothesis of antidepressant efficacy was proposed in the 1990s, although some would consider this hypothesis questionable today.[13]

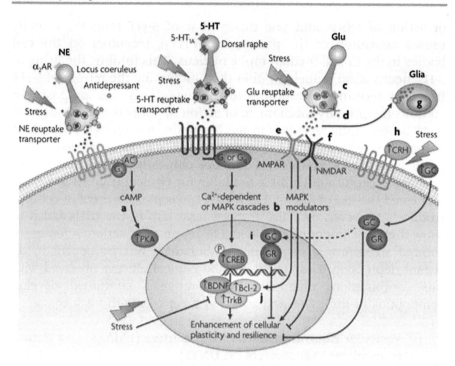

Figure 5.3 Cellular targets for the development of novel agents for the treatment of mood disorders. This figure shows the multiple targets by which transcription, neuroplasticity and cellular resilience can be increased in mood disorders. (a) Phosphodiesterase inhibitors increase levels of pCREB; (b) MAP kinase modulators increase the expression of the major neurotrophic protein Bcl-2; (c) mGluR II/III agonists modulate release of excessive levels of glutamate; (d) drugs such as lamotrigine and riluzole act on Na^+ channels to attenuate glutamate release; (e) AMPA potentiators upregulate the expression of BDNF; (f) NMDA antagonists like ketamine and memantine enhance plasticity and cell survival; (g) novel drugs to enhance glial release of trophic factors and clear excessive glutamate may have utility for the treatment of depressive disorders; (h) CRF antagonists and (i) glucocorticoid antagonists attenuate the deleterious effects of hypercortisolemia, and CRF antagonists may exert other beneficial effects in the treatment of depression *via* non-HPA mechanisms; (j) agents that upregulate Bcl-2 (*e.g.*, pramipexole, shown to be effective in dipolar depression). These distinct pathways have convergent effects on the cellular processes such as bioenergetics (energy metabolism), neuroplasticity, neurogenesis, cellular resilience, and survival.
Reprinted with permission from Macmillan Publishers Ltd: S. J. Mathew *et al.*, *Neuropsychopharmacology*, 2008, **33**, 2080–2092, copyright 2008.

To discover an antidepressant that has an effect within days rather than weeks, has challenged medicinal chemists and researchers for decades to understand the reasons for the delay in onset antidepressant action.[8,13] The focus has been mainly on the mechanism of action of SSRIs and one theory that of 5-HT reuptake initially causes activation of the presynaptic 5-HT$_{1A}$ receptors on the cell bodies in the dorsal median raphe nucleus. This inhibits the firing of 5-HT neurons, so reducing rather than increasing the release of 5-HT from the terminals. According to the hypothesis first proposed in 2001, as the primary mechanism of action of SSRIs is an activation of 5-HT postsynaptic receptors in the forebrain. A proposed action that is not achieved until the raphe 5-HT$_{1A}$ becomes desensitized.[14] The problem with putting this into practice clinically is that there are no selective receptor antagonists available for clinical use. Artigas *et al.* pioneered the use of pindolol as a 5-HT$_{1a}$ receptor for proof-of-concept studies.[15] However, evidence from at least half of the trials failed to show that the compound provided a faster onset of action. A few cases showed an improved response rate, but there was no benefit in resistant depression. Despite the limited clinical efficacy observed with this combination, pharmaceutical companies continued develop compounds in the 1990s and 2000s, at great cost with no success.

(i) **Vesicular Neurotransmitter Transporters (VMATs) and Neurotransmitter Transporters (SCDNTs)**

Neurotransmitters are synthesized in neurones and concentrated in vesicles for their subsequent Ca^{2+}-dependent release into the synaptic cleft, where they are inactivated by either enzymatic degradation or active transport neuronal and or glial cells by neurotransmitter transporters.[16] Molecular cloning has allowed the pharmacologic and structural characterization of a large family of related genes encoding Na^+/Cl^--dependent neurotransmitter transporters (SCDNTs). The identification of a superfamily of monoamine transporters (dopamine transporter (DAT), norepinephrine transporter (NET), and serotonin transporter (SERT)) has been the focus of recent research on the association of monoamine transporters with psychiatric illnesses, as discussed below.

(ii) **Norepinephrine Transporter Polymorphisms**

At least 13 polymorphisms of NET have been identified, the functional significance of which is unknown.[17] Alterations in the concentration of NE in the CNS have been hypothesized to cause, or contribute to, the development of psychiatric illnesses such as major depression and BPAD. Many studies have reported altered levels of NE and its metabolites NM and dihydroxyphenylglycol (DHPG) in the CSF, plasma, and urine of depressed patients as compared with normal controls. These variances could reflect different underlying phenotypes of depressive disorders with varying effects on NA activity. The melancholic subtype of depression (with positive vegetative features, agitation, and increased hypothalamic pituitary–adrenal (HPA) axis activity) is most often associated with increased NE. Alternatively, so-called atypical depression is associated with decreased NE and HPA axis and hypoactivation. In one study, urinary NE and its metabolite were found to be significantly higher in unipolar and bipolar depressed patients than in healthy volunteers, suggesting the unmedicated unipolar and bipolar depressed patients have a hyperresponsive noradrenergic system. Increased NE activity has also been observed to be a contributor to the borderline personality disorder traits of impulsive aggression and affective instability, high levels of risk taking, irritability, and verbal aggression. Furthermore, abnormal regulation or expression the human NET has been reported in major depression. In the postmortem human brain, [³H]-nisoxetine binding to NE was highest in dorsal *raphe nuclei* and *locus coeruleus.* Low levels of NET in the *locus coeruleus* in major depression may reflect a compensatory downregulation of this transporter protein in response to insufficient availability of its substrate (NE) at the synapse. These studies suggest that abnormalities that can cause impaired noradrenergic transmission could contribute to the pathophysiology of certain psychiatric disorders. However, results from other studies suggest that the investigated polymorphisms are not the main susceptibility factors in the etiology of major depression.[18] This was also the case for the NET DNA sequence variants identified in patients suffering from schizophrenia or BPAD. Subsequently control studies did not reveal any significant association between the variances and those diseases.[18]

(iii) **Dopamine Transporter Polymorphisms**
DAT terminates dopaminergic neurotransmission by reuptake of dopamine (DA) in presynaptic neurons and plays key role in DA recycling. DAT can also provide reverse transport of DA under certain circumstances. Psychostimulants such as cocaine and amphetamines and drugs used for attention deficit hyperactivity disorder (ADHD) such as methylphenidate exert their actions *via* DAT.

Altered DAT function or density has been implicated in various types of psychopathology, including depression,

BPAD, suicide, anxiety, aggression, and schizophrenia. Altered transporter properties associated with some of the coding variants of DAT suggest that individuals with these DAT variants could display an altered DA system.[16,19] Multiple human dopamine transporter (hDAT, *SLC6A3*) coding variants have been described, though to date they have been incompletely characterized. The antidepressant, bupropion (dose-dependently) increases vesicular DA uptake; an effect also associated with VMAT-2 protein redistribution. Another purported antidepressant that weakly block DATs is radafaxine ((2*S*,   3*S*)-2-(3-chlorophenyl)-3,S,S,-trimethyl   2-morpholinol hydrochloride). NCEs that block DATs, but clinical development was terminated in 2007.

(iv) **Serotonin Transporter Polymorphisms**
Reduced binding of imipramine and paroxetine to brain and platelet SERTs in patients with depression and suicide victims indicates that altered SERT function might contribute to aberrant behaviors. Polymorphic regions have been identified in the SERT promoter and implicated in anxiety, mood disorders, alcohol abuse, and in neuropsychiatric disorders. Thus, studies are emerging to support the notion that impaired regulation contribute to human disease conditions such as those seen in human variants of the SERT coding region.[20]

5.4.8   Experimental CNS Disease Models

Drug discovery in depression has been hampered by the lack of a universally accepted animal model(s) used to screen NCEs for antidepressant effects. Although there are several animal models that reproduced features of depression in the context of stress and/or separation, it is questionable as to whether these are related to human disorders, anxiety, MDD or BPAD. The advantages and disadvantages of animal models for depression are well documented in various reviews (see the recommended reading). In all cases, the behavioral features can be reversed by conventional antidepressant treatment. However, despite their intrinsic limitations, the full potential of these models is only now been challenged by genetically modified animal models and has not yet been fully evaluated and represents an underexplored opportunity to fully validated these well-established models. The heuristic value and the knowledge gained from the animal models in psychopharmacology is, explicitly or implicitly, the central preoccupation of psychopharmacologists.[21,23,50] There are a number of compelling reasons to believe in the legitimacy of animal models in the development new improved drugs for the treatment of mental disorders; however, these models need to be based on established criteria.[21]

- Predictive validity: The ability of a model to accurately predict clinical efficacy of a psychoactive pharmacological agent.
- Face validity: The similarity of the model to clinical manifestations of phenomenon/disorder in terms of behavioral and/or physiological symptoms and etiology.
- Construct validity: The strength of the theoretical rationale upon which the model is based.

Animal models have been defined as experimental preparations developed in one species for the purpose of understanding a phenomenon occurring in another species (*e.g.*, the serotonin (5-HT) syndrome translates across species). In animal models of human psychopathology, the aim is to develop specific syndromes that resemble those in humans and study selected aspects of psychopathology. The behavioral models are explicitly related to a broader body of behaviors that fulfill a valuable function in forcing the clinician and psychopharmacologists, alike to critically examine assumptions of these behaviors in relation to the pathophysiology of anxiety, depression and bipolar disorders.[22,23] To denigrate animal models of psychiatric disorders seems unwise when various examples can be clearly observed in animals and in humans, which argues for their validity in creating phenotype/genotype models of mental disorders. It is clear that the etiology of psychiatric disorders is still in its infancy; however, a healthy skepticism provides service in pointing out the many shortcomings when animal models are measured against the complexities of human behavior using qualitative data in both nonclinical and clinical practice.

5.4.9 Animal Models of Anxiety and Depression

Animal models of anxiety, depression and BPAD have proved to be of considerable value in elucidating pathophysiological mechanisms and in developing novel treatments. However, the challenges faced by neuroscientists in modeling human psychiatric disorders in experimental animals are fraught with difficulties and reproducibility. As new targets emerge through hypothesis-driven research or serendipity, the challenge is to link the mechanism to a clinically complex and heterogeneous disorder. Consequently, much of the animal research today is framed around physiological and neurobiological phenomena that may bear little resemblance to the disease state.[23]

However, the poverty of reliable clinical science feedback needs to be addressed first, which would aid future development. The pros and cons of the classical models of depression (*e.g.*, Porsolt forced (FST), tail suspension test, olfactory bulbectomy, learned helplessness, chronic mild stress, and resident intruder) that have stood the test of time in the development of novel antidepressants are now being challenged by genetically modified animal models (see the recommended reviews for more detailed discussion).[28]

(i) **Transgenic and Knockdown and -Out Mice: Models of Psychiatric Disease**

Transgenic animal technology (knockdown and -out), from DNA microinjection to gene targeting and cloning, resulting in "loss-of-function" mutants can be used to clarify the role of molecules thought to be involved in development and structural maintenance of the nervous system. Transgenic models of human disease are used extensively to assess the validity of therapeutic applications before clinical trials although there is an active debate as to the utility of transgenics in drug discovery.[24] The most compelling evidence of a link between genetic variation and the role of the SERT in depression and anxiety led to SERT knockout mice that show increased anxiety-like behaviors, reduced aggression, and exaggerated stress responses. Appropriate functioning of SERT and monoamine oxidase A (MAO-A) during early life appear critical to the normal development of these systems. MAO-A and SERT knockout mice mimic in some respects the consequences of reduced genetic expression in humans. MAO-A knockout mice exhibit high levels of aggression, similar to the elevated impulsive aggression seen in humans lacking this gene. SERT knockout mice may thus represent a more exaggerated version of the reduced SERT expression found in certain subjects, and a partial model of the increased vulnerability to anxiety and affective disorders seen in human subjects with the low expressing allele. Furthermore, some of the genetically modified mice have been reported to show depressive or antidepressant-like behavior in simple behavioral tests, *e.g.*, the Porsolt, or Forced Swim Test.[24]

(ii) **Ribonucleic Acid (RNA) Interference (Small Interfering RNA (siRNA))**

RNA interference (RNAi) allows post-transcriptional gene silencing where double-stranded RNA induces degradation of the homologous endogenous transcripts, mimicking the effect of the reduction or loss of gene activity. This technique, therefore, holds promise in understanding hippocampal autophagy and the role of the DRN in SERT activity. Recent siRNA-mediated knockdown of the SERT in the adult mouse and rat brain would support the concept, although selectivity and side effects remain issues.[25,26]

5.5 CNS Clinical Trial Issues

5.5.1 Clinical Efficacy

Clinical studies with antidepressants invariably involve self-reporting of symptoms (qualitative data) using standardized questionnaires including the Hamilton Depression Rating (HAM-D) scale for depression (17 or 21 items of a 23-item scale). A positive NCE treatment usually registers at greater than 50% in the baseline HAM-D score on a 17- or 21-item scale. The Montgomery–Asberg Depression Rating Scale (MADRAS) is a five-item scale that is used to identify anxiolytic-like activity, and is gaining prominence in the US and Europe. Whichever rating scale is used, a wealth of clinical data exist to show that all current antidepressants seem to be effective in 20–70% of those treated, with placebo responses occurring in 30–50% of treated individuals. Thus, the major issue with the current clinical "blunt" instruments and questionable trial design is that the overall efficacy of antidepressants may be less than 50%. Retrospective analysis of completed antidepressant trials has revealed that four out of six trials do not differentiate from placebo. The limitations and challenges of antianxiety, antidepressant and antischizophrenic clinical trials are well documented (see the further reading list for comprehensive reviews on the clinical challenges in these therapeutic areas[27]). More recent clinical trial approaches are focused on "adaptive design," and trail logistics models in CNS drug development.[27]

5.5.2 Current Treatments and Pharmacological Approaches for Psychiatric Diseases

The majority of psychiatric drugs in clinical use today act by enhancing neurotransmission of monoamines, serotonin (5-HT), NE, DA, or all three, either directly or indirectly (antidepressants).[29] This is by either blocking reuptake *via* monoamine transporters or blocking the metabolism of monoamine inhibiting the major catalytic degradation enzymes, monoamine oxidase (Figure 5.4) and catechol-0-methyltransferase (COMT). Other modes of action include direct or indirect modulation of receptors or signal transduction mechanism. Antidepressants that modulate neurotransmission of monoamine

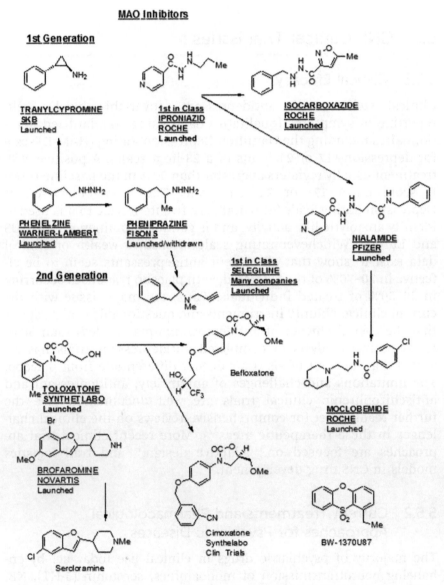

Figure 5.4 Examples of MAOIs chemical phenotypes in their chemical evolution.

uptake inhibition are divided into (i) nonselective (*e.g.*, TCAs, with dual action, see Figure 5.5), (ii) SSRis (see Figure 5.6) and (iii) selective norepinephrine reuptake inhibitors (SNRIs, see Figure 5.7) and recently an additional class of antidepressant (the so-called

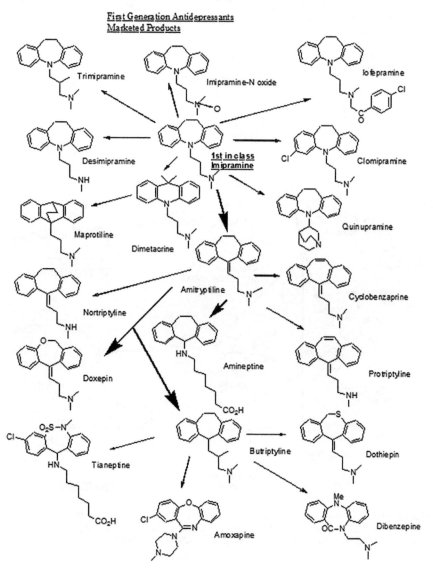

Figure 5.5   The early development of various first-generation chemotypes from the first clinically useful TCA, imipramine and the synthesis of other agents with selective α_2 antagonist and 5-HT uptake properties, with reported antidepressant-like properties.

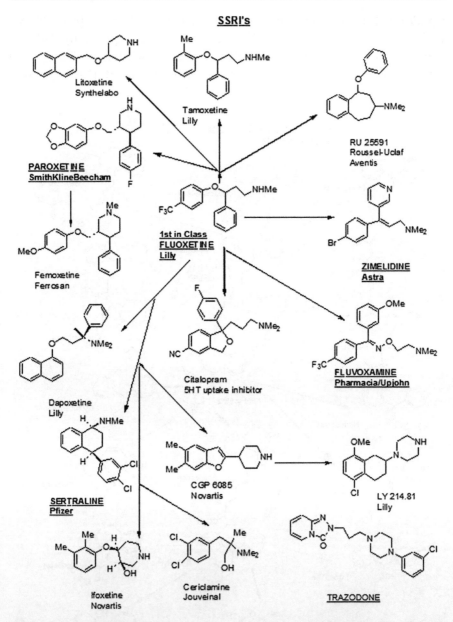

Figure 5.6 The early development of various chemotypes from the first clinically useful SSRI, fluoxetine and the synthesis of other agents with 5-HT uptake properties and 5-HT antagonist properties.

polypharmacology/multimodal) heterocyclics; *e.g.*, vortioxetine, acting at both reuptake sites and receptors. Other approaches include neuropeptide agents and transcription factors with potentially novel

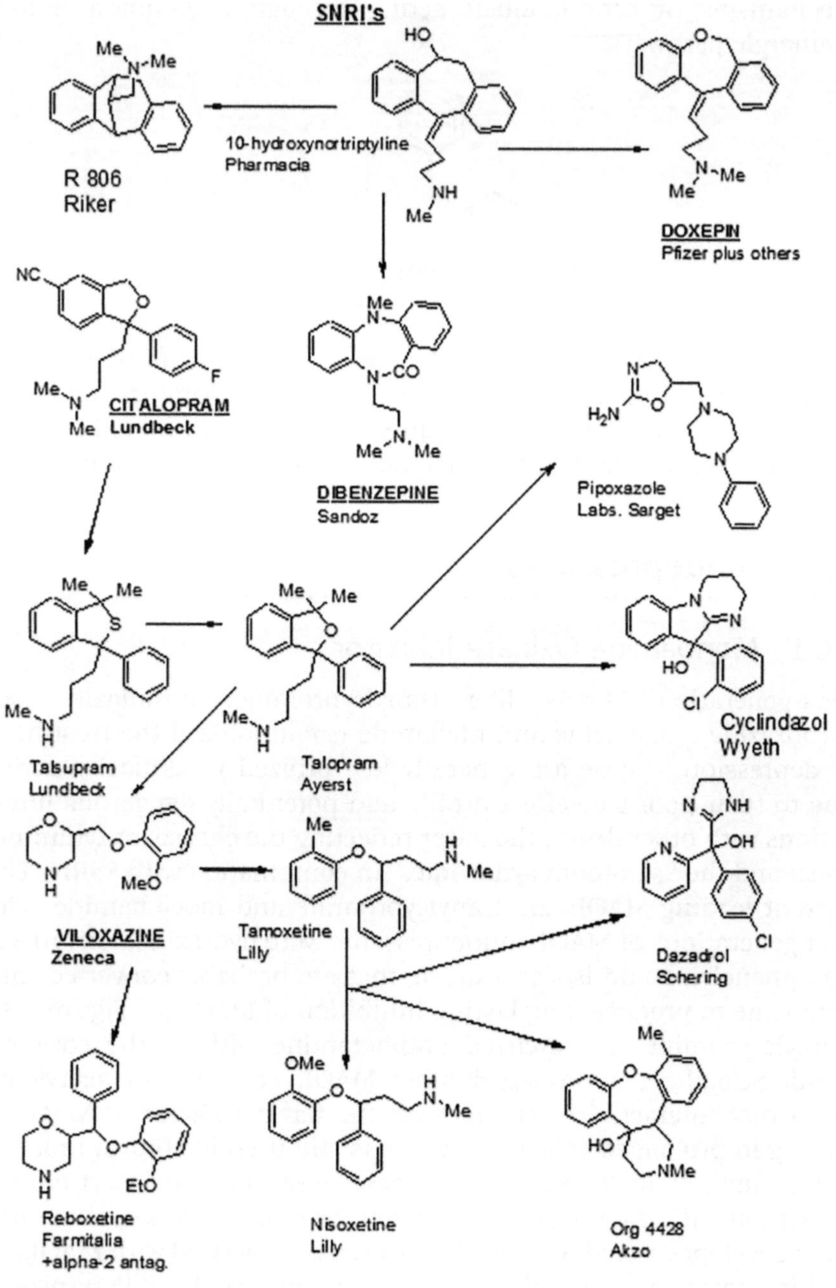

Figure 5.7 The early development of various chemotypes from SNRIs, and the synthesis of other agents with 5-HT uptake and α-adrenoceptor properties.

mechanisms of action, albeit acting through a common mono-aminergic pathways.

vortioxetine

The antipsychotic agents have the opposite mechanism of action by mainly antagonism of the neurotransmission of monoamines, serotonin (5-HT), NE, DA, or all three, either directly or indirectly, as well as other neurotransmitter sites (muscarinic) indiscriminately, *e.g.* clozapine.

5.6 Antidepressants

5.6.1 Monoamine Oxidase Inhibitor

First-generation MAOIs like tranylcypromine, iproniazid, iso-carboxazide, phenelzine and nialamide revolutionized the treatment of depression.[8] These are generally less utilized in clinical practice due to their poor side-effect profile and potentially dangerous inter-actions with other drugs, the latter reflecting the cheese, or tyramine, effect and the "serotonin syndrome," in combination with SSRIs. The current leading MAOIs are tranylcypromine and moclobemide. The first generations of MAOI antidepressants were hydrazine derivatives, *e.g.*, phenelzine and isocarboxazide that are probably converted into hydrazine to produce long-lasting inhibition of MAO (see Figure 5.3). Tranylcypromine is a cyclized amphetamine without the covalent bond. Selegiline, a propargylamine MAOI, contains an acetylenic bond that interacts irreversibly with the flavin cofactor of MAO re-sulting in prolonged MAOI. Selegiline is still used in clinical practice today, mainly in Parkinson's disease (see the later section on Parkinson's disease). However, a patch formulation of selegiline has been developed to reduce the adverse events associated with MAOIs in major depression. Rasagiline is currently marketed for Parkinson's disease in Europe. The major challenge over the last 60 years has been to synthesize short-acting, reversible MAOIs, moclobemide, rasagi-line, toloxatone, befloxatone, brofaromine, secloramine, cimoxatone are representatives of third-generation MAOIs.

   Isozyme-selective and partly reversible MAO have been developed over the last two decades and are reversible inhibitors of monoamine oxidase type A (RIMAs). In contrast to the irreversible nonselective selective RIMAs are isozyme selective – MAO-A or MAO-B. MAO-A inhibition results in increases in NE DA in the synaptic cleft, while selective MAO-B inhibitors increase only DA. The molecular mechanism of action of RIMAs is the same as in the nonselective NCEs: increase of monoamines, near to the receptor, leads, after a number of intermediate steps, to activation of functional proteins in the cell. The control of therapy with MAO-A inhibition is easier, because of their reversibility. Of the current RIMAs, only moclobemide has been shown to have moderate effects in depression.[29] The development of RIMAs, as antidepressants and with possible anti-Parkinson activity, has had limited success to date, and is constrained by the tyramine effect, since the amine can be displaced by the inhibitor from its binding site on the enzyme. Thus, RIMAs would inhibit both forms of enzymes to get the full functional activities of the amine neurotransmitters, without inducing a tyramine effect "cheese reaction." This was not possible until recently, with the development of the novel cholinesterase brain-selective MAO-A/B inhibitor ladostigil (a carbamate derivative of the irreversible MAO-B inhibitor rasagiline). Structure Ladostigil is a brain-selective MAO-A and MAO-B inhibitor; even after 2 months of daily administration little or no effect was observed on the enzymes in the intestinal tract and liver.[30] Pharmacologically, it has a limited tyramine potentiation effect, similar to moclobemide, but it has the anti-depressant, anti-Parkinson, and anti-Alzheimer's action of an MAO-A/B inhibitor in the respective animal models used to develop such NCEs. Ladostigil is currently reported to be in 3-year Phase IIb clinical studies for mild cognitive impairment (MCI).

ladostigil

5.6.2   Tricyclic Antidepressants

TCAs were introduced in the 1950s and became the gold standard treatment for depression before the launch of the first SSRI, fluoxetine, in 1987.[31] All current antidepressants rely upon the principle of enhancing monoamine neurotransmission interfering with the presynaptic transporter that reimports the neurotransmitter from the cleft once released from presynaptic nerve terminals. The classic tricyclic compounds were much less specific, representing a shotgun approach to several neurotransmitter receptors and transporters. Now they are increasingly used as a third-line therapy as the side-effect profile of second-generation SSRI antidepressants is far superior. Classical tricyclics include imipramine, clomipramine, and amitriptyline, along with many other tricyclic's and tetracyclic compounds (Figure 5.5). The TCAs such as imipramine exert their therapeutic actions by inhibition of both 5HT and NE transporter sites. Unfortunately, their polypharmacology target profile, which extends to antagonism adrenoceptors, muscarinic receptors, and histamine receptors, as well as cardiac ion channels, NMDA receptors, underlies their poor patient tolerance. Based on their complex pharmacology this class exhibits a variety of adverse life-threatening side effects. The most serious side effects of the classical tertiary compounds (*e.g.*, imipramine, clomipramine, amitriptyline, doxepin, trimipramine) and secondary amine (*e.g.*, amoxapin desipramine, maprotiline) tricyclics are attributed to direct quinidine-like actions on the heart, interfering with normal conduction and prolongation of the QRS or QT interval with associated cardiac arrhythmia. Other adverse effects include respiratory depression, delirium, seizures, shock, and coma. Anticholinergic side effects including dry mouth, blurred vision, urinary retention, and sinus tachycardia are perhaps the most common. The adverse event profile of tricyclics is also associated with a greater risk due to their interaction with cytochrome P450 (CYP) isoenzymes, in particular CYPZD6, which is a highly polymorphic gene. A minority of the population has CYPZD6 polymorphic variations, gene deletion, or gene duplication. Polymorphism results in ultrarapid or ultraslow metabolism, leading to varying degrees of drug bioavailability, which can lead to increased bioavailability and severe toxicity in some cases. All the "classical" TCAs have a basic three-ring pharmacophore (Figure 5.5). The therapeutic and commercial success aminoalkylphenothiazines such as promethazine, promazine, and chlorpromazine led to molecular modification of the polycyclic phenothiazine ring structure and its *N*-aminoalkyl side chain substitution of the sulfur bridge of the phenothiazine ring of promethazine with an ethylene bridge resulted in imipramine, the first clinically useful TCA (Figure 5.5). It did not take long for medicinal chemists focusing on the structure to substitute the additional N-CH2 group in imipramine for a C–CH group in amitriptyline (Figure 5.5). Following earlier dimethylamine work, monomethyl amine derivatives were then synthesized, from which desipramine and nortriptyline evolved.

5.6.3 Selective Serotonin Reuptake Inhibitors (SSRIs)

While the primary mode of action of TCAs was thought to be inhibition of NE reuptake, a reassessment of the actions of the diverse structures of antihistamines on the reuptake of various biogenic amines, especially 5HT, led to the hypothesis that an increase in brain noradrenergic function caused the energizing and motor stimulating effects of the TCAs, but an increase in 5HT function was responsible for their mood-elevating effects.[31] Structural analogs of diphenhydramine were sought as novel antidepressants. The phenoxyphenylpropylamine pharmacophore was used to identify fluoxetine (Figure 5.6), the first SSRI.[31] The phenomenal success of fluoxetine as an antidepressant led to the identification of SSRIs, *e.g.*, paroxetine, citalopram, fluvoxamine and sertraline. Most SSRIs are aryl or aryloxalkylamines. Within this chemical genus, fluoxetine, citalopram, and zimelidine are racemates, and sertraline and paroxetine are separate enantiomers. The (*S*)-enantiomer citalopram (escitalopram) and fluoxetine (norfluoxetine) are relatively more potent at SERT (see Figure 5.6 for SSRIs structures). Structural–activity relationships (SARs) are not well established for SSRIs, although the *para*-CF_3 substituent of fluoxetine critical for potency as removal and substitution at the *ortho-position* of a methoxy group yields nisoxetine, selective NE uptake (NET) inhibitor. The potency for 5-HT inhibition varies amongst this group, as does the selectivity for 5HT relative to NE reuptake inhibition. The relative potency of sertraline for DA reuptake inhibition differentiates it from other SSRIs. Affinity for neurotransmitter receptors such as sigma, muscarinic, and $5HT_{2C}$ also differs widely (for greater details see recommended reading[31]). Furthermore, the inhibition of nitric oxide synthetase by paroxetine, and possibly other SSRIs, may have significant pharmacodynamic effects. Fluoxetine has a long-acting and pharmacologically active metabolite (norfluoxetine) with high-affinity $5HT_{2C}$ receptor. Other important clinical differences among the SSRIs include differences in their half-lives, linear and nonlinear PK, effect of patient age on their clearance, and their potential to inhibit drug-metabolizing CYP isoenzymes. These differences underlie the increasingly apparent important clinical differences among the SSRIs. Of the very limited comparative clinical data available, a meta-analysis of 20 short-term comparative studies of SSRIs (citalopram, fluoxetine, fluvoxamine, paroxetine, and sertraline) showed no difference in efficacy between individual compounds but a slower onset of action for fluoxetine. The most common adverse reactions to these, according to FDA and MHRA websites, were gastrointestinal (especially nausea) and neuropsychiatric (particularly headache, tremor, discontinuation reactions, and sexual dysfunction).

5.6.4   Selective Norepinephrine Reuptake Inhibitors (SNRIs)

SNRIs are a class of antidepressants characterized by a mixed action on both major monoamines of depression: NE and serotonin (Figure 5.7). In essence, SNRIs are improved TCAs with less off-target activity, *e.g.*, muscarinic, histamine, α_1-adrenergic receptors, and MAOI. The combination of inhibition of 5-HT and NE uptake confers a profile of effectiveness comparable to TCAs and is reported to be higher than SSRIs, especially in severe depression. SNRIs are purported to be better tolerated than TCAs and more similar to SSRIs without the associated sexual dysfunction seen with the latter. Several other SNRIs have been approved to date (venlafaxine, milnacipran, levomilnacipan, desvenlafaxine, tofencin) (Figure 5.7). They are active on depressive symptoms, as well as on certain comorbid symptoms (anxiety, sleep disorders) frequently associated with depression. SNRIs appear to have an improved rate of response and a significant rate of remission, decreasing the risk of relapse and recurrence in the medium and long term and address two of the current goals of antidepressant treatment. Reboxetine is another SNRI that was approved in Europe and many countries for the treatment of depression but is not available in the USA (Figure 5.7). It is used off-label for ADHD and panic disorder.

5.6.5   Catechol-0-Methyltransferase Inhibitors

An increase in the functional monoamines NE, DA, and 5-HT can precipitate mania or rapid cycling in an estimated 30% of affectively ill patients. A strong association between velo-cardio-facial syndrome (VCFS) patients diagnosed with rapid-cycling bipolar disorder, and an allele encoding the low enzyme activity catechol-0-methyltransferase (COMT) has been identified. Between 85% and 90% of VCFS patients are hemizygous for COMT. There are expected to be higher levels of transynaptic catecholamines due to a reduced COMT degradation by COMT inhibitors (entacapone and tolcapone), which could therefore be beneficial as adjuncts to L-dopa Parkinson's disease, but also in the coincident depressive illness associated with rapid cycling.[8]

entacapone   **tolcapone**

5.6.6   Current Treatments for Bipolar Disorder

(a) Acute Mania

Lithium is generally the drug of choice to stabilize the person and is usually very effective in controlling and preventing new episodes. Response to lithium treatment may take several days initially.[8] If the patients are experiencing psychotic symptoms, antipsychotic medications, *e.g.* clozapine, olanzapine, and other antipsychotic agents, may be prescribed. Mood-stabilizing anticonvulsant drugs, such as valproate, lamotrigine, carbamazepine and topiramate may also be used. Often, these medications are combined with lithium for maximum effect with mixed benefit to the patient.[8]

(b) Acute Depression

Lithium can be a very effective treatment for the depression that occurs in bipolar disorder. Antidepressant SSRIs may also be prescribed. Antidepressant medications used to treat the depressive symptoms of bipolar when taken without a mood-stabilizing medication can increase the risk of switching into mania or developing rapid cycling, in people with bipolar disorder. Therefore, mood-stabilizing medications are urgently required, alone or in combination with antidepressants, to protect patients with bipolar disorder from this switch. Lithium and valproate are the most commonly used mood-stabilizing drugs today.[8]

Current medications used for the treatment of bipolar disorder are briefly summarized below.

(i) Lithium: Since the early 1800s, lithium has been a first-line medication in the treatment of bipolar disorder (manic depression). The mechanism of action by which lithium produces efficacy is unknown. Lithium inhibits production of monophosphate, which plays a role in gene expression.[32] Lithium is a well-established treatment for bipolar depression being effective in the treatment of both the manic and depressive phases.[33] It has an approximate response rate of 79%.

(ii) Anticonvulsants: The treatment spectrum for bipolar disorder has broadened since the use of anticonvulsants, such as carbamazepine, lamotrigine, and gabapentin. Patients with rapid cycling or mixed episode are more likely to benefit from treatment with anticonvulsants than patients with other types of bipolar disorder.[34]

(iii) Valproate: The mechanisms by which valproate exerts its therapeutic effects have not been established. It has been suggested that its activity in epilepsy is related to increased brain concentrations of GABA and also its sodium channel antagonist properties. It is widely used in the US but not in Europe or Japan.

(iv) Carbamazepine: Carbamazepine has antiepileptic psychotropic and neurotropic actions. It is believed that its primary mechanism of action is blockade of voltage-sensitive sodium and calcium channels, although the molecular mechanisms underlying the actions of carbamazepine and other mood-stabilizing drugs used in treatment of bipolar disorders are still largely unknown. Compared to a placebo, carbamazepine is effective for the long-term treatment of bipolar disorder, but is not approved worldwide. It has antimanic and antidepressant effects both as a monotherapy and in combination with lithium and antidepressants.

(v) Lamotrigine: The effects of lamotrigine are thought to include inhibition of excitatory amino acid and voltage-dependent sodium channels and blockade of 5-HT$_3$ receptors.[35] A number of studies have shown lamotrigine to be effective for the treatment of the depressive phase of bipolar disorder and rapid cycling bipolar disorder. The adverse effects of lamotrigine are similar to other anticonvulsants, with a slightly higher rate of headaches.[35] Lamotrigine is not practical for the treatment of the manic phase of bipolar disorder; this is partly due to the need for slow dose escalation. Lamotrigine has been reported to enhance the effectiveness of valproate in bipolar disorder; however, there is a risk of a rash with this dosing regimen. To reduce the risk of a rash a slow dose titration is recommended. In contrast, when coadministering lamotrigine with carbamazepine a more rapid increase is recommended.[34] Lamotrigine is emerging as a potentially useful agent for the treatment of bipolar disorder.[34]

(vi) Gabapentin: Gabapentin is an antiepileptic drug that is structurally related to GABA. Its mechanism of action is in part linked to voltage-gated calcium channels ($\alpha 2\delta$ subunit), but its distinct profile of anticonvulsant activity in animal seizure models and its activity at many drug binding sites associated with other antiepileptic drugs indicate that its mechanism of action is novel. Gabapentin was formed by the addition of a cyclohexyl group to GABA, which allowed this form of GABA to cross the blood–brain barrier. Despite its structural similarity to GABA, gabapentin does interact with GABA receptors in CNS. Its mechanism of action is unknown, but may involve enhanced neuronal GABA synthesis. Gabapentin differs from other mood-stabilizing drugs in two ways: (i) it is sometimes effective for patients who have failed to respond to antidepressants or mood stabilizers; and (ii) it has a relatively benign side-effect profile. Gabapentin has been successful in controlling rapid cycling and mixed bipolar states in a few people who did receive adequate relief from carbamazepine and/or valproate. It also appears that gabapentin has significantly more antianxiety and antiagitation potency than either carbamazepine or valproate. Gabapentin also may be useful treatment for people with antipsychotic-induced tardive dyskinesia. It is possible that gabapentin will prove to be a useful treatment for individuals with other mood disorders. The final dose of gabapentin is usually between 900 and 2000 $mg\,day^{-1}$ when used as an antidepressant or as a mood-stabilizing agent. Some individuals require doses as high as 4800 $mg\,day^{-1}$ to achieve efficacy.

(vii) Antidepressants: Standard antidepressants are effective for the treatment of BPAD-1 in combination with a mood stabilizer. The most commonly recommended antidepressants are SSRIs, TCAs, MAOIs, and bupropion.[33]

5.6.7 Unmet Medical Needs

It is well documented that the onset of action, drug efficacy, and side-effect profile are the three key unmet needs with current mood disorders therapies. An improvement in general efficacy of the drug was cited as the single most important factor physicians require in new antidepressant medication. It is widely acknowledged that approximately 20–35% of patients do not respond to any of the currently available agents and discontinuation are high (up to 50% within the first 4 months of treatment), primarily due to side effects related to the dysfunction associated with the SSRIs. ECT is useful, particularly for individuals whose depression is severe and life threatening or who cannot take antidepressant medication. It is often effective in cases where antidepressant medications do not provide sufficient relief of symptoms.[36] A variety of antidepressants are often tried before finding the most effective medication or combing medications, and often the dosage must be increased to be effective. Although some improvement may be seen in the first few weeks, antidepressant medications must be taken regularly for 3–5 weeks (in some instances 8 weeks the full therapeutic effect occurs).[36] Patient compliance is often the most cited reason for stopping medication due to lack of efficacy or adverse effects. It is evident that no significant advances have emerged over the last three decades for the treatment of bipolar disorders and new target are eagerly sought to overcome the dire unmet medical need in this area and for other psychiatric disorders (some new areas of research are listed below and discussed in Section 5.6.8).

5.6.8   New Areas for Future Research

The monoamine hypothesis of depression has been the cornerstone of antidepressant treatment for several decades. However, many questions remain unanswered as to the underlying pathophysiology of anxiety, depression and affective disorder, and whether monoamines are solely responsible for regulating psychiatric disorders. It is clear that the etiology of depression and bipolar disorder is still unknown. Arguably, however, the clinic preclinical data supporting the monoamine hypothesis is tried and tested. With this in mind, the fact remains that the clinical response is delayed several weeks following administration of monoaminergic antidepressant, suggesting that other mechanisms may well be involved in the efficacy of these agents. It has long been suggested that alterations in gene expression may be a contributing factor for the delayed clinical response, thereby resulting in changes in signal transduction mechanisms (Figure 5.3).[8,32,37] Several purported mechanisms may account for the clinical response: (i) receptor downregulation; (ii) other components of cellular signaling that are regulated cyclic AMP, which are prominent transcription factors in the brain (phosphorylated cAMP response element protein, CREB); and (iii) factors controlling cellular plasticity such as brain-derived neurotrophic factor (BDNF). At present, the number one challenge is to develop novel antidepressants with greater efficacy and rapid action. To this end, several pharmaceutical companies continue to bet on the tried and tested monoamine inhibitors approach, the latest entrant in the US depression market is brexpiprazole. It is a DA_2 partial agonist called serotonin–dopamine activity modulator (SDAM). The drug received FDA approval in 2015 for the treatment of schizophrenia, and as an adjunctive treatment for depression.

(i) *Serotonin Agents*

The fact that repeated SSRI treatment can downregulate a number of pre- and postsynaptic 5-HT receptors (see the section on serotonin) in healthy subjects and depressed patients has led to intensive research and numerous "selective" compounds synthesized to unravel the role of 5-HT receptor subtypes in the pathophysiology of anxiety, depression and BPAD.[8,15]

(ii) *Neuropeptide Approaches*

(a) NK1 Receptor Antagonists: Substance P is an 11 amino acid peptide belonging to the tachykinin family; it mediates its biological actions through trachykinin (NK1) receptors. Evidence to support a major role of the NK_1 receptor system in stress-related behavior has guided the clinical development of several NK_1 receptor antagonists, including aprepitant (MK-869), lanepitant, dapitant, vestiptant, PD-174424, and NBI 127914. The antidepressant effect of the first NK_1 receptor antagonist MK-0869 (aprepitant) was demonstrated in patients with major depression and high anxiety, and has recently been replicated with a second compound, L759274. Aprepitant improved depression/anxiety symptoms in a quantitatively similar manner to SSRIs. However, it failed to show efficacy in large Phase III clinical trials for depression.[38] Although other NK agents are also under investigation for their potential role in depression and anxiety disorders no positive finding have been reported to date.[8]

(b) *Corticotropin-releasing Factor (CRF)*

The 41 amino acid neuropeptide CRF initiates the HPA axis response to stress, and may have utility in the treatment of depression and anxiety. However, no clinical success has been reported with selective compounds.

(c) *Melanin Concentrating Hormone (MCH)*

MCH is an orexigenic hypothalamic neuropeptide that plays an important role in the complex regulation balance and body weight. SNAP-7941 is a selective, high-affinity MCH_1 receptor (MCH 1-R) antagonists and produced effects similar to clinic antidepressants and anxiolytics in animal models of depression/anxiety.[8] Given these observations, it has been suggested that MCH 1-R may be useful not only in the management of obesity but also as a treatment for depression and/or anxiety.[8]

(d) *Galanin*

Galanin is a biologically active neuropeptide that is widely distributed in the central and peripheral nervous and the endocrine systems. The amino acid sequence of galanin is very conserved (almost 90% among species) indicating the importance of the molecule. In the CNS, galanin is important in the release of several neurotransmitters. The $GALR_3$ receptor antagonist SNAP-37889 has been reported to show activity in several antidepressant tests and may have utility as a novel antidepressant.[8,39]

(e) *Neuropeptide Y*

Neuropeptide Y (NPY) is abundantly expressed in numerous brain areas, including the locus coeruleus, hypothalamus, amygdala, hippocampus, nucleus accumbens, and neocortex associated with mood and behavior. Rodent studies demonstrated antidepressant-like effects of centrally administered NPY_1 receptor agonists.[8]

(f) *Vasopressin VP$_{1b}$ Receptor Antagonists*

The nonpeptide vasopressin, which is synthesized in the PVN and supraoptic nucleus, acts *via* vasopressin receptor (V_{1a} and V_{1b}) expressed mainly in limbic areas and in the hypothalamus. Abnormalities in vasopressin expression receptor activity occur in both clinical depression and rodent genetic models of depression, whereas vasopressin release predicts anxiety reactions to stress provocation in healthy volunteers. The nonpeptide V_{1b} receptor antagonist S149415 exerts marked anxiolytic and antidepressant-like effects in rodents.[38]

(iii) *Glutamate Receptors and NMDA Receptors*

Several types of antidepressants can alter the expression of mRNA for the NMDA receptor, while glutamate levels are elevated in the occipital cortex of medication-free subjects with unipolar major depression.[40] Direct NMDA receptor antagonists have antidepressant effects in animal models including behavioral despair and chronic mild stress paradigms. Two drugs with putative glutamatergic properties, lamotrigine and riluzole, have been reported to show antidepressant activity in clinical trials, while a preliminary study using the noncompetitive NMDA receptor antagonist, ketamine, showed rapid and short-lived antidepressant actions. This has now become a very active area of research and more extensive clinical studies are in progress.

(iv) *Group 1 Glutamate Receptors (mGlu 1–8)*

Metabotropic glutamate (mGlu) receptors (mGlu 1–8) are a heterogeneous family of GPCRs that modulate excitability *via* presynaptic, postsynaptic, and glial mechanisms. Glutamate receptors regulate brain glutamate activity and may play an important role in the pathophysiology of depression. NMDA receptor antagonists, mGluRl and mGluRS antagonists, and positive modulators of AMPA receptors have antidepressant-like activity in a variety of preclinical models. Clinical data with subtype selective compounds is yet to yield any promise for this approach.

(v) *GABA$_A$ Receptor Antagonist*

A dysfunction of GABAergic system is implicated in the pathophysiology of anxiety and depression. Recent evidence points to a role of GABA$_A$ receptors in anxiety and depression. Metabotropic GABA$_B$ receptors predominantly as heterodimers of GABA$_{B1}$ and GABA$_{B2}$ subunits, but GABA$_{B1}$ can also form functional receptors in the GABA$_{B2}$ mice lacking the GABA$_{B1}$ subunit have altered behavioral responses in tests for anxiety and depress GABA$_{B1}$- and GABA$_{B2}$-deficient mice were found to be more anxious than wild type in the light-dark box paradigm.[41] In contrast, these mice exhibited an antidepressant-like behavior in the FST. Taken together, these data suggest heterodimeric GABA$_{B1,2}$ receptors are required for the normal regulation of emotional behavior and offer a novel approach for the treatment of mood disorders.[41] GABA$_B$ receptor antagonists in preclinical and clinical develop depression and cognitive disorders include AVE 1876 and SGS742.

(vi) *Glucocorticoids*

Glucocorticoids are important in the pathogenesis of depression, but potentially serious psychological side effects are often overlooked in clinical practice. The unwanted behavioral effects of anabolic steroids are widely known, but those of glucocorticoid therapy, though recognized for over 45 years, receive less attention. Placebo-controlled studies revealed that a third of patients taking glucocorticoids experience significant mood disturbance and sleep disruption. More importantly, up to 20% of patients on high-dose glucocorticoids report psychiatric disorders including mania, psychosis, or a mixed affective state. A recent double-blind, placebo-controlled trial of corticosteroid administration in healthy individuals showed that 75% of subjects developed disturbances in mood that reversed when steroids were stopped. Deregulation of the HPA axis in depression is one of the oldest consistent findings in biological psychiatry. A large-scale meta-analysis of over 140 studies using the dexamethasone suppression test illustrated that persistent adrenocortical hyperactivity is a robust indicator of poor prognosis and a weaker predictor of suicide in depression. Overall, data from conditions of both exogenous and endogenous steroid excess provide support for a glucocorticoid theory of depression. However, no further supportive data has been presented for this indication or other CNS indications for some time. ORG-34517, a glucocorticoid antagonist, has been in Phase II clinical trials for the treatment of depression. However, this study was discontinued.

(vii) *Transcription Factor*
cAMP Response Element-Binding Protein (CREB)
Antidepressants usually take weeks to exert significant therapeutic effects. This lag phase is suggested to be due to neural plasticity, which may be mediated by the coupling of receptors to their respective intracellular signal transduction pathways. Phosphorylated CREB, a downstream target of the cAMP-signaling pathway, is a molecular state marker for the response to antidepressant treatment in patients with MDD. The transcriptional activity of CREB is upregulated by antidepressant treatment. Therefore, it has been hypothesized that antidepressant treatment exert therapeutic effect through this mechanism, an hypothesis awaiting selective compounds.

(viii) *Trace Amines*
In addition to the classical monoamines NE, 5-HT, and DA, there exists a class of "trace amines" that are found in low levels in mammalian tissues, and include tyramine, P-phenylethylamine (β-PEA), tryptamine, and octopamine. TAAR1 is an amine-

activated G_s-coupled and G_q-coupled G protein-coupled receptor (GPCR) that is primarily located in several peripheral organs, lymphocytes, astrocytes, and in the intracellular compartments within the presynaptic plasma membrane (*i.e.*, axon terminal) of monoamine neurons in the central nervous system (CNS). The rapid turnover of trace amines, as evidenced by the dramatic increases in their levels following treatment with MAOIs or deletion of the MAO genes, suggests that the levels of trace amines at neuronal synapses may be considerably higher than predicted by steady-state measures. Although there is clinical data in the literature to support a role for trace amines in depression as well as other psychiatric disorders, the role of trace amine neurotransmitters in mammalian systems has not been thoroughly examined.[42] Because they share common structure with the classical amines and can displace other amines from their storage vesicles, trace amines are known as "false transmitters." Thus, many of the effects of trace amines are indirect and are caused by the release of endogenous classical amines. They have well-characterized presynaptic *amphetamine-like* effects on monoamine neurons *via* TAAR1 activation. The aim now is to prove TAAR1 is a mediator of some of amphetamine's actions *in vivo*, and the development of novel TAAR1-selective agonists and antagonists could provide a new approach for the treatment of amphetamine-related conditions such as addiction and/or disorders in which amphetamine is used therapeutically. In particular, because amphetamine has remained the most effective pharmacological treatment in ADHD for many years, the potential role of TAAR1 in the mechanism of the "paradoxical" effectiveness of amphetamine in this disorder should be explored.[43,44]

However, there is a growing body of evidence suggesting that trace amines function independently of classical amine transmitters and mediate some of their effects *via* specific receptors. Saturable, high-affinity binding sites for [^{3}H]-tryptamine, ρ-[^{3}H]-tyramine, and β-[^{3}H] PEA have been reported in rat brain, and the pharmacology and localization of these sites suggest that they are distinct from the amine transporters. However, although binding sites have been reported, no specific receptors for these trace amines have yet been identified conclusively. A family of 15 GPCRs has been described, two members of which are activated by trace amines. TAAR 1 activated most potently by tyramine and β-PEA, and TA2 by β-PEA. TAAR1 is one of six functional trace amine-associated receptors in humans, which are so named for their ability to bind endogenous amines that occur in tissues at trace concentrations. The localization of the TAAR1 receptor in human and rodent tissues, as well as the chromosomal localization of the human members of this family, has been characterized. The identification of this family of receptors should facilitate the understanding of the roles of trace amines in the mammalian nervous system and their role in mood and affective disorders.

Clinical studies have examined the levels of PEA in many conditions and discovered significant associations. Patients with depression have decreased PEA levels while levels are increased in schizophrenic and psychotic subjects. The administration of PEA has been found to reduce depression. Likewise, the administration of its precursor phenylalanine has been found to improve depression on its own and the therapy outcome when combined with antidepressants.[42]

Thus, the role of trace amines (TAAR1) in the pathophysiology of depression and other psychiatric disorders is an exciting area of research and selective drugs are now awaited to test the hypothesis.

5.6.9 Summary

The future trends in the development of drugs for the treatment of anxiety, depression and affective disorders have followed the high degree of anatomical overlap in the distribution and coexistence of monoamine and neuropeptide neurons in limbic regions of the human brain. The intimacy of the relationship between monoamines and neuropeptides suggests a common downstream effect on neural circuits that mediate stress. The relationship between stress-related interactions among glutamate, CRF, galanin, NPY, SP, and vasopressin VlB and monoamines in the brain is well documented. However, the role of neuropeptides in depression is still in its infancy and it is premature to speculate on whether and how the systems might exert common downstream effects on brain pathways/transduction mechanisms that mediate stress and emotion. Thus, although SP, CRF, vasopressin, NPY, MCH, and galanin demonstrate important functional interactions with monoamines implicated in the etiology and treatment of stress-related disorders, their effects almost certainly go beyond modulation of these neurotransmitters. Understanding these effects is a central goal of future research in this field. There are certain characteristics common to neuropeptides that might make them attractive targets for novel therapeutics.

As neuropeptides possess a more discrete neuroanatomical localization than monoamines, neuropeptide receptor ligands might be expected to produce relatively little disruption of normal physiological/pharmacological alteration of neuropeptide function, and might normalize aberrant activity in neuronal circuits such as the HPA axis, without producing unwanted side effects. Moreover, antagonists might be less likely than agonists to produce tolerance and dependence. Indeed, drugs that are antagonists at CRF, vasopressin, NPY, and receptors might have a particularly low side-effect burden because such compounds would not be expected to disrupt normal physiology in the absence of neuropeptide release. Preliminary clinical data appear to be encouraging in this regard. Several ligands that target neuropeptide receptors are currently undergoing clinical evaluation as to whether they provide efficacious alternatives to existing drug treatments for depression, affective disorders and anxiety.

Establishing the safety, therapeutic efficacy, and an acceptable tolerability profile of drugs that target neuropeptides or other neurotransmitter systems in depression and anxiety disorders would represent a major advance in the treatment of these diseases. The validation of future targets will be facilitated by the generation of new technologies and techniques *e.g.*, mutant rodents models to elucidate neurotransmitter function, where a paucity of selective, brain-penetrant ligands limits conventional psychopharmacological approach effects of constitutive neuropeptide mutations can be skewed by developmental alterations that compensate mutated neuropeptide or cause changes in other systems that confound interpretation of stress-related phenotypes. Therefore, engineering neuropeptide transgenic mutations that are limited to specific developmental stages and brain regions, improving other molecular manipulations (*e.g.*, CRISPR/Cas9 and DAT-Cre gene editing, optogenics, RNA interference, antisense, and viral vector delivery technique), will be very valuable in identifying novel targets. Once promising targets are identified, a further challenge is the generation of small-neuropeptide receptor ligands that are soluble, bioavailable, brain penetrant, and have a low potential for tachphylaxis. Although there are important obstacles to surmount, neuropeptide-based therapeutic strategies for anxiety, depression and affective disorders represents a highly promising approach to treating these dreadful, debilitating conditions.

5.7   Schizophrenia

Schizophrenia is a chronic, debilitating neurodevelopment psychiatric disorder that affects approximately 1% of the population. With >8 million people in the world affected by a disorder that requires daily drug treatment throughout their life. It appears to have genetic and epigenetic causality and is characterized by diminished drive and emotional problems in childhood.[45,46,58,59] Schizophrenia is defined as a wide spectrum of disorders consisting of positive, negative and cognitive symptoms. Positive symptoms include auditory and visual hallucinations, delusions, disorganized thought, and antisocial or violent behavior. Negative symptoms include dissociation, apathy, difficulty or absence of speech, and social withdrawal. The cognitive symptoms are disorganized thought, difficulty in attention or concentration, and poor memory. Schizophrenic symptoms are more prevalent in adolescence or early childhood, but can occur at any age. Current diagnostic criteria rely on DSM-IV-TR and ICD-10 for classification of schizophrenia (see the excellent references for further information).[45,46]

5.7.1   The Neuroanatomical and Neurochemical Basis of Schizophrenia

Schizophrenia appears to be a spectrum of disorders that are speculated to result from malfunctions in different neuronal circuits. It is well documented that changes in the mesolimbic dopamine pathway (the neuronal projection from the ventral tegmental area (VTA) to the nucleus accumbens, amygdala and hippocampus) being associated with psychosis, the positive symptoms, whereas negative symptoms are associated with changes in the mesocortical dopaminergic pathway (the projection from the VTA to areas of the prefrontal cortex).[48]

Dopamine receptors in this pathway and in the central nervous system fall into five subtypes and two functional classes: the D_1 type, comprising D_1 and D_5, and the D_2 type, comprising D_2, D_3 and D_4. Antipsychotic drugs owe their therapeutic effects mainly to blockade of D_2 receptors. It is well documented that antipsychotic effects require about 80% blockage of D_2 receptors.

5.7.2 Classification of Antipsychotic Drugs

Antipsychotic drugs have been divided into two groups;

- *first-generation, typical antipsychotic drugs:* chlorpromazine, haloperidol, fluphenazine, flupentixol, clopentixol;
- *second-generation* or *atypical antipsychotic drugs:* clozapine, risperidone, sertindole, quetiapine, amisulpride, aripiprazole, ziprasidone.

Pharmacological investigation showed that the first-generation antipsychotic agents (the phenothiazines, *e.g.* chlorpromazine), block many different mediators, including histamine, catecholamines, acetylcholine and 5-HT, and a multiplicity of other neurotransmitter effects. First- and second-generation chemotypes are shown below.

chlorpromazine   **clozapine**

However, the development of the second generation rested on their improved receptor profile, reduced incidence of extrapyramidal side-effects efficacy, particularly in the resistant group of patients with compounds like clozapine. The irony is that clozapine like many atypical antipsychotic, look much like the first-generation antipsychotic drugs with modest changes in receptor affinity for a multiplicity of receptor binding sites and modest improvement in efficacy and adverse events. Unfortunately, the long-awaited third generation of antipsychotics is still very much on the horizon.[49,50,61,62]

5.7.3 Pharmacological Effects of Antipsychotics

A number of neurotransmitters are purported to be involved in the pathogenesis of schizophrenia, most of the evidence points to dopamine, 5-hydroxytryptamine and glutamate.[47,48] The proposed role of dopamine in psychosis, is one school of thought based on the hyperactivity of dopamine neurocircuitry in these limbic areas. Supported by the development of phenothiazine derivatives and congeners potency at dopamine pre- and postsynaptic D_2 receptor antagonists. Suggesting a possible mechanism of action for these drugs. However, all of the known antipsychotic drugs have complex pharmacological profiles in addition to their dopaminergic receptor antagonism. The *atypical* neuroleptic compounds (*e.g.* respiridone and quetiapine) have significant affinity for the serotonergic 5-HT$_2$ receptors, as well as a number of other receptors (5-HT$_{1A}$, α_{1a}, mACh, H$_1$).[51,52] This indicates that the antipsychotic efficacy of these compounds may be a result of a summation of all these pharmacological properties – multimodal activity.[61,62]

This is not a new concept as the antipsychotic drug clozapine is unique in that it binds only weakly to D$_2$ receptors but also has affinity for other dopamine subtypes (D$_1$ and D$_4$). Based on the binding profile clozapine compounds with high affinity for the D$_4$ receptor were developed to determine if upregulation of G-protein coupled to D$_4$ receptors were important in the etiology and treatment of schizophrenia. However, a selective D$_4$ receptor antagonist L-745, 870 failed to show clinical efficacy in a Phase II study in schizophrenic patients.

L-745,870   quetiapine

Quetiapine and clozapine are the most widely used medications for the treatment of Parkinson's disease induced psychosis due to their very low extrapyramidal side-effect liability. Owing to the risks associated with clozapine (*e.g.* agranulocytosis, diabetes mellitus, *etc.*), quetiapine is now the drug of choice, although the 5-HT$_{2A}$ antagonist primavanserin has recently gained approval for this indication.

5.7.4 Current Therapy and Adverse Events

Antipsychotics induce a range of side effects that include extrapyramidal motor (see below), endocrine (galactorrhea), weight gain and sedative effects that can be severe and limit patient compliance. They may shorten survival through cardiac (proarrhythmic and increase in

QT-interval) effects. Abrupt cessation of antipsychotic drug administration may lead to a rapid onset psychotic episode distinct from the
underlying illness. The second-generation (atypical) antipsychotic
drugs held out much promise but similar side-effect issues prevailed,
although with some improvement in efficacy. However, advances in
new technologies, as described above may hold promise for the
development of new treatments for this dreadful disorder.[61,62]

5.7.5 Major Motor Disturbances Induced by Antipsychotic Drug Treatment.

Two main types of disturbance occur:

(i) An acute, reversible dystonia's and Parkinson-like symptoms.
Antipsychotic drugs generally worsen Parkinson's disease and
block the actions of drugs used to treat the disorder.[49,50,53,54]
The acute symptoms comprise involuntary movements, tremor
and rigidity, and are probably the direct consequence of
antagonism of nigrostriatal dopamine (D_2) receptors. The
incidence of acute dystonia's and tardive dyskinesia is less with
newer, second-generation antipsychotics, and particularly low
with clozapine, aripiprazole and zotepine.

(ii) A slowly developing tardive dyskinesia, which comprises mainly
involuntary movements of the face and limbs, often appearing
after months or years of antipsychotic treatment and may be
reversible. It is also associated with proliferation of dopamine
(D_2) receptors in the corpus striatum. Treatment is generally
unsuccessful.[49,50,53,54]

5.7.6 Translational Pharmacological Models

Translational of animal models of schizophrenia that relate to the
positive, negative and cognitive deficit components of this disorder are
imperfect.[55,56] Schizophrenia presents as a heterogeneous disorder
with patients exhibiting different combinations of symptoms that may
result from different neuronal abnormalities. Classical animal models
present behaviors resulting from heightened dopaminergic transmission in the brain that is antagonized by dopamine receptor antagonists.
More recent studies on inhibition of NMDA function by phencyclidine
(PCP) and related drugs are based on neurobehavioral endotypes types
observed in animals and humans that result in a schizophrenia-like
syndrome. A provisional consensus on a cognitive battery of animal

behavioral models for cognitive deficits and negative symptoms (MATRICS) has been developed to address aspects of schizophrenia.[46] The development of such models is a major challenge that to aid a better understanding of the pathophysiological processes that underlie different symptoms. For further details on the development of new animal models of schizophrenia (see the recommended reading).[56,62] Heterogeneity in the manifestations and course of schizophrenia has long been observed. It is now well recognized that the genetic basis of schizophrenia is multifactorial and environmental factors play an important part. The complex genotype of schizophrenia is evolving and focus on a single gene may only provide limited information (see further reading for more detailed information).[57]

5.8   Antiepileptics

5.8.1   Pathophysiology of Epilepsy

In this section we discuss our current understanding of the pathophysiology and mechanisms mediating the development of epilepsy. The use of antiepileptics and drug-resistance issues that have grown substantially over the past decade provide opportunities for the discovery and development of more efficacious antiepileptic and anti-epileptogenic drugs.[63,64] It is clear that all current medications for epilepsy fail to control seizures in 20–30% of patients. Historically, preclinical models and clinical trial designs may have limited progress in the discovery of better treatments.[63,64] Today, antiepileptic drug development has become more focused through the identification and application of tools for new target-driven approaches, and through comparative preclinical proof-of-concept studies and innovative clinical trials designs.

5.8.2   Prevalence of Epileptic Disorders

Epilepsy is a chronic brain disorder that is characterized by partial or generalized spontaneous (unprovoked) episodic epileptic seizures and is often associated with comorbidities such as anxiety and depression and increased mortality. Epilepsy is diverse, with over 15 different seizure types and over 30 epilepsy syndromes, affecting approximately 1% of the worldwide population.[63,66] It starts as a local abnormal focus of high frequency of impulses discharge from neurones in the brain.

Not all seizures are associated with convulsions and the site of the primary discharge and its spread will determine the full extent of the symptoms produced that range from brief lapse of attention to a full convulsive fit lasting several minutes, with accompanying changes in behavior and sensations. The involvement of specific brain areas will determine what behaviors manifest. For example, involvement of the motor cortex causes convulsions, whilst those emanating in the hypothalamus will cause peripheral autonomic discharge. The involvement of the reticular formation and upper brain stem will lead to loss of consciousness. Modern brain imaging techniques (MRI and PET) and advances in EEG mapping are now routinely providing the neurologists with more precise view of structural abnormalities and aiding diagnostic criteria that may lead to novel drug treatments based on more individual treatment regimes and targeted clinical trial design.[65,67]

5.8.3 Mechanism of Action of Antiepileptic Drugs

Antiepileptic drugs act through a number of different neurotransmitter mechanisms and a number of unknown mechanisms. Their main actions are as follows:

(i) reducing electrical excitability of cell membranes mainly through use-dependent block of sodium channels (*e.g.*, carbamazepine, phenytoin)

carbamazepine   **phenytoin**

(ii) enhancing GABA-mediated synaptic inhibition; at postsynaptic neurones (clonazepam)

(iii) Inhibition of GABA transaminase activity (valproate, vigabatrin)

valproate   vigabatrin

(iv) Inhibition of GABA uptake into neurones or glial cells (tiagabine, rufinamide)

tiagabine   rufinamide

(v) Inhibition of T-type calcium channels and controlling absence seizures (gabapentin, pregabalin)

gabapentin   pregabalin

Other mechanisms include: binding to SV2A proteins (levetiracetam), activation of KCNQ2 potassium channels (retgabine), perampanel (noncompetitive AMPA antagonist), inhibition of glutamate release (lamotrigine), AMPA receptor block (topiramate).[64]

5.8.4 Translational Animal Models of Seizure Disorders

Unfortunately, there are few aetiologically relevant animal models used in epilepsy today that have translational validity in humans. Below is a schematic representation of the development of animal models of epilepsy, an area sadly that has seen little investment from the pharmaceutical industry in the development of new models or drugs for this condition over the last two decades. Löscher *et al.* provided an excellent review in 2013 on this topic.[65]

5.8.5   Current Drug Treatment

The classical and newer antiepileptics (AEDs) has provided physicians and patients with more options for the treatment of many types of seizures.[63,64] No AED has been shown to prevent the development of epilepsy in patients prior to the first seizure and these drugs seem to act to symptomatically suppress seizures once they occur.[63,64]

The main antiepileptics in clinical use for the various types of seizures are briefly listed below, for more detailed reviews see the recommended reading list.[49]

> *Generalized tonic-clonic seizures*
> Carbamazepine, phenytonin and valproate, new drugs include vigabatrin, lamotrigine, topriamate, levetiracetam.
> *Partial (focal) seizure*
> Carbamazepine, valproate, clonazepam, gabapentin, pregabalin, lamotrigine, topiramate, levetircetam, zonisamide.
> *Absence seizures*
> Ethosuximide, lamotrigine, valproate (most useful when clonic/tonic coexist).
> *Myoclonic seizures and status epilepticus*
> Diazepam (iv).

5.8.6   Emerging Research Areas

There are a few new antiepileptics agents currently being evaluated in clinical trials:

(i)   GABA transporter (GAT) inhibitors.
(ii) Ganaxolone – structurally resembling endogenous neurosteriods, is a positive allosteric modulator of $GABA_A$ receptors containing the δ subunits.
(iii) Tonaberstat – gap-channel modulator.

5.8.7   Summary and Future Directions

It is abundantly clear that the treatment of seizure disorders is in need of new strategies for the discovery and development of novel anticonvulsants as it has become the forgotten child of the pharmaceutical industry, and alternative investment models need to be pursued in order to provide new and improved treatment options for patients with seizure disorders. With recent advances in the understanding of neurocircuitry and optogentic investigation to unmask the molecular and cellular events leading to subpopulations of seizure disorders and a focus on novel target-driven approaches for the discovery, it is hoped that more efficacious and better-tolerated antiepileptogenic drugs will become available.[66,68]

A major incentive for the industry is to adopt an alternative investment models/partnership strategy for valid and druggable targets, targeting translational preclinical proof-of-concept studies, disease and target-related biomarkers, diagnostic methodology for the identification of the specific patient subpopulations, and innovative clinical trial designs.[66] Fortunately, various initiatives from major public and private funding bodies in the United States and Europe have recently stimulated a focus on further identification of these tools, and this has led to new concerted efforts between academia and industry. This holds great potential for the revitalization of anticonvulsant drug discovery and development, in the goal to free children and adults from these seizure disorders.

5.9   Sedatives and Hypnotics

5.9.1   Sedatives

A sedative drug decreases CNS activity, diminishes neuronal excitement and calms the patient. The effects of sedatives are reversible and the subject can readily be aroused. A hypnotic induces drowsiness and facilitates the onset of sleep, from which the patient can be aroused. Sedation, hypnosis and general anesthesia may be regarded as a continuum of the depths of CNS depression that are representative of different mechanisms of action for each class of drug. In general anaesthesia, the patient loses the ability to feel pain until the effect of the drug wears off, unlike hypnotics and sedatives where the perception of pain is retained.

5.9.2   Drugs Used as Sedatives

(i) Ethanol
    Ethanol ingestion results in an initial stimulation followed by a general depression of inhibitory controls mechanisms occurs.

It has a complex pharmacology causing an enhancement of both GABA- and glycine-mediated inhibition, activation of certain types of K^+ channels and inhibition voltage-gated Ca^{2+} channels, inotropic glutamate receptors and adenosine transport. Only in extreme cases of ethanol intoxication does general anesthesia occur.

(ii) Barbiturates

The use of barbiturates as sedatives and hypnotics has diminished clinically, as most clinicians consider that, for sedation, the benzodiazepines are preferable from the standpoint of therapeutic efficacy and safety. Phenobarbital is still useful in emergency situations for the treatment of convulsions and relief of the symptoms of epilepsy. Short-acting barbiturates such as thiopental are still used in general anaesthesia (see Section 5.11).

(iii) Benzodiazepines

Benzodiazepines have a powerful sedative, anxiolytic, and anticonvulsant effect, selectively potentiating GABA on GABA$_A$ receptors. Depending on the subunit composition of the receptor they bind to an accessory site "benzodiazepine-receptor" on the GABA$_A$ receptor (Figure 5.8), thereby augmenting the agonist effect of GABA at its receptor.[69,70]

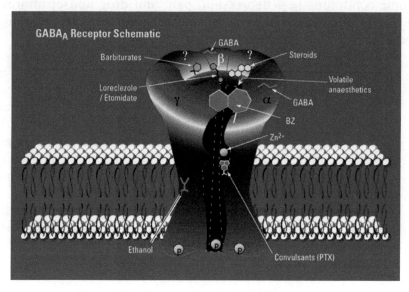

Figure 5.8 Drugs acting at the GABA$_A$ receptor.

The discovery of $GABA_A$ receptor subtypes and their profound regional differences in location provides the current new target for GABA acting anxiolytics; subtype-selective drugs. Currently, eleven native $GABA_A$ receptors are classed as conclusively identified (*i.e.*, $\alpha1\beta2\gamma2$, $\alpha1\beta\gamma2$, $\alpha3\beta\gamma2$, $\alpha4\beta\gamma2$, $\alpha4\beta2\delta$, $\alpha4\beta3\delta$, $\alpha5\beta\gamma2$, $\alpha6\beta\gamma2$, $\alpha6\beta2\delta$, $\alpha6\beta3\delta$ and ρ) with further receptor isoforms occurring with high probability.[71] The best evidence points to the α_1 and α_3 $GABA_A$ subtype mediating the anxiolytic effects of benzodiazepines so a number of agents with binding selectivity to these have been subunit synthesized (*e.g.*, midazolam, lorazepam, alprazolam, zolpidem).[71,72] It is clear that our understanding of precise subunit selectivity in terms of efficacy and reduced side effects for novel agents acting at the $GABA_A$ ion-channel receptor is awaiting synthesis of NCEs to test the hypothesis. Recently, compounds with $\alpha2/\alpha3$-selectivity $GABA_A$ receptor positive allosteric modulation have shown clinical efficacy in treating GAD that is differentiated from existing mechanisms (primarily benzodiazepines and SSRIs and SNRIs).[73]

5.9.3 Drugs Used to Treat Sleep Disorders (Hypnotic Drugs)

Whether hypnotics induce a natural sleep is still highly debatable.[70,74] Drugs used to treat insomnia following emotional trauma, underlying chronic conditions (*e.g.* depression, pain and drug abuse) improve sleep patterns and may alleviate the underlying conditions. The drugs used to treat these conditions are:

(i) Benzodiazepines. This class of drugs potentiate the action of GABA at $GABA_A$ receptors and are used to induce sleep. The short-acting benzodiazepines (*e.g.* lorazepam and temazepam) are used for treating insomnia as they produce little or no hangover effect in patients. On the other hand, diazepam and older benzodiazepines are longer acting and used to treat insomnia associated with anxiety disorders but are not recommended for long-term use and dependence liability.

The benzodiazepine so-called "Z" drugs are a special class of benzodiazepine-like $GABA_A$ positive allosteric modulators (PAMs), which have been developed for insomnia (see below).

| **Zolpidem** | **Zopiclone** | **Eszopiclone** |

The impetus for their development was to provide a faster acting and shorter half-life hypnotic, thereby providing better efficacy and less "hang-over" than the benzodiazepines in use up to the 1990s. One of these, zolpidem, was designed based upon simple binding assays to be preferentially effective at the cortical subtype of the benzodiazepine receptor.[74] Later molecular cloning studies revealed that this was due to its exceptionally high selectivity for the α_1 subtype, which is heavily expressed in the cortex. Subsequent research showed that the first generation of Z-drugs (zolpidem, zopiclone and zaleplon) were chemically distinct and all relatively α_1-selective, whereas the newest agent in this class, eszopiclone (the dextrorotary stereoisomer of zopiclone), shows relative selectivity for other receptor subtypes.[74] This probably explains why these drugs cause less tolerance and withdrawal than the benzodiazepines and eszopiclone can be used for up to a year without loss of efficacy.

(ii) Chlormethiazole. Is a positive allosteric modulator of GABA$_A$ receptors acting at a site distinct from the benzodiazapines.

(iii) Melatonin Receptor Agonists: *Tasimelteon* is a specific MT1 and MT2 receptor agonists developed for the treatment of circadian rhythm sleep disorders.

(iv) Orexin receptor antagonists: Suvorexant (see below) is now licensed to treat insomnia and acts on OX$_1$ and OX$_2$ receptors that mediates the action of the neuropeptides orexins important in in modulating diurnal rhythm. The neuropeptides are high in daylight and low at night, thus the orexin antagonists, suvorexant reduces wakefulness.[75]

Suvorexant

(v) Antihistamines: Diphenhydramine and promethazine are over-the-counter drugs used to induce sleep.

(vi) Antidepressants: Doxepin and other related antidepressants with histamine H_1- and H_2-receptor antagonist properties.

(vii) Discontinued drugs: Chloral hydrate, meprobamate and methaqualone are of little clinical significance today.

5.10 Psychostimulants and Psychotomimetic

5.10.1 Psychostimulants

Psychostimulant drugs produce a temporary increase in psychomotor activity or a temporary improvement in physical functions or mental processes or both. The most commonly used psychostimulants are caffeine, nicotine, amphetamines, methamphetamine, ecstasy, cocaine and methylphenidate.[76–78] Mild psychostimulants, such as caffeine, are used throughout the day to improve alertness, concentration, energy levels and performance in energy drinks and coffee.[77] The so-called "legal highs" (*e.g.*, derivatives of amphetamines) have also gained notoriety and are now banned in some countries. Except for caffeine, nicotine and some over-the-counter drugs, the majority of the psychostimulants are regulated by the government and classified as addictive.[76] This means they are only available legally through a physician's prescription for conditions such as obesity, attention deficit disorder/hyperactivity and narcolepsy. Psychostimulants are nevertheless widely abused illegally.[79] Psychostimulants can have negative long-term effects on a person's health because they raise blood pressure, increase heart beat and "speed up" other bodily functions and cause addiction.[79] When the drug wears off, the person experiences a "crash" with symptoms such as depression and fatigue, leading them to take more drugs.[79] Another problem is that users build up a tolerance for these drugs and have to take a higher dosage to get the same effects. People who are chemically dependent on psychostimulants and suffer from mental disorders, especially delusions about their abilities, partly due to sleep deprivation. Common side effects of psychostimulants can be nervousness, restlessness, headache, dry mouth, difficulty falling asleep or staying asleep, uncontrollable shaking of a part of the body and digestive problems such as diarrhea, nausea, vomiting, constipation and loss of appetite. Allergic reactions to these drugs, with sometimes life-threatening reactions such as hallucinations, seizures, rashes, chest pain, mania and shortness of breath occur in some people. Also, people with undiagnosed heart conditions have died suddenly after taking psychostimulants.[76,79]

5.10.2 Psychotomimetics

Psychotomimetics (so-called psychedelic or hallucinogenic drugs) affect thought perception and mood without causing psychomotor stimulation or depression.[76,79] Thoughts and perceptions are altered and distorted with dream-like qualities and mood changes that are more complex than heightened euphoria or depression. Drugs in this category are often called hallucinogenics, but some drugs fail to induce full hallucinations within this class of drugs. Also, a number psychotomimetics are nonaddicting, including lysergic acid diethylamide (LSD), psilocin and mescaline. For more details see ref. 80,81.

LSD

5.11 Anaesthetics

5.11.1 Introduction

The first clinical use of an inhalation anaesthetic (ether) was discovered in 1842 from which time most surgical operations became practically possible. In this section the main agents in current use are described, which fall into two categories: general anaesthetics and inhalation agents (gases and volatile liquids). Detailed information on the clinical pharmacology and use of anaesthetic agents can be found in the recommended reading list and specialized text books.[49,82–84]

5.11.2   General Anaesthetics

General anaesthetics render the patient unaware of and unresponsive to painful stimuli. They are given systemically either by inhalation or by intravenous injection and their main site of action is in the central nervous system.

Local anaesthetics reversibly block the action potentials involved in nerve impulse conduction at both sensory and motor neurones. Most drugs are given by parental administration, usually by subcutaneous administration or directly into the tissue or an organ.

5.11.3   Stages of Anaesthesia

The clinical criteria for assessment of anaesthesia induced by slow-acting agents as blood concentration increases, is briefly outlined below:

1. *Stage 1: Analgesia.* A mild depression of the higher centers in the cerebral cortex. The patient is conscious but drowsy and their response to painful stimuli is reduced.
2. *Stage II: Excitement.* Depression of motor centers in the cerebral cortex. The patient is unconscious and does not respond in a reflex manner to nonpainful stimuli but does respond to painful stimuli. The patient may be incoherent, hold breath, move and vomit. This stage is regarded as potentially dangerous and modern anaesthetic techniques and strategies have overcome these clinical issues.
3. *Stage III: Surgical anaesthesia.* This is the point of surgical intervention. Spontaneous movement ceases, and respiration becomes irregular. This stage is further subdivided into four planes, representing increasing depth of anaesthesia.
4. *Stage IV: Medullary paralysis.* Vasomotor control is lost. Depression of the CNS reaches the respiratory control apparatus in the medulla oblongata. Efferent impulses to the respiratory muscles (diaphragm and intercostal) stop. Respiration ceases and death ensues rapidly.

5.11.4   General Anaesthetics

(i) Inhalation Anaesthetics

Many of the early general inhalation anaesthetics; chloroform, diethyl ether cyclopropane, xenon and nitrous oxide are gases or low boiling point liquids and are the archetypal structurally nonspecific drugs. The molecular composition and configuration seem relatively unimportant but their physical chemical properties are critical. Their lipophilicity and hydrophobic properties, with the exception of nitrous oxide, are all of little clinical use today. The physiochemical properties of inhalation anaesthetics make them ideal gases: their solubility in different media is expressed as partition coefficients, defined as the ration of the concentration of the agent in two phases at equilibrium.[82,83]

The main factors that therefore determine the speed of induction of anaesthesia and recovery depend on the following physiochemical properties of the agent:

Properties of the anaesthetic:

– blood : gas partition coefficient (*i.e.* solubility in blood);
– oil : gas Overton and Meyer partition coefficient (*i.e.* solubility in fat).

Physiological factors:

– lung alveolar ventilation rate;
– cardiac output.

(ii) Current Inhalation Anaesthetics

The advent of less hazardous inflammable and explosive inhalation anaesthetics like halothane and polyfluorinated ethers enflurane, desflurane and isoflurane, has led to a more flexible control over the depth of anaesthesia during surgery and are often used in combination with nitrous oxide. The advantages of these agents is rapid induction and recovery, in most cases.[82,85] However, agents with a low blood : gas partition coefficients produce rapid induction and recovery (*e.g.* nitrous oxide, desflurane); agents with high blood : gas partition coefficients show slow induction and recovery. Agents with a high lipid solubility accumulate gradually in the body fat and may produce a hangover effect if used for a long operation. Other halogenated anaesthetics as shown below (halothane, desflurane and isoflurane). Halothane and methoxyflurane are no longer used due to safety issues with their metabolites on long-term use.[85,86]

halothane   desflurane   isoflurane

5.11.5   Mechanism of Action

The plethora of potential sites of action of anaesthetics is listed below and more detailed information is given in the recommended reading list.[49] Only in the last decade have advances in molecular mechanism identified effects of anaesthetics on neuronal cell membrane lipids and proteins and other multiple mechanisms often unique to specific anesthetics. Historically, their affects were deemed nonspecific due to molecular shape or electronic configurations of molecules.[83,85]

It is clear that many simple, unreactive compounds produce general anaesthesia, the most unlikely being the inert gas xenon. The Overton–Myers correlation of anaesthetic potency with lipid solubility and not with chemical structure has stood the test of time. Along with earlier theories of anaesthesia that postulated interaction with lipid membrane bilayer.[84,85] However, recent work favors interaction with membrane ion channels. This is apparently the case for most anaesthetics that enhance the activity of inhibitory $GABA_A$ receptors and other cys-loop ligand-gated ion channels (Figure 5.8). Meanwhile other important ion-channel effects occur following activation of a subfamily of potassium channels (the two-pore domain K^+ channels) and inhibition of excitatory NMDA receptors. Comprehensive reviews of the molecular and cellular actions of general anaesthetics can be found in the recommended reading list. Specific mechanisms of action for a number of intravenous anaesthetics will be discussed in the next section.[82,83,85]

- protein interactions;
- lipid interactions;
- ion channels;
- $GABA_A$ receptors – subtype selective;
- voltage-gated potassium channels;
- voltage-gated calcium channels;
- voltage-gated sodium channels;
- n-methyl-D-aspartate (NMDA) and glycine receptors;

- neuronal nicotinic receptors;
- hyperpolarization-activated cyclic nucleotide (HCN) or pacemaker channels.

5.11.6 Intravenous Anaesthetics

The so-called rapid onset inhalation anaesthetics all require a few minutes to act and cause a period of excitement (Stage II) before anaesthesia is induced. Intravenous anaesthetics act more rapidly, producing unconsciousness following CNS distribution from the site of injection in about 20 s. These drugs are often used to induce anaesthesia in conjunction with inhalation anaesthetics such as nitrous oxide. The structures shown below are of the more popular intravenous anaesthetics; propofol, thiopental and etomidate.[86]

propofol   thiopental   etomidate

The mechanism of action for these drugs and other intravenous anaesthetics are discussed briefly below, a more detailed pharmacology is given in the various reviews in the recommended reading list.[49,82,84–86]

- *Propofol* – Rapid onset, metabolism and recovery. The drug can be used to maintain anaesthesia during surgery.
- *Thiopental* – A high lipid soluble barbiturate, with a short duration of action (5 min), that is slowly metabolized and liable to accumulate in the body causing a prolonged effect if given repeatedly.
- *Etomidate* – Similar to thiopental but more rapidly metabolized. Less CVS depression but high incidence of nausea and may cause voluntary movements on induction.
- *Ketamine* – An analog of phencyclidine that inhibits NMDA-type glutamate receptors. The drug produces a "dissociative" slow-onset anaesthesia with high incidence of dysphoria, hallucinations, insensitive to pain and amnesic effects. It is currently being evaluated as an antidepressant.

- *Fentanyl* – The opioid is used in combination with inhalation anaesthetics. It lacks many of the undesirable cardiovascular effects of morphine and is 50–100 times more potent. It also produces a shorter duration of respiratory depression than morphine.
- *Midazolam* – Representative of several benzodiazepine drugs used in conjunction with some anaesthetics. It has a slower onset than other agents and has little respiratory or cardiovascular depression.
- *Alphaxalone* – A steroid anaesthetic used for short-term anaesthesia similar to barbiturates but with a wider safety margin. Its efficacy is *via* a neurosteroid binding site on the $GABA_A$ receptor and is primarily used in veterinary surgery (Figure 5.8).

5.11.7 Local and Topical Anaesthetics

Local anaesthetics are used to reduce sensation in a part of the body (*e.g.* skin, oral cavity) without the loss of either consciousness or control of vital functions.

They act by preventing the generation and conduction of a sensory (afferent) nerve impulse. It is widely accepted that the local anaesthetics act on specific binding sites within voltage-gated Na_v channels to block sensory transmission in a similar manner to inhalation anaesthetics. Many local anaesthetics (procaine, benzocaine and lignocaine) are esters or amides that contain a highly lipophilic moiety and a basic amino group that is often at the terminus of a side chain or, less commonly, part of a saturated heterocyclic ring (see the structures below).

| procaine | benzocaine | lignocaine |

In contrast to general anaesthetics, local anaesthetics are devoid of the physiological consequences of the former, *e.g.*, amnesia, and impairment of cardiac and respiratory function.[84] Mostly, they are formulated as water-soluble salts for parental administration. Local anaesthetic activity is pH dependent and it increases in an alkaline environment where the relative amount of uncharged amino groups is large.[87,88] The evidence suggests that the local anaesthetics molecules basic amino acid moiety takes on a proton from its environment and regains its cationic nature before binding to its Na_v ion-channel receptor.[89] Benzocaine is the exception to the rule, as it is poorly soluble, with extremely weak unprotonated basic amino groups under physiological conditions. Clinically, it is used as an over-the-counter drug because of its low toxicity, slow dissipation from the site of application due to its low water solubility, which makes it ideal for local application to wounds and ulcerated surfaces.[86,87]

5.12 Neurochemistry and Physiology of Neurotransmitters in the Central Nervous System

Chapter 4 has described nerve impulse transmission, phenomenon occurring at synapses and peripheral organs, and the roles of neurotransmitters in the autonomic nervous system. Similar neurotransmitter mechanisms and neurophysiological events also occur in the brain but with far greater complexity and interplay between several other nonpeptide and peptide neurotransmitter substances. With the development of early spectrophotoflurometry techniques advances in nuclear magnetic resonance spectroscopy (NMR) and the emergence of optigenetics in mapping brain neurotransmitter networks and several other applications offers great opportunities in advancing our knowledge of neuroscience and brain disorders (see the excellent review by Deisseroth, 2012).[68] This section will now review two classical neurotransmitters in the brain, 5-HT and DA, and their role in the pathophysiology of a number of brain disorders and their impact on drug discovery (for greater detail and other brain neurotransmitters systems the reader is referred to earlier sections and the references for several excellent reviews, and to other chapters of this book).

5.12.1 Serotonin (5-Hydroxytryptamine; 5-HT): Receptors and Therapeutic Targets

Seven-transmembrane G-protein-coupled and ion-channel receptors, through which 5-hydroxytryptamine (5-HT; serotonin) produces its wealth of physiological and pathological effects, have been the subject of intense investigation over the past 50 years.[90] Advances have been made since the pioneering days of isolated organ studies by Gaddum and Picarelli[91] and radioligand studies by Peroutka and Synder;[92] more recently, new data have been developed using molecular techniques, cloning, and *in situ* hybridization, and a superfamily of 5-HT receptor subtypes has emerged.[93,94] There now exist up to seven receptor classes, with isoforms existing for many of the receptor subtypes in the central and peripheral nervous system. To date, the success of the pharmaceutical industry in identifying 5-HT subtype-selective compounds has been remarkable in providing a tool-kit of compounds and drugs to further characterize these receptors; this has led to major clinical advances in migraine, psychiatric, neurological, and gastrointestinal disorders, with yet more clinical opportunities awaiting investigation.[95,96] This section focuses on the 5-HT receptor family members and their localization, pharmacological properties, functional role, and major contribution to clinical practice.[97,98]

Since its discovery in 1948, studies of 5-hydroxytryptamine (5-HT; serotonin) have provided considerable insight into the metabolism and physiological functions of this monoaminergic neurotransmitter in the peripheral and central nervous system (CNS).[97,99] Considerable interest in the biosynthesis, metabolism, and receptor pharmacology of 5-HT has elevated this simple indolealkylamine (see the structure below) into one of the most researched neurotransmitters of the twentieth century. The synthesis of 5-HT begins with the essential amino acid tryptophan, which can be found in foods such as bananas, turkey, and milk. The enzyme tryptophan hydroxylase hydroxylates tryptophan's benzene ring, to form 5-hydroxytryptophan. Aromatic amino acid decarboxylase then decarboxylates 5-hydroxytryptophan to produce 5-HT.[97,99] 5-HT is formed in nerve terminals, where it is held in storage granules; when nerve impulses release 5-HT into the synaptic cleft, it acts on the postsynaptic neuron to affect the process of neurotransmission across the synapse. Released 5-HT is inactivated primarily by an active serotonin reuptake transporter (SERT) process that moves 5-HT back into the neuron. Once inside the 5-HT neuron, the neurotransmitter is either repackaged into storage granules or degraded metabolically by monoamine oxidase (MAO). Pre- and postsynaptic 5-HT receptors exist on the 5-HT nerve terminals and cell bodies. Activation of these receptors leads to a disruption in the synthesis and release of serotonin. The role of the presynaptic autoreceptor is to modulate the concentration of serotonin in the synaptic cleft, thereby reducing further release and synthesis of 5-HT (see the sections "The 5-HT$_{1A(Gi/Go)}$ receptor" and "The 5-HT$_{1B(Gi/Go)}$ receptor").[94,98]

NH₂

HO

5-Hydroxytryptamine

It is probable that early work on the synthesis, neurochemistry, and distribution of 5-HT stimulated the expectation that 5-HT might be important in anxiety, depression, and schizophrenia.[6] There is increasing awareness of the 5-HT receptor subtypes associated with the multitude of disorders treated by drugs modulating 5-HT function; migraine, chemotherapy-induced emesis, depression, anxiety, schizophrenia, and irritable bowel syndrome are examples of such disorders. The promise of further therapeutic opportunities emerging awaits the development of more selective agents (combination agents) and the possibility of discovery of further novel 5-HT receptor subtypes.[97,99]

5.12.2   Evidence for Multiple Serotonin Receptor Genes

The classification and nomenclature of 5-HT receptors in the past 10 years has emerged from criteria established by the International Union of Basic and Clinical Pharmacology (IUPHAR) Subcommittee for the Classification and Nomenclature of 5-HT receptors Guide to Receptors and Channels (GRAC) 2015.[98,139]

To date, 13 distinct subtypes of human serotonin receptor are recognized on the basis of structural, transductional, and operational characteristics. A fourteenth purported receptor subtype (5-HT$_{5B}$) has been identified in rodents only (as shown in Figure 5.8).

5.12.3   G-protein-coupled 5-HT$_{(1-7)}$ Receptor Family

5-HT$_1$ serotonin receptors are expressed in a variety of neurons in the peripheral and central nervous systems. The 5-HT$_1$ receptor families – 5-HT1A, 5-HT1B (formerly termed 5-HT1Dβ), 5-HT1D (formerly 5-HT1Dα), 5-HT1E, and 5-HT1F – are all seven-transmembrane (7TM) G-protein-coupled receptors (*via* G$_i$ or G$_o$; see nomenclature in Figure 5.9). They are all encoded by intronless genes (of approximately 365–421 amino acids) and are negatively coupled to adenylyl cyclase, and principally cause hyperpolarization.[98]

(i) The 5-HT$_{1A(Gi/Go)}$ Receptor
The most studied of the 5-HT$_1$ receptor family is the 5-HT$_{1A}$
receptor, located on human chromosome HTR1/5q11.2–q13,
which encodes 421 amino acids. The receptor is found largely
on neuronal soma and dendrites and epithelial cells, where it
is found on both apical and basolateral membranes. Release
of 5-HT in the CNS is under the control of a 5-HT$_{1A}$ autoreceptor (and 5-HT$_{1B}$ autoreceptors, see later).[95,98] These
autoreceptors fall into two categories: cell-body autoreceptors
and terminal autoreceptors. The former inhibit 5-HT release
through inhibition of cell firing; the latter act through direct
inhibition of release at the terminal. Cell-body (or somatodendritic) autoreceptors belong to the 5-HT$_{1A}$ receptor subtype in all species studied so far. *In vivo* immunocytochemical
experiments have clearly demonstrated that, in striatal medium spiny neurons, the 5-HT$_{1A}$ receptor is restricted to the
somatodendritic level, while 5-HT$_{1B}$ receptors are shipped
exclusively toward axon terminals. In all systems examined to
date, there appears to be a differential sorting of the 5-HT$_{1A}$
and 5-HT$_{1B}$ receptors.[100,101] This observation is largely based
on an *in vivo* transgenic system and appears to represent
the only model that reconstitutes proper sorting of these
receptors.[103]

5-HT$_{1A}$ receptors in the raphe nuclei act as somatodendritic
autoreceptors and have little effect on extraneuronal 5-HT
levels alone, but potentiate the increase seen after selective
serotonin reuptake inhibitor (SSRI) administration. It is this
effect that has been purported to be the mechanism of action
of SSRIs, whereby desensitization of the 5-HT$_{1A}$ autoreceptors
may underlie the ability of chronic, but not acute, SSRI administration to increase synaptic cleft levels of 5-HT.[101]

Evidence from *in vivo* studies indicates that activation of
the postsynaptic 5-HT$_{1A}$ receptor in the rat results in a characteristic 5-HT syndrome consisting of flat body posture,
forepaw treading and head weaving, hypothermia, and ad-
renocorticotropic hormone (ACTH) release. Stimulation of
postsynaptic 5-HT$_{1A}$ receptors may also cause anxiogenic-like
responses. In contrast, activation of presynaptic 5-HT$_{1A}$ receptors induces both hyperphagia and anxiolytic-like effects
in rats and hence may account for the clinical anxiolytic efficacy of the 5-HT$_{1A}$ receptor agonists, buspirone and gepirone. 5-HT$_{1A}$ receptor agonists are also active in animal models

of depression, such as the rat forced swimming and mouse-tail suspension tests, consistent with their purported anti-depressant efficacy.[95,101]

Several agonists show selectivity for the 5-HT$_{1A}$ receptor, particularly $R(+)$-8-hydroxy-(di-n-propylamino) tetralin (8-OH-DPAT, U92016A, and $R(+)$-UH301, which act as high-efficacy agonists in most systems. In contrast, the nonbenzodiazepine anxiolytics (buspirone and gepirone) and other ligands (such as MDL72832) have been shown to be partial agonists (low efficacy) at the 5-HT$_{1A}$ receptor. Several apparent antagonists have also been characterized, such as NAN190, BMY7387, MDL73005EF, WAY100135, $S(-)$-UH301, $S(-)$-pindolol, spiroxatrine, spiperone, SDZ216525, and NAD299 (robalzotan).[95,98] However, all have demonstrated partial agonist properties in studies of somatodendritic autoreceptor function, perhaps due to the much larger receptor reserve associated with these, as opposed to postsynaptic receptors.[101] To date, two selective high-affinity silent antagonists at this receptor have been identified; WAY100635 (pK_B = 8.7) and lecozotan (SRA-333; K_i = 1.6 nM).[95,98]

The 5-HT$_{1A}$ receptor remains an enigma; however, continued great interest in this receptor persists, based on the ability of azapirone partial agonists (*e.g.*, gepirone, isapirone, and zalospirone) to modify 5-HT function by an action on the 5-HT autoreceptor. 5-HT$_{1A}$ agonists in this class include GR127935 and WAY163426, which are reported to raise serotonin levels more rapidly than occurs with an SSRI, and that exhibit activity in chronic models of depression that are consistent with a more rapid onset. A more recent compound, PRX-00023, with mixed 5-HT$_{1A}$ receptor antagonist and sigma agonist properties entered phase II studies for depression but was found not to be significantly superior to placebo and development was stopped.[102] A high placebo response in clinical trials and poor clinical design with antidepressants and other psychotropic drugs has been the nail in the coffin of far too many novel agents.

(ii) The 5-HT$_{1B(Gi/Go)}$ Receptor

5-HT$_{1B}$ receptors are expressed throughout the mammalian central nervous system. These receptors are located on the axon terminals of both serotonergic and nonserotonergic neurons, where they act as inhibitory autoreceptors or heteroreceptors, respectively.[101,102] 5-HT$_{1B}$ receptors inhibit the release of a range of neurotransmitters, including serotonin, noradrenaline, γ-aminobutyric acid (GABA), acetylcholine, and glutamate. In rats and mice, the terminal autoreceptor is known to be a 5-HT_{1B} receptor, whereas in humans, pigs, rabbits, and guinea pigs, the terminal autoreceptor is thought to belong to the 5-HT_{1D} receptor subtype (see later). The 5-HT_{1B} receptor is located on human chromosome HTR1/6q13 and the receptor gene is largely concentrated in the basal ganglia, striatum, and frontal cortex.[98] The receptor was originally defined by its pharmacology and also because of species differences in the binding affinity of key ligands, such as the β-adrenoceptor antagonist, cyanopindolol, which was thought to exist only in rodents.[12] More recently, the amino acid sequence of the receptor has been characterized (humans, 390 amino acids; rodents, 386 amino acids) and found to be 93% identical overall and 96% identical with the transmembrane domains of the $5\text{-HT}_{1D\beta}$ receptor, a close homolog found in higher species.[98,99] This has a distribution and function similar to that of the 5HT_{1B} receptor.[98] Indeed, the differences in the pharmacology of these two homologs are now attributed to the mutation of a single amino acid. Only the nonrodent form of the receptor was previously called 5-HT_{1D}: the human 5-HT_{1B} receptor displays a pharmacology different from that of the rodent forms of the receptor due to Thr335 of the human sequence being replaced by Asn in rodent receptors in the transmembrane-spanning region. In the rodent it is possible to differentiate between the pharmacology of the 5-HT_{1B} and 5-HT_{1D} receptors, as the 5-HT ortholog in these species exhibits a different pharmacology due to an Asp123/Arg123 switch in the rodent receptor.[98,99] Thus, it has recently been agreed to classify the receptors as species homologs of the same receptor, termed $\text{h}5\text{-HT}_{1B}$ (formerly $5\text{-HT}_{1D\beta}$) and $\text{r}5\text{-HT}_{1B}$, with the h and r prefixes referring to the human and rat species, respectively.[98]

Highly selective agonists and antagonists have now been identified to facilitate characterization of this receptor. There has been accumulating evidence, however, that 5-HT_{1B} receptors modulate drug reinforcement, stress sensitivity, mood, anxiety, and aggression. The general results of a number of studies and human genetics suggest that reduced 5-HT_{1B} heteroreceptor activity may increase impulsive behaviors, whereas reduced 5-HT$_{1B}$ autoreceptor activity may have an antidepressant-like effect.[101]

A large number of studies have established the 5-HT$_{1B}$ terminal autoreceptor in controlling 5-HT release in rats, guinea pigs, and humans. The notion that a selective 5-HT$_{1B}$ antagonist might prevent 5-HT negative feedback *via* this site, thereby increasing extraneuronal 5-HT and mimicking SSRI antidepressants, awaits clinical efficacy studies.[95]

RU24969 was the first reported full agonist at the 5-HT$_{1B}$ receptor, but, in binding studies, it is only fivefold selective over the 5-HT$_{1A}$ and 5-HT$_{1D}$ receptors. Since the first reports, several other compounds with dubious selectivity have emerged, at best having five-to-tenfold selectivity in binding and functional assays (CGS 12066B, anpirtoline tri-fluoromethylphenylpiperasine (TFM), *m*-cholorophenylpiper-azine (mCPP), MK 464, BW311C90, KSF99101H, and GR46611). The most potent agonist reported is L694247 (pK_D = 10.0). None of these, however, differentiates between the 5-HT$_{1B}$ and the 5-HT$_{1D}$ receptor subtypes. More recently, selective 5-HT$_{1B}$ receptor agonists (L694247) and antagonists (GR55562, pK_B = 7.4; SB224289, pK_B = 8.5; SB236657, pK_B = 8.9; SB236057, pK_B = 8.9) have been reported, generating further excitement around this receptor subtype and promising to shed light on the therapeutic utility of such agents.[95,98,102]

The first potent and selective 5-HT$_{1B/1D}$ antagonist to be reported, GR127935 (pK_D = 9.9), decreased extraneuronal 5-HT in the brain when administered systemically to guinea pigs. This may be due to its partial agonist activity at the 5-HT$_{1B}$ receptor or to the presence of 5-HT$_{1D}$ receptors on the

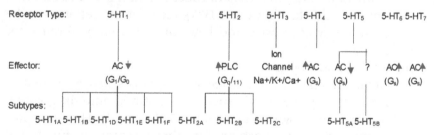

Figure 5.9 Classification of 5-HT receptor subtypes. 5-HT receptors are at present divided into seven classes, based upon their pharmacological profiles, cDNA-deduced primary sequences and signal-transduction mechanisms. With the exception of the 5-HT$_3$ receptor, which forms a ligand-gated ion channel, all 5-HT receptors containing a predicted seven-transmembrane domain structure.

raphe acting as somatodendritic autoreceptors, at which GR127935 also has agonist properties This interpretation is supported by the failure of the recently characterized selective silent 5-HT$_{1B}$ antagonist, SB224289 ($pK_B = 8.5$), >60-fold selective over 5-HT$_{1D}$ and other receptor to increase 5-HT release in the frontal cortex. NAS-181 has been reported to be the only selective antagonist of the rodent 5-HT$_{1B}$ receptor and it may add to our knowledge of the operational characteristics of the rat receptor subtype.[101]

Complementary studies in 5-HT$_{1B}$ null mutant ("knock-out") mice have proved unhelpful in understanding the pharmacology of the 5-HT$_{1B}$ receptor, as it has been suggested that a compensatory adaptation to the constitutive loss of 5-HT$_{1B}$ receptors becomes an important determinant of the altered response of 5-HT$_{1B}$ knockout mice to a variety of pharmacological challenges. Altered expression and functions of serotonin 5-HT$_{1A}$ and 5-HT$_{1B}$ receptors in knockout mice lacking the 5-HT transporter (SERT knockout mice) have also been reported.

Although the effects on 5-HT release are unclear from 5-HT$_{1B}$ knockout mice, *in vivo* studies have shown that stimulation of central postsynaptic 5-HT$_{1B}$ receptors in mice, but not rats, causes hyperlocomotion, and penile erection and hypophagia in rats are also reportedly 5-HT$_{1B}$ receptor mediated. Postsynaptic 5-HT$_{1B}$ receptor activation in another species, the guinea pig, is reported to induce hypothermia, while the hypothermic response to 5-HT$_{1B}$ agonists in the rat remains to be fully characterized. The putative 5-HT$_{1B}$ receptor agonist, anpirtoline, has analgesic, cognitive, and anti-depressant-like properties in rodents and it is of interest that mutant mice lacking the 5-HT$_{1B}$ receptor are reported to be both highly aggressive and have an increased preference for alcohol.[95,98,102]

(iii) 5-HT$_{1B}$ and Migraine

Great interest in 5-HT$_{1B}$ receptor agonists has been largely generated by the highly successful antimigraine drug sumatriptan (see below), a nonselective 5-HT$_{1D}$ and 5-HT$_{1B}$ receptor agonist with low selectivity for other receptors in functional studies. This compound may act either *via* constriction-mediating 5-HT$_{1B}$ receptors on cerebral arteries or by blocking neurogenic inflammation and nociceptive activity with trigeminovascular afferents. This latter action has been argued to be 5-HT$_{1B}$ receptor mediated, as protein extrava-sation induced by trigeminal ganglion stimulation is blocked by sumatriptan, the nonselective 5-HT$_{1B}$ receptor agonist, by CP-93129 (>100-fold selectivity over 5-HT$_{1A}$, 5-HT$_{1B}$, and 5-HT$_2$ receptors).[102]

Sumatriptan

(iv) The 5-HT$_{1D(Gi/Go)}$ Receptor

The 5-HT$_{1D}$ receptor has 63% overall structural homology with the 5-HT$_{1B}$ receptor (formerly 5-HT$_{1D\beta}$) and a 77% amino acid sequence homology in the seven-transmembrane do-mains. The receptor is located on human chromosome HR1D/1p34.3–p36.3 and contains 377 and 374 amino acids for the human and rodent gene, respectively. Low levels of the 5-HT$_{1D}$ receptor mRNA are found in the rat brain, pre-dominantly in the caudate putamen, nucleus accumbens, hippocampus, and cortex, but also in the dorsal raphe and locus coeruleus.[95,98,102]

It has been proposed that neurogenic inflammation and nociceptive activity within the trigeminovascular afferents may be 5-HT$_{1D}$ receptor mediated due to the presence of 5-HT$_{1D}$, but not 5-HT$_{1B}$, receptor mRNA in the guinea pig and human trigeminal ganglia. Thus antagonism of plasma ex-travasation, induced by electrical stimulation of the trigem-inal nerves, is observed with the antimigraine 5-HT$_{1B/1D}$ receptor agonists sumatriptan, naratriptan, rizatriptan, and zolmitriptan. Some advocates, however, claim that this is a 5-HT$_{1D}$ receptor-mediated response. Several other pharma-cological nonselective tool 5-HT$_{1B/1D}$ ligands have been used extensively to characterize the 5-HT$_{1D}$ receptor. These include sumatriptan, PU109291, L694247 (p$K_D = 10.0$), CP-122288, MK 464, and BW311C90, and the more brain penetrant agonists SKF 99101H, GR46611, and GR125743.

The location of 5-HT$_{1D}$ receptor mRNA in the raphe suggests that, like the 5-HT$_{1B}$ receptor, it too may function as a 5-HT autoreceptor. Indeed, there is electrophysiological, release, and voltammetric evidence to this effect. These data are further substantiated by the use of ketanserin and ritanserin, which, in addition their high affinity for the 5-HT_{2A} site, also have high affinity for the 5-HT_{1D}, but not for the 5-HT_{1B}, site. The antagonist SB714786 ($pK_D = 9.1$) and BRL-15572 ($pK_D = 7.9$) $5\text{-HT}_{1B/D}$ ligands mediate their effects at autoreceptors involved in the local inhibitory control of 5-HT release and may play a role in the pathogenesis of major depressive disorder (MDD) and in the antidepressant effects of the SSRIs in patients. A number of compounds in this class – including GR125743, SB224289, and L694247; AZD8129, a 5-HT_{1B} receptor antagonist; and CP-448187, a $5\text{-HT}_{1B/1D}$ receptor antagonist – a number of these agents entered phase II studies for depression, and their outcome like the 5-HT1$_A$ compounds led to further development being terminated.[95,98,102]

(v) The $5\text{-HT}_{1E(Gi/Go)}$ Receptor

The 5-HT_{1E} receptor was first characterized in humans as a [^{3}H]-5-HT binding site in the presence of 5-carboxyamidotryptamine (5-CT) that blocked binding to the 5-HT_{1A} and 5-HT_{1D} receptors. Human brain binding studies have reported that 5-HT_{1E} receptors (representing up to 60% of 5-HT_1 receptor binding) are concentrated in the caudate putamen, with lower levels in the amygdala, frontal cortex, and *globus pallidus*.[95,98] This is consistent with the observed distribution of 5-HT_{1E} mRNA. The receptor has been mapped to human chromosome HTR1E/6q14–q15, and the gene encodes a 365-amino-acid protein. There are no reported selective or high-affinity ligands for this receptor (except for 5-HT) and its function is currently unknown.[98] However, based on its distribution, one could argue that the amygdala is regarded as the "emotional engine", and dysfunction of the amygdala (and caudate putamen) has been associated with numerous psychiatric and neurological disorders, ranging from epilepsy to anxiety disorders, attention-deficit/hyperactivity disorder (ADHD), and social phobia, to Alzheimer's disease.[6] Some schools of thought have even proposed that the amygdala is involved in sociopathic and criminal behaviors. Only time will tell, with the development of selective agents for the 5-HT_{1E} receptor. To date, only one agent BRL-54443, a 5-HT_{1E} agonist ($pK_D = 8.7$) has been reported.[98]

(vi) The 5-HT$_{1F(Gi/Go)}$ Receptor

This receptor subtype is most closely related to the 5-HT$_{1E}$ receptor, with 70% sequence homology across the 7TM domains. The gene is located on chromosome HTR1F/3p11, encoding 366 amino acids across human and rodent species.[98,102] The mRNA coding for the receptor is concentrated in the dorsal raphe, hippocampus, and cortex of the rat and also in the striatum, thalamus, and hypothalamus of the mouse. 5-HT$_{1F}$ receptor mRNA has been detected in human brain and is also present in the mesentery and uterus. Sumatriptan has almost equal affinity for the 5-HT$_{1F}$ (pK_i = 7.6) and 5-HT$_{1B/1D}$ receptors (pK_i = 8.4 and 8.1, respectively). Thus, it has been hypothesized that the 5-HT$_{1F}$ receptor might be a target for drugs with antimigraine properties. 5-HT$_{1F}$ mRNA has been detected in the trigeminal ganglia, the stimulation of which leads to plasma extravasation in the dura, a component of neurogenic inflammation that is thought to be a possible cause of migraine. The first 5-HT$_{1F}$ receptor selective agonist, LY334370 (pK_D = 9.4) with >100-fold separation over the 5-HT$_{1B/1D}$ receptors, has been claimed to block the effects of trigeminal nerve stimulation, as does sumatriptan, naratriptan, rizatriptan, and zolmitriptan, but with no vascular effects.[102,104] LY334370 has also been used successfully as a radioligand (K_D = 0.5 nM) and demonstrated a reasonable correlation between the receptor protein and mRNA distribution, with the highest binding in the cortical areas, striatum, hippocampus, and olfactory bulb. LY334370 was withdrawn from development due to animal toxicity and antimigranaine studies are still ongoing with a follow-up compound lasmiditan, recently licensed by Lilly.[102,104]

(vii) The 5-HT$_{2(Gq11)}$ Receptor Family

5-HT$_2$ receptors are characterized by having a relatively lower affinity for indolealkylamines, including 5-HT, and are linked to the G$_q$/phospholipase C pathway of signal transduction. The structural organization of the 5-HT$_2$ receptor has largely been determined by mutagenesis analyses that have identified a number of primary binding residues located mainly on TMs 3, 5, and 6, which are believed to constitute the purported binding pocket. Other residues identified on TMs 2, 3, 6, and 7 have been implicated in receptor activation and G-protein coupling.[98] The 5-HT$_2$ receptor family mediates a large array of physiological and behavioral functions in humans *via* three distinct subtypes: 5-HT_{2A}, 5-HT_{2B}, and 5-HT_{2C}. While selective 5-HT_{2A} receptor antagonists have been known for some time, knowledge of the precise effects of 5-HT_{2C} and 5-HT_{2B} receptor antagonists has been hampered by the existence of only mixed receptor antagonists for 5-HT_{2A}, 5-HT_{2B}, and 5-HT_{2C}. However, selective 5-HT_{2A}, 5-HT_{2C}, and 5-HT_{2B} receptor antagonists have emerged over the past three decades. Indeed, several structural classes belonging to the various pharmacophores have been reported.[95,102]

(viii) The 5-HT_{2A} Receptor Subtype

The 5-HT_{2A} receptor subtype refers to the classical "D" 5-HT_2 receptor as defined by the neuroleptic [^{3}H]spiperone in rat brain frontal cortex.[93,102] Much of the early pharmacology of the 5-HT_{2A} receptor was pioneered using the selective ligand, [^{3}H]ketanserin, which made an important contribution to the characterization and localization of 5-HT_{2A} receptors. Since the early ligand-binding studies, only a few compounds have shown overall selectivity for the 5-HT_{2A} site. Of these compounds, MDL100097 possesses subnanomolar affinity for the 5-HT_{2A} receptor ($pK_B = 9.4$), with a minimum 300-fold selectivity for the 5-HT_{2A} site ($pK_B = 8.5\text{--}9.5$) over other receptors examined, including the 5-HT_{2C} receptor. Other less selective compounds include ICI 169369, ICI 170809, RP62203 (fananserin), and SR46349B (20- to 30-fold selective). The future introduction of more selective 5-HT_{2A} receptor antagonists will greatly accelerate the clinic potential of this receptor subtype. However, as yet, no selective agonists exist to provide a more precise classification of this receptor. Only α-methyl-5-HT has been shown to be a useful selective agonist for identifying $5\text{-HT}_{2A/2C/2B}$ receptors.[102] Other (nonselective) compounds with agonist properties at 5-HT_2 receptors include 1-(2,5-dimethoxy-4-methylphenyl)-2-amino propane (DOM) and its 4-bromophenyl (DOB) and 4-iodophenyl (DOI) congeners. DOI is the preferred radioligand, albeit that it is relatively nonselective.[98]

The actions elicited *via* 5-HT_{2A} receptors comprise mostly those previously considered to be mediated by postsynaptic "D" receptor and include vascular smooth muscle contraction in several preparations (*e.g.*, rabbit aorta, rat caudal artery, dog gastrosplenic vein); contraction of bronchial, uterine, bladder, and some gastrointestinal smooth muscle; platelet aggregation; increased capillary permeability; and some behavioral syndromes in rodents, such as head twitch, fore-paw treading, and coarse body tremors induced by 5-HT agonists.[102] In humans, hallucinations and vivid dreams have been reported to be induced by 5-HT$_{2A}$ receptor stimulation.[95] The contribution of 5-HT neurotoxins in understanding the role of 5-HT$_{2A}$ receptors and the mechanisms of action of hallucinogenic amphetamine analogs such as 3,4-methyl-enedioxymethamphetamine (MDMA) have implicated the 5-HT$_{2A/2C}$ receptors in psychiatric disorders. Several non-selective 5-HT$_{2A}$/5-HT$_{2C}$ antagonists under investigation have failed to show any clinical utility for the treatment of several psychiatric disorders, including schizophrenia, anxiety, neurotic depression, migraine, sleep disorders, and drug abuse. However, a number of compounds have been shown to have therapeutic potential in cardiovascular disorders and migraine prophylaxis.

As considered earlier, the striking similarities between 5-HT$_{2A}$ and 5-HT$_{2C}$ receptors with respect to amino acid sequence (70% in 7TM domains) holds true for the signal-transduction mechanism, as both receptors stimulate phosphatidylinositol (PI) turnover.[90] However, the receptors have distinct chromosomal localizations (HTR2A/13q14–q21 and HTR2C/Xq24). The characterization of the 5-HT$_{2A}$ receptor-coupled signal-transduction system was first studied in human platelets. It is likely that the biochemical events mediated by the 5-HT$_{2A}$ receptor are important for both the shape change and the aggregation of human platelets.[95] Controversy still exists, but a few studies have reported 5-HT-induced platelet aggregation in patients with peripheral cerebrovascular disease. Peripheral arterial vasospasms subsequent to platelet activation appear to be mediated in part by 5-HT and can be effectively reduced by ketanserin. In animal studies, post-thrombotic peripheral collateral circulation can be significantly restored by treatment with ketanserin ($pK_D = 8.5$–9.5) and MDL100907 ($pK_D = 9.4$), and 5-HT-induced reduction of blood supply through the collateral system can be effectively counteracted by pretreatment with ketanserin. In *ex vivo* human platelet studies, the 5-HT$_{2A}$/5-HT$_{2C}$/5-HT$_{2B}$ receptor antagonist ICI 170809 significantly inhibited 5-HT-induced aggregation (minimum effective dose, 0.1 mg kg^{-1} per os), thus lending support for the potential therapeutic role of 5-HT$_{2A}$ receptor antagonists in modulating the vasospastic effects of 5-HT released from platelets during vascular trauma.[102] The clinical significance of selective 5-HT$_{2A}$ (see also discussions of 5-HT$_{2B}$ and 5-HT$_{2C}$) receptor antagonists in this vascular disorder and various psychiatric disorders still awaits further conclusive evidence of clinical efficacy with selective agents. In psychiatric disorder studies, two recent 5-HT$_{2A}$ receptor antagonists have shown significant clinical efficacy in schizophrenia. Pimavanserin (ACP-103) significantly improved efficacy scores (positive and negative syndrome scale; PANSS), showing a faster onset and improved side-effect profile, following co-therapy with risperidone. Primavanserin has recently been approved for the treatment of psychosis associated with Parkinson's disease. Fananserin, which binds with high affinity to dopamine D$_4$, in addition to 5-HT$_{2A}$ receptors, showed less than expected clinically efficacy in schizophrenia, but with an improved side-effect profile. This drug is now approved for hyposexuality in women in 2015. This therapeutic target has been actively pursued by several pharmaceutical companies, and several compounds acting at 5-HT$_{2A}$ and dopamine D$_2$/D$_4$ receptors, were all at various stages of clinical development or clinical approval (*e.g.*, asenapine, blonanserin, eplivanserin, pruvanserin, paliperidone, uoperidone, and vabicaserin).[102] Other selective 5-HT$_{2A}$ inverse agonist, APD-125 (nelotanserin), showed a positive effect in the clinic for insomnia and sleep maintenance and is under investigation in Lewy body dementia (LBD) with associated REM sleep behavioral disorders. More recently, the most selective 5-HT$_{2A}$ agonist was identified as 2-(4-bromo-2,5-dimethoxyphenyl)-*N*-[(2-methoxyphenyl) methyl]ethanamine or 25B-NBOMe a derivative of the phenethylamine psychedelic 2C-B.[106,107]

(ix) The 5-HT$_{2B(Gq11)}$ Receptor

The 5-HT$_{2B}$ mRNA transcript has been now been identified in human and nonvertebrates and the receptor has been mapped to human chromosome HTR2B/2q36.3–2q37.1, with the human gene encoding 481 amino acids. It was first identified in the stomach of the rat and in the small intestine, kidney, heart, and cerebellum in the mouse. The presence of the transcript in the rat stomach and the pharmacology of the cloned 5-HT$_{2B}$ receptor have led ultimately to defining the rat stomach fundus receptor. In Gaddum and Picarelli's first classical attempt at subdivision of 5-HT receptors, the D subtype (5-HT$_{2A}$) preceded the first descriptions of the rat stomach fundus 5-HT receptor by only a few pages in the same issue of the *British Journal of Pharmacology*, whereas, the 5-HT$_{2C}$ receptor was not identified until some 37 years later.[91,108]

The 5-HT$_{2B}$ receptor represents the second member of the 5-HT$_2$ family, but it is still unclear how the clone relates pharmacologically to that 5-HT receptor that mediates contraction of the rat stomach fundus. Evidence indicates that the 5-HT$_{2B}$ contractile receptor in the rat stomach fundus is coupled to calcium influx through voltage-dependent calcium channels, intracellular calcium release, and activation of protein kinase C (PKC).[98,102] These actions may reflect a novel coupling mechanism unrelated to increases in PI hydrolysis, unlike the other 5-HT$_2$ receptors.

In the 1990s a series of potent, selective 5-HT$_{2B}$ receptor antagonists were identified, of which SB204741 (pK_i = 7.8) has proved to be an important tool for characterizing this receptor, along with the less selective 5-HT$_{2B}$ receptor agonists BW723C86 and α-Me-5-HT.[98,102] Recent additions to the 5-HT$_{2B}$ tool-kit have been the introduction of two highly selective 5-HT$_{2B}$ receptor antagonists, RS127445 (pK_i = 9.5) and EGIS-7625 (pK_i = 9.0).[98]

The clinical potential of compounds selective for the 5-HT$_{2B}$ receptor site has pointed to an alternative target for initiation of migraine and may account for the continued use of methysergide, pizotifen, and cyproheptadine for migraine prophylaxis. More selective compounds have been developed for this indication, but to date no selective 5-HT$_{2B}$ compound has shown clinical efficacy for migraine. The only 5-HT$_{2B}$ compound in clinical development for the treatment of pulmonary hypertension is PRX-08066, but no further clinical data has been reported with this compound for some time. The fact that activation of the 5-HT$_{2B}$ receptor results in heart valve damage, is a major concern with this therapeutic approach.[102]

(x) The 5-HT$_{2C(Gq11)}$ Receptor

The human 5-HT$_{2C}$ receptor gene HTR2C/Xq24 encodes 377 amino acid residues with seven hydrophobic domains of 20–30 residues.[1,109] The rat and mouse sequences suggest the presence of an eighth transmembrane region not seen in the human receptor.[90,97] The rat and human sequences are otherwise very similar, with a 90% overall homology. They are also positively coupled to phosphoinositide hydrolysis, providing a common secondary messenger system for 5-HT$_{2C}$ and 5-HT$_{2A}$ receptors.[94,98] It is well established that 5-HT$_{2C}$ receptors display a high degree of constitutive activity, and consequently inverse agonists have a large effect when they bind to the 5-HT$_{2C}$ receptor. Furthermore, the mRNA for the 5-HT$_{2C}$ receptor undergoes mRNA editing, which results in the expression of multiple 5-HT$_{2C}$ receptor isoforms. Correspondingly, INI (unedited) and VSV (a fully edited version) isoforms are abundant in rat brain. The VSV isoform lacks the high-affinity recognition site for 5-HT, which may be caused by low-efficiency coupling to G-proteins.[90,94] These edited isoforms appear to have reduced constitutive activity and reduced capacity to couple to G$_{q11}$ signaling proteins. This suggests that the INI isoform of the 5-HT$_{2C}$ receptor is pharmacologically similar to the VSV form of the 5-HT$_{2C}$ receptor, but that it couples more efficiently to G-proteins.[93,94] Importantly, the capacity of some cells in the brain to edit the 5-HT$_{2C}$ mRNA is reduced. It has been suggested that this may well be the underlying pathophysiology in schizophrenic patients and that this reduced capacity is evident in brains of patients who had committed suicide. Changes in the RNA-editing capacity of the brain will result in altered patterns of expression of the edited 5-HT$_{2C}$ isoforms and perhaps in altered responses to drugs.[93,94,109] Consequently, it is important to understand the signaling characteristics and mechanism of the edited 5-HT$_{2C}$ receptor isoforms in various psychiatric and neurological conditions, and selective agents are allowing this work to progress.

The pharmacological characteristics of the 5-HT$_{2A}$ and 5-HT$_{2C}$ (formally named 5-HT$_{1C}$) receptors are remarkably similar, with many of the "classical" 5-HT receptor antagonists showing high affinity for both sites.[93,94] Furthermore, some compounds may have different affinities for human as opposed to rat receptors. The first 5-HT$_{2C}$ selective antagonist SB200646A (50-fold) was reported in the early 1990s; however, few truly selective 5-HT$_{2C}$ receptor agonists have been identified. MK 212 (30-fold) is the most selective CNS active agonist, but is also a 5-HT$_{2A}$ agonist, while α-methyl-5-HT has equal affinity for all 5-HT$_2$ receptor subtypes. Of the various 5-HT$_2$ receptor agonists, (Ro 60-0175, lorcaserin, WAY-163909) *m*-chlorophenylpiperazine (mCPP) possess approximately tenfold selectivity over other binding sites in the rat but not in humans. However, mCPP is of particular interest because of its use as a clinical probe in the 1990s to study human 5-HT$_{2C}$ receptor function.[102]

mCPP is commonly observed to induce anxiety and panic attacks in humans. This mCPP-induced anxiogenesis is inhibited by the 5-HT$_{2C}$ and 5-HT$_{2A}$ receptor antagonist ritanserin, which is consistent with 5-HT$_{2C}$ receptor mediation. If 5-HT$_{2C}$ receptor activation is anxiogenic, the blockade of this receptor might induce anxiolysis, provided the receptors are tonically innervated. This argument is supported by the anxiolytic properties of SB243213, a selective 5-HT$_{2C}$ antagonist, in rat models of anxiety with quite different motivational and aversive bases (the rat social interaction and Geller–Seifter tests). To date this hypothesis awaits clinical testing.[102]

In support of this hypothesis, the selective 5-HT reuptake inhibitors (SSRIs) paroxetine, fluoxetine, clomipramine, sertraline, and citalopram, on chronic administration, have all been observed to desensitize behavioral responses to mCPP. The monoamine oxidase inhibitors phenelzine and nialamide, which also raise extraneuronal 5-HT levels, have a similar effect after chronic treatment. These results suggest that SSRIs and MAO inhibition antidepressant treatments may desensitize the 5-HT$_{2C}$ receptor, although there is no direct evidence from receptor binding, mRNA, or secondary messenger systems as yet.[102] Furthermore, unlike other SSRIs, fluoxetine and its metabolite norfluoxetine have weak affinity for the 5-HT$_{2C}$ receptor and are therefore likely to accumulate and directly block 5-HT$_{2C}$ sites.

Paradoxically, there is an emerging rationale and literature for the use of selective 5-HT$_{2C}$ receptor agonists in psychiatric and eating disorders that has been driven by recently reported compounds that have shown greater selectivity for the 5-HT$_{2C}$ receptor than has mCPP (lorcaserin). Early studies with mCPP pointed the way to far more selective agonists in eating and psychiatric disorders and locrcaserin is now in clinical use as an antiobesity agent.

Early clinical studies with mCPP in bulimic patients induced migraine-like headaches 8–12 h later. This response was correlated with plasma levels of mCPP and was more pronounced in patients with a personal or family history of migraine. Conversely, the nonselective $5\text{-HT}_{2C}/5\text{-HT}_{2A}$ receptor antagonists pizotifen, cyproheptadine, and methysergide are all clinically effective as migraine prophylactics. ICI 169369 and sergolexole, which have high affinity for 5-HT_{2C} and 5-HT_{2A} receptors, but little affinity for other sites, have subsequently been found to have modest antimigraine efficacy in an early trial.[102] As these drugs only share a high affinity for the 5-HT_{2C} and 5-HT_{2A} receptors, and as the selective 5-HT_{2A} receptor antagonist ketanserin is ineffective as a migraine prophylactic, it was proposed that blockade of 5-HT_{2C} receptors mediated this effect (although this has recently been challenged on the basis of their affinity for the 5-HT_{2B} receptor). Furthermore, downregulation of 5-HT_{2C} receptors may mediate the antimigraine properties of the antidepressants amitriptyline and fluoxetine; alternately, the latter compounds may act by direct antagonism of vascular 5-HT_{2C} receptor sites. Other possible therapeutic targets for drugs acting at 5-HT_{2C} receptors include eating disorders, sexual dysfunction, and head trauma, based on the actions of nonselective $5\text{-HT}_{2C/2A}$ agonists and antagonists.[95]

It is highly likely that the 5-HT_2 family will continue to grow. There is already considerable evidence for a number of other "orphan" receptors that may be adopted into the family. Several orphans are thought to exist, one of which is the endothelial 5-HT receptor, which is present in rabbit and rat jugular vein and pig pulmonary artery and vena cava. The rat jugular vein is now known to represent a 5-HT_{2B} receptor. More recently, a site identified in choroid plexus has been shown to bind the 5-HT_2-selective ligand RP62203 with high affinity, but does not possess identity with any of the three established 5-HT_2 receptors. However, it would appear that the endothelial receptors do not represent a pharmacologically homogeneous class. While possessing many of the characteristics of 5-HT_{2C} and 5-HT_{2B} receptors, subtle yet robust differences render any explicit affiliation impossible. These problems are confounded by the need, in many instances, to make comparisons across species. The potential for substantial species differences in pharmacology is highlighted on pharmacological comparison of mouse and rat homologs of the cloned 5-HT_{2B} receptor.[108] Thus, the eventual classification of the orphan endothelial 5-HT receptor awaits robust interspecies comparison of $5\text{-}HT_{2A}$, $5\text{-}HT_{2B}$, and $5\text{-}HT_{2C}$ receptors. In this regard, in rat jugular vein, the endothelial receptor, which was previously classified as $5\text{-}HT_{2C}$-like, has now been shown to be more like $5\text{-}HT_{2B}$ receptors, using ligands, which discriminate the rodent (rat) $5\text{-}HT_2$ receptor subtypes.

Another therapeutic area that has gained prominence recently arose from observations in humans that mCPP reduces total sleep time, sleep efficiency, slow-wave sleep (SWS), and rapid eye movement sleep (REMS).[102] The actions of $5\text{-}HT_{2A/2C}$ receptor antagonists on sleep architecture have since been widely studied. Thus, ritanserin has been reported to increase SWS, reduce sleep onset latency, and improve subjective sleep quality in volunteers, while REMS is reduced in some, but not all, studies. Ritanserin has also been observed to be of therapeutic use in insomniacs and in patients suffering from dysthymia. These effects were maintained with chronic treatment. Other nonselective $5\text{-}HT_{2A}/5\text{-}HT_{2C}$ receptor antagonists such as mianserin, cyproheptadine, and pizotifen have similar effects. In rats, ritanserin generally increases SWS and reduces REMS, although some studies report that the deepest phase of SWS (SWS2) is increased but total SWS is unaffected. Studies with three other $5\text{-}HT_{2A}/5\text{-}HT_{2C}$ receptor antagonists, ICI 169369 and ICI 170809 and SR46349B, report reduced REMS, but little effect on undifferentiated SWS.[102] These results, therefore, suggest that a $5\text{-}HT_{2C}$ receptor antagonist may improve sleep quality (see also agomelatine).[102]

Of great interest to pharmaceutical companies is the role of the $5\text{-}HT_{2C}$ receptor in anxiety and depression.[96] Compounds with $5\text{-}HT_{2C}$ receptor pharmacology are active in several animal models of depression, reflecting the ability of the $5\text{-}HT_{2C}$ receptor agonists and antagonists/inverse agonists to desensitize the $5\text{-}HT_{2C}$ receptor, as observed following chronic SSRI treatment. Agomelatine is a melanotinergic, MT_1 and MT_2 receptor agonist/$5\text{-}HT_{2C}$ antagonist with antidepressant properties and may be acting through a similar mechanism.[102] It is reported to improve disrupted sleep patterns in depressed patients (which also is likely, due to its $5\text{-}HT_{2C}$ receptor antagonist properties), without affecting daytime vigilance. The reported lack of effect on sexual function, tolerability problems, or discontinuation symptoms offers several advantages over current therapy for the treatment of depression.[102] It is clearly evident that therapeutic opportunities are still to be realized in the 5-HT$_2$ receptor family and will undoubtedly follow with clinical evaluation of selective, nonselective, and combination agents.[102,106]

(xi) The Ligand-Gated 5-HT$_3$ Receptor

The original naming of the "M" receptor by Gaddum and Picarelli was based on the ability of morphine to antagonize the serotonin receptor that mediates depolarization of cholinergic nerves in the guinea pig ileum.[90] Morphine was the first pharmacological tool to help characterize this receptor, now referred to as the 5-HT$_3$ receptor. Since the time of these early observations, 5-HT$_3$ receptors have been reported to be present on postganglionic autonomic neurons in the peripheral sympathetic and parasympathetic nervous systems, on enteric neurons, and on sensory neurons in various tissues.[95,110] The 5-HT$_3$ receptor is a member of the Cys–Cys loop ligand-gated ion channel superfamily and as such is composed of multiple subunits, containing 478 amino acids. Five human genes have been identified on chromosome HTR3/11, which encodes the 5-HT$_3$ subunit (5-HT$_{3A-E}$).[98] To date, conflicting reports as to the existence of central 5-HT$_{3B}$ subunits have been issued. However, immunohistochemical studies using sections of human temporal lobe demonstrated both 5-HT$_{3A}$ and 5-HT$_{3B}$ immunoreactivity associated with pyramidal neurons within all CA fields of the hippocampus, as well as with large neurons within the hilus. The expression of both h5-HT$_{3A}$ and h5-HT$_{3B}$ subunit mRNAs in human hippocampus would therefore support the presence of both homomeric h5-HT$_{3A}$ and heteromeric h-5-HT$_{3A/3B}$ receptors in human hippocampus.[110] The pathophysiological relevance of central h5-HT$_{3A}$ and h5-HT$_{3B}$ receptors remains to be fully evaluated with selective receptor antagonists in preclinical and clinical studies.[102] The pioneering animal studies in the 1980s with the nonselective antagonist ondansetron clearly point to a role for 5-HT$_3$ receptors in anxiety, depression, and schizophrenia.[95] However, this was not substantiated in human studies with several 5-HT$_3$ receptor antagonists, including, ondansetron, granisetron, BRL46470, and dolasetron.[95] 5-HT is found in both the brain and the gut, but it is now widely understood that 95% of the serotonin in the body resides in the gut. The two leading 5-HT$_3$ receptor antagonists, ondansetron and granisetron, have proved to be highly successful antiemetic agents, and they are widely used in the prophylactic treatment of chemotherapy-induced nausea and vomiting. This discovery arose from carefully reasoned experimental research. It was shown in animal studies that chemotherapeutic agents such as cisplatin released 5-HT in the gut in vast quantities, activating 5-HT$_3$ receptors in this region. Activation of 5-HT$_3$ receptors resulted in the firing of a vagally mediated response resulting in emesis. The discovery of the antiemetic properties of these 5-HT$_3$ receptor antagonists, along with several others that followed, opened the way for this class of compounds to become the gold standard in the clinical treatment of chemotherapy-induced nausea and emesis.[95] A second non-CNS therapeutic target that has gained prominence in recent years has been irritable bowel syndrome (IBS) and its association with the 5-HT$_3$ receptor and 5-HT$_4$ receptor. 5-HT$_3$ receptor antagonists were predicted to be effective in diarrhea-predominant IBS based on animal studies. Clinical studies with ondansetron and granisetron showed a reversal in rectal hypersensitivity in IBS, and the longer-acting agent alosetron showed significant clinical improvement in IBS symptoms.[102] However, alosetron has had a checkered clinical existence following early withdrawal after reports of rare serious gastrointestinal adverse events among patients taking the drug. Similar drugs (*i.e.*, cilansetron, ramosetron, and DDP-733) are also in clinical studies for IBS, with the hope that these side effects are not class related. The latter compound is a partial agonist that has recently been reported to be active in a Phase II IBS study. The complex symptomatology in IBS may be attributed to activation of other 5-HT receptor subtypes in the gastrointestinal tract. It was the identification of 5-HT$_4$ receptors that led many pharmaceutical companies to believe that agents acting as 5-HT$_4$ receptors would be useful in the treatment of this condition (see discussion of the 5-HT$_4$ receptor).[102,110]

(xii) The 5-HT$_{4(Gs)}$ Receptor

The 5-HT$_4$ receptor was originally identified in primary cell cultures of mouse embryo colliculli, where later it was found to have a broad tissue distribution and to be positively coupled to adenylate cyclase. In early functional assays studies the 5-HT$_4$ receptor was identified in the rat esophageal muscularis mucosae and guinea pig colon, where activation resulted in relaxation and contraction, respectively. The cloned human and rodent receptor was shown to consist of 388 and 406 amino acids, respectively. Various isoforms of the 5-HT$_4$ receptor have now been identified (5-HT$_{4b,c}$ and 5-HT$_{4d}$), encoded by a gene located on chromosome HTR4/5q31– q33.[98,111] In the brain, 5-HT$_4$ receptors appear to be located on neurons, whereas in the peripheral nervous system they are reported to facilitate the release of acetylcholine in smooth muscle of the intestinal tract.[111] It was these observations on the gastrointestinal tract that led to the notion that 5-HT$_4$ receptors may have a role in the pathophysiology of irritable bowel syndrome and gastroesophageal reflux disease (GERD).[102] The complex symptomatology and pharmacology of IBS led to the development of several highly selective 5-HT$_4$ receptor agonists and antagonists that are now established in clinical practice for this indication. Tegaserod (HTF-919) and purcalopride (R93877) are 5-HT$_4$ receptor partial agonists, and piboserod (SB207266) is an antagonist. All of these medications are reported to relieve the symptoms of constipation, the notion being that the 5-HT$_{3/4}$ agonists are intended to have prokinetic activity, like the mixed 5-HT$_4$ receptor agonist/5-HT$_3$ receptor agonist, cisapride. Such agents would therefore be beneficial for constipation-predominant IBS. Conversely, the antagonist piboserod would be indicated for diarrhea-predominant IBS. However, the therapeutic potential of piboserod was noted to be limited to a decrease in rectal sensitivity, and the compound was not advanced in further clinical studies. The disappointment with this class of drugs has been their lack of efficacy in IBS-related pain. However, other 5-HT$_3$ agonists (pumosetrag) and 5-HT$_4$ agonists (TD-2749 and TD-5108) are also being evaluated in IBS, which may address this important, unmet medical need. The potential CNS efficacy of 5-HT$_4$ receptor antagonists is less well developed, though the 5-HT$_4$ partial agonist SL65.0155 (5-(8-amino-7-chloro-2,3dihydro-1,4-benzodioxin-5yl)-3-[1-(2-phenylethyl)-4-piperidinyl]1,3,4-oxadiazol-2(3H)-one-monohydrochloride) has been reported to show cognitive-enhancing effects in animal models.[111] The high density of 5-HT$_4$ receptors in the nucleus accumbens has suggested that these receptors may be involved in reward mechanisms and may influence self-administration. Recent studies with GR113808 and its effects on reducing alcohol intake in rats would concur with this suggestion.[111] The clinical benefit of selective agents for the 5-HT$_4$ receptors in medical practice and the promise of more developments clearly represent one of many success stories of 5-HT drug development.[111]

(xiii) The 5-HT$_5$ Receptor (G-Protein Coupling, None Identified)

The 5-HT$_5$ receptor is highly localized in the rodent CNS; its distribution in the hippocampus (CA1, CA3, dentate gyrus), cortex, cerebellum (granular layer), olfactory bulb, habenula, and spinal cord would suggest a role in the pathophysiology of psychiatric and neurological disorders. A human 5-HT$_{5A}$, but not a 5-HT$_{5B}$, receptor has been identified; the human 5-HT$_{5A}$ receptor gene, like the rodent gene, contains two codon exons separated by a single large intron.[114] In the case of the human and rodent 5-HT$_{5A}$ receptor, the gene contains the same number of amino acids (357); similarly, the rodent 5ht$_{5b}$ gene contains 387.[98,103,104] To elucidate the pathophysiology of the 5-HT$_5$ gene, chromosomal localization studies have identified the receptor on HTR5A/7q36, whereas the 5ht$_{5b}$ gene is localized on htr5b/2q11–q13.[98] The signal-transduction mechanism for the 5-HT$_5$ receptor is still a matter for conjecture, as some suggest it may utilize a novel (perhaps ion channel) second-messenger system.[98] The physiological function of the 5-HT$_{5A}$ receptor is poorly understood. 5-HT$_{5A}$ null mutant mouse studies suggest a role in exploratory behavior, but the lack of selective pharmacological tools has hampered progress.[104,114] Localization studies have revealed that 5-HT$_{5A}$ receptors have widespread distribution in the CNS.[112,113] It has been speculated that, on the basis of their localization, the receptors may be involved in motor control, learning and memory consolidation, and anxiety and depression.[104,112,113] Recently, a selective 5-HT$_{5A}$ receptor antagonist, SB699551A, has been identified.[105] This compound has been reported to attenuate 5-carboxyamidotryptamine-induced inhibition of raphe neuronal cell firing *in vitro* and to increase 5-HT levels in prefrontal cortex *in vivo*, suggesting a role for the 5-HT$_{5A}$ receptor in the modulation of raphe 5-HT neuronal activity. This is another pharmacological opportunity whereby "tool" compounds will expose the receptor subtype operational and functional characteristics in native tissue, as it has been suggested that many of the operational characteristics of the 5-HT$_{1D}$ receptor may be attributed to the 5-HT$_{5A}$ receptor.[112,113]

(xiv) The 5-HT$_{6(Gs)}$ Receptor

5-HT$_6$ receptors have been identified in areas of the rat and human brain associated with learning and memory: hippocampus, CA1, CA3, dentate gyrus, olfactory tubercles, cerebral cortex, nucleus accumbens, and striatum.[98,103] The 5-HT$_6$ gene is localized on chromosome HTR6/1p35–p36, and studies to determine its linkage to several CNS disorders (memory dysfunction and schizophrenia) are actively being pursued.[98,115] However, the potential of 5-HT$_6$ receptor antagonists in the treatment of cognitive disorders has been an actively researched topic in recent years, and the availability of numerous tool compounds for this receptor, some of which are currently in clinical trials, will advance our understanding of the role of this receptor in cognitive process.[115–117] The notion that several atypical antipsychotic agents such as clozapine, quetiapine, and olanzapine possess high affinity for the 5-HT$_6$ receptor adds translational creditability to this approach.[106,115] The leading pharmacophores are indole-related compounds, of which several have emerged, including the recently described aryl sulfonamide indole chemotype, SB271046 (pK_i = 8.6), with excellent CNS bioavailability and *in vivo* efficacy, compared to several earlier compounds. More recently, 5-HT$_6$ receptors have been demonstrated to regulate central cholinergic neurotransmission. In this regard, administration of the 5-HT$_6$ receptor-selective antagonist Ro 04-6790 reversed scopolamine-induced rotation in 6-hydroxydopamine-lesioned rats.[115] Additionally, 5-HT$_6$ receptor antisense oligonucleotides or 5-HT$_6$ receptor-selective inhibitors enhanced retention by rats of the learned platform position in the Morris water maze.[106] These data suggest that 5-HT$_6$ receptor antagonists might boost cholinergic neurotransmission and reduce the cognitive impairments experienced by patients with dementia or schizophrenia.[107,108] Intriguingly, it has recently been determined that the 267C allele of the 5-HT$_6$ receptor is a significant risk factor for Alzheimer's disease.[26] Currently, there are a number of 5-HT$_6$ antagonists in clinical studies for cognitive disorders with no significant clinical efficacy shown to date (interpiridine (SB-74257), LU AE-58054 (indalopridine)). Taken together, these findings indicate that 5-HT$_6$ antagonists might prove useful in treating a number of common illnesses, including dementia and schizophrenia.[115,116] A major concern in drug discovery is that the cloned mouse receptor is significantly different in rat and human 5-HT$_6$ receptors. In addition to species differences in the binding of drugs to 5-HT$_6$ receptors, differences in the regional expression of 5-HT$_6$ receptors are also apparent.[115,116] Furthermore, quantitative polymerase chain reaction studies have demonstrated that the mouse 5-HT$_6$ receptor mRNA is at least tenfold less abundant than are the rat and human 5-HT$_6$ receptor mRNAs, in every brain region examined.[103,115] Surprisingly, whereas 5-HT$_6$ receptor mRNA and radioligand binding activity is enriched in the basal ganglia of rat and human brain, there is no such enrichment in the mouse brain.[98,115,116] Finally, using a combination of site-directed mutagenesis and molecular modeling studies, it has been shown that the peculiar mouse 5-HT$_6$ pharmacology is attributable to two amino acids – Tyr188 (in helix 5, which is Phe188 in rats and humans) and Ser290 (in helix 6, which is Asn290 in rats and humans) – and these account for the bulk of the differences in pharmacology.[116] These findings are a cautionary note but may have important implications for drug discovery.[115–117]

(xv) The 5-HT$_{7(Gs)}$ Receptor

Since the identification of the 5-HT$_7$ receptor in the early 1990s, evidence has accumulated supporting a role for 5-HT$_7$ receptors in various physiological functions, including sleep disorders, circadian rhythms, cognition, mood, and sleep.[118,119] A number of splice variants of the 5-HT$_7$ receptor have been identified and exhibit overlapping mRNA distributions, and all couple to G$_s$, though no differences in operational characteristics have been shown, which may have led to differentiation of behavioral phenotypes.[118] Work is still ongoing to elucidate the pathophysiology of these splice variants, and chromosomal localization studies have identified the receptor on HTR5A/7q36, encoding 445 and 488 amino acids for the human and rodent receptor, respectively.[98] The finding that several drugs used in clinical psychiatric practice had high affinity for 5-HT$_7$ receptor binding sites in the brain initiated an ongoing interest in evaluating the involvement of neuropsychiatric disorders. The distribution of 5-HT$_7$ receptors in areas of the brain associated with neuropsychiatric disorders – hippocampus (CA$_1$, CA$_2$), hypothalamus, thalamus, raphe nuclei, and superior colliculus – supported this notion. In concordance with its distribution in the hippocampus, the selective 5-HT$_7$ receptor antagonist SB269700 was shown to attenuate phencyclidine (PCP)-induced cognition dysfunction associated with schizophrenia.[118,119] The strongest case has been made for the importance of the 5-HT$_7$ receptor in depression. Inactivation or blockade of the receptor leads to an "antidepressant" state in behavioral models of depression. Furthermore, 5-HT$_7$ receptor downregulation following chronic administration of antidepressants in animal studies points to a possible molecular mechanism of action. This hypothesis is supported by recent animal experiments using 5-HT$_7$ receptor knockout mice that clearly show an antidepressant-like profile in the rat forced swim test and mouse tail suspension test. Selective antagonist tool compounds SB258717, SB269970, and SB656104-A have all been shown to exhibit antidepressant-like properties in animal studies and to modulate REM sleep in rats, results that concur with this line of investigation.[102,118] Several lines of evidence have shown that 5-HT-induced hypothermia is mediated by the 5-HT$_7$ receptor, as defined with selective antagonists and in 5-HT$_7$ knockout mice studies. What is interesting, however, is recent work with the endogenous fatty acid oleamide that suggests it may act through an independent mechanism as well as at an allosteric 5-HT$_7$ receptor site to regulate body temperature and perhaps circadian rhythms.[102] Clinical progress in this area has been limited by suboptimal clinical candidates, and clinical efficacy data for selective agents are awaited.[102,119,120]

5.12.4   The Future of 5-HT Research

It is now over 50 years since Gaddum first suggested "that a drug with a specific antagonistic action to 5-HT – might be used in therapeutics". Clearly, drugs acting at the 5-HT receptor subtypes have revolutionized therapeutics during the intervening years. However, it is only now that selective agents are appearing to differentiate many operational and physiological characteristics of 5-HT actions on the various receptor subtypes currently identified. The future therapeutic potential of drugs acting at these 5-HT subtypes and isoforms, or in combination with other neurotransmitter receptors, has opened up a new generation of novel serotonergic therapeutic agents as we differential the various subtypes and signal-transduction pathways mentioned above and their complex relationship between disease states.[60,67,102,121]

5.13 Dopamine and Parkinson's Disease and Huntington's Chorea

5.13.1 Dopamine: Parkinson's Disease and Huntington's Chorea

Parkinson's disease (PD) was the first brain disorder to be linked with a specific chemical deficiency – dopamine (DA).[122–124] Studies of brain structure showed degeneration in DA cell bodies in the *substantia nigra* (SN), caudate nucleus (CN), putamen (Pu) and the *globus pallidus* (Gp). These areas in the basal ganglia are interconnected by nerve fibers and work in harmony in the control and the initiation of movement.[122] Degeneration in these brain areas accounts for many of the symptoms of PD, where the overactivity of output pathways is believed to account for the clinical symptoms (see Figure 5.10). It is generally considered in PD, that DA loss in the striatum leads to overactivity of the so-called indirect pathway, from the striatum to the *globus pallidus* (Gpi) leading to a disinhibition of the subthalmic nucleus (STN). The subsequent overactivity in the STN and consequently the *globus pallidus* (GPm), inhibits cerebral motor cortices and possibly brain stem locomotor regions.[123,124]

Meanwhile, in dyskinesia (an involuntary movement; dystonia, athetosis, and chorea are specific types of dyskinesis) following prolonged ʟ-Dopa therapy there is strong evidence that the basal ganglia output activity is reduced. It has been proposed that overactivity in the direct pathway, and under activity of the STN and GPm, appears to be the underlying pathophysiology in dyskinetic movement.[125]

5.13.2 Neurochemical Changes

The neurochemical processes associated with selective loss of dopaminergic neurons in the nigral striatal neurones is believed to involve both mitochondrial dysfunction and oxidative stress (OS) in the pathogenesis of PD. Development of an effective causal therapy should be focused on preventing or at least retarding the neurodegenerative process underlying the disease.[124]

5.13.3 Pathogenesis of Parkinson's Disease

Parkinson's disease is a synucleinopathy with widespread degeneration within the peripheral nervous system and across a variety of brain regions.[122] The best studied and understood region of neural degeneration is the dopaminergic nigrostriatal system. Striatal dopamine insufficiency and nigral neuronal loss underlie the cardinal motor symptoms of PD (see Figure 5.10). Levodopa, dopamine agonists and deep-brain stimulation (DBS) are the most potent therapies.[126] However, they have side effects that warrant the investigation of new therapies.[127] Other pharmacological approaches included gene therapy for PD, which still has great promise but is still an emerging technology.[127] Delivery issues to maximize the spread of the vector to the target area, as well as moving to a less-impaired patient population, is yet to unlock the potential of this strategy. For more detail information on emerging therapies the reader is referred to the reviews indicated in the reference list.[127,128]

Parkinson's disease (PD), is a neurodegenerative disease, and is characterized by the progressive loss of dopamine neurons and the characteristic hallmark of the accumulation of Lewy bodies and neurites. The exact role of genetic and environmental factors in the pathogenesis of PD has frequently been debated. The association of MPTP (methyl-4-phenyl-1,2,3,6-tetrahydropyridine) and toxins (such as rotenone) with parkinsonism highlights the potential etiologic role of environmental toxins in disease causation.[129] The recent discoveries of monogenic (such as α-synuclein, PARK1-8, UCHL1, PINK1, DJ-1, LRRK2) forms of PD have provided considerable insights into its pathophysiology. Parkin, an ubiquitin protein ligase assists in the degradation of toxic substrates *via* the ubiquitin proteasome system. It can also mediate a nondegradative form of ubiquitination. PINK1 and LRRK2 are possibly involved in the phosphorylation of substrates important for various cellular functions.[127,128] Recent *in vitro* and *in vivo* studies suggest that deficits in mitochondrial function, oxidative and nitrosative stress, the accumulation of aberrant or misfolded proteins, and ubiquitin-proteasome system dysfunction underpin the pathogenesis of sporadic and familial forms of PD.[130] Elucidation of the functions of the proteins encoded by the disease-causing genes will provide an opportunity for identification of specific pathways that could be targeted in neurotherapeutics.[127,128]

5.13.4 Pharmacological Treatment of Parkinson's Disease

Today, there is only symptomatic relief from Parkinson's disease, medications, surgery, and multidisciplinary management are the only course of treatment. The main families of drugs useful for treating motor symptoms are levodopa (usually combined with a dopa decarboxylase or COMT inhibitor (*e.g.* tolcapone, entacapone), which

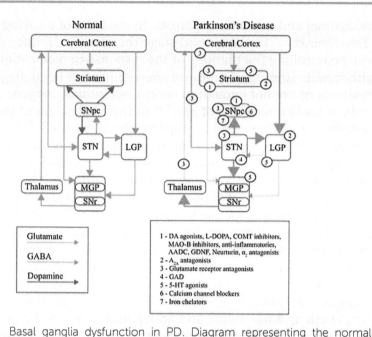

Figure 5.10   Basal ganglia dysfunction in PD. Diagram representing the normal function of the basal ganglia (left), the changes occurring in PD (right), and the site of primary action of therapeutic targets discussed in this review (numbered). Arrows represent the major neurotransmitters of glutamate (green), GABA (blue) and dopamine (red). Relative thickness of the arrows indicates level of activity of neurotransmitter. SNpc, *substantia nigra pars compacta*; SNr, *substantia nigra reticulata*; STN, *subthalamic nucleus*; MGP, *medial globus pallidus*; LGP, *lateral globus pallidus*.
From: Stayte S and Vissel B (2014) Advances in nondopaminergic treatments for Parkinson's disease. *Front. Neurosci.* **8**:113. doi: 10.3389/fnins.2014.00113. Reproduced with permission under Creative Commons Attribution License (CC BY) (https://creativecommons.org/licenses/by/3.0/).

does not cross the blood–brain barrier), dopamine agonists (ropinirole, bromocriptine, pergolide, pramipexole, apomorphine, cabergoline and lisuride). See Stayte and Vissel, 2014 (Figure 5.10) for the myriad drugs at various stages of clinical development for the treatment of PD.

The stage of the disease determines which group is most useful. Two stages are usually distinguished: an initial stage in which the individual with PD has already developed some disability for which pharmacological treatment is needed, then a second stage in which an individual develops motor complications related to levodopa usage. Treatment in the initial stage aims for an optimal tradeoff between good symptom control and side effects resulting from improvement of dopaminergic function. The start of L-Dopa treatment may be delayed by using other medications such as MAO-B inhibitors (rasagiline) and dopamine agonists, in the hope of delaying the onset of dyskinesia.[131] In the second stage the aim is to reduce symptoms while controlling fluctuations of the response to medication. Sudden withdrawals from medication or overuse have to be managed. When medications are not enough to control symptoms, surgery, and deep-brain stimulation can be of use.[126] In the final stages of the disease, palliative care is provided to improve quality of life.[128,131]

Unfortunately, L-Dopa preparations lead in the long term to the development of motor complications characterized by involuntary movements called dyskinesias and fluctuations in the response to medication.[125] When this occurs a person with PD can change from phases with good response to medication and few symptoms ("on" state), to phases with no response to medication and significant motor symptoms ("off" state).[125] For this reason, levodopa doses are kept as low as possible while maintaining functionality. Delaying the initiation of therapy with levodopa by using alternatives (dopamine agonists and MAO-B inhibitors) is common practice.[128,131] Other drugs such as amantadine and anticholinergics may be useful in the treatment of motor symptoms. Several other drugs are used to treat further complications of the disease, such as psychosis (quetiapine), daytime sleepiness (modafinil) and choilinesterase inhibitors for dementia.[127,128]

5.13.5 Summary

The future of neuroscience and drug discovery has turned a corner after decades in the doldrums and a lack of new innovative drugs. The rise of optogenetic, chemogenetic RNA-interferences technologies is now greatly influencing our understanding of traditional behavioral models and visualizing the unique spatiotemporal regional features that these tools unmask following behavioral changes induced by stress or psychotropic agents and visualizing the discrete control over neural circuits.[55–57] This is particularly so with Cre-recombinase/loxP technology to gain cell-type selective expression and thus precise excitation, inhibition, or modulation of specific circuits *in vivo*.[133,134] In a seminal paper by Tsai *et al.* (2009),[133] it was demonstrated that closed-loop control of behavioral responding was possible with *in vivo* optogenetics, such that the animal's presence in a certain context triggered phasic photostimulation of dopamine (DA) neurons to elicit place preference.[134]

Since then, numerous adaptations of closed loop, "real-time" *in vivo* optogenetic engagement of neural circuits have been used with behavioral models in Pavlovian, operant, and acute measures of valence.[135] The increased spatiotemporal control over neural circuits afforded by these modern approaches has rapidly advanced our understanding of the role of the limbic system in reward and aversion.[136–138]

Several recent studies utilizing adapted behavioral models and examining regional brain areas are advancing our knowledge of psychiatric and neurological disorders (for a comprehensive recent review of anxiety see Calhoon and Tye, 2015).[138] It is anticipated that further advances in optogenetic tools, brain microcircuitry, *in vivo* imaging, and mouse genetic models over the next decade will further expand the types of behavioral phenotypes that assess valence in the limbic system and other brain regions will pave the way for new treatments for mental health.[132,138]

References

1. World Health Report, *Mental Health: New Understanding New Hope*, World Health Organization, Geneva, Switzerland, 2001.
2. J. E. Lieberman, J. Greenhouse, J. Robert and M. Hamer, *et al.*, *Neuropsychopharmacology*, 2005, **30**, 445–460.
3. M. Anderson, D. J. Nutt and J. E. W. Deakin, *J. Psychopharmacol.*, 2000, **14**, 3–20.
4. G. M. J. Goodwin, *Psychopharmacology*, 2003, **17**, 149–173.
5. E. E. Duffy, W. Narrow, J. C. West, L. J. Fochtmann and D. A. Kahn, *et al.*, *Psychiatr. Q.*, 2005, **76**, 213–230.
6. American Psychiatric Association Diagnostic and Statistical Manual of Mental Disorders: DSM-IV, APA, Washington, DC, 1994.
7. World Health Organization. *ICD-10 Classification of Mental Disorders: Diagnostic Criteria for Research*, World Health Organization, Geneva, Switzerland, 1993.
8. T. P. Blackburn and J. Wasley, Affective Disorders: Depression and Bipolar Disorders, in *Comprehensive Medicinal Chemistry*, ed. W. Moos, 2nd edn, 2007, ch. 56, vol 4, Pub. Elsevier, pp. 46–81.
9. C. J. Murray and A. D. Lopez, *Lancet*, 1997, **349**, 1498–1504.
10. P. R. Bieck and W. Z. Potter, *Annu. Rev.Pharmacol. Toxicol.*, 2005, **45**, 227–246.
11. B. L. Roth, D. J. Sheffler and W. K. Kroeze, *Nat. Rev. Drug Discovery*, 2004, **3**, 353–359.
12. J. J. Schildkraut and J. J. Mooney, Toward a rapidly acting antidepressant: the normetanephrine and extraneuronal monoamine transporter (uptake 2) hypothesis, *Am. J. Psychiatry*, 2004, **161**(5), 909–911.
13. D. Petrik, D. C. Lagace and A. J. Eisch, The neurogenesis hypothesis of affective and anxiety disorders: are we mistaking the scaffolding for the building? *Neuropharmacology*, 2012, **62**(1), 21–34.
14. P. Blier, *J. Clin. Psychiatry*, 2001, **62**, 12–17.
15. F. Artigas, *et al.*, *Trends Pharmacol. Sci.*, 2001, **22**, 224–228.
16. T. Tellioglu and D. Robertson, *Expert Rev. Mol. Med.*, 2001, **19**, 1–10.
17. P. Zill, R. Engel and T. C. Baghai, *et al.*, *Neuropsychopharmacology*, 2002, **26**, 489–493.
18. D. Hadley, M. Hoff and J. Holik, *et al.*, *Hum. Hered.*, 1995, **45**, 165–168.

19. Z. Lin, G. R. Uhl, *Pharmacogenomics*, 2003, **3**, 159–168.
20. J. R. Wendland, B. J. Martin, M. R. Kruse, K.-P. Lesch and D. L. Murphy, *Mol. Psychiatry*, 2006, **274**(3), 1–3.
21. C. Belzung and L. Maël, Criteria of validity for animal models of psychiatric disorders: focus on anxiety disorders and depression, *Biol. Mood Anxiety Disord.*, 2011, **1**, 9.
22. M. L. Wong and J. Licino, *Nat. Rev. Drug Discovery*, 2005, **3**, 136–151.
23. N. A. Andrews, M. Papakosta and N. M. Barnes, *Neurobiol. Dis.*, 2014, **61**, 72–78.
24. M.-L. Wong and J. Lucino, *Nat. Rev. Drug Discovery*, 2004, **3**, 136–151.
25. D. R. Thakker, F. Natt and D. Husken, *et al.*, *Mol. Psychiatry*, 2005, **10**, 714.
26. Z. M. Wu, C. H. Zheng and Z. H. Zhu, *et al.*, *J. Neurol. Sci.*, 2016, **360**, 133–140.
27. S.-J. Wang, H. M. Hung and R. O'Neil, *Eur. Neuropsychopharmacol.*, 2011, **21**, 159–166.
28. M. J. Millan, *Eur. J. Pharmacol.*, 2004, **500**, 371–384.
29. M. B. H. Youdim and Y. S. Bakhle, *Br. J. Pharmacol.*, 2006, S287–S296.
30. M. Weinstock, L. Luques, C. Bejar and S. Shoham, *J. Neural Transm.*, 2006, **70**, 443–446.
31. E. T. Wong, K. W. Perry and F. P. Bymaster, *Nat. Rev. Drug Discovery*, 2005, **4**, 764–774.
32. H. K. Manji, G. J. Moore and G. Rajkowska, *et al.*, *Mol. Psychaitry*, 2000, **5**, 578–593.
33. P. E. Keck, S. L. McElroy and L. M. Arnold, *Med. Clin. North Am.*, 2001, **85**, 645–661.
34. R. W. Leicht, *Acta Psychiatr. Scand.*, 1998, **97**, 387–397.
35. C. L. Bowden, G. M. Asnis and I. D. Ginsburg, *Drug Saf.*, 2004, **27**, 173–184.
36. P. L. Delgado, H. L. Miller, R. M. Salomon and J. Licinio, *et al.*, 1993, 29, 389–396.
37. S. J. Mathew, H. K. Manji and D. S. Charney, Novel Drugs and Therapeutic Targets for Severe Mood Disorders, *Neuropsychopharmacology*, 2008, **33**, 2-80-2092.
38. A. Holmes, M. Heilig and N. M. J. Rupniak, *et al.*, *Trends Pharmacol. Sci.*, 2003, **24**, 580–595.
39. T. P. Blackburn, C. J. Swanson and T. D. Wolinsky, *et al.*, *Neuropsychopharmacology*, **30**, S99.
40. P. Skolnick, *Eur. J. Pharmacol.*, 1999, **375**, 31–40.
41. J. F. Cyran and K. Kaupmann, *Trends Pharmacol. Sci.*, 2005, **26**, 36–43.
42. T. Branchek and T. P. Blackburn, *Curr. Opin. Pharmacol.*, 2003, **3**, 90–97.
43. T. D. Sotnikova, M. G. Caron and R. R. Gainetdinov, Trace amine-associated receptors as emerging therapeutic targets, *Mol. Pharmacol.*, 2009, **76**(2), 229–235.
44. D. K. Grandy, Trace amine-associated receptor 1-Family archetype or iconoclast? *Pharmacol. Ther.*, 2007, **116**(3), 355–390.
45. S. M. Stahl, *Antipsychotics and Mood Stabilizers*, Cambridge University Press, New York, 3rd edn, 2008.
46. W. T. Carpenter and J. I. Koenig, The evolution of drug development in schizophrenia: Past issues and future opportunities, *Neuropsychopharmacology*, 2008, **33**, 2061–2079.
47. J. Pratt, C. Winchester, N. Dawson and B. Morris, Advancing schizophrenia drug discovery: optimizing rodent models to bridge the translational gap, *Nat. Rev. Drug Discovery*, 2012, **11**, 560–579.
48. M. Laruelle, A. Abi-Dargham and R. Gil, *et al.*, Increased dopamine transmission in schizophrenia: relationship to illness phases, *Biol. Psychiatry*, 1999, **46**, 56–72.
49. H. P. Rang, J. M. Ritter, R. J. Flower and G. Henderson, *Rang and Dale's Pharmacology*, Pub. Elsevier, Churchill Livingstone, 8th edn, 2015, vol. 46, pp. 559–569.
50. G. Gross, M. A. Geyer, 2012. Current antipsychotics. *Handb. Exp. Pharmacol.* 212.

51. G. F. Busatto and R. W. Kerwin, Perspectives on the role of serotonergic mechanisms in the pharmacology of schizophrenia, *J. Psychopharmacol.*, 1997, **11**, 3–12.

52. J. T. Coyle, A. Basu and M. Benneyworth, *et al.*, Glutamatergic synaptic dysregulation in schizophrenia: therapeutic implications, *Handb. Exp. Pharmacol.*, 2012, **213**, 267–295.

53. D. E. Casey, Tardive dyskinesia: pathophysiology, in *Psychopharmacology*, ed. F. E. Bloom and D. J. Kupfer, A Fourth Generation of Progress, Raven Press, New York, 1995.

54. H. L. Klawans, C. M. Tanner and C. G. Goetz, Epidemiology and pathophysiology of tardive dyskinesias, *Adv. Neurol.*, 1988, **49**, 185–197.

55. H. Steeds, R. L. Carhart-Harris and J. M. Stone, Drug models of schizophrenia, *Ther. Adv. Psychopharmacol.*, 2015, **5**(1), 43–58.

56. M. A. Geyer, G. Gross (2012). Novel antischizophrenia treatments. *Handb. Exp. Pharmacology*, 213.

57. S. Ripke, B. M. Neale and A. Corvin, *et al.*, Biological insights from 108 schizophrenia-associated genetic loci, *Nature*, 2014, **511**, 421–427.

58. K. A. Aberg, Y. Liu and J. Bukszár, *et al.*, A comprehensive family-based replication study of schizophrenia genes, *JAMA Psychiatry*, 2013, **70**, 1–9.

59. P. J. Harrison, Schizophrenia: a disorder of development, *Curr. Opin. Neurobiol.*, 1997, 7, 285–289.

60. D. J. Nutt and J. Attridge, CNS drug development in Europe–past progress and future challenges, *Neurobiol. Dis.*, 2014, **61**, 6–20.

61. A. Abbott, 2010. Schizophrenia: The Drug Deadlock. [Online] Available at: www.nature.com/news/2010/101110/full/468158a.html.

62. B. A. Ellenbroek, Psychopharmacological treatment of schizophrenia: What do we have, and what could we get? *Neuropharmacology*, 2012, **62**, 1371–1380.

63. T. R. Browne and G. L. Holmes (2008). *Handbook of Epilepsy.* Lippincott Williams & Wilkins, Philadelphia.

64. H. P. Rang, J. M. Ritter, R. J. Flower and G. Henderson, *Rang and Dale's Pharmacology*, Elsevier, Churchill Livingstone, 8th edn, 2015, vol. 46, pp. 559–569.

65. W. Löscher, H. Klitgaard, R. E. Twyman and D. Schmidt, *Nat. Drug Discovery Rev.*, 2013, **12**, 757–776.

66. M. Pandolfo, Genetics of epilepsy, *Semin. Neurol.*, 2011, **31**, 506–518.

67. M. Louise, I. Paterson, B. R. Kornum and D. J. Nutt, *et al.*, 5-HT radioligands for human brain imaging with PET and SPECT, *Med. Res. Rev.*, 2013, **33**(1), 54–111.

68. K. M. Tye and K. Deisseroth, Optogenetic investigation of neural circuits underlying brain disease in animal models, *Nat. Rev. Neurosci.*, 2012, **13**, 251–266.

69. H. P. Rang, J. M. Ritter, R. J. Flower and G. Henderson, *Rang and Dale's Pharmacology*, Elsevier, Churchill Livingstone, 8th edn, 2015, vol. 46, pp. 536–545.

70. J. A. Vida, Sedatives-Hypnotics, in *Burger's Medicinal Chemistry and Drug Discovery*, ed. D. J. Abraham, Wiley-Interscience, Hoboken, N.J., 6th edn, 2003, vol. 6, pp. 223–234.

71. D. R. Burt and G. L. Kamatchi, GABAA receptor subtypes: From pharmacology to molecular biology, *FASEB J.*, 1991, **5**(14), 2916–2923.

72. K. R. Tan, U. Rudolph and C. Lüscher, Hooked on benzodiazepines: GABAA receptor subtypes and addiction, *Trends Neurosci.*, 2011, **34**, 188–197.

73. J. R. Atack, GABAA receptor subtype-selective modulators. I. alpha2/alpha3-selective agonists as non-sedating anxiolytics, *Curr. Top. Med. Chem.*, 2010, **2**, 331–360.

74. D. J. Nutt and S. M. Stahl, Searching for perfect sleep: The continuing evolution of GABAA receptor modulators as hypnotics, *J. Psychopharmacol.*, 2010, **24**(11), 1601–1612.

75. D. Hoyer and L. H. Jacobson, Orexin in sleep, addiction and more: Is the perfect insomnia drug at hand, *Neuropeptides*, 2013, **47**, 477–488.

76. H. P. Rang, J. M. Ritter, R. J. Flower and G. Henderson, *Rang and Dale's Pharmacology*, Elsevier, Churchill Livingstone, 8th edn, 2015, vol. 46, pp. 589–597.

77. B. B. Fredholm, K. Battig and J. Holmes, *et al.*, Actions of caffeine in the brain with special reference to factors that contribute to its widespread use, *Pharmacol. Rev.*, 1999, **51**, 83–133.

78. D. Repantis, P. Schlattmann, O. Laisney and I. Heuser, Modafinil and methylphenidate for neuroenhancement in healthy individuals: a systematic review, *Pharmacol. Res.*, 2010, **62**, 187–206.

79. L. L. Iversen, *Speed, Ecstasy, Ritalin. The Science of Amphetamines*, Oxford University Press, Oxford and New York, 2006.

80. A. L. Halberstadt, Recent advances in the neuropsychopharmacology of serotonergic hallucinogens, *Behav. Brain Res.*, 2015, **277**, 99–120.

81. D. E. Nichols, Hallucinogens, *Pharmacol. Ther.*, 2004, **101**, 131–181.

82. A. S. Evers, C. M. Crowder and J. R. Balser, General Anesthetics, in *Goodman and Gilman's, The Pharmacological Basis of Therapeutics*, ed. L. L. Bruton, J. S. Lazo and K. L. Parker, McGraw-Hill, New York, 11th edn, 2008, vol. C13, pp. 341–368.

83. U. Rudolph and B. Antkowiak, *Nat. Rev. Neurosci.*, 2004, **5**, 709–720.

84. F. Yanagidate and G. R. Stricharz, *Handb. Exp. Pharmacol.*, 2007, **177**, 95–127.

85. H. P. Rang, J. M. Ritter, R. J. Flower and G. Henderson, *Rang and Dale's Pharmacology*, Elsevier, Churchill Livingstone, 8th edn, 2015, vol. 46, pp. 498–508.

86. *Pharmacology for Anesthessiologists*, ed. J. P. Fee, J. G. Howard and Bovill, Talyor & Francis, Boca Raton, FL, 2005.

87. H. P. Rang, J. M. Ritter, R. J. Flower and G. Henderson, *Rang and Dale's Pharmacology*, Elsevier, Churchill Livingstone, 8th edn, 2015, vol. 46, pp. 530–535.

88. D. E. Becker and K. L. Reed, Local anesthetics: review of pharmacological considerations, *Anesth. Prog.*, 2012, **59**, 90–102.

89. D. R. Ragsdale, J. C. McPhee, T. Scheuer and W. A. Catterall, *Science*, 1994, **265**, 1724–1728.

90. S. P. H. Alexander, A. Mathie and J. A. Peters, Guide to receptors and channels, *Br. J. Pharmacol.*, 2008, 153.

91. J. H. Gaddum and Z. P. Picarelli, Two kinds of tryptamine receptors, *Br. J. Pharmacol.*, 1957, **12**, 323–328.

92. S. J. Perotka and S. H. Synder, Multiple serotonin receptors, differential binding of [H3]-5-hydroxytryptamine, [H3]-lysergic acid diethylamide and [H3]-spiroperidol, *Mol. Pharmacol.*, 1979, **16**, 687–699.

93. D. Hoyer, D. E. Clarke and J. R. Fozard, *et al.*, International Union of Pharmacology classification of receptors for 5-hydroxytryptamine (serotonin), *Pharmacol. Rev.*, 1994, **46**, 157–203.

94. D. Hoyer and G. Martin, 5-HT receptor classification and nomenclature: Towards a harmonization with the human genome, *Neuropharmacology*, 1997, **36**, 419–428.

95. N. M. Barnes and T. Sharpe, A review of central 5-HT receptors and their function, *Neuropharmacology*, 1999, **38**, 1083–1152.

96. B. J. Jones and T. P. Blackburn, The medical benefit of 5-HT, *Pharmacol., Biochem. Behav.*, 2002, **71**, 555–568.

97. Rang & Dale's PHARMACOLOGY, *5-Hydroxytryptamine and the Pharmacology of Migraine*, ed. H. P. Rang, J. M. Ritter, R. J. Flower and G. Henderson, Elsevier, Churchill Livingstone, 8th edn, 2016, ch. 2, vol. 15, pp. 197–206.

98. S. P. H. Alexander, A. P. Davenport, E. Kelly and N. Marrion, *et al.*, 2015/2016. The Concise Guide to PHARMACOLOGY 2015/16: G protein-coupled receptors. Version of Record online: 10.1111/bph.13348.

99. L. L. Brunton, J. S. Lazo and K. L. Parker, *Goodman and Gilman's, The Pharmacological Basis of Therapeutics*, McGraw Hill, New York, 11th edn, 2005.

100. H. P. R. Hartig, D. Hoyer and P. P. Humphrey, *et al.*, Alignment of receptor nomenclature with the human genome: Classification of 5-HT1B and 5-HT1D receptor subtypes, *Trends Pharmacol. Sci.*, 1996, **17**, 103–110.

101. L. Lanfumey and M. Hamon, 5-HT1 receptors, *Curr. Drug Targets: CNS Neurol. Disord.*, 2004, **3**, 1–10.

102. T. P. Blackburn, Serotonin (5-Hydroxytryptamine; 5-HT) receptor, *Encyclopedia Neurosci.*, 2009, 701–714, DOI: 10.1016/B978-008045046-9.01162-1.

103. S. J. Bonasera and L. H. Tecott, Mouse models of serotonin receptor function: Toward a genetic dissection of serotonin systems, *Pharmacol. Ther.*, 2000, **88**, 133–142.

104. D. D. Mitsikostas and P. Tfelt-Hansen, Targeting to 5-HT1F receptor subtype for migraine treatment: lessons from the past, implications for the future, *Cent. Nerv. Syst. Agents Med Chem.*, 2012, **12**(4), 241–249.

105. E. Sanders-Bush, H. Fentress and L. Hazelwood, Serotonin 5-HT2 receptors: Molecular and genomic diversity, *Mol. Interventions*, 2003, **3**, 319–330.

106. J. E. Leysen, 5-HT2 receptors, *Curr. Drug Targets: CNS Neurol. Disord.*, 2004, **3**, 11–26.

107. J. I. Juncosa Jr., M. Hansen, L. A. Bonner and J. P. Cueva, Extensive rigid analogue design maps the binding conformation of potent *N*-benzylphenethylamine 5-HT2A serotonin receptor agonist ligands, *ACS Chem. Neurosci.*, 2013, **4**(1), 96–109.

108. K. Schmuck, C. Ullmer, P. Engels and H. Lübbert, Cloning and functional characterization of the human 5-HT2B serotonin receptor, *FEBS Lett.*, 1994, **342**(1), 85–90.

109. N. J. Stam, P. Vanderheyden, C. van Alebeek and J. Klomp, *et al.*, Genomic organisation and functional expression of the gene encoding the human serotonin 5-HT2C receptor, *Eur. J. Pharmacol.*, 1994, **269**(3), 339–348.

110. N. M. Barnes, T. G. Hales, S. C. Lummis and J. A. Peters, The 5-HT3 receptor–the relationship between structure and function, *Neuropharmacology*, 2009, **56**(1), 273–284.

111. J. Brockaert, S. Claeysen and V. Compan, *et al.*, 5-HT4 receptors, *Curr. Drug Targets: CNS Neurol. Disord.*, 2004, **3**, 39–51.

112. R. A. Glennon, Higher-end serotonin receptors: 5-HT5, 5-HT6, and 5- HT7, *J. Med. Chem.*, 2003, **46**, 2795–2812.

113. D. L. Nelson, 5-HT5 receptors., *Curr. Drug Targets: CNS Neurol. Disord.*, 2004, **3**, 53–58.

114. D. R. Thomas, 5-ht5A receptors as a therapeutic target, *Pharmacol. Ther.*, 2006, **111**(3), 707–714.

115. T. A. Branchek and T. P. Blackburn, 5-HT6 receptors as emerging targets for drug discovery, *Annu. Rev. Pharmacol. Toxicol.*, 2000, **40**, 319–334.

116. X. Codony, J. M. Vela and M. J. Ramirez, 5-HT(6) receptor and cognition, *Curr. Opin. Pharmacol.*, 2011, **11**, 94–100.

117. M. L. Woolley, C. A. Marsden and K. C. Fone, 5-ht6 receptors, *Curr. Drug Targets: CNS Neurol. Disord.*, 2004, **3**, 59–79.

118. D. R. Thomas and J. J. Hagen, 5-HT7 receptors, *Curr. Drug Targets: CNS Neurol. Disord.*, 2004, **3**, 81–90.

119. M. Leopoldo, E. Lacivita, F. Berardi, R. Perrone and P. B. Hedlund, Serotonin 5-HT7 receptor agents, structure–activity relationships and potential therapeutic applications in the central nervous system disorders, *Pharmacol. Ther.*, 2011, **129**, 120–148.

120. A. Frood, Serotonin receptors offer clues to new antidepressants: Shapes of binding sites could help drug discovery and the study of consciousness, *Nat. Rev.*, 2013, **21**, DOI: 10.1038/nature.2013.12659.

121. J. Lotharius and P. Brundin, Pathogenesis of Parkinson's disease: dopamine, vesicles and α-synuclein, *Nat. Rev. Neurosci.*, 2002, **3**, 833–842.

122. Rang & Dale's PHARMACOLOGY, *Parkinson's Disease*, **40**, 4, ed. H. P. Rang, J. M. Ritter, R. J. Flower and G. Henderson, Elsevier, Churchill, Livingstone, 8th edn, 2016, pp. 493–497.

123. A. H. V. Schapira, Neurobiology and treatment of Parkinson's disease, *Trends Pharmacol. Sci.*, 2009, **30**, 41–47.

124. E. Bezard, *et al.*, Pathophysiology of levodopa-induced dyskinesia in Parkinson's disease: opportunities for novel treatments, *Nat. Rev. Neurosci.*, 2001, **2**, 577–588.

125. M. S. Okun, Deep-Brain stimulation for Parkinson's disease and alpha synuclein in Parkinson's disease, *N. Engl. J. Med.*, 2012, **367**, 1529–1538.

126. W. Poewe, P. Mahlknecht and J. Jankovic, *Curr. Opin. Neurol.*, 2012, **25**, 448–459.

127. W. Meissner, M. Frasier, T. Gasser and C. G. Goetz, *Nat. Rev. Drug Discovery*, May 2011, **10**, 377–393.

128. W. J. Langston, *Trends Neurosci.*, 1985, **8**, 79–83.

129. K. K. Chung, *Methods Mol. Biol.*, 2015, **1292**, 195–201.

130. C. W. Olanow, O. Rascol and R. Hauser, *et al.*, *N. Engl. J. Med.*, 2009, **139**, 1268–1278.

131. A. H. V. Schapira, C. W. Olanow, J. T. Greenamyre and E. Bezard, (2014). The Lancet, pub. Online, June 19th, http://dx.doi.org/10.1016/S0140-6736(14)61010-2.

132. D. Atasoy, Y. Aponte, H. H. Su and S. M. Sternson, *J. Neurosci.*, 2008, **28**, 7025–7030.

133. H.-C. Tsai, F. Zhang, A. Adamantidis, G. D. Stuber and A. Bonci, *et al.*, *Science*, 2009, **324**, 1080–1084.

134. P. Namburi, R. Al-Hasani and G. G. Calhoon, *et al.*, *Neuropsychopharmacology*, 2016, **41**, 1697–1715.

135. R. Al-Hasani, J. G. McCall, G. Shin and A. M. Gomez, *et al.*, *Neuron*, 2015, **87**, 1063–1077.

136. E. H. Nieh, S.-Y. Kim, P. Namburi and K. M. Tye, *Brain Res.*, 2013, **1511**, 73–92.

137. K. M. Tye and K. Deisseroth, Optogenetic investigation of neural circuits underlying brain disease in animal models, *Nat. Rev. Neurosci.*, 2012, **13**, 251–266.

138. G. G. Calhoon and K. M. Tye, Resolving the neural circuits of anxiety., *Nat. Neurosci.*, 2015, **18**, 1394–1404.

139. http:. http://www.iuphar-db.org – IUPHAR Receptor Database.

6 Actions of Drugs on the Cardiovascular and Renal System

Thomas P. Blackburn

Translational Pharmacology BioVentures LLC, PO Box 3126, 56, 14th St., Hoboken, New Jersey, NJ 03070, USA
Email: tblackburn@tpbioventures.com

6.1 Anatomy and Physiology of the Heart

Drugs acting on the cardiovascular system may produce their primary effects on the heart or the vessels or both. To fully cover the complex integration of the cardiovascular system and the homeostatic processes involved is beyond the scope of this section (further reading can be found in the bibliography).[1–4] Thus, the focus of this section will be to briefly describe the main circulatory mechanisms, beginning with the heart. The human heart is a four-chambered organ, consisting of the right and left atria (auricles) and the right and left ventricles (Figure 6.1). The right auricle receives blood from all parts of the body *via* the venous system (superior *vena cava*). When the heart is filled with blood the atrial muscle contracts and forces blood through as one-way value (the tricuspid) into the lower right chamber, the right ventricle (Figure 6.1). The right ventricle now contracts and forces blood through another one-way value (the pulmonary) out into the

Pharmacology for Chemists: Drug Discovery in Context
Edited by Raymond Hill, Terry Kenakin and Tom Blackburn
© The Royal Society of Chemistry 2018
Published by the Royal Society of Chemistry, www.rsc.org

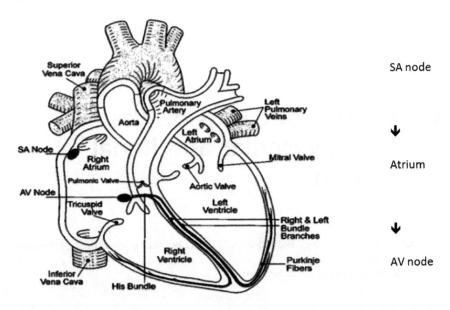

Figure 6.1 Functional anatomy and conduction/contraction impulse through the heart.

pulmonary artery, and into the right and left lungs. Once gas exchange takes place in the lung vasculature (oxygen entering the blood and carbon dioxide leaving the blood and being blown out in the expired air), newly oxygenated blood returns to the left atrium of the heart. A full left atrium now contracts and forces its contents through another one-way valve (the mitral) into the left ventricle (Figure 6.1). Then the left ventricle contracts, forcing its contents into the principal artery leaving the heart (the aorta), oxygenated blood is then delivered to all parts of the body. The left ventricle has a thicker muscular wall than the right, as it must create a hydrostatic pressure four times that produced in the right ventricle, to force blood out into the systemic circulation against a higher resistance. The cardiac output of blood from the left ventricle averages 4.2–5.6 $L min^{-1}$, and it rises to as much as 12 $L min^{-1}$ during strenuous activity.[2] Arteries carry oxygenated blood and the veins transport deoxygenated blood, with one exception the pulmonary vasculature, which carries blood between the heart and the lungs. The pulmonary vein carries oxygenated blood from the lungs to the heart and the pulmonary artery carries low-oxygen tension blood from the heart to the lungs. The heart's cardiac muscle receives blood from the coronary arteries, which form branches of the aorta.[1]

Cardiac muscle is striated like skeletal muscle, as the muscle proteins actin and myosin are arranged in sarcomeres, just as in skeletal muscle. However, its contraction is not under conscious control but is involuntary. Cardiac muscle cells usually have a single (central) nucleus. The cells are often branched, and are tightly connected by specialized junctions. The region where the ends of the cells are connected to another cell is called an intercalated disc.

The intercalated disc contains gap junctions, adhering junctions and desmosomes.

Gap junctions allow the muscle cells to be electrically coupled, so that they beat in synchrony. It has the same biochemical mechanism of contraction as skeletal and smooth muscle but differs in appearance. Cardiac muscle is unique in its ability to both participate in physiological functions and to discharge nerve impulses. This unique feature is controlled by the sinoatrial node ("SA node") found in the wall of the right atrium, where it generates nerve impulses that pass from cell to cell, travelling along the muscle cell membranes and inducing the heart to contract (Figure 6.1). The SA node or "pacemaker", therefore controls heart beats, and it is this small area, which needs to be stimulated before all of the heart muscle is excited. The heart maintains its self-stimulating properties, although the heart rate can change with various physiological stimuli – *e.g.* stress and anxiety.

A second part of the specialized conduction system of the heart is the AV node found at the top of the heart, in the so-called Koch's triangle, where it electrically connects the atrial and ventricular chambers, allowing the atria to constrict slightly before the ventricular contraction (Figure 6.1). This produces an effective pump mechanism to circulate blood around the human body (see the further reading for more detailed information on cardiac contraction and physiology).[1,2,4]

6.2 Cardiovascular Physiology and Pharmacology

The drugs discussed in this section fall into six categories and are used for their actions on the heart or other parts of the cardiovascular system, to control cardiac output and the distribution of blood throughout the cardiovascular system. The cardiovascular therapeutic areas reviewed are (1) hypertension, (2) angina, (3) cardiac arrhythmias, (4) hyperlipidaemias, (5) anticoagulants, (6) congestive heart failure (COPD) and diuretics.

6.3 Hypertension

6.3.1 Physiological Regulation of Blood Pressure

Physiological regulation of arterial blood pressure depends on two parameters:

1. *Cardiac output:* the volume of blood expelled by the heart into the arteries per unit of time. Factors determining cardiac output include the rate of heart rate and the force of the cardiac muscle contractions.

2. *Peripheral resistance:* resistance to the passage of blood through the arterial vessels. Several factors determine peripheral resistance, (i) the volume of circulating blood, (ii) the degree of dilation or contraction of blood vessel walls, (iii) the viscosity of the blood, more viscous blood requires higher pressure to circulate it, (iv) the degree of elasticity of the large arteries determines in part the rise in *systolic pressure* (when the heart is contracting) and the fall of *diastolic pressure* (between beats when the ventricles are relaxed, (v) a loss of elasticity in diseased or calcified arteries results in higher systolic and lower diastolic pressure.

Blood pressure homeostasis in normotensive individuals remains surprisingly constant given the continual, minute-to-minute large variations in heart rate and cardiac output. One of the body's homeostatic mechanisms that maintains blood pressure at nearly constant levels, using a negative feedback system, is the *baroreceptor reflex.* Baroreceptors (pressure receptors) are stretch receptors embedded in the walls of certain arteries (carotid artery and the aortic arch). These receptors, located in the aortic arch, carotid sinuses, monitor hydrostatic pressure through specialized nerve endings that are sensitive to mechanical deformation (stretching of the arterial wall). The greater the volume of blood expelled into the artery from the heart, the more the arterial wall dilates, which activates the baroreceptor, an afferent nerve impulse that travels to the brain, from which a series of efferent nerve impulses activate nerve fibers from the spinal cord to the arterial walls, signaling the smooth muscle fibers there to constrict. Thus, the *vasomotor tone* is reduced with relaxation of the muscles in the arterial walls and the arteries dilate, lowering peripheral resistance and lowering blood pressure. Heart rate and cardiac output are also lowered by other branches of the autonomic nervous system (parasympathetic and sympathetic nerves; see Section 4.1, Autonomic Nervous System), which contributes to the overall lowering of blood pressure. Baroreceptors are active even at normal blood pressures so that their activity informs the brain about both increases and decreases in blood pressure. Peripheral resistance and cardiac output also increases along with arterial blood pressure.[1,2]

Parasympathetic (cholinergic) fibers from the vagus nerve lead to the heart and stimulation of these nerves causes slowing of the heart rate. Significant sympathetic innervation of the arterial smooth muscle involves α_1 adrenoreceptors, whose stimulation produces vasoconstriction. Stimulation of β adrenoceptors in arterial smooth muscle produces vasodilation. In cardiac muscle, α adrenoceptors are of less significance. However, stimulation of the β adrenoceptors in the heart (mainly β_1) augments both *chronotropic* (heart rate) and *inotropic* (contractile force) effects. In general, stimulation of the peripheral noradrenergic system elevates blood pressure. The resulting increase in force with which the heart pumps blood and a less stretchable arterial system (caused by stimulation of α-adrenoceptors) augment the effects to raise both systolic and diastolic pressures. Parasympathetic nerves appear to have little importance in controlling vascular smooth muscle.[1,2]

However, the body's methods for overall control of blood pressure are much more complex than has been described thus far. The baroreceptor reflex and the noradrenergic and cholinergic systems act in concert with three slower-acting systems to regulate blood pressure: (i) the heart releases atrial natriuretic peptide when blood pressure is too high, and (ii) the kidneys sense and correct low blood pressure with the renin–angiotensin system (see below Figure 6.2), (iii) action of the steroidal hormone aldosterone and levels of sodium ion in the extra-cellular fluid. Increased sodium concentrations intensify the effects of released norepinephrine on the heart muscle and on the blood vessel walls, especially in borderline hypertensive patients. A low salt diet is recommended for this type of hypertensive patient.[1,2]

6.3.2 Control of Vascular Smooth Muscle

As discussed above, the renin–angiotensin system plays an important role in the physiological systems that regulate vascular tone; a second important control is the vascular endothelium. Endothelium cells release vasoactive mediators including prostacyclin (PGI_2), nitric oxide (NO) and a number of vasodilators (EDHF) and vasoconstrictors

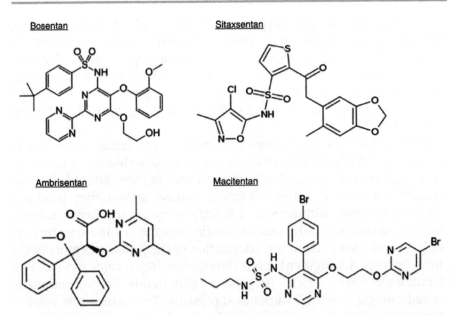

Figure 6.2 Chemical structures of clinically used endothelin receptor antagonists.

(endothelin, endoperoxide thromboxane receptor agonists). Several vasodilators (*e.g.* acetylcholine and bradykinin) act *via* endothelial NO production. The production of NO is derived from arginine, when intracellular $[Ca^{2+}]$ increases in the endothelial cell, or increased sensitivity of NO synthase to Ca^{2+}.[4]

NO causes a relaxation of smooth muscle by increasing cGMP formation.[6] The vasoconstrictor endothelin is a 21-amino acid peptide, which is produced in a variety of cell types in the body (including the endothelium, hence the name). Isolation and cloning of three isoforms of the endothelin genes (ET_1,ET_2,ET_3) identified only one in the endothelium ET_1 and was found to be the most potent vasoconstrictor substance known.[7,8] Stimulus for its release from endothelial cells in blood vessels results from many mediators released by trauma or in-flammation (*e.g.* endotoxins, cytokines). The resulting vasoconstriction prevents extensive bleeding. The pharmacology of the three isoforms of endothelins act at two G-protein-linked receptors ET_A and ET_B and similar transduction systems (Gq_{11}, G_s and Gq_{11}, G_{io}, respectively).[7,8] Among other physiological functions, endothelins participate in blood-pressure regulation. To date, a number of endothelin antagonists are used for the treatment of pulmonary hypertension (PH)[9] (Figure 6.2).

The release of ET_1 is poorly understood; it has a plasma half-life of less than 5 min, despite a much longer duration of action, and clearance occurs mainly in the lungs and kidney. It is presumed that the high concentrations are found in the extracellular space between endothelium and vascular smooth muscle, since endothelin receptor antagonists cause a vasodilation, implying tonic ET-1-mediated vasoconstrictor activity in resistance vasculature (see ref. 1 and 8 for more detailed description).

6.3.3   Clinical Categories of Hypertension

Globally, around 22% of adults aged 18 and over had raised blood pressure (SBP $\geq$ 140 or DBP $\geq$ 90) in 2014 (WHO 2014, http://apps.who.int/gho/data/node.main.A875?lang=en and BMJ Best Practices – http://bestpractice.bmj.com/best-practice/monograph/26/treatment/step-by-step.html). Two major types and one rare type of hypertension are recognized and the classification of hypertension are outlined in Table 6.1 (see also the recent references for further reading[13,14]).

6.3.4   Systemic (Primary) Hypertension

This is a common disorder and includes approximately 90–95% of all clinical cases, that if not effectively treated, increases the risk of compromising cerebral blood flow, cardiac output, kidney function and functions of the hormonal systems of the body and leading to cerebrovascular stroke, heart and kidney failure.[1] Previously termed, albeit incorrectly, essential hypertension ("essential" to maintain adequate blood perfusion), it is usually highly controllable with vasoactive drug therapy when that therapy is combined with programs of exercise, changes in dietary habits – DASH (Dietary Approaches to Stop Hypertension – https://www.ncbi.nlm.nih.gov/pubmed/27519166:) and weight loss.

Table 6.1   Classification of the various stages of hypertension See: https://www.nice.org.uk/guidance/cg127/chapter/1-guidance.[a]

Category	Blood pressure, mm Hg
Normal	SBP 90–119 and 60–79
Prehypertension	SBP 120–139 or DBP 80–89
Stage 1 HTN	SBP 140–159 or DBP 90–99
Stage 2 HTN	SBP $\geq$160 or DBP $\geq$100

[a]DBP = diastolic blood pressure; SBP = systolic blood pressure.

6.3.5   Secondary Hypertension

The remaining 5–10% of hypertensive patients are categorized as secondary hypertension, defined as high blood pressure due to an identifiable cause, such as chronic kidney disease, narrowing of the kidney arteries, hormonal imbalances, psychogenic causes, metabolic disorders and brain tumors.[4] Here, the etiology is understood, at least to some extent.[1] The most common form of secondary hypertension is *renal hypertension*, which is related to the action of a proteolytic enzyme *renin*, which is liberated into the blood stream from the kidney.[1]

Following release into the blood stream, renin acts on a specific substrate – a plasma globulin protein, *angiotensinogen (synthesised in the liver)*, which is a normal component of the blood – to release a decapeptide fragment, *angiotensin I. Angiotensin I is inactive but is converted by angiotensin converting enzyme,* which cleaves two amino acids from angiotensin I to produce angiotensin II, which is a potent vasoconstrictor. This substance increases the force of the heart beat (positive inotropic effect) and it constricts the small arteries and these combined actions increase blood pressure. Angiotensin II exerts these effects by interaction with G-protein-coupled receptors ($G_{q/11}$ and $G_{i/o}$).[7] The activated receptor stimulates phospholipase production and increases the cytosolic Ca^{2+} concentrations, which triggers cellular responses and stimulation of protein kinase C. The activated receptor also inhibits adenylate cyclase and various tyrosine kinase production.[1,7]

Angiotensin receptors are heterogeneous and four major subtypes have been characterized angiotensin AT_1–AT_4.[7] Angiotensin II is a substrate for aminopeptidase enzymes A and B, that remove single amino acid residues, giving rise, respectively, to AT_3 and AT_4. The known physiological role(s) of AT_{1-4} receptors are summarized below (Figures 6.3 and 6.4);

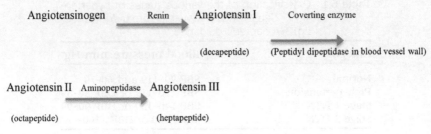

Figure 6.3   The renin–angiotensin system.

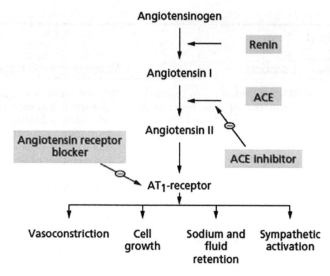

Figure 6.4 The angiotensin cascade and physiology.

The rapid onset and degradation of pressor effects of angiotensin II after the release of renin from the kidney occurs within minutes. Also, angiotensin II causes feedback inhibition of renin release. Angiotensin II is superpotent, approximately two orders of magnitude more potent than adrenaline. It is converted to a hepatpeptide, angiotensin III by an aminopeptidases as discussed above, which is also pharmacologically active but is considerably less potent than angiotensin II in elevating blood pressure. The role of angiotensin III is to stimulate release of aldosterone secretion. Further aminopeptidase enzymatic cleavage of the angiotensin III peptide chain results in the hexapeptide angiotensin IV, which is not directly involved in blood-pressure regulation, but may have other physiological functions. The actions of angiotensin II–IV at the angiotensin receptors and the renin–angiotensin system are briefly summarized in Figure 6.2 and Table 6.2. The clinical use of angiotensin converting enzyme inhibitors and angiotensin receptors blockers in the treatment of hypertension is discussed in detail in the reference list.[10]

6.3.6 Pulmonary Hypertension (PAH)

Pulmonary hypertension is a rare lung disorder in which the arteries that carry blood from the heart to the lungs become narrowed, making it difficult for blood to flow through the vessels and blood pressure in these arteries is abnormally high. In healthy, normal

Table 6.2 Angiotensin receptor subtypes and vasoconstrictor response.

Angiotensin II receptor interactions	Location	Vasoconstrictor response
AT1	Wide distribution in vascular endothelial cells, heart, brain, striated muscle and kidney.	Efferent arterioles of renal glomeruli, >noradrenaline release by inhibiting Uptake 1, reinforcing sympathetic stimulation, proximal tubular reabsorption of Na^+, increase in secretion of aldosterone from adrenal cortex and growth of cardiac and vascular cells
AT2	Expressed in fetus and neonate and distinctive brain regions in adults	Controversial effects – subtle and opposite effects to AT1 receptor stimulation (inhibition of cell growth and $<BP$), may be involved in CNS growth and development and exploratory behavior
AT3	Unknown	Still poorly characterized subtype, along with AT4
AT4	Distinctive distribution – hypothalmus	May be important in regulation of the CNS extracellular matrix, as well as modulation of oxytocin release

individuals, the pulmonary circulation is a low pressure, high flow system. PAH occurs in individuals at all ages, but is most common in young adults and leads to an overworked heart and right heart failure, which is frequently fatal. The disease is of unknown origin, progressive and has no cure. Therapeutic strategies for PAH, apart from the endothelin receptor antagonists previously discussed in Section 6.3.2, are still evolving, but current therapy and drugs are listed below:

- oxygen;
- oral anticoagulants;
- diuretics;
- digoxin;
- calcium-channel blockers;
- endothelin receptor antagonists (*e.g.* bosentan, ambrisentan, sitaxentan) for less-severe stages of the disease;
- prostanoid analogs (iloprost, treprostinil, beraprost);
- epoprostenol;

- inhaled NO is administered in intensive care;
- phosphodiesterase V inhibitor: sildenafil.

In contrast to systemic hypertension, pulmonary hypertension is also associated with other diseases (Raynaud's); more detailed information is given in the reference list.[2]

6.3.7 Advances in Therapeutic Strategies for Treating Hypertension

Current therapeutic strategies to treat hypertension rely on two possible ways of lowering blood pressure (lowering cardiac output or lowering peripheral resistance). Reducing cardiac output may result in adverse effects due to less oxygenated blood being transported to the tissues. The treatment of choice are drugs that lower peripheral resistance (by dilating the arteries), but also maintain adequate cardiac output to meet the oxygen demands of the body tissues.[1,10]

Systemic hypertension is a complex biochemical/pathophysiology event and beyond the scope of this section. The plethora and complexity of the molecular/biochemical mechanisms that regulate blood pressure depend on many factors, as discussed previously and summarized here; (i) the peripheral and central sympathetic nervous systems, (ii) the parasympathetic cholinergic system, (iii) the baroreceptor reflex, (iv) the rennin–angiotensin system, (v) aldosterone activity on sodium ion concentration in the extracellular fluid, (vi) endothelins and many more vasoactive agents.[1,4] The cascade of events involving one or more of these regulatory mechanisms and failure may lead to systemic hypertension. Thus, the daunting challenge faced by the physician in treating the hypertensive individual not knowing the underlying biochemical/physiological cause and which single or combinations of drugs gives the greater therapeutic benefit to the patient. The advent of evidence-based medicine and international regulatory guidelines (*e.g.* NICE – CG127, https://www.nice.org.uk/guidance/cg127), now aid the physician in the diagnosis and prognosis of the disease for each individual patient.

6.4 Antihypertensive Drugs

There are many classes of antihypertensives, which lower blood pressure by different pharmacological mechanism of action. Among the most important and most widely used drugs are thiazide diuretics,

Table 6.3 Evidence-based dosing for antihypertensive drugs.[a] Reprinted from The Lancet, Volume 386, N. R. Poulter, D. Prabhakaran and M. Caulfield, Hypertension, 801–812, Copyright 2015, with permission from Elsevier.

Antihypertensive medication	Initial daily dose, mg	Target dose in RCTs reviewed, mg	No. of doses per day
ACE inhibitors			
Captopril	50	150–200	2
Enalapril	5	20	1–2
Lisinopril	10	40	1
Angiotensin receptor blockers			
Eprosartan	400	600–800	1–2
Candesartan	4	12–32	1
Losartan	50	100	1–2
Valsartan	40–80	160–320	1
Irbesartan	75	300	1
β-Blockers			
Atenolol	25–50	100	1
Metoprolol	50	100–200	1–2
Calcium channel blockers			
Amlodipine	2.5	10	1
Diltiazem extended release	120–180	360	1
Nitrendipine	10	20	1–2
Thiazide-type diuretics			
Bendroflumethiazide	5	10	1
Chlorthalidone	12.5	12.5–25	1
Hydrochlorothiazide	12.5–25	25–100[b]	1–2
Indapamide	1.25	1.25–2.5	1

[a]Abbreviations: ACE, angiotensin-converting enzyme; RCT, randomized controlled trial.
[b]Current recommended evidence-based dose that balances efficacy and safety is 25–50 mg daily.

calcium channel blockers, ACE inhibitors, angiotensin II receptor antagonists (ARBs), and β-adrenorecepotor antagonists. A list of current therapeutic agents is presented in Table 6.3.

6.4.1 Angiotensin Converting Enzyme Inhibitors (ACEI)

Based on chemical knowledge of the target molecule and successful drug design the first ACEI, *Captopril* was marketed in 1981.[2] Since then several ACEI are now prescribed to treat mainly the early stages of hypertension (see Table 6.3).[1,11] ACEIs competitively inhibit the proteolytic enzyme ("converting enzyme") that catalyzes the conversion of angiotensin I into angiotensin II (Figure 6.1). *Captopril*, like other ACE inhibitors, has no effect on the pressor actions of angiotensin II;

their mode action strictly involves inhibition of formation of angiotensin II, resulting in a fall in blood pressure. Adverse effects have limited its clinical value (skin rash and temporary, reversible loss of taste sensation). Second-generation ACE inhibitors included *enalapril* and *lisinopril*. Enalapril, is a prodrug drug (a monomethyl ester that is rapidly metabolized in the liver to enalaprilat) and is a substrate for an active transport system to carry it across the intestinal wall.[2] Captopril is a classic example of a nonpeptide molecule that was designed to combine the steric properties of peptide antagonists into an orally active molecule.[2]

The second-generation drugs (Table 6.3) with a longer duration of action (once daily – OD) are now widely used in the clinic for several indications (hypertension, cardiac failure, myocardial infarction after ventricular dysfunction, ischaemic heart disease diabetic neuropathy and chronic renal insufficiency). The blood-pressure lowering mechanism(s) is due to a specific action on angiotensin-sensitive vascular beds, which include the kidney, heart and brain. This selectivity is important in sustaining the adequate perfusion pressure of these vital organs in the face of reduced perfusion pressure.[1,2] The critical exception to this is renal artery stenosis, where ACE inhibition results in a fall in glomerular perfusion rate.[1] Finally, the converting enzyme has other *in vivo* substrates in addition to angiotensin I (*e.g.* the converting enzyme accepts the nonapeptide *bradykinin* as a substrate and destroys its physiological activity). Bradykinin is a powerful vasodilator and it also stimulates prostaglandin biosynthesis. It has been proposed that some prostaglandin-related mechanisms (PGI_2, PGE_2, PGG_2, PGH_2 and thomoboxane, TP) released from the endothelium) are involved in lowering blood pressure.[1] Protection of bradykinin from metabolic destruction may be a direct or an indirect contributor to the hypotensive effect of the ACE inhibitor drugs. A persistent dry cough is one of the troublesome side effects of ACEIs and this has been attributed to kinin accumulation associated with the angioedema (painful swelling of tissues that can be life threatening if it involves the airways) that is likely to be due to an accumulation of bradykinin in the lungs. This and other life-threatening side effects discussed above should lead to an immediate discontinuation of ACEI treatment.[1,2]

6.4.2 Angiotensin II Receptor Antagonists

Following the introduction of ACEIs, angiotensin II receptor (AT_1) antagonists (ARBs) were synthesized based on the pharmacological knowledge derived from ACE inhibition. Although superficially similar in their mode of action ARBs did not cause a persistent cough, consistent with the lack of "bradykinin" accumulation as discussed above.[1] *Losartan, candesartan, valsartan and irbesarten* are all non-peptide, orally active AT_1 receptor antagonists. The first-in-class, losartan does not inhibit the converting enzyme, nor is it a partial agonist at AT_1 receptors; it is a competitive blocker at this receptor. Losartan has a rather short half-life, but the drug is metabolized to the carboxylic acid, which is also an AT_1 receptor blocker. Losartan is not categorized as a true prodrug, as both the parent compound and major metabolite are pharmacologically active.[1,2]

Finally, ACE is not the only enzyme capable of forming angiotensin II, *chymase* (which is not inhibited by the ACE inhibitors) provides an alternative route from angiotensin I to angiotensin II. The beneficial effects of this alternative pathway are not known but may be related to bradykinin/NO mediated effects.[1] As there is considerable overlap in the clinical indications for ARBs and ACEIs, it is still controversial that they share all the same therapeutic properties.

6.4.3 β-Adrenoceptor Antagonists

β-Adrenoceptor antagonists are widely used to control hypertension as well as a Class II drug for the treatment of myocardial infarction and cardiac arrhythmias (see Section 6.8.2). Typical examples of "β blockers" are *propranolol, atenolol, metoprolol, bisoprolol,* and *pindolol.* These drugs do not produce a fall in blood pressure in normotensive individuals but lower peripheral resistance and reduce cardiac output in hypertensive individuals. They work by blocking the effects of the hormone adrenaline (epinephrine). They act by reducing heart rate (beats and force) thereby reducing blood pressure. β-adrenoceptor antagonists also increase blood flow by increasing blood-vessel dilation. Selective $β_1$ beta-blockers mainly affect your heart, while nonselective $β_1/β_2$ affect the heart, lungs and blood vessels. Thus, each individual has to be carefully monitored and prescribed the right β blockers depending on the individual health and the condition being treated.[1]

Other mechanisms that may account for the hypotensive effects of β blockers include inhibition of the release of renin from the kidney, which is under noradrenergic ($β_1$ receptor) control. β blockers inhibit the release of renin, with resultant fall in angiotensin II levels and inhibition of its subsequent multiple effects on circulatory control as well as inhibition of aldosterone release. A contributory role of the central nervous system mechanisms for antihypertensive β blockers has also been suggested.[1,2] However, pharmacokinetic observations suggest that all β blockers are not alike, and the more hydrophilic the molecule, the less it is found in the brain tissue of both animals and man, although in the case of pindolol other factors may be important.[2] The clinical relevance of studies involving blood pressure or adverse effects (*e.g.* sleep disturbance) is still not clear.[12]

As well as the pharmacokinetic differences with these drugs, their selectivity for beta-adrenoceptor subtypes needs to be carefully considered, in particular nonselective drugs (β_1/β_2 – propranolol) are contraindicated in asthmatic patients in whom the β_2-adrenoceptor blocking effects constrict the bronchi and thus precipitate an asthmatic attack.[1]

The relatively selective β_1 blocker *metoprolol* is a safer alternative for asthmatic patients.[1]

6.4.4 Calcium Channel Blockers

Calcium channel antagonists block Ca^{2+} entry into voltage-gated L-type calcium channels. There are three main L-type antagonists, *verapamil, diltiazem* and dyhydropyridines (*e.g. amlodipine, nifedipine*).[1] *Amlodipine* and *nifendipine* are relatively smooth muscle selective and their vasodilators effect is mainly on resistance vessels, reducing afterload. The nondyhydropyridines act mainly on the heart and their clinical uses include dysrhythmias, to slow ventricular rate in rapid atrial fibrillation and recurrence of supraventricular tachycardia (SVT) and preventing variant angina. The clinical use of calcium antagonists for hypertension or their antidysrhythmic action, as with other therapeutic agents, should meet the clinical needs of the patient. Recent evidence-based guidelines for the management of high blood pressure provides recommendations for the management of hypertension and these recommendations are not a substitute for clinical judgment, and decisions about care must carefully consider and incorporate the clinical characteristics and circumstances of each individual patient.[13]

6.4.5 Diuretics

Therapeutically important calcium antagonists act on L-type calcium channels. There are three chemically distinct classes of L-type calcium channels; phenylalkylamines (*e.g. verapamil*), dihydropyridines (*e.g. nifedipine, amlodipine*), and benzothiazepines (*e.g.* diltiazem). All three classes bind to the α-subunit of the L-type calcium channel. The *thiazide-type diuretics* are the most commonly used drugs in hypertension (*e.g. hydrochlorothiazide,* chlorthalidone, bendroflumethiazide and indapamide, Table 6.3). They control hypertension in part by inhibiting reabsorption of sodium (Na^+) and chloride (Cl^-) ions from the distal convoluted tubules in the kidneys by blocking the thiazide-sensitive Na^+–Cl^- cotransporter (abbreviated as NCC or NCCT).[1]

Diuretics are usually administered in low doses and initially they decrease cardiac output and lower blood volume over several days, which is related to their mode of action. It has been proposed that these relatively long-acting diuretics produce a slowly developing decrease in extracellular sodium levels due to enhanced urinary excretion ("saluresis"), the likely cause of the hypotensive effect. Several calcium antagonists also show use dependence (*i.e.* they block more effectively cells in which the calcium channels are most active).[2] They also show voltage-dependent blocking action, blocking more strongly when the membrane is depolarized, causing calcium channel opening and inactivation (see also their related Class I antidysrhythmic activity).[1] The complexity and therapeutic uses of calcium channel blockers are beyond the scope of this chapter; please see the reference section for further reading. In summary, as discussed previously, diuretics are often used in combination with other hypotensive agents.[13,14] Further discussion on diuretics in cardiovascular disease can be found in Sections 6.11 and 6.12.

6.5   Vasodilator Drugs that Affect Noradrenergic Neurons

Emphasis in this section is on drugs that oppose the effects of noradrenergic (sympathetic) nerve stimulation and synthesis at peripheral and central sites. There are several specific mechanisms of hypotensive action discussed and a number of the drugs are also mentioned in Chapter 4, under the autonomic nervous system.

6.5.1   Drugs that Affect Noradrenaline Synthesis and Release

α-Methyldopa is still used as an antihypertensive agent but mainly during pregnancy. It is taken up by noradrenergic neurons where it is converted into a false transmitter α-methylnoradrenaline. As α-methylnoradrenaline is not deaminated by MAO (see Chapter 4), it accumulates, displacing noradrenaline from synaptic vesicles. It is released in the same way as noradrenaline but is less effective in causing vasoconstriction. However, it is more active at presynaptic α_2 adrenoceptors, resulting in an autoinhibitory feedback mechanism further reducing transmitter release below normal levels. This additional action combined with its central effects contribute to the drug's hypotensive action. The drug has typical side effects associated with centrally acting adrenergic drugs (*e.g.* sedation) with immune haemolytic reactions and liver toxicity reported.[1] *Reserpine* depletes catecholamine stores and blocks transmission, it is of little use today as an antihypertensive and is clinically obsolete. The drug causes side effects including depression. *Guanethidine* and *bethanidine* block transmitter release from vesicles partly by local anaesthetics effects (see Section 4.4.12), and are effective in hypertension but limited by adverse effects (postural hypertension, diarrhoea, *etc.*).[1]

6.5.2 Other Indirectly Acting Vasodilator drugs

Clonidine, an α_2-adrenoceptor agonist whose hypotensive effects are produced in part at the same central nervous system locations and is now largely clinically obsolete.[1,11]

Moxonidine, a centrally acting agonist at the imidazoline I_1 receptors, that causes less drowsiness and is licensed for mild to moderate hypertension. However, clinically there is little evidence to support its use.[1]

Minoxidil, is a prodrug for a relatively minor hepatic metabolite, the N–O–sulfate ester, which is the hypotensively active entity.[1] The drug activates K_{ATP}-modulated potassium channels in vascular smooth muscle. The resulting hyperpolarized cells, switch off voltage-gated calcium channels resulting in a relaxation of the vascular smooth muscle and a hypotensive effect. Potassium activators, like minoxidil and diazoxide work by antagonizing intracellular ATP on these channels. Minoxidil is mainly used in combination with a diuretic and β-adrenoceptor antagonist, and is sometimes effective in resistant hypertension (see Section 4.4.12). Other K-channel activators with additional properties include, Nicorandil (+NO donor activity) and Levosimendan (binds to Ca^{2+} binding protein troponin C – used in decompensated heart failure).[1,11]

6.5.3 Vasodilators that Act *via* Cyclic Nucleotides

There are many examples of drugs that relax vascular smooth muscle by increasing the cellular concentration of either cGMP or cAMP.[1]

Nesiritide(rB-type NP) and other natruiuretic peptides, NO and nitrates all act through cGMP, whereas, β-adrenoceptor agonists, adenosine, dopamine and PGI_2 increase cAMP. Phosphodiesterase inhibitors have mixed activity, PDE III (*e.g.* milrinone) increase cAMP and the PDE V inhibitors (*e.g.* sildenafil) inhibits the breakdown of cGMP, with the later compound showing efficacy in pulmonary hypertension, along with its known activity for erectile dysfunction. The methylxanthines (*e.g.* theophylline), papavarine and dipyridamole are nonselective PDE inhibitors.[1,11]

Sodium nitroprusside (Na^{2+} nitroferricynanide) is a powerful vaso-dilator that dilates small arteries and veins and has pronounced hypotensive effects.[1,11] It is metabolized in the endothelium of vascular smooth muscle to produce (NO), the pharmacologically active species.[6] NO's action in stimulating cGMP, which has been described previously (in Chapter 4), produces the powerful vasodilation.[6] However, the hypotensive effect of sodium nitroprusside is of very short duration (approximately 3 min) and its clinical use is limited due to intravenous injection in hypertensive emergencies and to produce controlled hypotension during cardiopulmonary bypass surgery. Further reading on vasodilators in cardiovascular disease see the ref. 5, 6, 11.

6.5.4 Other Vasodilators

Hydralazine, causes a relaxation of smooth muscles of the arteries and arterioles and a fall in blood pressure. This is accompanied by a reflex tachycardia and an increase in cardiac output. The molecular mechanism of this effect is to modulate the action of inositol triphosphate on Ca^{2+} release from the sarcoplasmic reticulum. It is metabolically acetylated at the terminal nitrogen of the hydrazine moiety and the metabolite is pharmacologically inactive. The rate of *N*-acetylation is genetically determined: fast acetylators require larger doses than do slow acetylators. N-acetyltransferase 2 (arylamine *N*-acetyltransferase), also known as NAT2, is an enzyme that in humans is encoded by the NAT2 gene. The drug is still used for short-term treatment of severe hypertension in pregnancy, but it frequently causes a variety of potentially serious side effects involving immunological reactions (*e.g.* lupus erythematous), which appear to be due to the production of HLA-DR antigens.[1] Finally, in heart failure patients of African origin, hydralazine is used combination with a long-acting organic nitrate.[1]

6.6 Atherosclerosis and Lipoprotein Metabolism

In this section we discuss the pathophysiological process of atherogenesis and therapeutic approaches to prevent atherosclerotic disease. Lipoprotein synthesis and transport form the basis for understanding drugs used to treat dyslipidaemia.

To cover all drugs used to treat abnormal lipoprotein metabolism is beyond the scope of this section and the reader is referred to the reference list and further reading section for more detailed information. Figure 6.5 shows a schematic representation of the role of lipid metabolism and atherosclerosis and the atheroprotective effects of HDL in the development of atherosclerotic lesions. Other diseases associated with aberrant lipid metabolism are also shown.

Atheroma is a focal disease of the intima of large and medium-sized arteries. Over time, presymptomatic lesions in the vessel wall begins to form fatty streaks or deposition of lipid material in the smooth muscle wall that eventually develops into a fibrous plaque that becomes laden with lipids (cholesterol and triglycerides) and cell debris and calcium. The blood vessels lose their elasticity due to arterial calcification and plaques, narrowing the diameter of the artery. These damaged sites in arteries due to plaque build-up are where a thrombus (blood clot) can develop.[1] These plaques are often difficult to detect with noninvasive techniques, although ultrasound of carotid vessels blood vessels is a useful method for determining reduced aortic compliance and arterial calcification from aortic pulse wave velocity. With the ensuing decrease in their elasticity due to arterial calcification and plaques narrowing the diameter of the artery, these severely damaged sites in arteries may ultimately lead to thrombus formation (blood clot) and death. Artherosclerotic lesions are induced to form in two ways: by enriching the diet in cholesterol or by repeated injury to an arterial wall. In either case the earliest lesions are focal and they occur at vessel orifices and branches, points where normal arterial shear forces are disrupted by turbulent flow conditions.

Epidemiological studies have identified numerous for atheromatous disease risk factors both familial and modifiable risk factors (*e.g.* hypertension, dyslipidemia, smoking and obesity, *etc.*), that offer therapeutic intervention or life-style changes. The therapeutic focus in recent years for the pharmaceutical industry are drugs that alter lipoprotein transport and dyslipidemia (as discussed below). The emergence of good translational animal models in transgenic mice

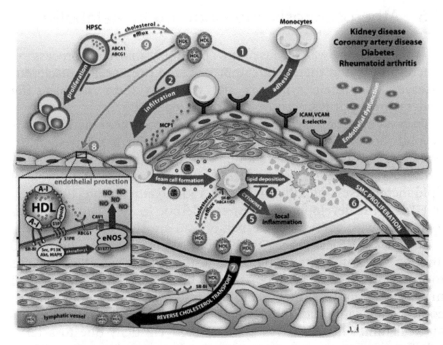

Figure 6.5 Potential atheroprotective activities of HDL in the development of atherosclerotic lesions. Inhibitory activities of HDL are shown in red and activating/promoting HDL activities in green. (1) HDL has the ability to inhibit monocyte adhesion by inhibiting vascular cell adhesion molecule (VCAM-1), intercellular cell adhesion molecule (ICAM-1), and E-selectin expression and (2) suppresses monocyte chemotaxis by inhibiting chemokine secretion, such as monocytic chemotactic protein-1 (MCP1). (3) HDL promotes cholesterol efflux from macrophage foam cells *via* the cholesterol exporter ATP-binding cassette transporter A1 (ABCA1), ABCG1 and scavenger receptor type 1 (SR-BI) thereby (4) preventing excessive cholesterol accumulation and (5) the secretion of proinflammatory cytokines from lipid-laden macrophages. (6) HDL enters the lymphatic vessels *via* SR-BI to promote reverse cholesterol transport. (7) HDL inhibits smooth muscle cells (SMC) proliferation and migration into the intima. (8) HDL binding to endothelial SR-BI and presenting HDL associated sphingosine-1-phosphate (S1P) to endothelial S1P receptors (S1PR) activates endothelial nitric oxide synthase (eNOS). In addition, HDL induced cholesterol efflux *via* ABCG1 reduces inhibitory interaction of eNOS with cavelon-1 (CAV1). (9) Activation of cholesterol efflux pathways by HDL *via* ABCA1 and ABCG1 controls the proliferation of hematopoietic stem cells (HPSC). Adapted from R. Birner-Gruenberger, M. Schittmayer, M. Holzer and G. Marche, *Progress in Lipid Research*, 2014, **56**, 36–46.[15] Creative Commons Attribution CC BY license (http://creativecommons.org/licenses/by/3.0/).

deficient in apolipoproteins or receptors has given pharmacologists new insights into lipoprotein metabolism. Such studies along with clinical investigation and progress in our current understanding of atherogenesis form human epidemiology and pathology studies have identified numerous risk factors for atheromatous disease and potential therapeutic targets (see Section 6.6.2).

6.6.1 Physiology and Pharmacology of Lipoproteins

Lipoproteins are complexes containing lipids and cholesterol that are transported in the bloodstream. They consist of a central core of hydrophobic core lipid (containing triglycerides and cholesterol). The so-called atherogenic lipoproteins are the *low-density* (LDL) *intermediate-density* (IDL) and very low-density (VLDL) lipoproteins.[15]

Each class lipoproteins has a specific role in lipid transport using different pathways;

(i) Exogenous – cholesterol and triglycerides are absorbed from the intestine and transported by chylomicrons in lymph/blood capillaries to muscle and adipose tissue.

Triglycerides are removed from the chylomicrons by a hydrolytic mechanism involving the lipoprotein lipase system, the tissues take up the resulting free fatty acids and glycerol. The chylomicron, still containing cholesteryl easters, pass to liver, bind to receptors on hepatocytes and undergo endocytosis. The remnant of the chylomicron is taken up into the liver cells; the cholesteryl esters are hydrolyzed; and the free cholesterol is in part excreted in the bile or can enter the endogenous pathway.[1,2]

(ii) Endogenous Pathway – cholesterol and newly synthesized triglycerides are transported from the liver as VLDL to muscle and adipose tissues, where triglyceride is hydrolyzed to fatty acids and glycerol that enter the circulation, as described for the exogenous pathway. The lipoprotein particles become smaller in this process, but retain the full complement of cholesteryl esters and become LDL particles. The LDL particles provide the source of cholesterol for incorporation into the cell membranes, the synthesis of steroids, and is a key component in atherogenesis.[1] LDL is taken up by cells by endocytocysis *via* LDL receptors that recognize apoB-100. In this process, cholesterol is esterified with long-chain fatty acids in HDL particles, with the resulting cholesteryl particles transferred to VLDL or LDL particles by the action of a plasma transfer protein, *cholesteryl ester transfer protein* (CETP). Lipoprotein(a) or Lp(a), is a species of LDL that is localized in atherosclerotic lesions and contains a unique protein, apo(a), with structural similarities to plasminogen. Lp(a) competes with plasminogen for its receptor on endothelial cells. Thus, plasminogen is the substrate for plasminogen-activating factor, which is secreted and bound to endothial cells, generating the fibrinolytic enzyme plasmin. The resulting pathophysiology is the binding of Lp(a) produces less plasmin, and fibrinolysis is inhibited and thrombosis is promoted.[1,15]

(iii) Reverse Cholesterol Transport – is when cholesterol is returned to plasma from tissues in HDL particles. Recent studies based on combinatorial chemistry and structure-based design have identified potent and subtype-selective peroxisome proliferator-activated receptor agonists (PPARs) that act at dietary lipid sensors that regulate fatty acid and carbohydrate metabolism. PPAR agonists (*e.g.* GW501515) increase expression of the reverse cholesterol transporter ATP-binding cassette A1 and induces apolipoprotein A1-specific cholesterol efflux. Additionally, when dosed to insulin-resistant middle-aged obese rhesus monkeys, GW501516 causes a dramatic dose-dependent rise in serum HDL cholesterol while lowering the levels of low density lipoprotein, fasting triglycerides, and fasting insulin. These results suggest that PPARd agonists may be effective drugs to increase reverse cholesterol transport and decrease cardiovascular disease associated with the metabolic syndrome X.[1,15]

6.6.2 Dyslipaemia

Dyslipaemia can be primary (combination of diet and genetics), or secondary to a disease (*e.g.* hypothyroidism, diabetes mellitus, alcoholism, chronic renal failure).

They are classified according to which lipoprotein is abnormal in six phenotypes (WHO – Frederickson Classification). The higher the LDL cholesterol, the higher the risk of ischaemic heart disease (Table 6.4). A greater risk of ischaemic heart disease is associated with a subset of the primary type IIa hyperlipoproteinaemia caused by single-gene defects of LDL receptors, known as familial hypercholesteroaemia (FH).[1,15]

The human body consists of a complex lipid transport system with five main classes of lipoproteins that, in order of molecular size, largest to smallest, are chylomicrons, very low density lipoproteins, intermediate-density lipoproteins, low-density lipoproteins (LDL), and high-density lipoproteins (HDL). Plasma levels of LDL-cholesterol

Table 6.4 World Health Organization/Frederickson classification of hyperlipoproteinaemia.[a]

Type	Lipoprotein elevated	Cholesterol	Triglycerides	Atherosclerosis risk	Drug treatment
I	Chylomicrons	↑	↑↑↑	NE	None
IIa	LDL	↑↑	NE	High	Statin ± ezetimibe
IIb	LDL + VLDL	↑↑	↑↑	High	Fibrate + statin + nicotinic acid
III	βVLDL	↑↑	↑↑	Moderate	Fibrates
IV	VLDL	↑←	↑↑	Moderate	Fibrates
V	Chylomicrons + VLDL	↑	↑↑	NE	Fibrates, niacin, fish oil and statin combinations

[a] ↑, increased concentration; LDL, low-density lipoprotein; NE, not elevated; VLDL, very low density lipoprotein; VLDL. A qualitatively abnormal form of VLDL identified by its pattern on electrophoresis.

are one of many factors implicated in the development of athero-sclerosis and cardiovascular disease. LDL transfers cholesterol to peripheral tissue and lowering LDL-cholesterol by statins in people with pre-existing cardiovascular disease is effective in decreasing mortality. HDL is the smallest of the lipoprotein particles containing the highest proportion of apolipoproteins to lipids. The major HDL associated apolipoproteins A-I (apoA-I) and apoA-II are secreted into plasma by the liver and the intestine, where they are lipidated to form lipid-poor, discoidal, nascent HDL.[14,15]

All plasma lipoproteins are spherical and have a hydrophobic core region containing triglycerides and colesteryl esters. Surrounding this core is a monomolecular layer of phospholipid and unesterified cholesterol.[2] The *apolipoproteins* (proteins associated with lipids) are located on the surface. Some lipoproteins (B-proteins) contain very high molecular weight apolipoproteins that, unlike some lower molecular weight apolipoproteins, do not migrate from one particle to another. There are two important forms of apolipoprotein-B: (i) *Apo B-48* is formed in the intestine and is found in the chylomicrons. (ii) *Apo B-100* is formed in the liver and is found in VLDL, IDL and LDL.[2]

The chylomicrons are the largest of the lipoproteins. They are formed in the intestine and they carry dietary triglycerides, together with some esterified cholesterol. Phospholipids, free cholesterol and newly synthesized apolipoprotein B-48 form the surface layer of the chylomicron. The chylomicron is absorbed from the intestine into the lymphatic system and thence into the blood stream. Triglycerides are removed from the chylomicrons by a hydrolytic mechanism involving the lipoprotein lipase system.[1,2] As the core triglycerides are removed, the size of the chylomicron diminishes and the surface lipids are transferred to HDL. The remnant of the chylomicron is taken up into the liver cells; the cholesteryl esters are hydrolyzed; and the free cholesterol is in part excreted in the bile and in part is metabolized. Chylomicrons are not normally present in the serum of individuals who have fasted.[1] Figure 6.5 illustrates this role of HDL in mobilizing and excreting cholesterol. HDL is composed of phospholipids, unesterified cholesterol and several different proteins. In its nascent form HDL exists as discs and the lipid bilayer forms an oily, hydro-phobic interior. While moving through the bloodstream and absorbing cholesterol, HDL becomes rounder, larger and more sphere like. The liver recognizes and takes up this sphere-like form of HDL by an as-yet unknown mechanism.[2]

VLDL are secreted by the liver and they carry triglycerides syn-thesized in the liver. After leaving the liver, the triglycerides are hydrolyzed by lipoprotein lipase and the freed fatty acids are oxidized in the tissues or are stored in the adipose tissue. Now, the depleted VLDL are termed intermediate-density lipoproteins (IDL). Some of these are taken up by the liver and the rest lose more of their triglycerides to form LDL.[2] The proteins of HDL are secreted by the intestine and the liver. Much of the lipid content of HDL derives from the surface monolayers of the chylomicrons and form lipolysis of the VLDL. The high-density lipoproteins also acquire cholesterol from the peripheral tissues.[1,2]

The risk of atherosclerotic heart disease is associated with abnormally high levels of LDL and cholesterol. Hypertriglyceridemia has long been associated with coronary heart disease, atherosclerosis with a strong link to hypertriglyceridemia in some families. Some types of hypercholesterolemia are genetically transmitted as a dominant trait.[2] Studies in animals have consistently shown that HDL is protective on several processes involved in atherosclerosis, at least in part by mediating the removal of cholesterol from lipid-laden macrophages. In mice, genetic lowering of plasma HDL decreases the appearance of macrophage-derived cholesterol in the feces and transgenic expression of apoA-I increases HDL and suppresses atherosclerosis in the apoE-deficient mouse. In humans, regressive changes in human atherosclerotic plaques were reported in relatively small studies when reconstituted HDL or apoA-I was provided exogenously. However, the mechanisms by which HDL may impact cardiovascular health and disease are complex and remain to be fully understood (see reviews in further reading for more discussion on this topic).

6.6.3 Dyslipidemia and Therapeutic Intervention

Dyslipidemia exists in two forms, primary and secondary:

- *Primary*: which is due to a combination of diet and genetics (often but not always polygenic) and are classified as shown in Table 6.4, with drug treatment for the various types.
- *Secondary*: which is associated with other diseases (*e.g.* diabetes, alcoholism, nephrotic syndrome, chronic renal failure, hyperthyroidism, liver disease.

Diet is a necessary adjunct for all types of drug therapy and in some instances diet may be sufficient to correct the hyperlipidemic condition.

6.6.4 Lipid-lowering Drugs

Drug therapy is based either on the specific metabolic defect, or in some instances diet may be sufficient to correct the hyperlipidemic condition. The main drugs used to modify cardiovascular risk factors are:

- statins, 3 hydroxy-3-methylglutaryl-coenzyme A (HMG-CoA) reductase inhibitors;
- fibrates;
- inhibitors of cholesterol adsorption;
- nicotinic acid or its derivatives;
- fish oil derivatives.

(i) *Statins: HMG-CoA Reductase Inhibitors*

The rate-limiting enzyme in cholesterol synthesis is HMG-CoA, which catalyzes the conversion of HMG-CoA to mevalonic acid. *Lovastatin, simvastatin* and *pravastatin* are specific, reversible, competitive inhibitors of HMG-CoA reductase. Atorvastatin and rosuvastatin are representative of a series of longer-acting competitive inhibitors of HMG-CoA reductase. They directly compete with the endogenous substrate for the active site cavity of HMGR (Figure 6.6). The statin pharmacophore binds to the same active site as the substrate HMG-CoA and inhibits the HMGR enzyme, resulting in a decrease in the synthesis of cholesterol via the mevalonate pathway (Figure 6.6). Thereby, decreasing LDL and TG (triglycerides) and total cholesterol levels as well as increasing HDL in plasma.[5] Statins are also noncompetitive with the cosubstrate NADPH (nicotinamide adenine dinucleotide phosphate).[23] The essential structural components of all statins are a dihydroxyheptanoic acid unit and a ring system with different substituents. The statin pharmacophore is a modified hydroxyglutaric acid component, which is structurally similar to the endogenous substrate HMG CoA and the mevaldyl CoA transition state intermediate (Figure 6.7). It has also been shown that the HMGR is stereoselective and as a result all statins need to have the required 3R, 5R stereochemistry.

All statins have the liver as their target organ and induce an increase in high-affinity LDL receptors, increasing clearance of LDL, which lowers the plasma pool of LDL-cholesterol. The liver is also the major route for excreting statins and its metabolites (enterohepatic circulation). Only a very small percentage of them

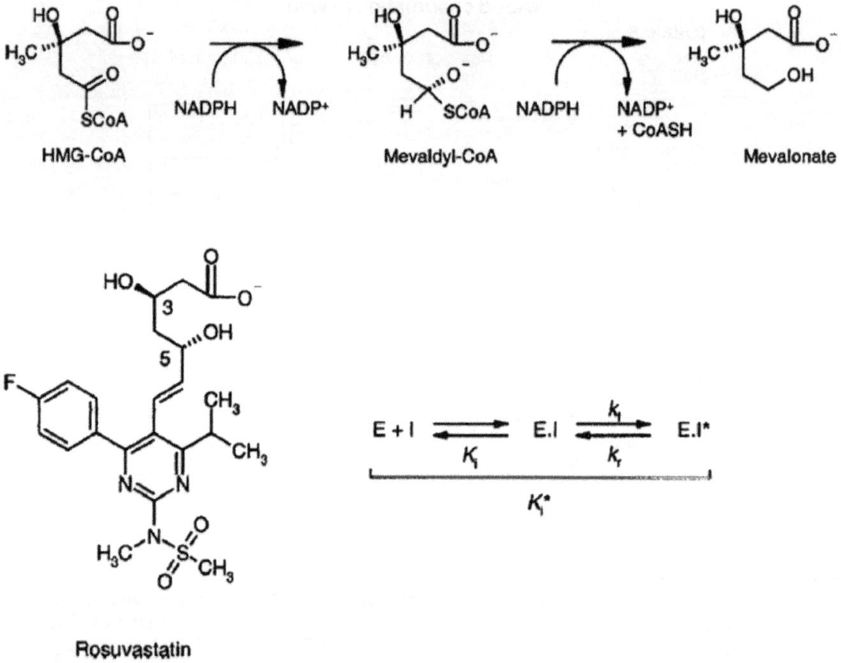

Figure 6.6 Catalysis by HMG-CoAR and inhibition by rosuvastatin. Top panel: the reaction catalyzed by HMG-CoAR. Bottom panel: the structure of rosuvastatin and kinetic scheme for inhibition. There is rapid formation of an initial E · I complex, with a dissociation constant, K_i, of ≈ 1 nM, followed by a slow transition (forward rate constant $k_f = 0.019$ s^{-1}; reverse rate constant $k_r = 0.009$ s^{-1}) to give a more tightly bound complex, and an overall steady-state inhibition constant, K_i^*, of ≈ 0.1 nM.
Reproduced from: G. A. Holdgate, W. H. J. Ward and F. McTaggart, *Biochemical Society Transactions*, 2003, **31**(3), 528–531.

are found in the urine. The phenomenal global success of the statins in treating hyperlipidemia is not without its controversy, both for and against their potential benefit in cardiovascular diseases/survival studies and their adverse side effects.[16,18]

However, the pharmaceutical industry is working on the next generation of antihyperlipedemia agents. These recent studies based on PCSK9 inhibition have provided for unprecedented lowering of plasma LDL-C (up to 70%) along with an excellent safety profile in ongoing clinical trials (see the bibliography for reviews).[17] Current knowledge indicates that PCSK9 and LDLR coexpression is regulated at the transcriptional level by SREBP-2. The biology/pharmacology in this field is still evolving and the reader is referred to the bibliography for further reading in this important area of drug development.[17,18]

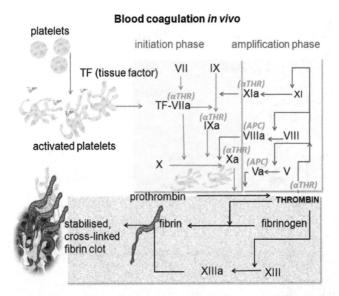

Figure 6.7   Principal cascade of blood coagulation.
Dr Graham Beards (UTC) Blood coagulation pathway *in vivo* (2014) 25
October 2014 (UTC) Reproduced with kind permission from Dr Graham
Beards.

(ii) *Fibrates*

Gemfibrozil is one of several fibric acid derivatives or fibrates
(*e.g.* bezafibrate, ciprofibrate, fenofibrate and clofibrate). These
agents markedly reduce circulating VLDL, TG with modest
reductions (10%) in LDL and HDL.[1] It was once thought that
the mechanism of action of fibric acid derivatives involves
increased lipolysis of lipoprotein lipase but their mechanism of
action is more complex. They are agonists of PPARα nuclear
receptor and their main effects are to increase transcription
genes for lipoprotein lipase, apoA1 and apoA5.[1]

Besides their effects on lipoproteins and increasing hepatic
LDL uptake, they have a wide range of additional properties they
(i) increase hepatic LDL uptake, (ii) reduce plasma C-reactive
protein and fibrinogen, (iii) improve glucose tolerance, (iv) inhibit vascular smooth muscle inflammation by inhibiting the
expression of the transcription factor NF-κB.[1]

Clinically, gemfibrozil is useful in hypertriglyceridemias, in
which VLDL predominates. Fenofibrate is uricosuric, which
may be useful where hyperuricaemia coexists with mixed
dyslipidaemia. The fibrates are efficiently absorbed from the
intestine when given with a meal, but it is less efficiently

absorbed when it is given on an empty stomach. They are often combined with other lipid-lowering drugs in patients with severe treatment resistant dyslipidaemia. This, on rare occasions, may lead to a risk of rhabodomyolysis.[1]

6.6.5 Inhibitors of Cholesterol Absorption

Historically, orally administered bile acid-binding resins (*e.g.* cholestyramine, colestipol) were the only agents available to reduce plasma cholesterol absorption. The resins are high molecular weight copolymers of polystyrene and divinylbenzene. They are large molecule, water-insoluble anion-exchange resins and sequester endogenous bile acids in the lumen of the intestine and prevent their being reabsorbed and enterohepatic recirculation. The concentration of HDL, however, remains unchanged and they cause an unwanted increase in TGs. They also increase fecal excretion of bile acids ($\geq$10-fold). Therefore, there is increasing demand for more of the body's cholesterol for replacement synthesis of bile acids by the liver. This process is normally controlled by the bile acids through negative feedback. Resins are bulky, unpalatable and often cause diarrhoea. They are nonselective ion-exchange resins and interfere with the absorption of fat-soluble vitamins, and several drugs to treat cardiovascular disorders (*e.g.* thiazide diuretics, digoxin, warfarin, propranolol and also some other hypolipidemic agents such as gemfibrizil).

Thus, with the introduction of statins, their use in treating dyslipidaemia is very limited. *Colesevelam* is a recent addition to the resin class of drugs but with little improvement apart a reduction in daily dose up to 4 g *vs.* 36 g for colestyramine.[2]

Ezetimibe is another cholesterol-absorption inhibitor that is used as an adjunct to diet and with statins in hypercholesterolaemia. It is a member of a class of compounds called azetidinones that inhibit the absorption of cholesterol from the duodenum by blocking a transport protein (NPC1L1) in the brush border cells of the enteroctyes, without affecting the absorption of fat-soluble vitamins and TGs or bile acids.[1] *Ezetimibe* represents a useful advance over the resins, however, in combination with a statin (simvastatin) it was shown to have little additional efficacy in the early 90s, although more recent clinical trials are long-awaited. Today, ezetimibe is still used along with a statin in severe dyslipidaemia. It is generally well tolerated, and is also used to treat pruritus in patients with partial biliary obstruction and bile acid diarrhoea resulting from diabetic neuropathy.[1]

(iii) *Nicotinic Acid or its Derivatives*

Nicotinic acid is a vitamin and is essential for many metabolic processes.[1] It is converted into nicotinamide, which inhibits hepatic VLDL secretion, resulting in reductions in circulating TGs and LDL, Lp(a), and an increase in HDL. The mechanism is poorly understood but is may be related to an effect on lipolysis *via* a G-protein-coupled orphan receptor HM74A found in adipocyte membranes. Chronic intake of therapeutic level doses of nicotinic acid for hyperlipidemia can cause severe hepatotoxicity, flushing, palpitations and gastrointestinal disturbances. Sustained release dosage forms of nicotinic acid are said to be somewhat less prone to producing severe liver toxicity, but this dosage form is not without controversy. Recent studies report that dose levels of nicotinic acid as used in humans have been associated with birth defects in experimental animals and some authorities suggest that pregnant women should not take this drug.[2] Equally concerning, the addition of nicotinic acid to a statin has no additional benefit but does increase serious side effects. Therefore, nicotinic acid treatment in dyslipidaemia is of little clinical value.[1]

(iv) *Fish Oil Derivatives*

Omega-3 polyunsaturated fatty acids are found in fish oil and there is epidemiological evidence to show they mitigate the risk of cardiovascular disease. They reduce plasma TGs, but increase cholesterol. Omega-3 fatty acids are essential fatty acids because they cannot be synthesized *de novo* and must be consumed from dietary sources such as marine fish. High levels of plasma TGs are less associated with coronary heart disease than cholesterol, although the actions on TG are less understood. Evidence-based medicine from large epidemiological studies show that eating fish regularly does reduce ischaemic heart disease, and dietary supplementation with ω-3 polyunsaturated fatty acids (PUFAs) improves survival in patients who have recently had a myocardial infarction (GISSI-Prevenzione Investigators, 1999). Fish oil is rich in PUFAs, including eicosapentaenoic and docosahexaenoic acid. PUFAs have also been shown to have beneficial effects on arrhythmias, inflammation, and heart failure. They may also decrease platelet aggregation and induce vasodilation. PUFAs also reduce atherosclerotic plaque formation and stabilize plaques preventing plaque rupture leading to acute coronary syndrome. Moreover, omega-3 fatty acids may have antioxidant properties that improve endothelial function and may contribute to its antiatherosclerotic benefits. Eicosapentaenoic acid substitutes for arachidonic acid in cell membranes and gives rise to 3-series prostaglandins ($PGI_{2\&3}$), thromboxanes ($A_{2\&3}$) and 5-series leukotrienes.[1] The haemostasis and antiplatelet aggregation properties of thomboxanes and the anti-inflammatory properties of leukotrienes (and the production of resolvins) largely account for the therapeutic efficacy of PUFAs. However, fish oil is contraindicated in type IIa hyperlipoproteinaemia because of the increase in LDL it causes.[1]

It is well established that PUFAs lower plasma TGs in patients with severe hypertriglyceridemia who are refractory to statins, they help to augment triglyceride reduction, they also increase HDL and decrease LDL levels.

As a prescription medicine, omega-3 acids are widely used and usually well tolerated with minimal adverse effects, and no known interactions or rhabdomyolysis-like symptoms.[1]

(v) *Other Novel Therapies*

To treat orphan indications as adjuncts in treating patients with the rare homozygous form of familial hypercholesterolaemia:

- *Mipomerson* is an antisenseoligonucleotide complementary to coding region for apoB-100 of mRNA, which thereby inhibits synthesis of apoB-100 and LDL. The drug, however, has ongoing regulatory issues with liver toxicity being a concern.

- *Lomitapide* is a small molecular weight inhibitor of microsomal triglycerides transfer protein (MTP or MTTP). MTP plays an important role in the assembly and release of apoB-containing lipoproteins (VLDL) into the circulation and inhibition of this protein significantly lowers plasma protein levels. This action is contrary to other lipid-lowering drugs, which work by increasing LDL uptake rather than reducing hepatic lipoprotein secretion. In Phase III studies raised aminoacidtransferase and fat deposits in the liver were reported.[1]

It is beyond the scope of this section to cover all the aspects of dyslipidemia and therapeutic intervention and the list of reviews in the further reading section is a resource of additional information.

6.7   Antiarryrythmic Drugs and Ischameic Heart Disease

6.7.1   Pathophysiology and Pharmacology

Cardiac arrhythmias result from irregular heart rhythm associated with disorders of heart rate and beating rhythm resulting in serious heart conditions. They include:

- *Atrial fibrillation*, the most common type, which consists of irregular fast heart beats, as individual atrial heart muscle fibers act independently and not in synchrony with the rest of the cardiac muscle fibers.
- *Ventricular fibrillation*, a rare, rapid and disorganized rhythm of ventricular muscle that leads to irregular heartbeats and sudden death if not treated immediately.
- *Supraventricular tachycardia*, present as episodes of abnormally fast heart rate.
- *Heart block*, a disease or inherited condition that causes a fault within the heart's natural pacemaker due to some kind of obstruction (or "block") in the electrical conduction system of the heart.
- *Bradycardia*, where the heart beats more slowly than normal, resulting from several factors, including aging, conditions that can slow electrical impulses through the heart (low thyroid level – hypothyroidism), electrolyte imbalance, medicines for treating heart problems or high blood pressure, such as beta-blockers, antiarrhythmics, and digoxin.

Arrhythmias occur as a result of (i) altered formation of impulses in the SA node (the pacemaker of the heart) Figure 6.1; (ii) altered conduction of impulses from the SA node through the heart muscle; or (ii) a combination of the two. It is the nerve impulse originating in the SA node that travels from cell to cell across the entire heart, triggering muscle contraction. Prior to this contraction, the cardiac muscle relaxes so that the chambers of the heart can dilate and again fill with blood. As a consequence of the sinus node-generated impulse moving from cell to cell, nerve impulses may circle back and reactivate cells that were activated earlier ("re-entry"). This re-entry is not normal because the cells, once stimulated, become refractory for a period of time long enough for the original signal fail to die out before a new signal from the SA node appear. In cases where re-entry does occur, nerve impulse shortens the refractory period or increases the rate of impulse conduction re-entry. However, re-entry is purported to be one of three mechanisms believed to result in cardiac arrhythmias, others include enhanced automaticity and triggered activity and further reading of this complex field can be found in the reference list.[1]

6.7.2　Antidysrhythmic Drugs

The following is a classification of antiarrhythmic drugs based on their electrophysiological classification as proposed by Vaughan Williams 1970 criteria (Table 6.5), according to their mode of action:

- Class I: drugs that block voltage-regulated sodium channels thus reducing the excitability of the heart and subdivided into: Ia, Ib, and Ic.
- Class II: *β-adrenoceptor antagonists.*
- Class III: drugs that substantially prolong the cardiac action potential.
- Class IV: drugs that block voltage-gated calcium channels and thus impair electrical impulse propagation in the SA and AV nodes.

(i) **Class I Drugs**

Disopyramide has now replaced quinidine as a Class 1a drug of choice in blocking sodium channels, see Table 6.5. Disopyramide, like quinidine, has local anaesthetic activity, binding on the α-subunit of the channel and inhibiting action potential propagation in many excitable cells (intermediate dissociation). The characteristic effect on the action potential is to reduce the maximum rate of depolarization during Phase 0 (fast depolarization) leading to Phase 1 (partial repolarization), Phase 2 (a plateau), Phase 3 (repolarization) and finally Phase 4 (pacemaker action potential).

Class Ia drugs bind to sodium channels that exist in three distinct functioning states: resting, open and inactivated.

Table 6.5　Summary of antidysrhthmic drugs (according to Vaughan Williams classification).

Class	Examples	Mechanism
Ia	Disopyramide Quinidine	Na^+ block of a subunit (intermediate dissociation)
Ib	Lidocanine	Na^+ block (fast dissociation)
Ic	Flecainide	Na^+ block (slow dissociation)
II	Propranolol	β-adrenoceptor antagonism
III	Amiodarone Sotalol	K^+-channel block
IV	Verapamil	Ca^{++}-channel block

Channels switch rapidly from resting to open in response to depolarization (activation), prolonged depolarization results in ischaemic cardiac muscle, which causes the channels to change more slowly from open to inactivated, and the membrane, which is refractory, must then be repolarized before it can be activated again.[1,11]

This class of drugs bind to channels most strongly when they are either in the open or inactivated state, less strongly in the resting state. This type of action is called *use dependence* (*i.e.* the more frequently the channels are activated, the greater the block that is produced).

Disopyramide mechanism of action is to depress the increase in sodium permeability in the open channel during Phase 0 of the cardiac action potential, in turn decreasing the inward sodium current. This results in an increased threshold for excitation and a decreased upstroke velocity.[1,2] Disopyramide prolongs the heart's PR interval by lengthening both the QRS and P wave duration. This effect is particularly well suited to the treatment of ventricular tachycardia as it slows the action potential propagation through the atria to the ventricles.[5] Disopyramide also has a significant negative inotropic effect on the ventricular myocardium, that may reduce contractile force up to 42% at low doses and up to 100% in higher doses, thus leading to heart failure. Finally, Class 1a drugs (quinidine, procainamide and disopyramide) have properties lying midway between Ib and Ic but in addition prolong repolarization less markedly than Class III drugs (see below). Disopyramide is currently used to treat ventricular dysrhythmias, prevention of recurrent paraoxysmal atrial fibrillation triggered by vagal overactivity and obstructive hypertonic cardiomyopathy. Furthermore, disopyramide has a number of anticholinergic effects on the heart, which accounts for many adverse side effects (see Chapter 4).[1,5]

Disopyramide

(ii) **Class Ib Drugs**

The local anaesthetic *procaine* is a Class Ib drug with pronounced antiarrhythmic activity when given intravenously. However, it is rapidly metabolized by blood and other tissue esterases accounting for its short duration of action and limited therapeutic utility.

Another Class Ib local anaesthetic, *lidocaine* is also an effective antiarrhythmic agent; it is a fast dissociation sodium channel blocker. *Lidocaine* blocks both open and refractory channels. The utility of *lidocaine* is also limited by its short duration of action and it is mainly administered intravenously for the treatment and prevention of ventricular tachycardia and fibrillation during and immediately after myocardial infarction. When given by mouth, it is efficiently absorbed across the wall of the gut, but it undergoes extensive first-pass metabolism. Several analogs of *lidocaine*, for example *mexiletene* were designed to provide enhanced resistance to first-pass metabolic inactivation, to make chronic oral therapy effective. This drug is effective orally; it is used to control ventricular arrhythmias. *Mexiletene* eventually undergoes metabolism by liver enzymes that are inducible by several other drugs, including the anticonvulsant drug phenytoin.[1,2]

lidocaine   mexiletene

(iii) **Class Ic Drugs**

Flecainide and *encainide* are representatives of this class of antiarrhythmic drugs that are associated with slow dissociation, reaching a steady-state of blockade during the cardiac cycle by strongly inhibiting the electrical conduction in the His–Purkinje fibers that innervate the ventricles. This class of drugs are used mainly to prevent paroxysmal atrial fibrillation and recurrent tachyarrhythmias associated with abnormal conducting pathways (*e.g.* Wolf–Parkinson–White syndrome).[1]

flecainide   **encainide**

(iv) Class II Drugs

The β-adrenoceptor blockers such as *propranolol, metoprolol* are useful in preventing mortality in patients recovering from myocardial infarction and it is claimed that the protective action of these drugs reflects their ability to prevent ventricular arrhythmias.

Ventricular arrhythmias following myocardial infarction may be partly the consequence of an increase in sympathetic activity and adrenaline release. β-adrenoceptor antagonists also act by increasing the refractory period of the AV node conduction (see Chapter 4), thereby preventing recurrent attacks of supraventricular tachycardia (SVT). Additionally, β-adrenoceptor antagonists are used to prevent paroxysmal attacks of atrial fibrillation during sympathetic action.[1]

propranolol   **metoprolol**

(v) Class III Antiarrhythmic Drugs

This class of drugs was originally based on the rich pharmacology of *amiodarone* and more recently *sotalol*. The prolongation of the refractory period of the heart produced by drugs in this category is believed to be an effect on ion channels; inhibition of repolarizing K^+ currents and/or activation of slow inward Na^+ current. These drugs have been described as sodium and/or potassium channel blockers. However, mechanisms in this category are complex and are not well understood.

Amiodarone   Sotalol

Amiodarone exhibits a multiplicity of pharmacological effects, none of which is clearly linked to its antiarrhythmic action. The drug could be viewed as an analog of the thyroid hormone and some of its actions on the heart may be a reflection of its interaction with thyroid hormone receptors. *Amiodarone* is extensively bound in the tissues. It has a long elimination half-life (10–100 days) and it accumulates in the body during dosing. It normally requires days or weeks for its action to develop. *Sotalol*, besides its nonselective β-adrenoceptor antagonists properties, also acts on potassium channels and causes a delay in relaxation of the ventricles. By blocking these potassium channels, *sotalol* inhibits efflux of K^+ ions, which results in an increase in the time before another electrical signal can be generated in ventricular myocytes (see Figure 6.12). This increase in the period before a new signal for contraction is generated helps to correct arrhythmias by reducing the potential for premature or abnormal contraction of the ventricles but also prolongs the frequency of ventricular contraction to help treat tachycardia. However, adverse effects due to prolonged cardiac action potential may result in proarrhythmic effects, notably a polymorphic form of ventricular tachycardia known as *tosade de pointes*. This occurs frequently in patients treated with drugs that prolong the QT interval, including antipsychotic drugs (see Rang and Dale for more detailed information and drug interactions).[1]

(vi) **Class IV drugs**

These drugs act by blocking voltage-gated calcium channels, preventing calcium ion passage, reducing the rate of conductance in the AV and SA nodes and action potential propagation. This action depends on a slow inward Ca^+ current slowing the heart and terminating SVT by causing partial AV block, thereby shortening the plateau of the action potential and reducing the force of contraction, resulting in a resumption of normal heart rhythmicity and reducing premature ectopic beats. Pharmacologically distinct classes of L-type voltage-gated calcium channel blockers are expressed in the heart and vascular smooth muscle.[1,5]

Three commonly employed antiarrhythmic calcium channel blockers are *verapamil*, *nifedipine* and *diltiazem*. *Nifedipine* and *diltiazem's* pharmacology is similar to that of *verapamil*, but both these drugs act mainly on vascular smooth muscle indirectly increasing sympathetic tone *via* their hypotensive effects, causing reflex tachycardia.[1,5]

verapamil   **nifedipine**   **diltiazem**

6.8   Ischameic Heart Disease

6.8.1   Myocardial Ischemia

The pathophysiological consequences of coronary atherosclerosis are two-fold;

- angina (chest pain caused by cardiac ischemia);
- myocardial infarction.

Angina is usually caused by *myocardial ischemia* (*"ischemia"* is a deficiency in blood supply to some body part caused by some obstruction or constriction of blood vessels). Anginal pain is often associated with coronary artery disease. The ischemic myocardial pain results from an imbalance between oxygen supply and oxygen demand by the heart muscle. Three principal types of angina are recognized; (i) *stable angina* – exercise-induced (the most common) where a variety of treatments are used (*e.g.* organic nitrates, β blockers, calcium antagonists, statins and aspirin to treat various aspects of the underling atheromatous disease (see Chapter 4, Section 4.5). (ii) *unstable angina* – no or little exertion. Intravenous organic nitrates, *glycerly trinitrate* and/or antiplatelet drugs (aspirin or ADP antagonists such as *clopidogrel* or *prasugrel*) are effective in reducing the risk of myocardial infraction in this indication. (iii) *variant angina*, relatively uncommon and occurs at rest due to coronary artery spasm. Coronary artery vasodilators are the drugs of choice (*e.g.* organic nitrates, calcium antagonists).[1,3,5]

The pathophysiological mechanism by which ischemia evokes pain in the heart is far from clear (see the references for more detailed reviews). Briefly, inadequate oxygen-carrying capacity of the blood, restricted blood flow resulting from atheromatous plaque/thrombus formation reducing blood flow and blood vessel viscosity of the blood, all decrease the caliber of the coronary vessels. The resulting increase in myocardial oxygen demand is driven by an increased heart rate and sympathetic tone due to an excess release of inotropic hormones (*e.g.* epinephrine) or by increased myocardial wall tension. A sudden anginal attack/ischemia is often preceded by a rise in systemic blood pressure and/or heart rate, resulting in increased oxygen demand by the heart muscle and subsequent muscle ischemia. Thus, pharmacological intervention is aimed at reducing ischemia by increasing oxygen supply to the heart muscle or by decreasing cardiac work (oxygen demand).[3]

Initially, therapeutic agents were developed for their ability to dilate the coronary arteries in animals and humans to increase oxygen supply (*e.g.* nitrates). However, this strategy failed to identify agents with both vasodilitatory and analgesic properties, as anginal pain is influenced by so many factors and a variety of drug treatments. The main therapeutic approaches are drugs that reduce cardiac work and improve function by maintaining oxygenation, as well as treating pain and preventing further thrombosis.[1,3] They are used in combination, and include:

- combinations of thrombolytic and antiplatelet agents (heparin, aspirin and clopidogrel) to open occluded artery and prevent reocclusion;

- oxygen given during arterial hypoxia;
- opioids to reduce pain and excessive sympathetic activity;
- organic nitrates (vasodilatory);
- calcium antagonists (vasodilatory);
- β-adrenoceptor antagonists (once patient is stable);
- angiotensin-converting enzyme inhibitors (ACEIs) or angiotensin AT1 receptor antagonists (ARBs) (reduce cardiac work and survival).

6.8.2 Drugs that Increase Myocardial Contraction

Myocardial contraction involves stimulation by acetylcholine resulting in a depolarization of the myocardial myofibrils (see Chapter 4). Calcium cations are present in the muscle myofibrils, where the myosin and actin interact along with two regulatory proteins, troponin and tropomyosin, which allow the muscle to shorten in the presence of Ca^{2+}. Ca^{2+} ions enter the muscle cell through calcium channels in the cellular membrane in response to depolarization of the membrane by acetylcholine. This causes a further release of additional Ca^{2+} stored in the cell. ATP and ADP act as sources of energy in the cross-bridge contraction cycle with the Ca^{2+} ions enabling the actin and myosin strands to form a cross-bridge and the muscle myofibrils to contract. In the relaxation stage calcium ions are detached from the protein filaments. The filaments expand and the muscle fiber relaxes. Following this process some calcium cations leave the cell through the calcium channels and the remainder is stored in the muscle cell.

There are four different kinds of calcium channels in the human body and they are classed according to their location and physiological function:[26]

1. *L-type,* found in smooth, skeletal and cardiac muscle;
2. *T-type,* found in pacemaker cells (in the heart);
3. *N-type,* fond in neurons and participate in release of neurotransmitter;
4. *P-type,* found in Purkinje fibers in the heart – physiological role not understood.

Therapeutically important calcium antagonists in cardiovascular disease act on L-type channels and comprise of three distinct classes: phenylalkylamines (*e.g. verapamil*), dihydropyridines (*e.g. nifedipine, anmlodipine*), and benzodiazapines (*e.g. diltiazem*).

Calcium-channel antagonists also selectively interfere with the entry of Ca^{2+} into the smooth muscle cells of the blood vessel walls and resulting arterial/arteriolar dilation across all vascular beds, thereby reducing blood pressure, with no effect on veins/venules.

They cause coronary vasodilation, increasing the supply of blood to the heart muscle and are used in patients with coronary heart spasm (*variant angina*).[3] The positive inotropic effect of raised Ca^{2+} levels in heart muscle fibers further increases the contractile force of the myocardium, thereby aggravating the angina symptoms. Thus, blockade of the calcium channels prevents entry of a significant amount of Ca^{2+} into the heart muscle cells and alleviates the symptoms. Therefore, the discovery and development of calcium channel-blocking drugs led to a dual mechanism of action in treating cardiovascular disease (i) dilation of arterioles combined with (ii) negative inotropic effect, establish their value in treating essential hypertension as well as myocardial ischemia.

6.8.3 Other Antianginal Agents

Recent advances in the discovery and development of antianginal agents are reviewed in the reference and further reading lists. This section will briefly review the historical and more recent advances in the treatment of angina. The early development of antianginal drugs was based on organic nitrate esters and to a lesser extent, organic nitrite esters. These esters are powerful vasodilators, acting on veins to reduce cardiac reducing preload and afterload. The desired pharmacological effects of organic nitrate and nitrite esters result from their enzyme-mediated release of the free-radical form of nitric oxide in the vascular smooth muscle. Nitric oxide stimulates the formation of cGMP that activates protein kinase G, affecting both the contractile proteins (myosin light chains, as cited previously above) and Ca^{2+} regulation.

Early examples include; *glyceryl trinitrate*, a short-acting antianginal drug (due to rapid first-pass metabolism), administered sublingual under the tongue. It is highly lipophilitic, which permits rapid absorption across the skin and mucous membrane and effective relief of angina pain is evident within 2–3 min. However, tolerance occurs quickly (depletion of stores of the cofactor(s) required for enzymatic conversion of the drug into nitric oxide) and it is important clinically therefore for the frequent use of long-acting or sustained release preparations. More recent agents include *isosorbide mononitrate*, which is a long-acting glycerol trinitrate, as it is absorbed and metabolized more slowly with a similar pharmacological action. It is orally administered for prophylaxis or once a day treatment. Various strategies to increase nitric oxide signaling in cardiovascular disease are currently being developed.[17]

Finally, organic nitrates and calcium antagonists have already been briefly reviewed above, along with β-adrenoceptor antagonists. Of the more recent agents, *Ivabradine* (β-adrenoceptor antagonists) acts on sinus node I_i current that is used in patients tolerant or contraindicated to existing β-adrenoceptor antagonists. *Ranolazine,* inhibits late sodium currents and thus indirectly reduces intracellular calcium and force of contraction without affecting heart rate; other more potent agents are currently in development for this class of compound. *Nicorandil,* is a potassium channel activator with additional nitrovasodialtor activity that is used in combination with existing antianginal agent in treatment-resistant patients.[1]

Glyceryl trinitrate   **Isosorbide mononitrate**

Ivabradine   **ranolazine**

nicorandil

6.9 Haemostasis and Thrombosis

6.9.1 Physiology of Haemostasis and Thrombosis

Hemostasis is the spontaneous arrest of blood from a damaged blood vessel and is essential for survival. The immediate hemostatic response of a damaged vessel is a vasospasm and resulting vasoconstriction, accompanied by:

- adhesion and activation of platelets;
- formation of fibrin.

Drug therapy to promote haemostasis (*e.g.* antifibrinolytic and haemostatic drugs) is used when a defective in the coagulation cascade (*e.g.* coagulation factors in haemophilia or following excessive anticoagulant therapy), or when it proves difficult following a haemorrhage during surgery or for menorrhagia. Drug therapy to treat or prevent thrombosis or thromboembolism is common and used extensively in emergency cases (*e.g.* stroke). Drugs affect haemostasis and thrombosis in three distinct ways, by influencing:

- blood coagulation (fibrin formation);
- platelet function;
- fibrin removal (fibrinolysis).

6.9.2 Blood Coagulation

Blood coagulation is the conversion of liquid blood to a clot. The first step in the conversion by thrombin of soluble fibrinogen to insoluble stands of fibrin, the final step in a complex cascade (Figure 6.7 for complete cascade).[1] Several blood components or factors are present in the blood as inactive precursors (zymogens) of proteolytic enzymes (serine proteases) and cofactors (see Figure 6.7 for the complete cascade). For example, a clotting factor zymogen (*factor VII*) undergoes proteolysis and becomes an active protease (*factor VIIa*). This protease in turn activates the next clotting factor in the cascade (*factor IX*) until eventually a solid fibrin clot is formed (see Figure 6.7 for the complete cascade). This accelerating cascade has to be controlled by inhibitors otherwise all the blood in the body would solidify within minutes of the initiation of haemostasis. Of the various inhibitors, antithrombin III, appears to be the most important and quickly neutralizes all serine proteases in the coagulation cascade. There are two pathways in the cascade: (i) the *in vivo* (extrinsic) pathway, (ii) the contact (intrinsic) pathway (*e.g.* contact with glass in a test tube). Both pathways result in activation of factor X to Xa that converts prothrombin to thrombin. Other components, including calcium and negatively charged phospholipid (PL) are essential for three steps, namely the actions of factor IXa on X, factor VIIa on X, and factor Xa on II. PL is formed by activated platelets adhering to the damaged vessel. Some factors promote coagulation by binding to PL and a serine protease factor (*e.g.* factor Va in the activation of II by Xa, or VIIIa in the action of X by IXa). However, the ultimate control of blood coagulation is by enzyme inhibitors (*e.g.* antithrombin III) and fibrinolysis (see Figure 6.7 for the complete cascade).

6.9.3 Thrombosis

Thrombosis is the pathological formation of a haemostatic plug within the vaculature in the absence of bleeding ("haemostasis in the wrong place").[1] It is a life-threatening disease, the consequences of which can lead to myocardial infarction, stroke, deep vein thrombosis (DVT) or pulmonary embolus. It occurs within seconds of the vasospasm, the blood platelets in the vicinity of the damaged vessel endothelium lumina become sticky and platelets adhere and aggregate to the injured surface of the blood vessel. The platelets physically change shape and lose their individual membranes and form a gelatinous plug that quickly arrests bleeding (see Figure 6.7 for the complete cascade). The plug is then reinforced and strengthened by *fibrin* synthesis. From the resulting coagulation cascade, thrombin (factor IIa) cleaves fibrinogen, producing fragments that polymerize to form fibrin. It also activates factor XIII, a fibrinoligase, which reinforces fibrin-to-fibrin threads cross-linking and forming covalent bonds to stabilize the plug (coagulum) (see Figure 6.7 for the complete cascade).

Thrombin also causes platelet aggregation, stimulates cell proliferation and modulates smooth muscle by an interaction with specific protease-activated receptors (PARs). These important receptors belong to a class of GPCRs and not only contribute to haemostasis and thrombosis, but also may be involved in inflammation and angiogenesis. Their pharmacological action and signal-transduction mechanism is unusual in that receptor activation requires proteolysis by thrombin of the extracellular domain of the receptor, revealing a new *N*-terminal sequence that acts as a "tethered agonist" (see Chapter 2, on Receptor Pharmacology).

6.9.4 Drugs that Act on Coagulation

Two classes of drugs affect coagulation: those that modify the cascade when there is a defect and drugs that act when there is an unwanted coagulation.

(i) Anticoagulant Drugs

Heparin, a heterogeneous mixture of sulfated glycosaminogly-cans (mucopolysaccharides), is a powerful anticoagulant whose mechanism of the action on thrombin is related to the action of an endogenous plasma protease inhibitor, *antithrombin III* (heparin cofactor); to inhibit factor Xa.

Heparin only needs to bind to antithrombin III to inhibit thrombin. Heparin forms an equimolar stable complex with antithrombin III. The LMWHs increase the action of antithrombin III on factor Xa but not its action on thrombin. This complex induces conformational change that makes antithrombin's active site more complementary to the topography of thrombin.[2] Once an antithrombin–thrombin complex is formed (which inactivates the catalytic activity of the thrombin), heparin is released and is available for binding to more antithrombin III. It has been known for a long time that the proactivation effects of heparin on platelets leads to an increase in bleeding time. This action is mediated through signaling *via* the glycoprotein (GP) IIb–IIIa complex (integrin αIIbβ3) action induced by heparin, and LMWHs.[1,2]

Heparin is administered by intravenous infusion, but for prophylaxis of thrombosis or embolisms, it can be administered subcutaneously. The drug has a rapid onset of action and is ineffective orally. In severe overdosage of heparin where bleeding occurs (*e.g.* haemorrhage or heparin-induced thrombocytopenia), a specific antidote, *protamine sulfate,* is administered. Protamine is a peptide that forms a stable complex with heparin, the complex of which is devoid of anticoagulant activity. Protamine itself has *anticoagulant* activity due to its effect on blood platelets, fibrinogen and other plasma proteins. Therefore, it is essential to administer the minimum amount of protamine required to neutralize the plasma heparin and an *in vitro* neutralization test is performed on the patient to provide an accurate dosing regimen. *Heparin* fragments (*e.g.* enoxaparin, dalteparin) or a synthetic pentasaccaride (fondaparinux), referred to as low molecular weight heparins (LMWHs) are longer acting than unfractionated heparin and are usually preferred. The unfractionated product is used only in renal failure where LMWHs are contraindicated. These agents are clinically superior to native heparin and are replacing heparin in clinical practice.[1,2]

(ii) Direct Thrombin Inhibitors and Related Drugs

Hirudins are polypeptides derived from the saliva of the medicinal leech. Unlike heparin their action does not depend on activation of antithrombin. *Lepirudin*, a recombinant form of hirudin binds irreversibly to both the fibrin-binding and catalytic site of thrombin and is used for thromboembolic disorders (Type II heparin-induced thrombocytopenia). *Bivalirudin* is an hirudin analog used mainly in combination with aspirin and clopidogerel in patients undergoing coronary artery surgery. A number of recent direct acting agents have shown promise as antithrombin drugs; including *dibigatran* (serine protease inhibitor), rivaroxaban and apixiban both direct orally active factor Xa inhibitors. The future success of these drugs will dramatically change the clinical management of patients currently treated with warfarin.[1]

Warfarin (Coumadin) is the most widely used oral anticoagulant. The oral anticoagulants are antagonists of vitamin K. Vitamin K antagonists only act *in vivo* and interfere with post-translational *γ-carboxylation* of glutamic acid residues of the coagulation factors (II, VII, IX, X).[18] Their action is to inhibit vitamin *K epoxide reductase component 1* (VKORC1), and thus inhibit the reduction of vitamin K epoxide to its active hydroquinone form.[1,18,20]

Carboxylation is directly related to the oxidation of vitamin K hydroquinone to the epoxide. The hydroquinone form must be regenerated from the epoxide to maintain the peptide carboxylation reaction. This process proceeds through the quinone. The reductase enzyme(s) that effect this reaction are targets for warfarin, which blocks their action. The body contains other reductases that catalyze reduction of vitamin K oxide, but these require higher concentrations of the substrate. These alternative reductases are less sensitive to warfarin, which may be the reason that the anticoagulant effects of large doses of warfarin are reversed with similarly large doses of vitamin K.[20]

Oral anticoagulants are a class of drugs that have major safety concerns (*e.g.* teratogenic, drug–drug interactions) and require constant individual blood monitoring for their effect on prothrombin time, the time required for clotting of citrated plasma (citrate sequesters Ca^{2+}) after adding specified standard amounts of Ca^{2+} and reference standard thromboplastin. Future routine screening using genotyping of the VKORC1 gene, which is polymorphic with different haplotypes, may represent a better screening assay to reduce the variability in response to warfarin in individuals. This combined with genotyping the cytochrome responsible for most of the drug–drug interactions, CYP2C9 would be a significant step in reducing the many safety issues associated with warfarin.[1,2]

6.9.5 Platelet Adhesion and Activation

Platelets maintain the integrity of the circulation and their activation results in a sequence of reactions that are essential for haemostatsis, the healing of damaged blood vessels along with the formation of a haemostatic plug to stop bleeding that is reinforced by fibrin at the injured site.

The complex sequence of reactions and pathways provide several redundant or autocatalytic routes if one or more become blocked, these include.[1,18]

- *Adhesion* – following vascular damage (*e.g.* von Willebrand factor (vWF) bridging between subendothelial macromolecules and glycoproteins [GP]1b receptors on the platelet surface (see Figure 6.9).
- *Shape change* – smooth discs to spiny spheres with protruding pseudopodia.
- *Secretion* – of granule contents (*e.g.* platelet agonists, ADP, 5-HT, epinepherine and coagulation/ growth factors such as platelet-derived growth factors).
- *Biosynthesis of labile mediators* – such as platelet-activating factor and thromboxane (TXA_2).
- *Aggregation* – which is promoted by various agonists (*e.g.* ADP, 5-HT and TXA_2, acting on specific receptors on the platelet surface; activation by agonists, ADP action on $P2Y_{12}$ leads to expression of GPIIb/IIIa receptors that bind fibrinogen, which links adjacent platelets to form aggregates.
- *Exposure of acidic phospholipids* – on the platelet surface, promoting thrombin formation and results in further platelet activation *via* thrombin receptors and fibrin formation *via* cleavage of fibrinogen (as discussed previously).

6.9.6   Antiplatelet Drugs

Platelets play a critical role in thromboembolic disease and the physiology and pharmacology of platelet function led to the development of several classes of antiplatelet drugs each acting on one or more of the several endogenous mediators of platelet activation and aggregation. Early studies on the inhibition of the arachidonic acid cascade formed the basis for the use of aspirin as a prophylactic agent to prevent heart attack, which radically altered clinical practice. The blood platelets, which are involved in an early phase of the clotting phenomenon, contain the components of the arachidonic acid cascade. Aspirin inhibits thromboxane (TXA_2) synthesis, by irreversible acetylation of a serine residue in the active site of the cyclo-oxygenase 1 (COX-1) enzyme isoform, which retards platelet aggregation and disrupts the coagulation process. The inability of platelets to synthesize new proteins due to aspirin's action on platelet cyclo-oxygenase is permanent, lasting the lifetime of the platelet (7–10 days). Thus, repeated doses of aspirin produce a cumulative effect on platelet function. Aspirin is maximally effective as an antithrombotic agent at doses much lower than those required for other actions of the drug. There are several other drugs marketed as antiplatelet agents, and these will be briefly discussed:

   (i) *Dipridamole* – inhibits phosphodiesterase, TXA_2 synthesis and adenosine uptake. It is used in addition to aspirin in some patients with stroke or transient ischaemic attack and its effects are additive.

  (ii) *Adenosine ($P2Y_{12}$) Receptor Antagonists* – $P2Y_{12}$ antagonists include: *ticlopidine* (first-in-class but causes neutropenia and thrombocytopenia).

 (iii) *Clopidogrel* (a prodrug), *prasugrel* and *tricagrelor* – each of which can be combined with low dose in patients with unstable coronary artery disease.

     Clopidogrel and *parasugrel* inhibit platelet aggregation by irreversible (covalent) inhibition of $P2Y_{12}$ receptors to which they link *via* a disulfide bond, whereas *ticagrelor* is a reversible, noncompetitive inhibitor of the $P2Y_{12}$ receptor. Care must be taken with these drugs in patients with variant alleles of CYP2C19 (poor metabolizers) who are at risk of therapeutic failure, or potential drug–drug interactions with proton pump inhibitors (*omeprazole*) that are metabolized by CYP2C19.

clopidogrel   **ticagrelor**

(iv) *Glycoprotein IIB/IIIA Receptor Antagonists* – these drugs inhibit all pathways of platelet activation as they all converge on activation of GPIIb/IIIa. Abciximab, is a hybrid murine–human monoclonal antibody Fab fragment directed against the GPIIb/IIIa receptor. Tirofiban and eptifibatide are a synthetic non-peptide and a cyclic peptide, respectively, that have a high risk of bleeding and therefore are sparingly used in high-risk patients undergoing coronary angioplasty or coronary syndrome.

(v) *Synthetic PGI$_2$* – *epoprostenol* is an agonist at prostanoid IP receptors causing vasodilation and inhibiting platelet aggregation. It is administered intravenously with a half-life of 3 min to patients undergoing dialysis to prevent thrombosis during haemodialysis or haemofiltration, especially in patients in whom heparin is contraindicated. Other uses include; severe pulmonary hypertension, circulatory shock associated with meningococcal septicaemia. Adverse effects include flushing, headache and hypotension.[1]

6.10   Fibrinolysis (Thrombolysis)

As discussed earlier, the fibrinolytic system comes into play once the coagulation cascade is activated *via* several endogenous plasminogen activators (tPA), urokinase-type plasminogenactivator, kallikrein and neutrophilelastase. tPA is inhibited by a structurally related lipoprotein(a). Plasminogen is deposited on the fibrin strands within the thrombus plug and the plasminogen activators (serine proteases) diffuse into the thrombus and cleave plasminogen to release plasmin. *Plasmin (trypsin-like protease)* digests fibrin by hydrolytic removal of one amino acid from its inactive precursor, *plasminogen,* catalyzed by *tissue plasminogen activator.* The fribrinolytic system is regulated so that unwanted fibrin thrombi are removed as well as factors II, V and VIII but fibrin in wounds persists to maintain haemostasis. Thrombolytic drugs dissolve the fibrin deposits at sites of vascular injury and pathological thrombi. However, these drugs can produce serious haemorrhage, as discussed previously. Current fibrinolytic drugs that act by increasing or inhibiting fibrinolysis are now briefly discussed.

6.10.1   Fibrinolyltic and Haemostatic Drugs

(i) *Streptokinase* is a protein produced by and extracted from β-hemolytic streptococci. It has no enzymatic activity, but it forms a stable noncovalent 1:1 complex with plasminogen that induces a conformational change on the plasminogen molecule, facilitating its conversion into active plasmin. Streptokinase is infused to reduce mortality in acute myocardial infarction and its beneficial effect is additive with coadministered aspirin. As it is a foreign protein in the body its side effects include allergic reactions and (rarely) anaphylaxis.

(ii) *Tissue plasminogen activators*: such drugs include *alteplase, duteplase and reteplase*, the first two drugs are, respectively, single- and double-stranded tPA. Reteplase is similar but with an extended half-life and can be given as a bolus injection. All three drugs are given clinically for myocardial infarction. Unlike streptokinase they are not antigenic but like all fibrinolytic agents the main adverse side effects are gastrointestinal bleeding and haemorrhage stroke. The plasminogen inactivator, tranexamic acid is used in patients with serious internal bleeding to inhibit plasminogen activation and in the rare disorder of hereditary angio-oedema.[1]

6.11   Heart Failure

6.11.1   Physiology and Pharmacology

The pathophysiology of chronic heart failure (HF)/congestive heart failure (CHF) is a result of the heart's inability to pump sufficient blood to meet the demands of the body. Many conditions culmimate in HF, a multiorgan systemic syndrome with an intrinsically poor prognosis. The heart malfunctioning as an efficient pump compromises the vascular circulation. The pathophysiology of HF/CHF is still largely unknown, although causality is related to the inability of the heart to translate ATP (the body's prime source of chemical energy) into mechanical work. In HF, compensatory mechanisms characterized by haemodynamic impairment and progressive neurohormonal dysregulation, with an increase in sympathetic activation, elevated peripheral vascular and cardiac remodelling resistance (largely mediated by the autonomic nervous system, see Chapter 4), resulting in cardiac output being compromised.[21] As discussed in Section 6.1, a complex interaction occurs following an increased sympathetic nervous activity due to a compromised arterial baroreceptor reflex and a decline in the body's ability to suppress the activity of centrally mediated sympathetic activity. This, combined with a retention of sodium ions and water by the kidney, causes an increased blood volume (see Section 6.12). As a consequence, the heart increases in size (cardiac hypertrophy), with a rapid increase in heart rate (tachycardia). Thus, the heart can no longer function as an effective pump, resulting in veins becoming engorged with blood and fluid escaping from the capillaries into the tissues. Heart failure quickly becomes apparent with oedema of the lower legs, ankles, feet and the lungs (pulmonary edema).[2] Left ventricular failure occurs and results in a dramatic decrease in peripheral blood flow and blood leaking from pulmonary vessel into the lung tissue. Difficulty in breathing quickly ensues (*dyspnea*) requiring therapeutic intervention to improve cardiac function without undermining the compensatory mechanisms of the body.

6.11.2 Drugs and their Pharmacological Effects on HF and CHF

Pharmacological intervention in the treatment of HF has focused on correcting the neurohormonal dysregulation, occupying the forefront of drug development since the 1980s, resulting in a large and robust evidence base supporting their use clinically (see Figure 6.8). This evidence underpins the European Society of Cardiology (ESC) guidelines for the diagnosis and treatment of acute and chronic heart failure.[19,20] There follows a brief account of current drugs used in the treatment of HF/CHF, many are discussed in earlier sections and further information is found in the reference list.[1,19,20]

(i) *ACE Inhibitors and ARBs*

ACE inhibitors and ARBs are the drugs of choice for the treatment of heart failure (see Section 6.3). ACE inhibitors promote dilation of the blood vessels and thus improve blood flow and slow the progression of HF/CHD. They can also reduce blood pressure and lighten the workload on the heart (see Section 6.3, Antihypertensive Drugs). Adverse effects include dry cough while taking an ACE inhibitor, an ARB, such as candesartan or valsartan is usually prescribed as an alternative treatment.

(ii) *Beta-Blockers*: These drugs such as carvedilol and metoprolol can also decrease the workload of the heart. Carvedilol was reported to reduced deaths and total hospitalizations by up to 50 percent when added to standard treatments for heart failure (see Section 6.3) (Figure 6.8).[1]

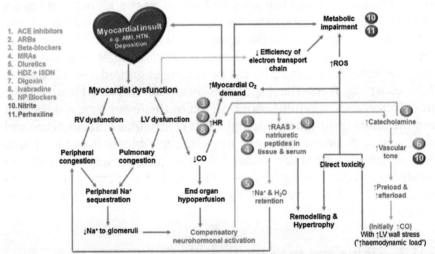

Figure 6.8 Overview of compensatory mechanisms in HF and complementary pharmacotherapeutic agents. Compensatory mechanisms and key pathophysiological changes take place in HF. Multiple pharmacotherapeutic agents have been developed to target these pathways and improve disease burden in HFrEF. ACE, angiotensin-converting enzyme; AMI, acute myocardial infarction; ARBs, angiotensin receptor blockers; CO, cardiac output; H2O, water; HDZ, hydralazine; HR, heart rate; HTN, hypertension; ISDN, isosorbide dinitrate; LV, left ventricle; MRAs, mineralocorticoid receptor antagonists; Na, sodium; NP, neprilysin; RAAS, renin–angiotensin–aldosterone system; ROS, reactive oxygen species; RV, right ventricle.

Adapted from: ref. 19, B. L. Loudon, H. Noordali, N. D. Gollop, M. P. Frenneaux and M. Madhani, *Br. J. Pharmacol.*, 2016, **173**, 1911–1924. © 2016 The Authors. *British Journal of Pharmacology* published by John Wiley & Sons Ltd on behalf of British Pharmacological Society.

(iii) *Aldosterone Blockers:* Spironolactone and eplerenone block the activity of aldosterone, an adrenal hormone that causes sodium retention. Aldosterone blockers are prescribed to individuals who develop heart failure after a heart attack, and studies show that these drugs can reduce the risk of hospitalization and death from cardiovascular disease (see Section 4.5.3).

(iv) *Diuretics:* Diuretics enable the body to rid itself of excess fluids (fluid overload) and sodium through urination and thus relieve the heart's workload (see Section 6.3). Also, they decrease the build-up of fluid overload in the lungs causing pulmonary congestion and other parts of the body, such as the gastrointestinal tract, hepatic and peripheral congestion. Different diuretics remove fluid at varied rates and through different methods and a careful balance in fluid homeostasis has be titrated against adverse properties of some diuretics (*e.g.* hypolakaemia, sympathetic hypotension and renal decline). The effects of this class of drugs will be discussed in greater detail in the section on the kidney and urinary system (see Section 6.12).

(v) *Calcium Antagonists*

Therapeutically important calcium antagonists act on L-type calcium channels. There are three chemically distinct classes of L-type calcium channels; phenylalkylamines (*e.g.* verapamil), dihydropyridines (*e.g.* nifedipine, amlodipine), and benzothiazepines (*e.g.* diltiazem). All three classes bind to the α-subunit of the L-type calcium channel. The *thiazide-type diuretics* are the most commonly used drugs and are known as "loop diuretics" (*e.g. hydrochlorothiazide,* chlorthalidone, bendrofluethiazide and indapamide). Their diuretic action is in part by inhibiting reabsorption of sodium (Na^+) and chloride (Cl^-) ions from the distal convoluted tubules in the kidneys by blocking the thiazide-sensitive Na^+–Cl^- cotransporter.

Diuretics are usually administered in low doses and initially they decrease cardiac output and lower blood volume over several days, is related to their mode of action. It has been proposed that these relatively long-acting diuretics produce a slowly developing decrease in extracellular sodium levels due to enhanced urinary excretion ("saluresis"), the likely cause of the hypotensive effect. Several calcium antagonists also show use dependence (*i.e.* they block more effectively cells in which the calcium channels are most active). They also show voltage-dependent blocking action, blocking more strongly when the membrane is depolarized, causing calcium channel opening and inactivation (see also their related class I antidysrhythmic activity).[1,2] The complexity and therapeutic uses of calcium-channel blockers are beyond the scope of this section and further reading is listed in the reference section.[1,20,22,23] In summary, diuretics are often used in combination with other agents based on symptoms and patient experience for the treatment of HF/CHF.

(vi) *Cardiac Glycosides*

The mode of action of the cardiac glycosides is extremely complex and is not completely understood. In the normal heart, Na^+/K^+ pumps in the cardiac myoctes pump potassium ions in and sodium ions out. Cardiac glycosides inhibit this pumping by stabilizing its transition state, so that sodium cannot be extruded: intracellular sodium concentration therefore increases. A second membrane ion exchanger, $Na+/Ca2+$ exchanger is responsible for "pumping" calcium ions out of the cell and sodium ions in. The raised intracellular sodium levels inhibit this pump, so calcium ions are also not extruded and will begin to build up inside the cell, as well. Increased cytoplasmic calcium concentrations cause increased calcium uptake into the sarcoplasmic reticulum (SR) via Ca^{2+} ATPase transporter within the myoctes (Figure 6.9). Raised calcium stores in the SR allow for greater calcium release on stimulation, so the myocyte can achieve faster and more powerful contraction by cross-bridge cycling (Figure 6.9). The refractory period of the AV node is increased, so cardiac glycosides also function to regulate heart rate. Binding of cardiac glycoside to Na^+-K^+ ATP-ase is slow, and after binding, intracellular calcium increases gradually. If SR calcium stores become too high, some ions are released spontaneously through SR ryanodine receptors. This effect leads initially to bigeminy (continuous alternation of long and short heart beats): regular ectopic beats following each ventricular contraction. If higher glycoside doses are given, rhythm is lost and ventricular tachycardia ensues, followed by fibrillation. It is likely that all of them, regardless of their specific chemical structures, have the same mechanism of action.

A positive inotropic effect combined with vagally mediated negatively chronotropic effects bring about increased cardiac

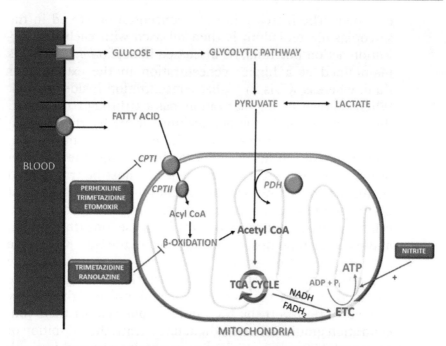

Figure 6.9 Schematic of the key metabolic pathways in cardiac myocytes. Fatty acid and glucose metabolism are key metabolic pathways within cardiac myocytes that are responsible for generating large amounts of ATP. Perhexiline and nitrite are therapeutic agents that have the ability to modulate and enhance cardiac metabolism. Acetyl CoA, acetyl coenzyme A; ADP, adenosine diphosphate; ATP, adenosine triphosphate; CPT, carnitine palmitoyltransferase; ETC, electron-transport chain; $FADH_2$, flavin adenine dinucleotide; NADH, nicotine adenine dinucleotide; Pi, inorganic phosphate; TCA, tricarboxylic acid.

Adapted from: ref. 19, B. L. Loudon, H. Noordali, N. D. Gollop, M. P. Frenneaux and M. Madhani, 2016, *Br. J. Pharmacol.*, 2016, **173**, 1911–1924. © 2016 The Authors. *British Journal of Pharmacology* published by John Wiley & Sons Ltd on behalf of British Pharmacological Society.

output, decreased heart size, lowered venous pressure, lowered blood volume, mobilization of edema fluid from the tissues and increased urine output. The cardiac glycosides themselves have no direct effect on the kidney and they do not produce diuresis in the individual who has a normally functioning heart. Moreover, in the individual who has a normally functioning heart, the cardiac glycosides produce only a very moderate positive inotropic effect.[1,2]

These drugs have a complex pharmacology, they inhibit Na^+/K^+ pump and increase Na^+ concentration, thereby reducing the inwardly directed gradient for Na^+. The smaller the gradient the slower the extrusion of Ca^{2+} by the Na^+/Ca^{2+}

exchange. The increase in Ca^{2+} concentration stored in the sarcoplasmic reticulum is then released with each depolarization action potential. As discussed in Section 4.5.1, Na^+ is maintained at a higher concentration in the extracellular fluid, whereas K^+ is in higher concentration inside the cell. When a wave of depolarization passes through the heart, there is a change in the permeability of the heart cell membranes Na^+ rapidly moves into the cell and K^+ moves out, both by passive diffusion. After the heart beats, the direction of ion migration must be reversed; K^+ must be transported against its concentration gradient into the cell and Na^+ must be transported against its concentration gradient out of the cell.[2] The active transport systems for these ions (the sodium and potassium pumps) intimately involve the Na^+/K^+ ATPase system. The hydrolysis of ATP to ADP by the enzyme and the accompanying release of energy by the cleavage of the so-called high-energy phosphate bond provide the energy needed to transport the Na^+ and K^+ ions against their concentration gradients. In the digitalized heart, the inhibition of transport ATPase results in and maintains elevated intracellular Na^+ concentrations. This in turn enhances the function of the Na^+–Ca^{2+} exchange transport system that results in an increase in available intracellular Ca^{2+}, to produce the positive inotropic effect, the desired effect of cardiac glycoside therapy. A low level of K^+ in the extracellular fluid compartment *enhances* and a high extracellular level of K^+ *antagonizes* the effects of the cardiac glycosides on the heart muscle. Thus, a common procedure for treating acute cardiac glycoside intoxication is the administration of a potassium salt to increase extracellular levels. These increased extracellular levels of K^+ stimulate the Na^+-K^+ ATPase pump system and this promotes removal of excess Na^+ from inside the cells and subsequent lowering of intracellular Ca^{2+}. The elevated extracellular K^+ also apparently decreases the binding affinity of the cardiac glycoside molecule to the heart tissue. These factors operate in concert to antagonize the effect of the cardiac glycoside on the heart. The cardiac glycosides slow conduction through the AV node by a central nervous system action that stimulates the vagus nerve. It will be recalled that stimulation of the vagus nerve slows heart rate (see Chapter 4). The resultant slowing of the rate of ventricular contraction is a desirable effect. Digoxin also suppresses the renin–angiotensin system in congestive heart failure patients and inhibits noradrenergic nervous activity.[1,2] However, all of the digitalis glycosides have about the same extremely small therapeutic index and the role of digoxin in HF remain heatedly controversial given the risk of toxicity.[2,20] A typical therapeutic dose is approximately 50–60% of the toxic dose level. All of these are potentially dangerous drugs. In most individuals, digoxin is excreted in the urine virtually 100% chemically unaltered, a result of combined glomerular filtration and direct tubular secretion.[1]

(vii) *Hydralazine and Isosorbide Dinitrate*

Hydralazine and isosorbide dinitrate (ISDN) act by reducing both resistance vessel and venous tone on the failing heart hydralazine and ISDN in combination was also shown to reduce nitrate tolerance by previously unknown mechanisms. Organic nitrates activate NADPH oxidase, which results in deleterious ROS generation, partially contributing to nitrate tolerance *via* endothelial dysfunction (uncoupling of NOS) and inhibition of mitochondrial aldehyde dehydrogenase 2 (responsible for a large part of organic nitrate enzymatic bioactivation to NO).[19] Hydralazine prevents NADPH oxidase activation, reducing nitrate toxicity, tolerance and free-radical production.[19]

(viii) *Ivabradine*

Ivabradine is a selective *If* current inhibitor in the SA node causing exclusively negative chronotropic effects. In the clinical setting, it is increasingly used for patients in whom β blockers are not tolerated or is contraindicated. A number of clinical studies have shown promise but not without controversy. The debate on the role of ivabradine in cardiovascular disease is therefore ongoing.[20]

(ix) *Neprilysin Inhibition*

Heart failure stimulates the compensatory natriuretic peptide system, which is composed of atrial natriuretic peptide, brain natriuretic peptide (BNP), C-type natriuretic peptide and other vasoactive substances such as angiotensin II and bradykinin.[21] Neprilysin degrades these vasoactive peptides, and so targeted inhibitors of neprilysin increase circulating levels of these substances and counteract the effects of neurohormonal overactivation. Sole inhibitors of neprilysin, however, failed to demonstrate a significant impact on blood pressure in hypertensive cohorts and so were discontinued.[21,23] This may have been in part due to an attendant increase in angiotensin, necessitating the combination of neprilysin inhibitors with drugs that target the RAAS. Omapatrilat, the first dual neprilysin and ACE inhibitor, was shown to be superior to enalapril alone. Unfortunately, there was a significant increase in the incidence of angioedema in the treatment arm (likely to be due to the substantial inhibition of bradykinin degradation), and the drug was discontinued. LCZ696, a combination of sacubitril (neprilysin inhibitor) and valsartan (ARB), as previously stated, showed a similar reduction in all-cause mortality compared with an ACE inhibitor combination, with better safety data due to an attenuated effect on bradykinin breakdown.[21,22]

In summary, a variety of drugs may be used to manage heart failure and have been discussed in previous sections for other cardiovascular disorders. As discussed, these therapeutic interventions will depend on the patient's specific condition, including the severity of heart failure, its cause, and the presence of other health problems.[22,23]

6.12   Kidney and Urinary System

There follows a brief outline of renal physiology based on the functional unit of the kidney – the nephron and the various drugs that affect renal function. It is beyond the scope of this section to fully review both the kidney's role in cardiovascular disease and drug treatment. Several drug classes have previously been discussed, *i.e.* diuretics and their role in the excretion of Na^+ ions and water, and their hypotensive activity (see also Section 6.6), antihypertensive agents other than diuretics, see sections on Autonomic and Cardiovascular Pharmacology. Drugs used to treat patients with renal failure and urinary tract disorders will also be briefly discussed.

6.12.1   Physiology and Pharmacology

The kidneys are the main organ by which drugs and their metabolites are eliminated from the body and the adverse effects of drug therapy are critical in patients with impaired renal function and specific diseases (*e.g.* hypertension and immunosuppressant agents), along with age-related decline. The various kinds of drug toxicity are due in part to the very high concentrations of drugs and drug metabolites in some renal tissues.[1] Each kidney consists of an outer cortex, an inner medulla and a hollow pelvis, which empties into the ureter. The functional unit is the nephron that regulates fluid volume, electrolyte and pH of the extracellular fluid, resulting from dietary intake and other environmental demands (*e.g.* climatic).[1] The kidneys are also the main organ by which drugs and their metabolites are eliminated from the body.

The functional unit of the kidney, the nephron, is shown in Figure 6.10. Each kidney contains approximately 1 million nephrons; their basic structure consists of a glomerulus, proximal tubule, loop of Henle, distal convoluted tubule and collecting ducts (Figure 6.10). The glomerulus network of small capillaries form tufts or fenestration that project into the renal tubule and act as pores to permit the passage and filtration of water and small soluble molecules (approximately 20% of the blood). Glomerular filtration acts on the same principle as systemic circulation where blood enters the nephron and is filtered under hydrostatic pressure through the afferent renal arterioles that flow through the network of thin-walled glomerulus capillaries into the dilated end of the renal tubule (Figure 6.10). The blood supply to each nephron consists of two capillary beds. The

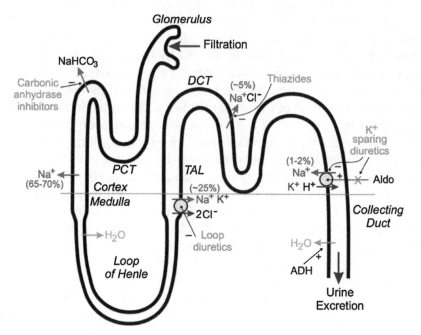

Figure 6.10 The kidney and renal handling of sodium and water. Reproduced with permission from Richard Klabunde, www.cvpharmacology.com.

afferent arteriole of each cortical nephron branches to form a network of glomerulus capillaries that coexist in the glomerulus with the efferent arterioles that form Bowman's capsule (Figure 6.10). The filtrate that enters Bowman's capsule is driven by hydrodynamic force opposed by the oncotic pressure of plasma proteins. It largely contains water, inorganic solutes and organic solutes that include drug molecules that are not bound to plasma proteins. Larger proteins (above a molecular weight of 20 000), such as albumin and structures such as red and white blood cells and platelets are not passed through the filtration process. These blood constituents and formed elements are returned to the general circulation *via* a second capillary network in the cortex surrounding the convoluted tubule and loops of Henle that converge onto venules before returning to the renal veins. The efferent arterioles of juxtamedullary nephrons descend to form the *vasa recta* that returns blood into the renal arterioles in the medulla (Figure 6.10). Over 80% of the blood from the glomerulous network is returned into the general circulation as a result of glomerular filtration pressure, estimated to be 50 mm Hg. The body's essential elements, such as sodium and potassium cations, glucose and amino acids are reabsorbed into the blood from the tubular filtrate by active transport mechanism. The remaining glomerular filtrate is concentrated in the collecting ducts and passes into the urinary bladder and excreted as urine (Figure 6.10). It has been estimated that 170 L of fluid is filtered by the human kidneys each day, of which urine excretion accounts for 1.5 L per day. Thus, leading to 99% of the glomerular filtrate being reabsorbed by the body.

The various regions of the renal tubule are presented in Figure 6.10. In the *proximal convoluted tubule*, potassium and sodium ions, glucose, amino acids, phosphate and bicarbonate ions and large amounts of water are reabsorbed into the general circulation. The fluid remaining in the tubule moves into the *loop of Henle*, first to the *descending limb* thence to the *ascending limb*. The loop of Henle is believed to be another important site for reabsorption of water. Next, the remaining fluid moves into the *distal convoluted tubule* where additional sodium is reabsorbed into the general circulation to exchange for some potassium ions and protons. This ion exchange in the distal tubule is stimulated by aldosterone. In the next region, the "collecting tubule", more water is reabsorbed and this effect is regulated by a peptide released from the posterior pituitary gland, the *antidiuretic hormone* (arginine vasopressin) whose function is to promote increased reabsorption of water from the tubular fluid.

Thus, the body conserves ions that it utilizes in its normal physiology. These ion-reabsorption processes (notably for Na^+, K^+ and Cl^-) involve active transport mechanisms. Waste materials of the body's metabolism, such as urea, are not reabsorbed but they remain in the tubular fluid and they are transported to the urinary bladder, whence they are excreted.[1,2]

6.12.2 Tubular Function and Drug Excretion

The kidneys tubular cells like all epithelia cells consist of tight junctions. It is these specialized regions of the cell membrane that separate the intercellular space from the lumen. The movement of ions and water across the epithelium between cells occurs through these tight junctions driven by various Na^+–K^+-ATPase transporters in different parts of the nephron, as discussed below.[24]

(i) Glomerulus/Bowman's Capsule: Osmotic diuretics – Most of the osmotic diuretics are sugars (*e.g.* glucose, sucrose) or polyols (*e.g.* mannitol, sorbitol) and urea. They elevate the osmotic pressure in the tubules, impeding the reabsorption of water, largely from the loop of Henle and, to a lesser extent, from the proximal convoluted tubule. There is an accompanying modest increase in urinary excretion of almost all ionic species: Na^+, K^+, Ca^{2+}, Mg^{2+}, Cl^-, phosphate and HCO_3^-. Osmotic diuretics are rarely indicated, *e.g.* lowering intraocular pressure in the eye and reducing cerebral edema before and after neurosurgery (Figure 6.10).

(ii) The Proximal Convoluted Tubule
The Na^+/H^+ exchange transporter is the most important transport mechanism for Na^+ entry into proximal tubular cells. Na^+ is reabsorbed from tubular fluid into the cytoplasm of proximal tubular cells in exchange for cytoplasmic H^+, produced by intracellular carbonic anhydrase. Na^+ is then transported out of the cells into the interstitium by a Na^+–K^+-ATPase (sodium pump) in the basolateral membrane. This is the main active transport mechanism of the nephron. Reabsorbed Na^+ then diffuses into blood vessels (Figure 6.10). Proximal tubular fluid (approximately 30–40% of the total filtrate) now passes into the loop of Henle, containing organic acids and bases that are actively secreted into the tubule from the blood by specific transporters.[1,2]

(iii) The Loop of Henle

The loop of Henle consists of two limbs, (i) *descending* and (ii) *ascending* (Figure 6.10) that enable the kidney to excrete urine and regulate urine osmolality and the osmotic balance within the body. The loops of Henle of the juxtamedullary nephrons act as countercurrent multipliers, and the *vasa recta* as countercurrent exchangers. NaCl is actively reabsorbed in the thick ascending limb, causing hypertonicity of the interstitium. In the descending limb, water moves out and the tubular fluid becomes progressively more concentrated as it approaches the bend.[1]

The *descending limb* is permeable to water and kept hypertonic by the countercurrent concentrating system. The hypertonic milieu of the kidney medulla, through which the collecting ducts of all nephrons pass, is important in providing a mechanism by which the osmolarity of the urine is controlled. The *ascending limb* has very low permeability to water, thus enabling the build-up of a substantial concentration gradient across the wall of the tubule. Approximately, 20–30% of filtered Na^+ is actively reabsorbed along with NaCl in the thick tubular wall, thus, reducing the osmolarity of the tubular fluid and making the interstitial fluid of the medulla hypertonic. Cations move into cells in the ascending limb of the loop of Henle across the apical membrane by a $Na^+/K^+/2Cl^-$ cotransporter, driven by the Na^+ gradient produced by Na^+-K^+-ATPase in the basolateral membrane. The K^+ taken into the cell by the $Na^+/K^+/2Cl^-$ cotransporter returns to the lumen through apical potassium channels, but some K^+ is reabsorbed, along with Mg^{2+} and Ca^{2+}.[1]

The countercurrent multiplier system plays an important role in maintaining the osmotic gradient in the nephron. Other filtrate constituents (*e.g.* urea) that contribute to the gradient increase in concentration and pass into the collecting tubules (Figure 6.10). Tubular fluid is hypotonic in the loop of Henle with respect to plasma as it enters the distal convoluted tubule and salt reabsorption is not balanced by reabsorption of water.[1]

(iv) The Distal Tubule

Again, active transport is driven by the Na^+-K^+-ATPase pump in the basolateral membrane. This lowers cytoplasmic Na^+ concentration, and consequently Na^+ enters the cell from the lumen down its concentration gradient, accompanied by Cl^-, by means of a Na^+/Cl^- cotransporter and results in a further dilution of the tubular filtrate (Figure 6.10). The excretion of Ca^{2+} is regulated in this part of the nephron, *parathormone* and *calcitriol* both increasing Ca^{2+} reabsorption.[1]

(v) The Collecting Tubule and Collecting Tubule

The filtrate from the distal convoluted tubules empty into collecting tubules, that include the principal cells, which re-absorb Na^+ and secrete K^+ along with two populations of intercalated cells, α and β, which secrete acid and base, re-spectively. It is important to note that the tight junctions in this portion of the nephron are impermeable to water and ions. The movement of ions and water in this segment is under independent hormonal control: absorption of NaCl by *al-dosterone* of water by *antidiuretic hormone* (ADH), also termed *vasopressin*.

Aldosterone enhances Na^+ reabsorption and promotes K^+ excretion. It promotes Na^+ reabsorption by: a rapid effect, stimulating Na^+/H^+ exchange by an action on membrane al-dosterone receptors and a delayed effect, *via* nuclear receptors directing the synthesis of a specific protein mediator that ac-tivates sodium channels in the apical membrane.[1]

6.12.3 Other Drugs Acting on the Kidney

The kidneys are a defence mechanism against foreign substances ("xenobiotics") and the target for various kinds of drug toxicity re-sulting from high concentrations of the parent compound or drug metabolites. The body's organs responsible for drug metabolism (*e.g.* liver) act to prepare the kidney to excrete drug molecules in urine.[1] Several antihypertensive agents are commonly indicated in kidney disease, along with immunosuppressant drugs (*e.g.* following renal transplantation), antibacterial drugs (used to treat renal and urinary tract infections), and drugs used to treat lower urinary tract disorders that commonly cause urinary retention or incontinence.[11,25] The physical properties of a drug determine its metabolism and excretion, 99% of lipophilic drugs are reabsorbed by passive reabsoption.[1,25] Such drugs are actively converted by the body to more hydrophilic, lipophobic molecules to reduce their reabsorption by the tubules and eventual excretion. Overall, such drugs tend to have a long duration of action in the body.[1,25] In contrast, hydrophilic molecules are rapidly and efficiently excreted by the kidneys in the urine. However, a drug molecule that is bound to plasma proteins is not filtered by the glomerulus and remains in the circulation, until such time when the protein bound and free-drug equilibrium in the aqueous phase is able to pass through the tubular glomerulus network for eventual excretion (see Figure 6.10).

There are many examples of the complexity of tubular excretion of drugs that are beyond the scope of this chapter (see references for further reading). There follows the various classes of drugs acting at the kidney tubules.

(i) Drugs that Alter the Acid–Base Balance of the Body

A drug that inhibits tubular reabsorption of sodium and bicarbonate ions, thus making the tubular fluid pH more alkaline, results in a systemic acidosis (elevated acidity in the blood and extracellular fluid). Acetazolamide whose mechanism of action involves inhibition of the enzyme carbonic anhydrase (see Figure 6.10) increases urinary pH by blocking bicarbonate reabsorption. The strategy of altering of acid–base balance to produce diuresis is considered to be obsolete. The carbonic anhydrase inhibitors overall show a low level of diuretic efficacy.[1,25]

(ii) Potassium Balance

The kidney is primarily responsible for maintaining total-body K^+ content by matching K^+ intake with K^+ excretion. Adjustments in renal K^+ excretion occur over several hours; therefore, changes in extracellular K^+ concentration are initially buffered by movement of K^+ into or out of skeletal muscle. The regulation of K^+ distribution between the intracellular and extracellular space is referred to as internal K^+ balance (see Figure 6.10). The most important factors regulating this movement under normal conditions are insulin and catecholamines.

(iii) Natriuretic Peptides

NPs (A, B and C) released from the heart or endothelium affect the kidneys by increasing glomerular filtration rate (GFR) and filtration fraction, which produces natriuresis and diuresis. These renal effects of NPs are K^+ sparing, unlike most diuretic drugs that are used to induce natriuresis and diuresis in patients.

Additionally, NPs decrease renin release, thereby decreasing circulating levels of angiotensin II and aldosterone, leading to further natriuresis and diuresis. Decreased angiotensin II release also contributes to systemic vasodilation and decreased systemic vascular resistance. The mechanism of systemic vasodilation involves NP receptor-mediated increases in vascular smooth muscle cGMP as well as a reduction of sympathetic vascular tone. Nesiritide and endopepitases inhibitors are antagonists of NPs.

(iv) Prostaglandins and Renal Function

Prostaglandins generated in the kidney influence its haemodynamic and excretory functions.[1] Their actions result in vasodilation and natriuruesis. They act at different renal sites: PGE_2 in the medulla and PGI_2 (prostacyclin) in glomeruli. Their synthesis and release occurs following ischaemia, angiotensin II, ADH and bradykinin stimulus. Local release of PGE_2 and PGI_2, maintains renal blood flow to compensate for vasodilator action.

Drugs that influence of renal prostaglandins on salt balance include nonsteroidal anti-inflammatory drugs (NSAIDs), which inhibit prostaglandin production by inhibiting cyclooxygenase but are contraindicated in patients with acute renal failure in which renal blood flow depends on vasodilator prostaglandin biosynthesis (*e.g.* cirrhosis, heart failure, nephrotic syndrome, glomerulonephritis and extracellular volume contraction).[1,25]

(v) Loop Diuretics: Frusemide, bumetanide, torasemide and ethacrynic acid are potent diuretics, which inhibit $Na^+K^+2Cl^-$ carrier in the thick ascending limb of the loop of Henle (see Figure 6.10). Thus, these agents are effective even when GFR is low. Frusemide is started in low dose of 20–40 mg per day and dose increased to achieve the desired volume reduction. Frusemide and bumetanide are short acting and there is postdose retention of Na^+ from all segments of nephron. This Na^+ retention causes loss of diuretic efficacy and can be avoided by giving frusemide twice a day.[1,25] Hypokalemia can be prevented by combining loop diuretics with spironolactone, amiloride or triamterene. Frusemide is the diuretic of choice in acute left ventricular failure/pulmonary oedema. It rapidly reduces dyspnoea and pulmonary congestion by its venodilatory action that is observed earlier to diuretic effect. It is given as 20–40 mg i.v. and may also be given by continuous infusion at a rate of 10 mg h^{-1}. This method reduces the risk of ototoxicity and there is sustained natriuretic action. Adverse effects of loop diuretics lead to increased renin and angiotensin-II secretion because of a reduction in blood volume. Chronic use causes electrolyte disturbances (hypokalemia), raise LDL and uric acid levels and worsens diabetes mellitus. There is no reduction in mortality on long-term use. Diuretic resistance may occur. In nephrotic syndrome, loop diuretics become bound to albumin in the tubular fluid, and consequently are not available to act on the $Na^+/K^+/2Cl^-$ carrier – another cause of diuretic resistance.[1,25] Molecular variation in the $Na^+/K^+/2Cl^-$ carrier may also be important in some cases of diuretic resistance. In nephrotic syndrome, loop diuretics become bound to albumin in the tubular fluid, and consequently are not available to act on the $Na^+/K^+/2Cl^-$ carrier – another cause of diuretic resistance. Molecular variation in the $Na^+/K^+/2Cl^-$ carrier may also be important in some cases of diuretic resistance.

(vi) Drugs Acting at the Distal Tubule

Thiazides are the most commonly used diuretic and inhibit the Na^+/Cl^- transporter in the distal tubule, inhibiting its action and causing natriuresis with loss of sodium and chloride ions in the urine. Because this transporter normally only reabsorbs about 5% of filtered sodium, these diuretics are less efficacious than loop diuretics in producing diuresis and natriuresis. Nevertheless, they are sufficiently powerful to satisfy most therapeutic needs requiring a diuretic and are preferred in treating uncomplicated hypertension. Thiazides are less powerful than loop diuretics, at least in terms of a peak increase in the rate of urine formation.[1] Their mechanism depends on renal prostaglandin production. They are better tolerated than loop diuretics, and in clinical trials have been shown to reduce risks of stroke and heart attack associated with hypertension (see Section 6.3). Both loop and thiazide diuretics increase Na^+ delivery to the distal segment of the distal tubule, this increases K^+ loss (potentially causing *hypokalemia*) because the increase in distal tubular Na^+ concentration stimulates the aldosterone-sensitive Na^+ pump to increase Na^+ reabsorption in exchange for K^+ and H^+ ion, which are lost to the urine (see Figure 6.10). The resulting contraction in blood volume stimulates renin secretion, leading to angiotensin formation and aldosterone secretion (see Section 6.3). This homeostatic mechanism limits the effect of the diuretic on the blood pressure, resulting in an *in vivo* dose–hypotensive response relationship with only a gentle gradient during chronic dosing. The effects of thiazides on Na^+, K^+, H^+ and Mg^{2+} balance are qualitatively similar to those of loop diuretics (see Figure 6.10), but less effective. In contrast to loop diuretics, however, thiazides reduce Ca^{2+} excretion, which may be advantageous in older patients at risk of osteoporosis.[1] Although thiazides are milder than loop diuretics when used alone, coadministration with loop diuretics has a synergistic effect, because the loop diuretic delivers a greater fraction of the filtered load of Na^+ to the site of action of the thiazide in the distal tubule (see Figure 6.10). Thiazide diuretics have a vasodilator action and their action in hypertension are discussed in detail in Section 6.4. Thiazide diuretics have a paradoxical effect in diabetes insipidus, where they reduce the volume of urine by interfering with the production of hypotonic fluid in the distal tubule, and hence reduce the ability of the kidney to excrete hypotonic urine (*i.e.* they reduce free-water clearance). Thiazides and thiazide-related drugs are effective orally. All are excreted in the urine, mainly by tubular secretion (see Figure 6.10). and they compete with uric acid for the organic anion transporter.[1,25] The clinical uses of thiazide diuretics are listed below;

- *Hypertension.*
- Mild *heart failure* (loop diuretics are usually preferred).
- Severe resistant *oedema* (metolazone, especially, is used, together with loop diuretics).
- To prevent recurrent stone formation in *idiopathic hypercalciuria.*
- *Nephrogenic diabetes insipidus.*

Apart from an increase in urinary frequency, the commonest unwanted effect of thiazides are: erectile dysfunction, potassium loss can be important, as can loss of Mg^{2+}. Excretion of uric acid is decreased, and hypochloraemic alkalosis can occur. Impaired glucose tolerance due to inhibition of insulin secretion, is thought to result from activation of K_{ATP} channels in pancreatic islet cells. Diazoxide, a non-diuretic thiazide, also activates K_{ATP} channels, causing vasodilatation and impaired insulin secretion. Indapamide is said to lower blood pressure with less metabolic disturbance than related drugs, possibly because it is marketed at a lower equivalent dose.[1,25] Hyponatraemia is potentially serious, especially in the elderly. Hypokalaemia can be a cause of adverse drug interaction (see above under loop diuretics) and can precipitate encephalopathy in patients with severe liver disease.[1,25,27]

(vii) Carbonic Anhydrase Inhibitors (*e.g.* azetazolamide, citrate) inhibit the transport of bicarbonate out of the proximal convoluted tubule into the interstitium, which leads to less Na^+ reabsorption at this site and therefore greater sodium, bicarbonate and water loss in the urine. These agents alter tubular fluid pH resulting in alkalinization that is important in preventing certain weak acid drugs with limited aqueous solubility, such as *sulfonamides*, from crystallizing in the urine; it also decreases the formation of uric acid and cystine stones by favoring the charged anionic form that is more water soluble (see Figure 6.10). Drugs that are weak acids (*e.g.* salicylates and some barbiturates) are excreted in the urine by this process.[1,25,27]

These are the weakest of the diuretics and are seldom used in cardiovascular disease. Their main use is in the treatment of glaucoma, inhibiting carbonic anhydrase in the eye to decrease the amount of fluid (aqueous humor) and thus lowering intraocular pressure.[2]

(viii) Xanthines

Caffeine, theophylline and theobromine are nonselective adenosine receptor antagonists with mild diuretic activity, which is competitively antagonized by theophylline. Their clinical utility is limited by their low potency, tolerance and CVS and CNS stimulation.[1,25,27]

(ix) Aldosterone Antagonists

There is a third class of diuretic that is referred to as potassium-sparing diuretics. Unlike loop and thiazide diuretics, some of these drugs do not act directly on sodium transport. Some drugs in this class antagonize the actions of aldosterone (aldosterone receptor antagonists) at the distal segment of the distal tubule. This causes more sodium (and water) to pass into the collecting duct and be excreted in the urine. They are called K^+-sparing diuretics because they do not produce hypokalemia like the loop and thiazide diuretics. The reason for this is that by inhibiting aldosterone-sensitive sodium reabsorption, K^+ and H^+ ion are exchanged for sodium by this transporter and therefore less K^+ and H^+ are lost to the urine. Other K^+-sparing diuretics directly inhibit Na^+ channels associated with the aldosterone-sensitive sodium pump, and therefore have similar effects on K^+ and H^+

ion as the aldosterone antagonists (see Figure 6.10). Their mechanism depends on renal prostaglandin production. Because this class of diuretic has relatively weak effects on overall sodium balance, they are often used in conjunction with thiazide or loop diuretics to help prevent hypokalemia.[1,27]

Spironolactone and its long-acting metabolite canrenone are mild K^+ conserving/sparing diuretics, which given with thiazides or loop acting diuretics antagonize their hypokalemic effects. Eplerenone is another K^+-sparing diuretic. The advantage with eplerenone is that it does not have endocrine adverse effects (gynecomastia, impotence, menstrual disturbance) of spironolactone. They compete with aldosterone for its intracellular receptor, thereby inhibiting distal Na^+ retention and K^+ secretion.[1,25,27]

6.12.4 Electrolyte Disorders and Renal Failure

(i) Hyperphoshataemia: Antacids (*e.g.* aluminium hydroxide and phosphate binding agents – calcium carbonate) affect phosphate and calcium absorption and are contraindicated in dialysis patients and in patients predisposed to tissue calcification (hardening of the arteries).

Sevelamer is an anion-exchange resin that lowers plasma phosphate and is less likely to cause tissue calcification, but its use is limited by gastrointestinal disturbances and contraindicated in bowel obstruction.[1,27]

(ii) Hyperkalaemia: Severe hyperkalaemia is life-threatening and treated with intravenous calcium gluconate with concurrent measurement of the cellular uptake of K^+ into intracellular compartments.[1,25] Intravenous salbutamol (albuterol) or sodium bicarbonate stimulate cellular uptake of K^+ and is used in this indication. Excessive K^+ can also be removed by cation-exchange resins (*e.g.* sodium or calcium polystyrene sulfonate).[1,25,27]

6.12.5 Excretion of Organic Molecules

Uricosuric drugs are used to treat abnormalities of uric acid metabolism and secretion (*e.g.* gout). Such drugs increase the elimination of urate, include probenecid and sulfinpyrazone but have largely been replaced in the clinic by allopurinol that inhibits urate synthesis.[1,25,27]

6.12.6   Drugs Used in Urinary Tract Disorder

Drug treatment falls in to four categories:

- Enuresis (bed wetting) in children is treated with desmopressin (ADH analog).
- Benign prostatic hyperplasia – A_1-adrrenoceptor antagonists (*e.g.* doxazosin, tamsulosin) or the androgen synthesis inhibitors (*e.g.* finasteride).
- Neurogenic detrusor muscle instability – muscarinic receptor antagonist (*e.g.* oxybutynin).
- Overactive bladder – selective β_3 agonist (*e.g.* mirabegron).

References

1. *Rang & Dale's Pharmacology*, ed. H. P. Rang, J. M. Ritter, R. J. Flower and G. Henderson, Elsevier, Churchill Livingstone, 8th edn, 2016.
2. J. C. Cannon, *Pharmacology for Chemists*, Oxford University Press, 2nd edn, 2007.
3. D. A. Jones, A. Timmis and A. Wragg, Novel drugs for treating angina, *Br. Med. J.*, 2013, **347**, 34–37.
4. D. L. Mann, D. P. Zipes, P. Libby and R. O. Bonow, *Braunwald's Heart Disease: A Textbook of Cardiovascular Medicine*, Saunders/Elsevier, Philadelphia, 10th edn, 2014.
5. L. H. Opie and B. J. Gersh, *Drugs for the Heart*, Saunders/Elsevier, Philadelphia, 8th edn, 2013.
6. M. Seddon, N. Melikian and R. Dworakowski, *et al.*, *Circulation*, 2009, **119**, 2656–2662.
7. S. P. H. Alexander, H. E. Benson and E. Faccenda, *et al.*, 2013/14, Concise Guide to Pharmacology (Angiotensin receptors, p. 1484).
8. S. P. H. Alexander, H. E. Benson and E. Faccenda, *et al.*, 2013/14, Concise Guide to Pharmacology (Endothelin receptors, p. 1505).
9. A. H. Chester and M. H. Yacoub, *Glob. Cardiol. Sci. Pract.*, 2014, **29**, 70–78.
10. J. M. Ritter, *Br. Med. J.*, 2011, **342**, 868–873.
11. J. A. Oates and N. J. Brown, in *Goodman and Gilman's The Pharmacological Basis of Therapeutics*, ed. J. G. Hardman, L. E. Limbird and A. G. Gilman, McGraw-Hill, New York, 11th edn, 2001, pp. 871–900.
12. J. McAinsh and J. M. Cruickshank, *Pharmacol. Ther.*, 1990, **46**(2), 163–197.
13. P. A. James, S. Oparil and B. L. Carter, *et al.*, *J. Am. Med. Assoc.*, 2014, **311**(5), 507–520.
14. N. R. Poulter, D. Prabhakaran and M. Caulfield, *Lancet*, 22 August 2015, **386**(9995), 801–812.
15. R. Birner-Gruenberger, M. Schittmayer, M. Holzer and G. Marche, *Prog. Lipid Res.*, 2014, **56**, 36–46.
16. T. Alonzi, C. Mancone, L. Amicone and M. Tripodi, *Expert Rev. Proteomics*, 2008, **5**, 91–104.
17. J. O. Lundberg, M. T. Gladwin and E. Weitzberg, *Nat. Rev. Drug Discovery*, 2015, **14**, 623–641.
18. M. J. Shearer and P. Newman, *Thromb. Haemostasis*, 2008, **100**, 530–547.
19. B. L. Loudon, H. Hannah Noordali, N. D. Gollop, M. P. Frenneaux and M. Madhani, *Br. J. Pharmacol.*, 2016, **173**, 1911–1924.

20. J. J. McMurray, S. Adamopoulos, S. D. Anker, A. Auricchio, M. Bohm and K. Dickstein, *et al.*, *Eur. Heart J.*, 2012, **33**, 1787–1847.
21. O. Vardeny, R. Miller and S. D. Solomon, *JACC: Heart Failure*, 2014, **2**, 663–670.
22. M. Parker, J. J. McMurray, A. S. Desai and J. Gong, *et al.*, *Circulation*, 2015, **131**, 54–61.
23. C. M. White, *J. Clin. Pharmacol.*, 2002, **42**(9), 963–970.
24. W. Lee and R. B. Kim, *Annu. Rev. Pharmacol. Toxicol.*, 2003, **44**, 137–166.
25. G. S. Pazhayattil and A. C. Shirali, *Int. J. Nephrol. Renovasc. Dis.*, 2014, **12**(7), 457–468.
26. S. P. H. Alexander, H. E. Benson and E. Faccenda, *et al.*, 2013/14, Concise Guide to Pharmacology (calcium-sensing receptors, p. 1491).
27. A. C. Guyton and J. E. Hall, 2000, *Urine Formation in the Kidney: 1. Glomerular Filtration, Renal Blood Flow, and their Control. 2. Tubular Processing of the Glomerular Filtrate*, Textbook of Human Physiology, W.B. Saunders, Philadelphia, Pa., pp. 279–312.

7 Actions of Drugs in the Gastrointestinal Tract

Gareth J. Sanger[†]

Blizard Institute, Barts & The London School of Medicine and Dentistry,
Queen Mary University of London, UK
Email: g.sanger@qmul.ac.uk

7.1 Introduction

After eating and chewing, food is moved into the stomach *via* the esophagus. There, it is exposed to acid, peptidases and mechanical forces that begin to break it into smaller, partly digested fragments before being moved into the intestine for further digestion and absorption of water and nutrients. Different mechanisms operate to induce satiety, minimise overeating or ingestion of food high in one type of nutrient (*e.g.* amino acids, glucose, fatty material) and if necessary, induce nausea and/or vomiting to protect against ingestion of harmful or toxic material. After passing through the small intestine further absorption of water occurs in the proximal colon, along with digestion of plant cellulose by bacteria within the lumen. In the distal colon and rectum, faeces are stored prior to defecation. Throughout the gastrointestinal (GI) tract an immune system protects the mucosa from invasion by bacteria, parasites and

[†]Postal address: National Bowel Research Centre, Abernathy Building, 2 Newark Street, London, E1 2AT, UK.

Pharmacology for Chemists: Drug Discovery in Context
Edited by Raymond Hill, Terry Kenakin and Tom Blackburn
© The Royal Society of Chemistry 2018
Published by the Royal Society of Chemistry, www.rsc.org

antigenic proteins. Diarrhoea will occur normally to expel toxins or ingested parasites.

GI disorders and symptoms occur, for example, when there is a localised irritation or inflammation of the mucosa (*e.g.* ulceration), gastric acid "in the wrong place" (gastroesophageal reflux disease), inappropriate activation of vagal nerve and other pathways that lead to nausea and vomiting, changes in GI movements caused by different factors including neurodegeneration and/or loss of interstitial cells of Cajal, the presence of toxins and parasites leading to diarrhoea, the effects of many different drugs used for treatment of non-GI disorders and a range of "idiopathic" mechanisms hypothesised to explain functional bowel disorders such as functional dyspepsia (FD) and irritable bowel syndrome (IBS). The following summarises the drugs used to treat GI diseases and identifies current gaps in research and treatments. Inflammation bowel disorders are the focus of extensive research into the generic mechanisms of inflammation and are discussed elsewhere.

7.2 Protection and Healing of the Oesophagus and Stomach

This is required for patients with gastroesophageal reflux disease (GERD), hiatus hernia, gastric and duodenal ulcers, gastritis and dyspepsia. GERD is further classified into patients with nonerosive reflux disease (NERD), erosive esophagitis and Barrett's esophagus, each promoted by chronic exposure to acid. Barrett's esophagus is characterised by metaplasia in cells of the distal esophagus, with replacement of stratified squamous epithelium by simple columnar epithelium with goblet cells (usually found lower in the GI tract) and often leading to an esophageal adenocarcinoma.

7.2.1 Inhibition of Gastric Acid

Hydrochloric acid (HCl) is formed by the stomach during ingestion and digestion of food, helping to denature proteins and by activating pepsinogen to form the enzyme pepsin, facilitate protein breakdown into chains of amino acids. HCl is produced by parietal cells within the epithelium of gastric glands in the fundus and body of the stomach. In humans, the acid is secreted mostly "on demand", during anticipation of eating (*via* vagal nerve activation) and by the presence of amino acids within the stomach. In animals that "graze" for long periods of time (*e.g.* rodents) the secretion of gastric acid is more prolonged.

HCl is synthesised by:

- Dissociation of water to generate H^+ and OH^-, the latter combining with CO_2 (cataylsed by carbonic anhydrase) to form HCO_3^-.
- Transport of HCO_3^- out of the parietal cell into the venous blood in exchange for Cl^-.
- Transport of Cl^- and K^+ out of the parietal cell into the stomach lumen.
- Pumping of H^+ out of the parietal cell and into the gastric lumen, in exchange for K^+, *via* the H^+/K^+ ATPase (the proton pump). These pumps are increased in number when required and removed during deactivation.
- Diffusion of water into the lumen along the osmotic gradient created by the H^+ (HCl now part of the "gastric juice").

Acetylcholine (ACh), released from the vagus and enteric neurons, activates parietal cells directly and also indirectly by stimulating enterochromaffin-like cells near the parietal cells to release histamine. Histamine and also gastrin, released from endocrine cells within the mucosa in response to a number of different stimuli, directly stimulate the parietal cells by acting at, respectively, histamine H_2 and cholecystokinin CCK_2 receptors.

An early use of muscarinic receptor antagonists to inhibit gastric acid secretion (*e.g.* pirenzepine) was discontinued in favour of the H_2 receptor antagonists (cimetidine, famotidine, nizatidine, ranitidine) and proton pump inhibitors (omeprazole, lansoprazole, pantoprazole, rabeprazole, esomeprazole, tenatoprazole). H_2 receptor antagonists competitively block the ability of histamine to stimulate acid secretion. PPIs are prodrugs and weak bases that accumulate in the acid of the parietal cell, before acid-dependent conversion to molecules that react covalently with cysteines on the H^+/K^+ ATPase, blocking its function. Compared with the H_2 receptor antagonists PPIs are slower in onset of activity but more effectively inhibit gastric acid secretion. Other treatments include alginate-based formulations to minimise reflux (by creating a foamy raft in the stomach).

PPIs and H_2 receptor antagonists are also successfully used to treat many patients with functional dyspepsia, a functional bowel disorder with unknown mechanisms, characterised symptomatically by postprandial fullness, early satiation, epigastric pain and burning, often accompanied by nausea, belching or abdominal bloating (see section on Functional Bowel Disorders). In these patients, it has been suggested that a raised perception of gastric acid (perhaps associated with the presence of *Helicobacter pylori*; see below) may explain the effectiveness of treatments which reduce acid secretion.

7.2.2 Eradication of *Helicobacter Pylori*

These bacteria live within the gastric mucosa and epithelial cells, producing irritant/harmful substances that in some individuals can result in gastritis. Inflammation reduces the mechanisms that protect the stomach and duodenal mucosa from gastric acid and pepsin, increasing the likelihood of developing ulcers. Eradication of *H. pylori* by antibacterial treatment in combination with inhibitors of gastric acid secretion reduces their recurrence.

7.2.3 Mucosal Protection

Misoprostol is a synthetic prostaglandin analogue that inhibits gastric acid secretion and also stimulates mucus secretion and mucosal blood flow to increase mucosal integrity. Together, these actions promote healing of gastric and duodenal ulcers, particularly nonsteroidal anti-inflammatory drug (NSAID)-associated ulcers in frail or elderly people.

7.2.4 Future

In spite of the use of PPI therapy, GERD remains a common disorder and a significant number of patients continue to experience night-time heartburn or noncardiac chest pain, do not achieve complete healing of esophageal erosion and experience symptoms and extra-esophageal consequences of gastroesophageal reflux such as cough or asthma.

1. *Alternative methods of inhibiting gastric acid secretion.* The development of long-acting, noncompetitive H_2 receptor antagonists has so far been unsuccessful. Attempts to achieve "on demand" therapy by developing reversible, potassium-competitive antagonists that inhibit gastric H^+/K^+-ATPase have also so far been unsuccessful; these agents would have faster onset of activity as they do not require prior proton pump activation to achieve an antisecretory effect.
2. *Inhibition of reflux of gastric contents into the esophagus.* These include the use of $5\text{-}HT_4$ or motilin receptor agonists to stimulate esophageal propulsion, increase lower esophageal sphincter

tone and/or stimulate gastric emptying and thereby minimise exposure of the esophagus to gastric acid. This approach has achieved variable success but other than the 5-HT$_4$ receptor agonist mosapride (registered in Japan and other Asian countries for various disorders associated with chronic gastritis, including heartburn), none are registered world-wide for this indication (see "Stimulation of gastric motility and emptying", below). A different approach is to use cannabinoid receptor agonists, GABA-B receptor agonists, antagonists at metabotropic glutamate receptor 5 or partial 5-HT$_3$ receptor agonists to reduce the number and duration of transient lower esophageal sphincter relaxations (a normal vagus-mediated mechanism by which gas is released from the stomach as a "burp" or "belch"). Developments in this area have, for different reasons, met with limited success.

3. *Inhibition of esophageal sensitivity.* Resistance to treatment may be related to a continued hypersensitivity of the esophagus and/or to an impairment of mucosal integrity, perhaps associated with increased permeability between the esophageal epithelium. This has led to proposals to treat with, respectively, analgesics or agents that promote mucosal healing. For those patients where noncardiac chest pain remains a problem after treatment with PPIs, the use of antagonists at the transient receptor vanilloid 1 (TRPV1) has been tried without success.

7.3 Inhibition and Prevention of Nausea and Vomiting

Vomiting is coordinated by structures in the hindbrain. Sensory inputs are to the *area postrema* (AP) and the *nucleus tractus solitarius* (NTS). The AP is highly vascularised and not protected by the blood–brain barrier. It is often known as the "chemoreceptor trigger zone" and responds to different emetogenic substances circulating in the blood. The NTS receives direct or indirect inputs from the abdominal and thoracic vagus, pharyngeal, glossopharyngeal and trigeminal nerves, the spinal tract, the AP, hypothalamus, cerebellum and vestibular/labyrinthine systems, and the cerebral cortex. One or other of these pathways may be stimulated by the toxins (*e.g.* from bacteria) or locally released substances (*e.g.* 5-HT) in the lumen of the gut, by toxic materials (including drugs) or endogenous agents in the blood, by a GI pathology (*e.g.* pyloric stenosis or intestinal blockage) or other visceral organ pathology (*e.g.* renal failure leading to uremia, myocardial infarction), by surgery, a stimulus within the CNS (fear, anticipation, psychiatric disorders, brain trauma, sudden raised intracranial pressure), by pain and migraine, a disturbance of the vestibular system (motion sickness, space sickness, Meniereís disease) and by end-of-life scenarios for cancer and other causes, combining any of the above and more. The induction of nausea and vomiting by new chemical entities is also a major reason for their discontinued development as drugs.

- *Medications.* This includes, for example, opioid-based analgesics and dopamine receptor agonists used to treat Parkinson's disease.
- *Pregnancy.* The precise aetiology is unclear but metabolic and hormonal factors (*e.g.* estrogen, progesterone) are implicated.
- *Surgery.* The incidence depends on factors such as the type of surgery and physical manipulation, the use of inhalation anaesthetics or opioid-based drugs, and the presence/absence of established sensitivity to nausea and vomiting.
- *Pain and Migraine.* Activation of trigeminal and facial nerves during intense headaches/migraine/dental pain activates parasympathetic reflexes that regulate cardiovascular, GI and respiratory functions, including emesis.
- *Abnormal motion (motion sickness).* Driven by a mismatch between converging information received by the NTS from visual and sensory systems.
- *Cancer chemo- and radiotherapy.* In terms of emetogenic potential, cancer treatments have "low risk" (*e.g.* fluorouracil, etoposide, methotrexate, vinca alkaloids, abdominal radiotherapy), "medium risk" (*e.g.* the taxanes, doxorubicin, intermediate/low doses of cyclophosphamide, mitoxantrone, high doses of methotrexate) or "high risk" (*e.g.* cisplatin, dacarbazine, high doses of cyclophosphamide). The nausea and vomiting may be acute (occurring within 24 h of treatment), delayed (occurring more than 24 h after treatment) or anticipatory (prior to subsequent doses). Acute emesis is thought to be largely driven by a treatment-induced production of reactive oxygen species that stimulate the release of 5-HT from the enterochromaffin cells within the duodenum and antrum. The released 5-HT acts at 5-HT$_3$ receptors on the nearby vagal nerve terminals to sensitise these neurons to a large range of other substances generated during therapy (*e.g.* eicosanoids and inflammatory cytokines),

and initiate a prolonged emetic signal *via* the vagal nerve projection to the NTS. The causes of the delayed emesis are thought to be mediated *via* the area postrema, perhaps involving eicosanoids and inflammatory mediators.

- *Palliative care.* Nausea and vomiting are common in patients with advanced cancer and the causes vary, including, for example, drug-induced or metabolic causes (*e.g.* hypercalcaemia, renal failure), gastritis, gastric stasis and bowel obstruction.

7.3.1 Drug Treatments

Many antiemetic drugs were discovered a long time ago, before some of the receptors on which they act were identified. It is now understood that the older drugs can have multiple actions, which can be disadvantageous or advantageous for different groups of patients.

In summary:

- Muscarinic receptor antagonists (*e.g.* hyoscine hydrobromide) act within the vestibular nuclei to control motion sickness; notably, dopamine D_2 and 5-HT$_3$ receptor antagonists are ineffective treatments of motion sickness.
- Drugs that antagonise at histamine H_1 (*e.g.* cinnarizine, cyclizine, dimenhydrate, meclizine, promethazine) and/or D_2 receptors (*e.g.* the phenothiazines prochlorperazine, perphenazine, trifluoperazine, chlorpromazine, droperidol, haloperidol, domperidone and metoclopramide) act, respectively, within brainstem nuclei and at dopamine D_2 receptors within the area postrema, and are commonly used for most forms of emesis. For example, D_2 receptor antagonists can prevent emesis caused by opioid-based analgesics or dopamine-based treatments given to patients with Parkinson's disease. Pain/migraine may be relieved by metoclopramide, domperidone, phenothiazine or antihistamine antiemetic's. Some of these drugs have affinity for D_3 and other receptors.
- Metoclopramide is also an agonist at 5-HT$_4$ receptors, increasing acetylcholine (ACh) release from enteric cholinergic neurons (a major motor-neuron within the GI tract) to increase gastric emptying. This additional GI prokinetic activity of metoclopramide may be used to increase absorption of orally administered analgesics in patients with migraine, whilst simultaneously reducing emesis. It also finds use in overcoming partial bowel obstructions in patients receiving palliative care (see below).

- There is increasing use of olanzapine (affinity for D_1, D_2, D_4, 5-HT_{2A}, 5-HT_{2C}, 5-HT_3, α_1-adrenoceptor, H_1, muscarinic receptors) and levomepromazine (H_1, muscarinic, D_2, α_1 adrenoceptor and 5HT_2 receptor antagonist), the latter for patients with intractable nausea and vomiting receiving palliative care where it is also used to treat severe delirium or agitation at the end of life.
- Glucocorticoids such as dexamethasone reduce the ability of various inflammatory mediators to sensitise different emetic pathways (*e.g.* vagal nerve terminals), may have direct actions at different points along the emetic pathways (*e.g.* the NTS and at NK_1 receptors) and reduce pain and the concomitant use of opioids. They are used to prevent low-level emesis (*e.g.* after surgery) but most importantly are given together with 5-HT_3 and NK_1 receptor antagonists to provide a synergistic antiemetic activity.
- 5-HT_3 receptor antagonists (*e.g.* granisetron, ondansetron, tropisetron, palonosetron) were developed to prevent the severe emesis caused by anticancer treatments such as cisplatin, and their introduction led to a revolution in the care of cancer patients. They block the action of 5-HT at the peripheral vagal nerve terminals, although a central site of action may sometimes operate. 5-HT_3 receptor antagonists are used to treat mild-to-moderate emesis in cancer patients, in combination with dexamethasone. These drugs can also reduce postoperative nausea and vomiting, perhaps together with droperidol, dexamethasone, some phenothiazines (*e.g.* prochlorperazine), and antihistamines (*e.g.* cyclizine).
- NK_1 receptor antagonists (aprepitant, fosaprepitant) were also developed to prevent the severe emesis caused by anticancer treatments such as cisplatin; they act within the central vagal neurocircuits. They are used in patients with severe emetogenic challenges (*e.g.* following treatment with cisplatin) as part of a "triple therapy" of 5-HT_3 and NK_1 receptor antagonism plus dexamethasone.
- Cannabinoid receptor agonists (nabilone, dronabinol) act within the vestibular and brainstem nuclei and have restricted use for treatment of chemotherapy-induced emesis in cancer patients.

The first generation of 5-HT_3 receptor antagonists are competitive antagonists. Palonosetron is a 5-HT_3 receptor antagonist that appears to perform better than previous 5-HT_3 receptor antagonists; the drug has good efficacy against both "acute" and "delayed" emesis. Palonosetron has relatively high binding affinity and a persistent ability to inhibit receptor function after the drug is removed, caused by triggering receptor internalisation of the drug–receptor complex into the cell. The drug has no affinity for the NK_1 receptor yet it can reduce certain NK_1 receptor-mediated functions; the combination of palonsetron with NK_1 receptor antagonists such as netupitant appears to have synergistic activity. It has been speculated that since different types of receptor can "cross-talk", changes in $5\text{-}HT_3$ receptor internalisation can reduce NK_1 receptor functions.

Adjunct treatments with no direct antiemetic activity may control the cause of nausea and vomiting. The benzodiazepine lorazepam can reduce anticipatory vomiting because of its amnesic, sedative, and anxiolytic activity. In patients with intestinal obstruction caused by cancer a somatostatin receptor agonist (octreotide) inhibits fluid secretion in the intestinal lumen, minimising build-up of pressure that would otherwise cause pain and vomiting.

7.3.2 Future

Paradoxically, antiemetic drugs (including the $5\text{-}HT_3$ and NK_1 receptor antagonists) appear to be more effective at controlling vomiting than controlling nausea. The reasons are unclear but suggest different, albeit related, mechanisms of action for nausea and vomiting. Major pathways currently thought to be involved in mechanisms of nausea include:

1. *Satiety and nausea.* Drugs that promote appetite might reduce nausea. An example is ghrelin, known to promote appetite (and perhaps "hedonistic eating" *via* a constitutively active receptor) and alleviate anorexia and vomiting in animal models and reduce cachexia and nausea in cancer patients. The ghrelin receptor agonist relamorelin reduced nausea, vomiting and sense of fullness in patients with diabetic gastroparesis. A different ghrelin receptor agonist, TZP-102, was not effective in a trial in patients with diabetic gastroparesis but there was some reduction in nausea. Conversely, agonists at the glucagon-like peptide receptor, stimulate insulin secretion, inhibit glucagon secretion, delay gastric emptying and reduce appetite and food intake, but may also induce a relatively high incidence of nausea.

2. *Gastric dysrhythmia and nausea.* In various disorders, including gastroparesis, nausea has been associated with disruptions of the normal slow-wave rhythm of gastric motility, a dysrhythmia linked with changes in numbers and/or functions of interstitial

cells of Cajal (ICC). The ICCs are a specialised type of smooth muscle cell that generate rhythmic oscillations in membrane potential and act as pacemakers to promote gastric emptying in an anal direction and facilitate the functions of motor neurons; this dysrhythmia is, in part, responsible for the delay in gastric emptying and may represent the interface between nausea and the feeling of impending vomiting, an act associated with transition of movement into retropulsion.

3. *Vasopressin and nausea.* During nausea, blood concentrations of vasopressin (from the posterior pituitary) are thought to increase and this has been implicated in reduction of gut blood flow, disruption of gastric motility and the genesis of nausea, possibly by activation of V_1 receptors within the central emetic pathways *via* the area postrema. This release of vasopressin is consistent with its role in fluid homeostasis, regulating retention of water.

It should be remembered that nausea is a subjective human sensation, so studying this in animals is difficult without some degree of uncertainty over the translational value of the data. Examples of animal models of nausea-like behaviours include conditioned taste aversion, pica consumption, conditioned gaping or signs of disgust.

7.4 Stimulation of Gastric Motility and Emptying

Gastric prokinetic drugs are useful treatments for some patients with PPI-resistant GERD (see above), gastroparesis and FD. These drugs also find use in patients requiring rapid intubation and/or emptying of stomach (prior to emergency surgery or endoscopy), patients requiring enteral feeding during intensive care (the illness causes a delay in gastric emptying, limiting the delivery of nutrition and thereby compromising clinical outcomes) and diabetic patients needing improved blood glucose control achieved by better regulation of gastric emptying.

Drugs currently used to increase gastric emptying include the nonselective $5\text{-}HT_4$ receptor agonist and D_2 receptor antagonists metoclopramide (also a $5\text{-}HT_3$ receptor antagonist when used at higher doses), levosulpiride and clebopride, a $5\text{-}HT_4$ receptor agonist with poor intrinsic efficacy (mosapride) and the peripherally restricted D_2 receptor antagonist, domperidone. Activation of $5\text{-}HT_4$ receptors within the enteric nervous system (ENS) of the stomach facilitates release of ACh from cholinergic motor neurons to promote coordinated GI movements (Figure 7.1). Antagonism at the D_2

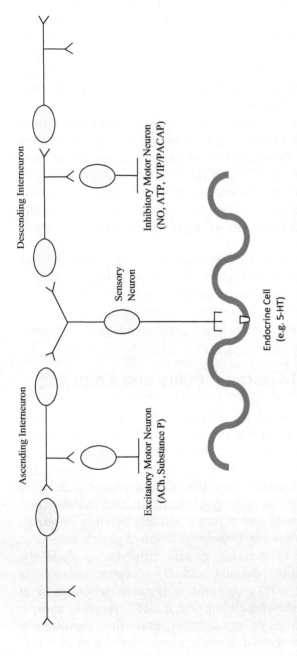

□ In brief, the ENS in the intestine possesses sensory neurons projecting from the mucosa to nerve plexuses running the length of the bowel. Sensory nerves (and other cells) are sensitive to movement and stimulated or sensitised by substances from endocrine and other cells within the epithelial layer (*e.g.* 5-HT from enterochromaffin cells).

□ The myenteric plexus (between the outer longitudinal and inner circular muscle) contains excitatory and inhibitory motor-neurons that control muscle movements. The submucosal plexus (between the circular muscle and mucosa) contains motor-neurons controlling intestinal secretion.

□ Not shown are the interstitial cells of Cajal (ICCs). Their spontaneous depolarisation creates a 'slow wave' current transmitting to the muscle and leading to regular, propagating low-amplitude contractions, further controlled by the ENS.

□ Drugs directly affecting GI movements (*e.g.* stimulation by 5-HT$_4$ or motilin receptor agonists, and inhibition by opioid receptor agonists) generally affect motor-neuron transmission, although sensory nerve functions may also be affected.

Figure 7.1 Schematic and simplified diagram of the enteric nervous system (ENS).

receptor within the AP exerts antiemetic activity and in patients with delayed gastric emptying (*e.g.* gastroparesis; not in healthy volunteers) this activity may increase gastric emptying. D_2 receptor antagonism in the pituitary also increases prolactin secretion and for metoclopramide which crosses the blood–brain barrier, akathisia or other extrapyramidal movement disorders; the drug is therefore used only for short-term treatments. Gastric emptying can also be increased by erythromycin and azithromycin, which are antibiotic drugs that also activate the motilin receptor within the ENS to promote ACh release and increase gastric emptying. For this activity, these drugs are used off-label.

Gastric pacing or gastric electrical stimulation is a technique that has been approved to treat medically refractory diabetic and idiopathic gastroparesis. Electrodes are placed in the serosa and a stimulation device implanted subcutaneously in the abdominal wall. Treatment significantly reduces gastric dysrhythmia and gastric emptying, improving GI symptoms such as vomiting and increasing quality of life.

7.4.1 Future

1. *New gastric prokinetic drugs.* Each of the available treatments are unsatisfactory as gastric prokinetics, in terms of their short-lived activity (*e.g.* domperidone), side effects (metoclopramide), expense (gastric pacing), inappropriate use (erythromycin, azithromycin may accelerate the rise of antibiotic drug resistance) and a link to sudden cardiac death, exacerbated when given concomitantly with cytochrome 3A4 inhibitors (erythromycin). However, there has been only limited success in developing alternatives. Acotiamide (a muscarinic receptor inhibitor licenced in Japan for treatment of FD) promotes the release of ACh by inhibiting acetylcholinesterase (AChE) activity and also antagonising at muscarinic M_1 and M_2 receptors but not at M_3 receptors; this combination of activity stimulates GI motility without significant side effects. The AChE inhibitor and D_2 receptor antagonist itopride, marketed in Japan and certain other countries, has not been a successful treatment of FD. The selective 5-HT$_4$ receptor agonist prucalopride (developed for chronic constipation; see below) has been shown to increase gastric emptying and improve acid clearance from the esophagus in healthy volunteers but further development for treatment of PPI-resistant GERD or other indications seems slow. The selective

motilin receptor agonist GSK962040 has been shown to increase gastric emptying in healthy volunteers. Interestingly, a CCK_1 receptor antagonist has been shown to promote gastric emptying in critically ill patients receiving lipid-enriched enteral feeding but only slow progress is being made in further development.

2. *Better patient phenotyping.* There is little correlation between symptoms and delayed gastric emptying so different therapeutic approaches are required. Impaired gastric accommodation may be present in a significant number of patients with FD but treatments aimed at increasing accommodation (*e.g.* $5\text{-}HT_{1A}$ receptor agonism by buspirone) have so far been unsuccessful. For patients with gastroparesis and FD there is early promise in treatments that reduce nausea (see above for ghrelin receptor agonists). Another proposal is that in patients with diabetic gastroparesis, damage to ICCs and loss of enteric nerves leads to arrhythmic movements, confusing the signal transmitted to the brainstem by the vagus and leading to nausea; this group of patients may also find some similarities with patients with functional dyspepsia and chronic unexplained nausea and vomiting. In patients with idiopathic gastroparesis, accompanied by nausea, the loss of ICCs is associated with a higher myenteric immune and macrophage infiltrate. If confirmed, these observations point to a need to develop strategies in the control of ICC functions and/or immune activity. Finally, an extensive trial with amitriptyline has shown some benefit in those patients with the subgroup of patients with FD defined as the epigastric pain syndrome.

7.5 Treatment of Chronic Constipation, Megacolon and Intestinal Pseudo-obstruction

Chronic constipation is common. It may be caused by neuropathies of the CNS (*e.g.* Parkinson's disease, spinal cord damage, multiple sclerosis), certain metabolic conditions, by disorders within the gut (*e.g.* scleroderma) or as a side effect of several classes of medication (*e.g.* opioid-based analgesics, certain tricyclic antidepressant medications and calcium channel blockers). For multiple reasons the prevalence of chronic constipation increases significantly in the elderly population. Other conditions with intestinal hypomotility and acute or chronic small and/or large intestinal dilation, include postoperative ileus, intestinal pseudo-obstruction, "megacolon" and "acute colonic pseudo-obstruction". The latter can occur after major surgery or trauma, severe illness or during intensive care, especially in elderly patients leading to dilatation of the colon, perforation and death.

7.5.1 Laxatives and Faecal Softeners

Laxatives treat drug-induced constipation, clear the bowel before surgery and radiological procedures, expel parasites and manage patients with colostomy, ileostomy, haemorrhoids, anal fissure, diverticular disease, IBS and ulcerative colitis. They are bulk-forming laxatives (*e.g.* unprocessed wheat bran or methylcellulose, which increase faecal mass to stimulate peristalsis), stimulant laxatives (*e.g.* bisacodyl, sodium picosulfate and the anthraquinones, senna and dantron), faecal softeners (*e.g.* liquid paraffin and enemas containing arachis oil) and osmotic laxatives (*e.g.* lactulose that draws fluid from the body into the bowel).

7.5.2 Peripheral Opioid-receptor Antagonists

Alvimopan, methylnaltrexone and naloxegol are peripherally acting opioid-receptor antagonists used to treat opioid-induced constipation (especially during palliative care) without altering the central analgesic effect of opioids. These drugs may also treat postoperative ileus, a condition in which the recovery of GI movements is delayed by the procedure of surgery, prolonging the time before patients can be discharged from hospital.

7.5.3 Prucalopride

Several compounds were originally developed to retain the gastroprokinetic activity of metoclopramide but not the side effects associated with D_2 receptor antagonism. Much of this work occurred before the 5-HT_4 receptor was shown to mediate the GI prokinetic activity of metoclopramide and before several 5-HT receptors were identified. The early compounds were therefore nonselective in their activity (*e.g.* cisapride: inhibition of the inwardly rectifying K^+ channel encoded by the human ether-á-go-go gene (hERG), increasing the risk of cardiac ventricular arrhythmia; tegaserod: antagonism at the 5-HT_{2B} receptor acting against the prokinetic activity of the 5-HT_4 receptor) or have poor intrinsic activity at the receptor (*e.g.* mosapride, tegaserod). Later, the discovery of different 5-HT$_4$ receptor COOH-terminal splice variants provided a possible explanation for why 5-HT$_4$ receptor agonists act as full agonists at the receptors expressed by enteric cholinergic neurons (explaining their GI prokinetic activity; Figure 7.1) but as weak or partial agonists in other tissues, including cardiac muscle. Prucalopride is a selective 5-HT$_4$ receptor agonist (at the human 5-HT$_{4(a)}$ and 5-HT$_{4(b)}$ receptor isoforms) with high efficacy in GI tissues and without clinically meaningful actions in the heart. The drug is registered for treatment of chronic constipation in adults.

7.5.4 Lubiprostone

This compound activates ClC2 chloride channels on the luminal surface of epithelial cells lining the gut (Figure 7.2). Stimulation of chloride efflux increases movement of water into the lumen of the intestine, softening the stools and improving bowel movements. Lubiprostone is also approved for treatment of constipation-predominant IBS in women. Critically, the drug is poorly absorbed and cannot exit the gut to access ClC2 channels in the lung. It is a derivative of prostaglandin E$_1$ and retains an ability to activate prostaglandin EP receptors, perhaps explaining the occurrence of nausea as an adverse event and the need for a pregnancy category C label.

7.5.5 Linaclotide

The luminal surface of intestinal epithelial cells expresses guanylate cyclase C (GC-C) receptors that when activated by the guanylin family of peptides opens the cystic fibrosis transmembrane conductance regulator (Figure 7.2). This leads to secretion of chloride and bicarbonate ions into the lumen followed by water, a decrease in colonic fluid absorption and increased GI transit. The peptide linaclotide is a selective GC-C receptor agonist for treatment of idiopathic constipation and constipation-predominant IBS. In IBS trials linaclotide also reduced abdominal pain and discomfort. The mechanism of this latter activity is not clear, but in rodents, the drug can reduce visceral hypersensitivity, an activity argued to be caused by extracellular cGMP generated by GC-C activation, reducing nociceptor mechanosensitivity. Linaclotide is metabolised within the GI tract.

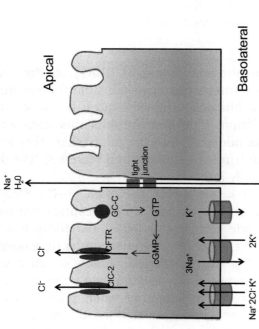

ClC-2: Chloride Channel-2

CFTR: Cystic Fibrosis Transmembrane Conductance Regulator

GC-C: guanylate cyclase-C (enzyme normally activated by heat stable enterotoxins to cause watery diarrhoea)

Intracellular composition regulated by channels at the basolateral boarder, drawing ions from blood circulation

Lubiprostone stimulates ClC-2, Linaclotide stimulates the GC-C receptor, leading to activation of CFTR

Cl⁻ enters the lumen via ClC-2 or CFTR

Na⁺ diffuses through intracellular spaces to balance Cl⁻ and water follows passively into the lumen

Intestinal motility is increased and the stool is softened

Figure 7.2 Epithelial cells lining the gut.

7.5.6   Cholinesterase Inhibitors

AChE inhibitors (*e.g.* neostigmine or pyridostigmine) are sometimes used to enhance intestinal propulsion in patients with severe hypo-motility, especially those with acute colonic pseudo-obstruction, patients with cancer and opioid-induced chronic constipation, chronic constipation associated with spinal injury and autonomic neuropathies. Clearly, these drugs have limited use because of their potential to cause cardiobronchial side effects, nausea and/or vomiting or abdominal pain. However, the fact that AChE inhibitors are sometimes used testifies to the need for new, more effective agents to treat chronic constipation.

7.5.7   Future

It is important to find new methods of treating chronic constipation, especially for those with spinal cord damage, Parkinson's disease or multiple sclerosis.

1. *Elobixibat.* The ileal bile acid transport (IBAT) on the brush border membrane of enterocytes is responsible for the initial uptake of conjugated bile acids from the intestine during their enterohepatic circulation. Elobixibat inhibits this transporter, increasing delivery of bile acids into the colon where they act at bile acid receptors to stimulate motility and secretion. The drug is excreted substantially intact.

2. *New prokinetic agents acting within the colon.* 5-HT_4 receptor agonists (with good intrinsic activity at the receptor) reportedly in development include naronapride (ATI-7505) and velusetrag (TD-5108). Similarly, GC-C receptor agonists under development include plecanatide. The use of donepezil, a more selective AChE inhibitor with less intense ability to stimulate colonic motility may be an alternative to neostigmine when given alone or together with a 5-HT_4 receptor agonist when the two activities are likely to synergise. Recently, orlistat, a lipase inhibitor that reduces absorption of dietary fat has been shown to reduce symptoms of chronic constipation.

3. *Prokinetic agents acting within the spinal cord.* Descending dopaminergic pathways (originating in the brain) release dopamine into the lumbosacral defecation centre of the spinal cord (L6-S1 region), activating dopamine D_2-like receptors (D_2, D_3 and D_4 receptors) to increase sacral parasympathetic nerve

firing to stimulate colorectal propulsive motility. Theoretically, therefore, this pathway may be involved in the chronic constipation of patients with Parkinson's disease, where brain areas supplying the dopaminergic innervation of the spinal cord have degenerated. Alternative approaches that may explain the development of chronic constipation in these patients include the involvement of vagal nerve degeneration and/or the presence of phosphorylated α-synuclein within the enteric nervous system. Regardless of cause, treatment may be achieved by application of ghrelin receptor agonists that act at neurons in the lumbosacral region of the spinal cord to stimulate pelvic nerve activity and promote propulsive contractions of the colon. The ghrelin receptor agonists ulimorelin, relamorelin and capromorelin increase defecation in humans.

7.6 Inhibition of Diarrhoea

In acute diarrhoea, the priority is to prevent or reverse fluid and electrolyte depletion. This is particularly important in infants and in frail and elderly patients.

- *Antimotility drugs.* Include (a) loperamide, a mu-opioid receptor agonist that does not cross the blood–brain barrier and slows intestinal transit by reducing fluid secretion and motility, and (b) racedadotril, a prodrug of the enkephalinase inhibitor thiorphan that inhibits breakdown of endogenous opioids, thereby reducing intestinal secretion.
- *Bile acid sequestrants.* Cholestyramine is an anion-exchange resin not absorbed from the GI tract, which forms an insoluble complex with unabsorbed bile salts to provide symptomatic relief of diarrhoea (and pruritus) following ileal disease or resection.
- *Pancreatin.* Coadministered with an inhibitor of gastric acid secretion to assist digestion of starch, fat and protein, compensating for reduced/absent exocrine secretion in cystic fibrosis or following pancreatectomy, gastrectomy or chronic pancreatitis.

7.7 Functional Bowel Disorders

These are highly prevalent disorders with unknown aetiology and no agreed structural pathology. These are several functional bowel disorders, but perhaps the most common are those known as functional dyspepsia (FD) and irritable bowel syndrome (IBS).

FD is characterised by bothersome postprandial fullness, early satiation, epigastric pain, and/or epigastric burning.

- Benefit may be achieved by eradication of *H. pylori* and/or by treatment with PPIs (see earlier).
- The tricyclic antidepressant, amitriptyline (unlicensed indication), has some benefit in patients with epigastric pain as the predominant symptom. Notably, in addition to inhibiting 5-hydroxytryptamine and noradrenaline transporter mechanisms, amitriptyline has good affinity for the H_1 receptor, the muscarinic receptor, α_1-adrenoceptor and the 5-HT_{2A} receptor, when K_i values fall within the ranges observed at binding sites for the 5-HT and noradrenaline transporter sites.
- Gastric prokinetic drugs may be of use although efficacy is not predicted by the presence of delayed gastric emptying. Acotiamide is registered in Japan and may improve symptoms and increase gastric emptying by antagonism at muscarinic M_1 and M_2 receptors together with inhibition of acetylcholinesterase (see earlier).

IBS symptoms include abdominal pain or discomfort, disordered defaecation (either diarrhoea, or constipation with straining, urgency, and incomplete evacuation), passage of mucus, and bloating. To facilitate advances in treatments, patients have been further characterised as those in which the predominant GI disturbance is diarrhoea (diarrhoea-predominant IBS), constipation or alternating.

- Antispasmodic drugs (*e.g.* alverine citrate, mebeverine hydrochloride, otilonium, hyoscine, cimetropium, pinavarium, dicyclomine, peppermint oil) may provide some short-term symptomatic relief, although the evidence is not always robust.
- A laxative can be used to treat constipation but not other symptoms of IBS.
- Loperamide may provide relief from diarrhoea but not other symptoms of IBS.
- The selective 5-HT_3 receptor antagonist, alosetron, is available in the USA for treatment of severe diarrhoea-predominant IBS in females (there is also evidence for efficacy in males). Alosetron is thought to act primarily by reducing inappropriate colonic motility and perhaps also by reducing colonic and rectal hypersensitivity.

Constipation is a recognised side effect and a small incidence of reversible ischaemic colitis restricts widespread use.

- Linaclotide has been registered for treatment of chronic constipation and constipation-predominant IBS (see above).
- Lubiprostone has been registered for treatment of chronic constipation and constipation-predominant IBS (see above).
- Eluxadoline, a mu-opioid receptor agonist/delta opioid receptor antagonist/kappa receptor agonist is now available for treatment of diarrhoea-predominant IBS. This drug reduces visceral hypersensitivity without completely inhibiting GI motility, primarily due to inhibition of pain and GI motility by the mu-opioid agonism, but with the degree of inhibition of GI motility moderated by the delta-opioid antagonist activity that also increases central analgesia.
- The tricyclic antidepressant, amitriptyline, is used for abdominal pain or discomfort, although this is not a registered use for the drug.
- Rifaximin, a minimally absorbed antibiotic for treatment of nonconstipation IBS.

7.7.1 Future

1. *Improved patient phenotyping.* In FD, patients may now be subgrouped as those with postprandial distress syndrome or epigastric pain syndrome. Some, but not all, may have delayed gastric emptying or poor fundus accommodation. There are reports of increased eosinophils within the duodenum, leading to speculation that inflammatory cytokines generate symptoms by activating local or extrinsic nerve control pathways. In IBS, a subtle intestinal inflammation has also been observed, along with changes in gut microbiota, disordered mucosal immune activation and increased numbers of mast cells in the mucosa. A mutation has been found in the $Na_v1.5$ sodium channel, potentially contributing to the pathology in up to 2% of the IBS population.
2. *Patient genotyping.* Includes the identification of genetic and epigenetic associations, accompanied by better stratification of the patient population.
3. *New drugs.* The NK_2 receptor antagonist ibodutant has shown improvement in symptoms in diarrhoea-predominant IBS and is currently undergoing Phase III trials. Several other trials have previously been conducted with proposed new treatments of

abdominal pain but without success (*e.g.* the Kappa opioid receptor agonists). Better understanding of the pathways of visceral pain is therefore required. Currently, there is extensive exploration of the functions of the different transient receptor potential channels and also the potential role for a vagal anti-inflammatory pathway in which the released ACh interacts with macrophages and other immune cells. A peripherally restricted inhibitor of tryptophan hydroxylase (LX-1031) has been shown to reduce 5-HT synthesis and in a Phase II trial in patients with diarrhoea-predominant IBS, improved stool consistency and provided adequate relief from symptoms.

7.8 Factors that Could Change the Landscape

1. Understanding the importance of the nutrient receptors expressed by GI endocrine cells.
2. Translated associations of functional GI disorders with epigenetic mechanisms.
3. The development of biased agonists that maintain therapeutic benefit while minimising side-effects (*e.g.* acting at the opioid receptors).
4. Control of neurogenesis and the application of stem cell biology for GI repair.

Further Reading

1. A. Acosta and M. Camilleri, Elobixibat and its potential role in chronic idiopathic constipation, *Ther. Adv. Gastroenterol.*, 2014, 7, 167–175.
2. T. R. Angeli, L. K. Cheng, P. Du, T. H. Wang, C. E. Bernard, M. G. Vannucchi, M. S. Faussone-Pellegrini, C. Lahr, R. Vather, J. A. Windsor, G. Farrugia, T. L. Abell and G. O'Grady, Loss of interstitial cells of Cajal and patterns of gastric dysrhythmia in patients with chronic unexplained nausea and vomiting, *Gastroenterology*, 2015, **149**, 56–66.
3. D. T. Beattie and J. A. M. Smith, Serotonin pharmacology in the gastrointestinal tract, *Naunyn-Schmiedebergs Arch. Pharmacol.*, 2008, **377**, 181–203.
4. J. Broad, V. W. S. Kung, G. Boundouki, Q. Aziz, J. H. De Maeyer, C. H. Knowles and G. J. Sanger, Cholinergic interactions between donepezil and prucalopride in human colon: potential to treat

severe intestinal dysmotility, *Br. J. Pharmacol.*, 2013, **170**, 1253–1261.

5. M. Camilleri, Pharmacological agents currently in clinical trials for disorders in neurogastroenterology, *J. Clin. Invest.*, 2013, **123**, 4111–4120.

6. C. C. Chu, C. H. Hsing, J. P. Shieh, C. C. Chien, C. M. Ho and J. J. Wang, The cellular mechanisms of the antiemetic action of dexamethasone and related glucocorticoids against vomiting, *Eur. J. Pharmacol.*, 2014, **722**, 48–54.

7. B. Y. De Winter, A. Deiteren and J. G. De Man, Novel nervous system mechanisms in visceral pain, *Neurogastroenterol. Motil.*, 2016, **28**, 309–315.

8. A. Emmanuel, Current management strategies and therapeutic targets in chronic constipation, *Ther. Adv. Gastroenterol.*, 2011, **4**, 37–48.

9. J. B. Furness, The enteric nervous system and Neurogastroenterology, *Nat. Rev. Gastroenterol. Hepatol.*, 2012, **9**, 286–294.

10. J. B. Furness, L. R. Rivera, H.-J. Cho, D. M. Bravo and B. Callaghan, The gut as a sensory organ, *Nat. Rev. Gastroenterol. Hepatol.*, 2013, **10**, 729–740.

11. M. Grover, C. E. Bernard, P. J. Pasricha, M. S. Lurken, M. S. Faussone-Pellegrini, T. C. Smyrk, H. P. Parkman and T. L. Abell, *et al.*, Clinical-histological associations in gastroparesis: results from the Gastroparesis Clinical Research Consortium, *Neurogastroenterol. Motil.*, 2012, **24**, 531–539.

12. S. Gupta, V. Kapoor and B. Kapoor, Itopride: A Novel Prokinetic Agent, *Drug Rev.*, 2004, **6**, 106–108.

13. W. L. Hasler and K. L. Koch, Amitriptyline for functional dyspepsia: Importance of symptom profile and making a case for gastric emptying testing, *Gastroenterology*, 2015, **149**, 270–289.

14. R. C. Heel, R. N. Brogden, T. M. Speight and G. S. Avery, Loperamide: a review of its pharmacological properties and therapeutic efficacy in diarrhea, *Drugs*, 1978, **15**, 33–52.

15. A. M. Holmes, J. A. Rudd, F. D. Tattersall, Q. Aziz and P. L. Andrews, Opportunities for the replacement of animals in the study of nausea and vomiting, *Br. J. Pharmacol.*, 2009, **157**, 865–880.

16. C. C. Horn, Is there a need to identify new anti-emetic drugs? *Drug Discovery Today: Ther. Strat.*, 2007, **4**, 183–187.

17. K. Jordan, H. J. Schmoll and M. S. Aapro, Comparative activity of antiemetic drugs. *Crit Rev. Oncol. Hematol.*, 2007, **61**, 162–175.

18. C. Maradey-Romero and R. Fass, New and future drug development for gastroesophageal reflux disease, *J. Neurogastroenterol. Motil.*, 2014, **20**, 6–16.

19. B. F. Kessing, A. J. P. M. Smout, R. J. Bennink, N. Kraaijpoel, J. M. Oors and A. J. Bredenoord, Prucalopride decreases esophageal acid exposure and accelerates gastric emptying in healthy subjects, *Neurogastroenterol. Motil.,* 2014, **26**(8), 1079–1086.
20. B. E. Lacy, Emerging treatments in neurogastroenterology: eluxadoline – a new therapeutic option for diarrhea-predominant IBS, *Neurogastroenterol. Motil.*, 2016, **28**, 26–35.
21. R. Latorre, C. Sternini, R. De Giorgio and B. Greenwood-Van Meerveld, Enteroendocrine cells: a review of their role in brain-gut communication, *Neurogastroenterol. Motil.*, 2016, **28**, 620–630.
22. P. Layer and V. Stanghellini, Review article: linaclotide for the management of irritable bowel syndrome with constipation, *Aliment. Pharmacol. Ther.*, 2014, **39**, 371–384.
23. K. J. Lee, Pharmacologic agents for chronic diarrhea, *Intest. Res.*, 2015, **13**, 306–312.
24. M. Matsushita, T. Masaoka and H. Suzuki, Emerging treatments in neurogastroenterology: Acotiamide, a novel treatment option for functional dyspepsia, *Neurogastroenterol. Motil.*, 2016, **28**, 631–638.
25. S. Menees, R. Saad and W. D. Chey, Agents that act luminally to treat diarrhoea and constipation, *Nat. Rev. Gastroenterol. Hepatol.*, 2012, **9**, 661–674.
26. K. Naitou, H. Nakamori, T. Shiina, A. Ikeda, Y. Nozue, Y. Sano, T. Yokoyama, Y. Yamamoto, A. Yamade, N. Akimoto and Y. Shimizu, Stimulation of dopamine D_2-like receptors in the lumbosacral defecation centre causes propulsive colorectal contractions in rats, *J. Physiol.*, 2016, **594**, 4339–4350
27. M. L. Nolan and L. J. Scott, Acotiamide: First global approval, *Drugs* 2013, **73**, 1377–1383.
28. J. H. Oh and P. J. Pasricha, Recent advances in the pathophysiology and treatment of gastroparesis, *J. Neurogastroenterol. Motil.*, 2013, **19**, 18–24.
29. P. J. Pasricha, R. Colvin, K. Yates, W. L. Hasler, T. L. Abell, A. Unalp-Arida, L. Nguyen, G. Farrugia, K. L. Koch, H. P. Parkman, W. J. Snape, L. Lee, J. Tonascia and F. Hamilton, Characteristics of patients with chronic unexplained nausea and vomiting and normal gastric emptying, *Clin. Gastroenterol. Hepatol.*, 2011, **9**, 567–576.
30. N. Percie du Sert, J. A. Rudd, R. Moss and P. L. R. Andrews, The delayed phase of cisplatin-induced emesis is mediated by the area postrema and not the abdominal visceral innervation in the ferret, *Neurosci. Lett.*, 2009, **465**, 16–20.

31. R. V. Pustovit, J. B. Furness and L. R. Rivera, A ghrelin receptor agonist is an effective colokinetic in rats with diet-induced constipation, *Neurogastroenterol. Motil.*, 2015, **27**, 610–617.

32. E. M. M. Quigley, Prokinetics in the management of functional gastrointestinal disorders, *J. Neurogastroenterol. Motil.*, 2015, **21**, 330–336.

33. S. S. C. Rao, K. Rattanakovit and T. Patcharatrakul, Diagnosis and management of chronic constipation in adults, *Nat. Rev. Gastroenterol. Hepatol.*, 2016, **13**, 295–305.

34. C. Rojas, M. Raje and T. Tsukamoto, *et al.*, Molecular mechanisms of 5-HT$_3$ and NK$_1$ receptor antagonists in prevention of emesis, *Eur. J. Pharmacol.*, 2014, **722**, 26–37.

35. G. J. Sanger, Translating 5-HT$_4$ receptor pharmacology, *Neurogastroenterol. Motil.*, 2009, **21**, 1235–1238.

36. G. J. Sanger and J. B. Furness, Ghrelin and motilin receptors as drug targets for gastrointestinal disorders, *Nat. Rev. Gastroenterol. Hepatol.*, 2016, **13**, 38–48.

37. A. Shin, M. Camilleri, G. Kolar, P. Erwin, C. P. West and M. H. Murad, Systematic review with meta-analysis: highly selective 5-HT$_4$ agonists (prucalopride, velusetrag or naronapride) in chronic constipation, *Aliment. Pharmacol. Ther.*, 2014, **39**, 239–253.

38. P. Singh, S. S. Yoon and B. Kuo, Nausea: a review of pathophysiology and therapeutics, *Ther. Adv. Gastroenterol.*, 2016, **9**, 98–112.

39. R. M. Stern, K. L. Koch and P. L. R. Andrews, Nausea: Mechanisms and Management, Oxford University Press, NY, USA, 2011.

40. J. Tack and N. J. Talley, Functional dyspepsia–symptoms, definitions and validity of the Rome III criteria, *Nat. Rev. Gastroenterol. Hepatol.*, 2013, **10**, 134–141.

41. N. J. Talley, Functional dyspepsia and the Rome criteria: a success story, *Neurogastroenterol. Motil.*, 2015, **27**, 1052–1056.

42. N. J. Talley, G. Holtmann and M. M. Walker, Therapeutic strategies for functional dyspepsia and irritable bowel syndrome based on pathophysiology, *J. Gastroenterol.*, 2015, **50**, 601–613.

43. N. P. Visanji, C. Marras, D. S. Kern, A. Al Dakheel, A. Gao, L. W. Liu, A. E. Lang and L. N. Hazrati, Colonic mucosal a-synuclein lacks specificity as a biomarker for Parkinson disease, *Neurology*, 2015, **84**, 609–616.

8 The Future of ADME in Drug Design and Development

Phil Jeffrey*[a] and Scott Summerfield[b]

[a] Pfizer, Cambridge, UK; [b] GlaxoSmithKline, Stevenage, UK
*Email: phil.jeffrey@hygamp.com

8.1 Introduction

Small-molecule drug discovery and development is a lengthy and complex process, associated with high financial costs and low success rates. The average new molecular entity (NME) takes around 13 years to develop from drug target identification to marketable drug, with fully loaded costs from discovery (find the molecule) through full development (make the medicine) of up to $1 billion.[1] Historically, regulatory approvals have declined but over recent years this trend is being reversed with a record 33 NME approvals in 2015.[2] A key source of attrition is the inability to translate and predict successful, efficacious clinical outcomes from preclinical data. NME survival continues to be adversely affected by the issue of insufficient or lack of efficacy observed in early clinical studies (Phase II) and is widely recognised as one of the major causes of NME failure. Despite the fact that large amounts of data are available; either pre-existing, *i.e.* based on similar drug targets, mechanisms, pathways and/or chemical class(es) or generated as part of the overall drug discovery/ development process attrition rates are high and the success rate remains steadfastly low.[1]

Pharmacology for Chemists: Drug Discovery in Context
Edited by Raymond Hill, Terry Kenakin and Tom Blackburn
© The Royal Society of Chemistry 2018
Published by the Royal Society of Chemistry, www.rsc.org

NMEs, especially those that are orally available, are concentrated in a relatively narrow range of physicochemical properties known as "drug-like space". These fundamental properties, such as molecular size, shape, hydrogen-bonding capability and polarity are directly correlated with solubility, membrane permeability, metabolic stability, receptor promiscuity, biological activity (safety and efficacy) and attrition in drug discovery/development. Viewed at an aggregate level across the drug discovery/development pipeline, the physicochemical properties of NMEs have an important influence on overall attrition rates and therefore ultimately on the return of investment.[3] *De novo* NME design is a balancing act since the control of physicochemical properties is dependent on the specific target (*e.g.* G protein-coupled receptors [GPCRs or 7TMs], enzymes [protein kinases, proteases, esterases], ion channels, nuclear hormone receptors, structural and membrane transporter proteins), the mode of perturbation (*e.g.* inhibition, agonism, antagonism) and the target product profile (*i.e.* the disease, the patient population, route and frequency of NME administration).

Over the last 20 or so years the importance of this balancing act has become abundantly clear but the challenge of optimising NMEs with the required "biological" properties *versus* "chemical" properties has been exacerbated by several factors. First – the practice of high-throughput screening (HTS) in the early discovery of initial "hit" compounds with high receptor binding affinity has resulted in compounds with less optimal physicochemical properties, *i.e.* highly potent compounds are larger, bulkier, more lipophilic and are subject to high attrition. Secondly – the proliferation of multiple ancillary assays to screen large numbers of compounds and enabled the process of lead optimisation to inflate poor physicochemical properties. Thirdly – current and future drug targets are less druggable than those pursued historically and future strategies now necessitate operating in a less optimal "drug-like space".[1]

Drug metabolism and pharmacokinetics (DMPK) cements the disciplines of biology and medicinal chemistry and is fundamental to the understanding of the biological and physical processes associated with the administration of NMEs to animals and humans. DMPK underpins the safety and efficacy framework by which key preclinical or clinical findings (desirable or adverse) can be related to quantifiable drug-related properties such as NME concentrations at the active site, at both a molecular level (*e.g.* V_{max} and K_m) and macroscopic level (*e.g.* tissue partitioning, organ specific drug elimination). Drug concentrations and their rate of change can be related to physical chemistry (Figure 8.1), biochemistry (Michaelis–Menten kinetics) and

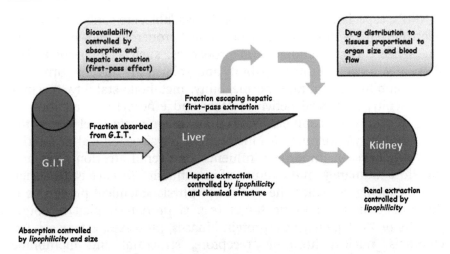

Figure 8.1 The fundamental processes and the relationship and influence of physicochemical properties on the absorption, distribution, metabolism and excretion of drugs flowing oral administration.

biological activity.[4] This pharmacokinetic/pharmacodynamic (PKPD) relationship is the backbone of informed NME discovery and development. By integrating PKPD analyses it is possible to characterise onset, intensity and duration of the pharmacological effects of an NME and relate these to the mechanisms of action in terms of safety and efficacy at both the clinical and preclinical level. The objective of this chapter is to describe and highlight the pivotal role that DMPK plays in underpinning a more integrated approach to early phase drug discovery where medicinal chemistry is at its most impactful.

8.2   Understanding ADME

The term ADME describes the absorption, distribution, metabolism and excretion of drug (and/or metabolites where appropriate) following drug administration. Absorption may be used to describe permeation across a membrane (*e.g.* the GI tract following oral administration) or, in the case of pharmacokinetics, is used to describe all the processes involved from the site of administration to the site of drug concentration measurement, usually blood or plasma. Distribution describes the reversible transfer of drug between the site of measurement and other site(s) within the body, metabolism is the irreversible loss of drug form the body by biochemical conversion and excretion is the irreversible loss of drug from the body. Disposition is a hybrid term describing the combination of both distribution and elimination, where elimination is the irreversible loss of drug from the site of measurement within the body (*e.g.* hepatic or biliary elimination) rather than excretion.

Pharmacokinetics, seeks to provide a mathematical basis for the description and prediction of the time course of drugs (and their metabolites) in the body and refers to the quantitative study of the process of ADME. In simple terms, pharmacokinetics may be defined as the effect of the body on the drug in contrast with pharmacodynamics which is the effect of the drug on the body. The fundamental hypothesis of pharmacokinetics is that the relationship between pharmacologic effect(s), be they efficacious or toxic, can be described by measuring drug concentrations in a readily and accessible site within the body (*e.g.*, the blood or plasma). The basic pharmacokinetic parameters are:

- *Clearance* – the measure of the ability of the body to eliminate drug. It is a proportionality factor that links the volume of blood/plasma that is cleared of drug removed per unit time, *i.e.* the rate of drug removal.
- *Volume of distribution* – relates the amount of drug in the body to the concentration of drug in the blood or plasma, depending on the matrix in which drug concentration is measured.
- *Terminal half-life* – is the time taken for the for drug concentration in the blood or plasma to reduce by 50%.
- *Bioavailability* – is the fraction of unchanged drug that is absorbed intact and reaches the site of action, or reaches the systemic circulation after administration by any route.

A drug-dosing regimen, "how often the drug needs to be administered", is dependent on the terminal half-life of the drug and so is interdependent on clearance and volume of distribution, while clearance and fraction of dose absorbed will impact the dose or "amount of drug required". Optimisation of these parameters continues to be the cornerstone of successful drug discovery[5] and an understanding of the basic principles of pharmacokinetics is therefore essential in order to describe and predict drug behaviour in both *in vivo* and *in vitro* models of ADME, drug efficacy and safety. A more indepth mathematical derivation and explanation of pharmacokinetic terms and interrelationships can be found elsewhere[6,7] but as the vast majority of small-molecule NMEs continue to be intended for oral use the prediction of human ADME characteristics continues to be of considerable interest and development.

For small-molecule NMEs estimating *in vivo* drug clearance using *in vitro* systems is critical for successful clinical translation. Understanding the clearance mechanism of novel compounds in a drug-discovery environment is the first step in achieving a successful prediction of human clearance. Historically, *in vitro* drug clearance assays have focused on drug metabolism[8] evaluating both Phase I drug metabolism involving cytochrome P450 (CYP), aldehyde oxidase (AOX), xanthine oxidase (XO), monoamine oxidase (MAO), flavin-containing monooxygenase (FMO) and hydrolases and the Phase II drug metabolism involving conjugation reactions such as glucuronidation and sulfation. However, in recent years there has been increased recognition that in addition to drug-metabolising enzymes, drug transporters play an important role in small-molecule ADME. A number of key drug transporters, both active uptake and efflux, have now been identified that play a pivotal role not only in drug disposition but in the transport of exogenous compounds.[9] In contrast to drug-metabolising enzymes that are predominantly located in the liver and intestine, drug transporters are present in all tissues in the body thus requiring more complex and varied *in vitro* tools with which to evaluate NMEs and the role of transporter proteins in drug and/or metabolite(s) disposition in humans.[10]

8.3 Screening Assays

Since many dynamic and competing processes influence the disposition of drugs *in vivo*, screening studies play an important role in triaging molecules throughout drug discovery and development (Table 8.1). In essence, these screens measure specific parameters in a reduced system, such as V_{max}/K_m values from a biological assay or partition coefficients from plasma or tissue binding experiments. Data derived from these screens is placed into context with the pharmacokinetics of the whole *in vivo* system (*e.g.* clearance, metabolism, distribution, drug accumulation) and ultimately informs the structure optimisation process.

Desirable features in a screening assay are exemplified by the measurement of intrinsic clearance (CL_i) from liver fractions.

1. Screening assays should be biologically relevant to the *in vivo* situation. CL_i is a fundamental property of the drug–enzyme interaction and measures the maximal rate of enzyme metabolism, such as the action of CYP450 isoenzymes present in the endoplasmic reticulum of the liver.

Table 8.1 *In vitro* DMPK screening approaches commonly applied to compound selection, lead optimisation or during early drug development.

DMPK screening approach	Example *in vitro* system	Ref.
Hepatic clearance	Intrinsic clearance (CL_i) measurements in liver microsomes, hepatocytes	91
Permeability	Rate of drug passage through Caco-2 cells (heterogeneous human epithelial colorectal adenocarcinoma) or MDCK cells (Madin–Darby canine kidney) cell monolayers	92,93
Drug distribution and partitioning into tissues	Tissue binding (slices or homogenates)	
Drug transporters	MDR-MDCK cells supplemented with human transporters	94,95
Enzyme inhibition	Aldehyde oxidase CYP inhibition	96
Induction of liver enzyme levels	AhR (aryl hydrocarbon receptor – regulated genes include CYP1A1, CYP1A2 and CYP1B1), CAR (constitutive androstane receptor – regulated genes include CYP2B, CYP2C, and CYP3A subfamilies, sulfotransferases and glutathione-S-transferase), PXR (pregnane X receptor – regulates CYP3A4).	97
Toxicity potential of reactive drug metabolites	Reactive metabolite trapping (*e.g.* glutathione or cyanide adducts)	98,99
Mechanism dependent inhibition	Time-dependent inhibition of CYP enzymes	100,101
Metabolite identification		102

2. Screening assays should provide high-fidelity measurements. Experimentally CL_i can be determined by monitoring the rate of loss of parent molecule on coincubation with liver fractions containing CYP450s (*e.g.* microsomes) and additional metabolising enzymes (*e.g.* S9, hepatocytes). LC-MS/MS as an endpoint ensures a good assay throughput and high analyte selectivity.

3. Screening assay data should correlate to measurable parameters *in vivo* or observable effects. CL_i can be transformed to an *in vivo* hepatic clearance (CL_H) using *in vitro–in vivo* extrapolation (IVIVE), such as the well-stirred liver model, eqn (8.1);[11] This model incorporates the rate of metabolism (CL_i), the rate of delivery of drug to the organ *via* blood flow (Q) and the extent of drug binding in blood (fu_b).

4. Screening assay data should be decision making. Molecules characterised by high intrinsic clearance values in liver

microsomes (>5 mL min^{-1}g^{-1}) would undergo extensive liver metabolism (*via* CYP450 enzymes), which in turn would lead to low oral bioavailability.

5. Screening assay methodology should be translational. Human CL$_i$ values can be generated alongside preclinical CL$_i$ values as a guide to the expected behaviour in the clinic.

$$CL_H = \frac{Q \cdot fu_b \cdot CL_i}{Q + fu_b \cdot CL_i} \qquad (8.1)$$

8.3.1   Assumptions Underlying Screening Assays

Although screening assays provide valuable insights into specific facets of drug disposition, these are reduced systems and their underlying assumptions and limitations must be borne in mind.

8.3.2   Impact of Nonspecific Binding

The screening assays described in Table 8.1 are configured in the absence of supplemented protein and therefore the reported PK parameters should be related to the unbound concentration in the incubation medium. However, the absence of supplemented protein does not preclude nonspecific binding to cellular components or experimental apparatus. If significant, this nonspecific binding reduces the unbound concentration available in the incubation and may confound the interpretation of results. At the very least, the impact of nonspecific binding should be understood.

In permeability assays, calculating the system's mass balance assesses the reliability of the reported data. Since the concentration (C) in each transwell is measured and the volumes (V) of the basolateral and apical compartments are known, then the apparent amount (M) of drug within the *in vitro* system can be determined (since $M = C \times V$). If this agrees closely with the amount administered at the start of the incubation then the mass balance is acceptable (*e.g.* $>80\%$). Substantially lower values than this would suggest significant losses are occurring within the system, such as *via* adsorption to the plate walls or retention within the cellular matrix. Alternatively, Tran *et al.* reported a correction methodology taking account of system losses, referred to as P_{exact}.[12]

In the case of intrinsic clearance measurements, equilibrium dialysis may be used to determine to extent of nonspecific binding to metabolically inactivated liver microsomes,[13] *e.g.* in the absence of the cofactors that promote enzyme activity. Complementary *in silico* approaches to estimate microsomal binding have also been reported,[14] where 92% of the variance (100 test compounds) was accounted for and log *P* was highlighted as the most important molecular property in the model.

Nor are protein binding measurements themselves exempt from the impact of nonspecific adsorptive losses, especially if the means of separating the total and unbound fractions is a dynamic process. For ultrafiltration, nonspecific binding to the filter or the ultrafiltrate container would lead to a reduction in the unbound concentration being transferred; consequently the fraction unbound in plasma would appear lower than reality.[15] Indeed, nonspecific adsorptive losses may still occur in an equilibrium dialysis method of measuring protein or tissues binding, however, since the system attains a state of equilibrium then the total concentrations and unbound concentrations will remain at a constant ratio.

8.3.3 Dynamic Range of the Measurement

The dynamic range and sensitivity of most screens is defined by the method of analytical measurement. For example, in the *in vitro* intrinsic clearance screening model, incubations are often run over a short time course (*e.g.* 30 min) with sampling of the incubation at selected time points (*e.g.* 0, 3, 6, 12, 15, 21, 30 min). The lowest intrinsic clearance value measurable (equating to low hepatic clearance *in vivo*) is determined by the assay variability across the concentration–time profile and whether the slope of the curve can be defined reliably (Figure 8.1). The rapid LC-MS/MS assays deployed in this screen are optimised to cover a wide range of physicochemical space expected in the lead selection and optimisation phases. Assays optimised in this way are often characterised by wider acceptance (*ca.* 20% assay bias and precision) rather than those used to support regulated studies variability in the range of ($\leq$20% assay and precision).[16] Thus, when the difference in analyte signal across the time profile is lower than 20% then significant caution should be placed on any derived intrinsic clearance values, unless additional control measurements, statistical analyses or reproducibility experiments can provide additional confidence. Although this limit of detection may be adequate for triaging compounds where the hepatic clearance is likely to be high or low, not being able to define CL_i may preclude IVIVE to predict human clearance; seriously limiting the translatability of the assay data.

Some analytical measurements are performed on peak area ratios with matrix matching rather than constructing individual calibration curves; an example being protein or tissue binding measurements.[17,18] This approach is valid provided the analytical signal being measured is within the linear dynamic range of the assay and no detector saturation is impacting the response. Understanding the linearity of the detector should be part of validating the screening methodology.

8.3.4 Unbound Concentrations and the Site of Action

The movement and physiological action of drugs within the body is driven by the unbound concentration. Unbound drug is available to diffuse through cellular membranes and bathes the target site in order to elicit efficacy (*e.g.* at the drug target), adverse pharmacology/toxicity (*e.g.* off-target), drug transport (*e.g.* carrier proteins) or metabolism. Although there are a few examples where drug efficacy appears to correlate better with total rather than unbound drug concentrations,[19–22] in the majority of instances there is strong support for unbound drug concentration being the driver for target engagement[23] or effect.[24]

Drug concentrations are linked to both target engagement and pharmacological activity through the pharmacokinetic–pharmacodynamic (PKPD) relationship. Testing the mechanistic hypothesis for treating a disease in humans starts in earnest with proof-of-concept (PoC) studies,[25] where success establishes a causal link between the drug, its target and the mechanism of drug action. However, failure in a PoC study is less clear cut; the proposed biological mechanism could be wrong, or at the clinical doses tested, there could be insufficient drug exposure in the biophase to elicit adequate target engagement.[25,26] Being able to measure or model the unbound drug concentration in the biophase provides a rational framework for setting the doses in a PoC study, but, also guides after action review if a suboptimal outcome were observed. In essence, target engagement and pharmacological effect both rely on sufficient exposure at the target site.

8.3.5 Factors Affecting the Unbound Concentration at the Site of Action

Depending on the processes acting on a given drug molecule, the unbound drug concentration may be comparable throughout the body or vary at a subcellular level relative to blood (or plasma). The main factors that drive the local unbound drug concentration relative to that in the systemic circulation are passive diffusion, carrier-mediated transport and the local pHs between aqueous fluid compartments.

8.3.6 Unbound Drug Concentrations Relating to Passive Diffusion

The rate of diffusion across a cellular barrier (dn/dt) is governed by Fick's first law, being a function of permeability (P), surface area (S) and the concentration gradient dC/dx.[27] This relationship holds for solutes (such as unionised lipid soluble drug molecules) moving across the lipid-rich chemical environment of the membrane from one aqueous compartment to another.

$$\left(\frac{dn}{dt}\right) = P \cdot S \cdot \left(\frac{dC}{dx}\right)$$ (8.2)

For molecules subject to passive diffusion, the unbound concentration on either side of the diffusion barrier would be the same at steady state since the concentration gradient would have been abolished. Passive diffusion is important for neutral molecules, where pH differences between aqueous compartments do not alter the fraction of ionised drug significantly or in the absence of drug transport. At steady state the dC/dx term would be equivalent to the equilibrium unbound partition coefficient, $K_{p,uu}$, which has become an important parameter to measure when attempting to understand difference in unbound concentrations in the body.[28]

For molecules whose distribution is governed by passive diffusion then the unbound concentration at the target site will reflect that in systemic circulation, given adequate permeability. In this instance, the plasma pharmacokinetics may provide an acceptable surrogate for the unbound concentration in the biophase.

8.3.7 Unbound Drug Concentrations Relating to Carrier-mediated Transport

The carrier-mediated transport of molecules across cell membranes is a key element of cellular homeostasis, leading to the potential for xenobiotic molecules to be transferred alongside endobiotics. If the carrier-mediated process is energy dependent then a concentration gradient may be established. For active efflux, steady-state $K_{p,uu}$ would be less than unity, whereas for active influx $K_{p,uu}$ would be greater than unity. The consequence of active drug transport is that the unbound blood concentration is unlikely to reflect the unbound concentration at the target site. Therefore, due consideration is required to either measure or model the pharmacologically relevant drug concentration.

8.3.8 Unbound Drug Concentrations Relating to pH

Differences in pH may also affect the unbound concentration gradient across cellular barriers, leading to a phenomenon often referred to as ion trapping.[29,30] Again, the unbound concentration of drug at the target site would differ from the unbound concentration in blood.

For weak acids and bases, both the ionised and unionised forms coexist in solution at physiological pH 7.4, the relative proportions of which are dependent on the pK_a and pK_b values. pH is moderated in cellular environments[31] and hence the extent of ionisation can differ for weak acids and bases across a cell membrane. Passive diffusion of the unionised form is driven by entropic changes arising from the concentration gradient (dC/dx), whereas permeation of the ionised form includes enthalpic factors;[32] as a consequence the unionised form is the major species undergoing passive diffusion.[33] In organelles such as the lysosome where the pH is reduced (pH 5–6), the proportion of ionised drug increases markedly for weak bases, which drives an increase in $K_{p,uu}$.

8.3.9 Unbound Drug Concentrations Relating to Drug–Drug Interactions

Drug–drug interactions (DDI) occur when the unbound concentrations of one coadministered drug perturb the ADME processes of another. These DDIs arise from altered enzymology of the "victim" drug (*e.g.* metabolism), reduced carrier-mediated transport (*e.g.* excretion) or downstream off-target binding to a regulatory element that modulates protein expression (*e.g.* of an enzyme or transporter). By inference, DDIs lead to a change in the unbound concentration of the victim drug or its metabolites, which in turn may drive reduced efficacy (reduced concentrations of victim drug) or raised safety risks (increased concentrations of victim drug). Assessments of DDI potential often include the common isoforms of major drug metabolism enzymes such as the Cytochrome P450s,[34] induction *via* regulatory elements such as hPXR[35,36] and hCAR[37,38] or inhibition of hepatic and renal transporters including OATP1B1,[39] OAT1 and OAT3[40] and OCT2.[41] One of the key successes of ADME in this field is *in vitro– in vivo* extrapolation, for example, taking inhibitory concentration measurements (unbound IC_{50} values) and predicting the clinical impact by mathematical modelling approaches.[42–44] The clinical consequences of DDIs on safety, efficacy and drug-development costs far outweigh the attrition in drug discovery to balance potency and selectivity with reduced DDI potential.

Many DDIs are competitive in nature and driven by the binding characteristics (V_{max}, K_m) alongside the unbound concentration of the perpetrator drug. However, noncompetitive or irreversible binding can also occur leading to a reduced overall pool of enzyme or transporter that persists longer than would direct inhibition (*e.g.* until protein is turned over and replenished). In the case of drug metabolism, chemically reactive species may be produced whose presence propagate a range of toxicities.[45,46] Of greatest concern are idiosyncratic adverse events that are unlikely to manifest themselves in early development but lead to serious clinical complications or deaths in late-phase development or postlaunch[47] when patient numbers are large enough to make the occurrence more probable. From a reputational perspective maximum damage is caused by these late-stage safety-related failures arising and the outlay in development costs are at their highest. Ironically, the idiosyncratic nature also presents the greatest challenge to drug discovery and so there is still much to be done to understand the biochemical and biological pathways. Currently, the best approaches available to the medicinal chemist and ADME scientist are to avoid isosteres that form chemically reactive species following drug metabolism (such as electrophiles) or to reduce dose.[45]

8.3.10   What is the Drug Concentration at the Target Site?

Two approaches provide insight into the drug concentration at the target site: (1) direct measurements based on imaging or subcellular fractionation, and (2) mathematical modelling based on physicochemical drug properties and physiological properties of the system.

Imaging techniques require some molecular property that facilitates high-fidelity detection at the subcellular level; examples of this include fluorescent chemical moieties[48] or readily ionisable functional groups coupled with the high resolution mass-spectrometric analysis.[49] Fluorescence lends itself well to nondestructive imaging approaches, *e.g.* those amenable to *in vivo* or *in situ* measurements.

Intravital microscopy enables biological processes to be viewed within the tissues of a living animal at high spatial resolution (mouse) and has provided imaging and pharmacokinetic information at the single-cell and subcellular level *in vivo*.[50] Although a fluorophore is required with this approach, fluorescent analogues may provide useful insights, such as for the microtubule inhibitor eribulin, where drug efflux and the three-dimensional tumour vasculature were shown to strongly determine the extent of drug accumulation.[51]

Subcellular fractionation provides a complementary measure on drug concentrations within a tissue fraction or organelle. Isolation of functional and pure mitochondria has been demonstrated using magnetic immunoprecipitation[52,53] and in principle this approach could be applied to any cell type, given a suitably selective immunocapture reagent. Once the fractionation has been completed, then a wide range of analytical techniques would be applicable, including fluorescence or mass spectrometric detection. Capturing the subcellular concentration of drug by destructive methods does, however, come with some caveats. Arguably the most important is postmortem redistribution of drug; this occurs at the tissue level[54] and by inference would occur at the subcellular level too. Disruption of the local concentration gradients during tissue excision and processing occurs and therefore molecules subject to energy-dependent carrier-mediated transport may be most susceptible.

Modelling employs molecule-specific PK properties and system specific properties to construct a mathematical description of the unbound drug concentrations across various compartments or in relation to drug effect.[55] For example, a framework for calculation of $K_{p,uu}$ values between plasma and cell cytosol or indeed cell cytosol and organelles has been described,[32] which accounts for the passive diffusion of both the unionised species (Fick's first law) and ionised species (Nernst–Planck equation). The relevant molecule-specific properties required to calculate $K_{p,uu}$ are pK_a and the permeability of the unionised and ionised species, while system specific properties are pH and electric potential across the membrane.

8.4 Dose

The therapeutic dose of a drug is intrinsically linked to its physico-chemical properties. The molecular characteristics must enable the drug to move from the administration site and reach the target at sustained active concentrations. Furthermore, those molecular characteristics should minimise safety concerns arising from promiscuous biological interactions with other receptors, transporters or enzymes. Such is the importance of dose that discussions relating to dose selection, prediction and refinement are hardwired across the entire breadth of the drug discovery and development process.

Paracelsus (1493–1541)[56] captured the salient point when commenting that "poison is in everything, and no thing is without poison. The dosage makes it either a poison or a remedy." In essence, increasing the dose will, at some point, lead to toxicity with the body being subjected to a higher burden of drug and metabolites. Given the extreme costs associated with late-phase clinical development of new medicines,[57] a variety of meta-analyses have related clinical dose to safety concerns. Of the drug withdrawals in the United States between 1980 and 2006, direct organ toxicity was identified as the major cause for many drugs where the dose was in excess of 100 mg.[58] Clinical doses greater than 100 mg have also been associated with increased risk of drug-induced hepatoxicity[59,60] and drug-induced liver injury.[61,62] At first sight, these doses may appear somewhat low (on a mg kg^{-1} basis) relative to the doses tested in preclinical safety studies. However, toxicological findings are a function of both dose and time and therapeutic indices narrow markedly over the development phase of a drug; this is because longer-term studies draw out the effects of prolonged drug (or metabolite exposure) exposure.[63]

Drug solubility in intestinal fluid may also become a concern with higher doses. The Developability Classification System (DCS) stratifies drugs according to both their solubility and permeability characteristics,[64] as shown in Figure 8.2, while some reports have highlighted a substantially larger proportion of class 2b molecules progressing through drug development compared to marketed drugs accompanied by a lengthier development process.[65] Molecules falling into the class 2b category are characterised by good permeability but limited drug solubility, *e.g.* there is insufficient liquid volume available to dissolve the entire dose during its transit through the small intestine and hence enable absorption. Poor solubility impacts both discovery and development process, including *in vitro* measurements of potency, low preclinical bioavailability leading to an inability to achieve adequate safety exposure margins in preclinical toxicology species, increased time devoted to formulation development and ultimately low oral/variable bioavailability in humans leading to difficulties in testing the clinical mechanism in (PoC) studies (cannot achieve adequate dose response).

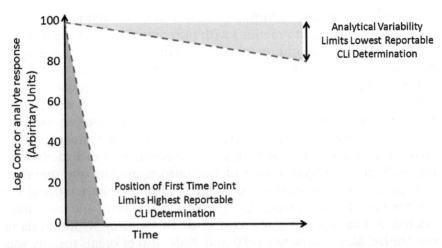

Figure 8.2 Example of how assay conditions limit the measurement range, *e.g.* analyte disappearance for the intrinsic clearance (CL_i) assay. Determining CL_i for low metabolic turnover (blue zone) is limited by the analytical variability of the experimental method. Determining CL_i for high metabolic turnover (orange zone) is limited by earliest time point collected (and assay limit of quantification).

Classification systems may also help identify key metabolic or transport processes that impact the adsorption, distribution, metabolism or elimination of drugs and drug candidates. For example, the dominant mechanism of drug passage across the blood–brain barrier (BBB) may be assessed by comparing *in vitro* efflux ratios (*e.g.* P-glycoprotein) and $K_{p,uu}$.[66,67] Similarly, the extended clearance classification system (ECCS) attempts to stratify clearance mechanisms (renal, hepatic uptake, metabolism) by combining permeability measurements and physicochemical properties, namely molecular weight and whether the molecule is an acid, base, zwitterion or neutral.[68] Recently, Wenlock published a study profiling the estimated human *in vivo* free plasma concentrations of 215 marketed oral drugs from a simple one-compartment pk model built using published values of oral bioavailability, clearance and volume of distribution. Predicted maximum and minimum steady-state plasma concentrations at clinically relevant doses were correlated with the minimum effective therapeutic concentration and the no observed adverse effect level.[69] Based on the empirical observations that the ratio of maximum to minimum plasma concentration was found to be less than 30 and that (for a subset of 104 drugs) adverse effects were observed at free plasma concentrations above 100 nM a new definition of oral drug-likeness was proposed whereby "ideal" compounds should preferably be designed within these limits. Furthermore, it was shown that *in silico* human ADME model predictions can be used in the absence of experimental data for the necessary input parameters and augment and improve the predictions.

8.4.1 Early Human Pharmacokinetic Predictions

It is now widely recognised that the correct balance of DMPK properties, optimised for suitable pharmacokinetics and an appropriate safety profile (including minimal drug–drug interactions), is a prerequisite for enabling efficient early-phase drug discovery and ultimately a successful and efficacious therapy. Consideration of basic physicochemical and molecular properties, integrated with *in vitro* ADME screening data (including human material) and *in vivo* preclinical pharmacology studies are core activities. However, over recent years, the need for thorough and comprehensive data integration and the ability to predict the efficacious clinical dose to man are key areas of emerging science. The primary objective of any medicinal chemistry research programme is to prosecute and create molecules that possess suitable potency and selectivity against the pharmacological target and are able to achieve sufficient clinical exposure at the site of action long enough to exert a clinical pharmacological effect. Understanding the dose required and the dosing regimen is underpinned by the pharmacokinetic profile of the compound and the importance and timing of human pharmacokinetic predictions is now an integral and essential part of compound optimisation.[70]

Predicting the human pharmacokinetic profile should be conducted at all stages in drug discovery, not just at the candidate selection stage. Judicious use of the data will not only inform on molecular design, optimisation strategy and ultimately drug candidate selection but, when combined with appropriate *in vitro* and *in vivo* pharmacology data, provide an estimate of clinical efficacious dose and a prediction of the appropriate dosing regimen. The key pharmacokinetics parameters of clearance, volume of distribution and oral bioavailability can be predicted using a variety of empirical and mechanistic techniques including allometric scaling[71,72] and physiological based pharmacokinetic modelling (PBPK).[73] Recognising the need for improved human pharmacokinetic predictions in early drug discovery and development and a clear description and comparison of the predictability of the different approaches, a recent initiative was co-ordinated by the Pharmaceutical Research and Manufacturers of America (PhRMA). In this comprehensive and cross-industry collaborative exercise, data on a total of 108 clinical compounds were shared from 12 PhRMA member companies.[74] The data set contained intravenous ($n = 19$) and oral pharmacokinetic data ($n = 107$) in humans, physicochemical data and associated preclinical *in vivo* and *in vitro* ADME data. All data were blinded to all parties, thereby protecting both the anonymity of the data and the company submitting the data. The dataset was evaluated using both empirical and physiological methodologies, derived from the literature, in order to predict human pharmacokinetic properties and plasma concentration–time profiles. In turn these predictions were then compared with the observed clinical data to assess predictive accuracy based on robust statistical analysis and criteria. A key advantage of this study was that the compounds chosen were (at the time of analysis) current development compounds rather than relying on a retrospective analysis of historical data which is often incomplete. All compounds had the associated physicochemical and *in vitro* and *in vivo* ADME data and, interestingly, a broad distribution with the Biopharmaceutics Classification System (see earlier) with 15%, 57%, 12% and 23% in classes 1, 2, 3 and 4, respectively. The approaches that were taken could be divided into three core areas: allometric scaling, *in vitro–in vivo* extrapolation (IVIVE) and mechanistic PBPK modelling. Interestingly, no one method emerged as consistently better than the rest and predicting human disposition pharmacokinetics after intravenous administration was far better than after oral administration. No one preclinical species was found to be a better predictor for human than the others, nor was the inclusion of multiple species seen to improve the predictive ability. Whilst allometry remains a popular approach, it does rely on *in vivo* preclinical data that has both cost and ethical implications and fundamentally the science of allometric scaling cannot progress much further. IVIVE has been applied extensively to predict hepatic clearance utilising both human and other preclinical species to understand known species differences and separate metabolic clearance from active transport processes. Moreover, such an approach is amenable to higher throughput without the use of preclinical species and therefore can be employed successfully in compound optimisation.[75] PBPK modelling, though an emerging science, has much to offer.[76] Using a mechanistic, bottom-up approach building on structural molecular properties and physicochemical data requires fewer resources (including preclinical species) and whilst supporting drug discovery, PBPK models have the potential to extend far beyond this, impacting and guiding clinical development on many fronts such as drug–drug interactions, effects of age and disease and patient stratification.

This approach is very much in vogue with regulatory agencies and authorities although a recent review has highlighted that whilst PBPK modelling is an exciting and emerging science full of potential, there is a considerable lack of consistency with respect to model development and quality assessment guidelines, demonstrating a real need for the development of best-practice and consistent approaches.[77]

8.4.2 Integrated Design – the Increasing Role of Pharmacodynamics–Pharmacokinetics Modelling and Simulation

As stated earlier, pharmacokinetics (pk) is the time-course of drug in the plasma (blood) or site of action (tissues, cells, *etc.*) and is effectively what the body does to the drug. Pharmacodynamics (PD) is the study of the pharmacological effects exhibited by the drug by whatever mechanism of action and is therefore what the drug does to the body. Pharmacokinetics is therefore the forcing function behind pharmacodynamics response. PKPD analysis is the integration of these two disciplines and enables exploration and characterisation of "concentration–effect relationships" built around the onset, duration and intensity of biological effect linked with drug concentration.

PKPD modelling and simulation is very much the cornerstone of successful drug discovery[78–80] but also spans the entire drug development landscape enabling "learning and confirming" at key stages (Figure 8.3). Quantitative PKPD modelling promotes the simultaneous optimisation of multiple parameters rather than a single one such as clearance or tissue penetration, thereby enabling medicinal chemists to relate pharmacodynamics properties such as potency and selectivity to structural properties associated with the compound of interest.[79] Such an approach is not necessarily "off the shelf" and often requires significant investment in both architectural infrastructure and staff training in order to build an integrated modelling and simulation capability that can impact drug discovery by improving success through reduced attrition,[80] more informed decision making and reducing the burden on *in vivo* studies.[78]

In those instances when there is either no validated translational *in vivo* or *in vitro* PKPD approach or uncertainty around the target efficacious drug concentrations then a "no regrets approach" targeting a minimum effective concentration (MEC) over the entire dosing interval is a pragmatic alternative.[81] Using a data set of 63 marketed drugs with known human pharmacokinetics and a clinically validated

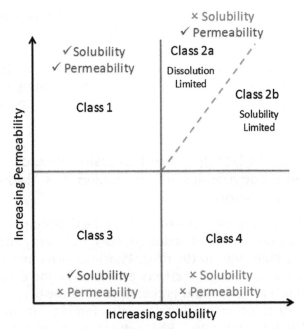

Figure 8.3   Depiction of the Drug Classification System (DCS, Butler and Dressman, 2010) that uses a 4-box design to stratify drugs in terms of permeability and solubility characteristics. Poor solubility may be a consequence of reaching the solubility limit (*e.g.* with increasing dose) or as a result of the dissolution rate; hence Class 2 is generally subdivided into 2a and 2b subcategories.

MEC, the clinical dose was shown to be well predicted using a simple pharmacokinetic modelling approach based on *in vitro* and minimal *in vivo* ADME data. By adjusting for plasma protein binding, the MEC of the majority of compounds was ≤3-fold over the respective *in vitro* potency for a significant proportion of the test set and for approximately 90% of the compounds the predicted clearance based on IVIVE combined with estimates of renal clearance were within 2-fold of the observed clinical values. This holistic approach although nascent continues to be further enhanced,[82] thereby enhancing data-driven decisions to support rational drug design strategies focused on delivering drug candidates that can test potential therapeutic targets in patients.

8.4.3   Systems Concepts in ADME

So what of the future for DMPK? As a discipline DMPK draws upon the full range of biological sciences (biology, biochemistry, physiology) and chemical sciences (organic, physical, analytical) with a mathematical context being placed upon the how the body acts on the drug (*e.g.* rates of change, partition coefficients). Although a screening assay might focus on a single element of the entire "DMPK-ome" the real value to the subject is when all of the data are brought together holistically. Arguably, the true success of drug discovery does not lie in the myriad of high-throughput screening data placed into columns of an Excel spreadsheet and colour-coded according to a traffic light approach of red, amber or green; the success is when the mathematical concepts and measurements (native to DMPK) are moulded with potency and PD to yield a systems view of drug properties in the context of clinical dose.[83] In a systems view, facets of the three pillars of drug survival[4,25] are presented with their most exquisite contact, where knowledge is refined throughout the drug discovery and development process (Figure 8.4). Renewed interest in phenotypic screening[84,85] and the dawn of 3D *in vitro* systems within the drug discovery setting[86] may also aid a more holistic mindset. Mathematical modelling approaches lend themselves well to contextualising *in vitro* data[87] as well as building more complex tasks, such as adverse outcome pathways.[88]

However, the frame of reference for scientific development and career progression is still heavily biased to a "wet science" as opposed

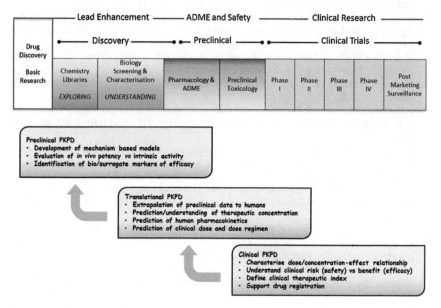

Figure 8.4 An integrated PKPD modelling approach; establishing exposure–response relationships at all stages and using all available data to learn and confirm across drug discovery and development.

to a "dry science" approach. Drug-discovery sciences, especially in the ADME arena are notoriously "information rich....but knowledge poor" so perhaps the greatest barrier to embracing a systems approach is more of an *in cerebro* challenge than an *in silico* mathematical modelling one. Furthermore, poor experimental reproducibility in science[89] begs the question as to whether a greater use of modelling and simulation would actually enhance our scientific abilities to design better experiments and improve decision making based on theoretical "what if" scenario testing rather than actual experimentation.[90] Knowing what not do is often just as impactful as knowing what to do. Whatever the outcome, there is no doubt that a substantially greater amount of data integration and contextualisation is conceivable with the aid of a systems- and modelling-based approach and therein lies the likely road ahead.

References

1. M. E. Bunnage, Getting pharmaceutical R&D back on target, *Nat. Chem. Biol.*, 2011, **7**(6), 335–339.
2. A. Mullard, FDA drug approvals, *Nat. Rev. Drug Discovery*, 2016, **15**(2), 73–76.
3. M. J. Waring, J. Arrowsmith, A. R. Leach, P. D. Leeson, S. Mandrell, R. M. Owen, *et al.*, An analysis of the attrition of drug candidates from four major pharmaceutical companies, *Nat. Rev. Drug Discovery*, 2015, **14**(7), 475–486.
4. S. Summerfield and P. Jeffrey, Discovery DMPK: changing paradigms in the eighties, nineties and noughties, *Expert Opin. Drug Discovery*, 2009, **4**(3), 207–218.
5. M. E. Dowty, D. M. Messing, Y. Lai and L. Kirkovsky, *ADME. ADMET for Medicinal Chemists*, John Wiley & Sons, Inc., 2011, pp. 145–200.
6. L. Z. Benet and P. Zia-Amirhosseini, Basic principles of pharmacokinetics, *Toxicol. Pathol.*, 1995, **23**(2), 115–123.
7. P. Jeffrey, Pharmacokinetics, in *Medicinal Chemistry Principles and Practice*, ed. F. D. King, Royal Society of Chemistry, Cambridge, UK, 2nd edn, 2002, pp. 118–137.
8. L. Di, The role of drug metabolizing enzymes in clearance, *Expert Opin. Drug Metab. Toxicol.*, 2014, **10**(3), 379–393.
9. K. M. Hillgren, D. Keppler, A. A. Zur, K. M. Giacomini, B. Stieger, C. E. Cass, *et al.*, Emerging transporters of clinical importance: an update from the International Transporter Consortium, *Clin. Pharmacol. Ther.*, 2013, **94**(1), 52–63.
10. K. L. Brouwer, D. Keppler, K. A. Hoffmaster, D. A. Bow, Y. Cheng, Y. Lai, *et al.*, In vitro methods to support transporter evaluation in drug discovery and development, *Clin. Pharmacol. Ther.*, 2013, **94**(1), 95–112.
11. G. R. Wilkinson and D. G. Shand, Commentary: a physiological approach to hepatic drug clearance, *Clin. Pharmacol. Ther.*, 1975, **18**(4), 377–390.
12. T. T. Tran, A. Mittal, T. Gales, B. Maleeff, T. Aldinger, J. W. Polli, *et al.*, Exact kinetic analysis of passive transport across a polarized confluent MDCK cell monolayer modeled as a single barrier, *J. Pharm. Sci.*, 2004, **93**(8), 2108–2123.

13. R. S. Obach, Prediction of human clearance of twenty-nine drugs from hepatic microsomal intrinsic clearance data: An examination of *in vitro* half-life approach and nonspecific binding to microsomes, *Drug Metab. Dispos.*, 1999, **27**(11), 1350–1359.

14. H. Gao, L. Yao, H. W. Mathieu, Y. Zhang, T. S. Maurer, M. D. Troutman, *et al.*, In silico modeling of nonspecific binding to human liver microsomes, *Drug Metab. Dispos.*, 2008, **36**(10), 2130–2135.

15. M. A. Zeitlinger, H. Derendorf, J. W. Mouton, O. Cars, W. A. Craig, D. Andes, *et al.*, Protein binding: do we ever learn?, *Antimicrob. Agents Chemother.*, 2011, **55**(7), 3067–3074.

16. S. Lowes, R. Hucker, M. Jemal, J. C. Marini, V. M. Rezende, R. Shoup, *et al.*, Tiered approaches to chromatographic bioanalytical method performance evaluation: recommendation for best practices and harmonization from the Global Bioanalysis Consortium harmonization team, *AAPS J.*, 2015, **17**(1), 17–23.

17. R. E. Curran, C. R. Claxton, L. Hutchison, P. J. Harradine, I. J. Martin and P. Littlewood, Control and measurement of plasma pH in equilibrium dialysis: influence on drug plasma protein binding, *Drug Metab. Dispos.*, 2011, **39**(3), 551–557.

18. H. Wang, M. Zrada, K. Anderson, R. Katwaru, P. Harradine, B. Choi, *et al.*, Understanding and reducing the experimental variability of *in vitro* plasma protein binding measurements, *J. Pharm. Sci.*, 2014, **103**(10), 3302–3309.

19. V. P. Gerskowitch, J. Hodge, R. A. Hull, N. P. Shankley, S. B. Kalindjian, J. McEwen, *et al.*, Unexpected relationship between plasma protein binding and the pharmacodynamics of 2-NAP, a CCK1-receptor antagonist, *Br. J. Clin. Pharmacol.*, 2007, **63**(5), 618–622.

20. P. H. Van Der Graaf, E. A. Van Schaick, R. A. Math-ot, A. P. Ijzerman and M. Danhof, Mechanism-based pharmacokinetic-pharmacodynamic modeling of the effects of N6-cyclopentyladenosine analogs on heart rate in rat: estimation of *in vivo* operational affinity and efficacy at adenosine A1 receptors, *J. Pharmacol. Exp. Ther.*, 1997, **283**(2), 809–816.

21. R. A. Hull, N. P. Shankley, E. A. Harper, V. P. Gerkowitch and J. W. Black, 2-Naphthalenesulphonyl L-aspartyl-(2-phenethyl)amide (2-NAP)–a selective cholecystokinin CCKA-receptor antagonist, *Br. J. Pharmacol.*, 1993, **108**(3), 734–740.

22. J. A. Wald, R. M. Law, E. A. Ludwig, R. R. Sloan, E. Middleton Jr. and W. J. Jusko, Evaluation of dose-related pharmacokinetics and pharmacodynamics of prednisolone in man, *J. Pharmacokin. Biopharm.*, 1992, **20**(6), 567–589.

23. J. Watson, S. Wright, A. Lucas, K. L. Clarke, J. Viggers, S. Cheetham, *et al.*, Receptor occupancy and brain free fraction, *Drug Metab. Dispos.*, 2009, **37**(4), 753–760.

24. C. H. Large, M. Kalinichev, A. Lucas, C. Carignani, A. Bradford, N. Garbati, *et al.*, The relationship between sodium channel inhibition and anticonvulsant activity in a model of generalised seizure in the rat, *Epilepsy Res.*, 2009, **85**(1), 96–106.

25. P. Morgan, P. H. Van Der Graaf, J. Arrowsmith, D. E. Feltner, K. S. Drummond, C. D. Wegner, *et al.*, Can the flow of medicines be improved? Fundamental pharmacokinetic and pharmacological principles toward improving Phase II survival, *Drug Discovery Today*, 2012, **17**(9–10), 419–424.

26. D. Cook, D. Brown, R. Alexander, R. March, P. Morgan, G. Satterthwaite, *et al.*, Lessons learned from the fate of AstraZeneca's drug pipeline: a five-dimensional framework, *Nat. Rev. Drug Discovery*, 2014, **13**(6), 419–431.

27. S. Summerfield, P. Jeffrey, J. Sahi and L. Chen, Passive Diffusion Permeability of the BBB—Examples and SAR, in Blood-Brain Barrier in Drug Discovery, in *Blood-brain barrier in drug discovery : optimizing brain exposure of CNS drugs and*

minimizing brain side effects for peripheral drugs, ed. L. Di and E. H. Kerns, John Wiley & Sons, Inc., Hoboken, NJ, 2015, pp. 97–112.

28. M. Hammarlund-Udenaes, M. Friden, S. Syvanen and A. Gupta, On the rate and extent of drug delivery to the brain, *Pharm. Res.*, 2008, **25**(8), 1737–1750.

29. W. A. Daniel, M. H. Bickel and U. E. Honegger, The contribution of lysosomal trapping in the uptake of desipramine and chloroquine by different tissues, *Pharm. Toxicol.*, 1995, **77**(6), 402–406.

30. G. A. Siebert, D. Y. Hung, P. Chang and M. S. Roberts, Ion-trapping, microsomal binding, and unbound drug distribution in the hepatic retention of basic drugs, *J. Pharm. Exp. Ther.*, 2004, **308**(1), 228–235.

31. J. R. Casey, S. Grinstein and J. Orlowski, Sensors and regulators of intracellular pH, *Nat. Rev. Mol. Cell Biol.*, 2010, **11**(1), 50–61.

32. A. Ghosh, T. S. Maurer, J. Litchfield, M. V. Varma, C. Rotter, R. Scialis, *et al.*, Toward a unified model of passive drug permeation II: the physiochemical determinants of unbound tissue distribution with applications to the design of hepatoselective glucokinase activators, *Drug Metab. Dispos.*, 2014, **42**(10), 1599–1610.

33. D. K. Walker, R. M. Jones, A. N. R. Nedderman and P. A. Wright, Primary Secondary and Tertiary Amines and their Isosteres Metabolism, Pharmacokinetics, in *Metabolism, Pharmacokinetics, and Toxicity of Functional Groups: Impact of the Building Blocks of Medicinal Chemistry in ADMET*, ed. D. A. Smith, Cambridge, UK, Royal Society of Chemistry, 2010, pp. 168–175.

34. M. Turpeinen, L. E. Korhonen, A. Tolonen, J. Uusitalo, R. Juvonen, H. Raunio, *et al.*, Cytochrome P450 (CYP) inhibition screening: comparison of three tests, *Eur. J. Pharm. Sci.*, 2006, **29**(2), 130–138.

35. S. J. Shukla, S. Sakamuru, R. Huang, T. A. Moeller, P. Shinn, D. Vanleer, *et al.*, Identification of clinically used drugs that activate pregnane X receptors, *Drug Metab. Dispos.*, 2011, **39**(1), 151–159.

36. X. Cui, A. Thomas, V. Gerlach, R. E. White, R. A. Morrison and K. C. Cheng, Application and interpretation of hPXR screening data: Validation of reporter signal requirements for prediction of clinically relevant CYP3A4 inducers, *Biochem. Pharm.*, 2008, **76**(5), 680–689.

37. S. S. Ferguson, E. L. LeCluyse, M. Negishi and J. A. Goldstein, Regulation of human CYP2C9 by the constitutive androstane receptor: discovery of a new distal binding site, *Mol. Pharm.*, 2002, **62**(3), 737–746.

38. J. Kublbeck, T. Laitinen, J. Jyrkkarinne, T. Rousu, A. Tolonen, T. Abel, *et al.*, Use of comprehensive screening methods to detect selective human CAR activators, *Biochem. Pharm.*, 2011, **82**(12), 1994–2007.

39. P. Sharma, V. E. Holmes, R. Elsby, C. Lambert and D. Surry, Validation of cell-based OATP1B1 assays to assess drug transport and the potential for drug-drug interaction to support regulatory submissions, *Xenobiotica*, 2010, **40**(1), 24–37.

40. M. G. Soars, P. Barton, L. L. Elkin, K. W. Mosure, J. L. Sproston and R. J. Riley, Application of an *in vitro* OAT assay in drug design and optimization of renal clearance, *Xenobiotica*, 2014, **44**(7), 657–665.

41. K. Hacker, R. Maas, J. Kornhuber, M. F. Fromm and O. Zolk, Substrate-Dependent Inhibition of the Human Organic Cation Transporter OCT2: A Comparison of Metformin with Experimental Substrates, *PLoS One*, 2015, **10**(9), e0136451.

42. G. Baneyx, Y. Fukushima and N. Parrott, Use of physiologically based pharmacokinetic modeling for assessment of drug-drug interactions, *Future Med. Chem.*, 2012, **4**(5), 681–693.

43. C. Wagner, Y. Pan, V. Hsu, J. A. Grillo, L. Zhang, K. S. Reynolds, *et al.*, Predicting the effect of cytochrome P450 inhibitors on substrate drugs: analysis of

physiologically based pharmacokinetic modeling submissions to the US Food and Drug Administration, *Clin. Pharm.*, 2015, **54**(1), 117–127.

44. M. M. Posada, J. A. Bacon, K. B. Schneck, R. G. Tirona, R. B. Kim, J. W. Higgins, *et al.*, Prediction of renal transporter mediated drug-drug interactions for pemetrexed using physiologically based pharmacokinetic modeling, *Drug Metab. Dispos.*, 2015, **43**(3), 325–334.

45. A. Kamel and S. Harriman, Inhibition of cytochrome P450 enzymes and biochemical aspects of mechanism-based inactivation (MBI), *Drug Discovery Today Technol.*, 2013, **10**(1), e177–e189.

46. R. J. Riley and C. E. Wilson, Cytochrome P450 time-dependent inhibition and induction: advances in assays, risk analysis and modelling, *Expert Opin. Drug Metab. Toxicol.*, 2015, **11**(4), 557–572.

47. J. Uetrecht and D. J. Naisbitt, Idiosyncratic adverse drug reactions: current concepts, *Pharmacol. Rev.*, 2013, **65**(2), 779–808.

48. G. M. Thurber, K. S. Yang, T. Reiner, R. H. Kohler, P. Sorger, T. Mitchison, *et al.*, Single-cell and subcellular pharmacokinetic imaging allows insight into drug action *in vivo*, *Nat. Commun.*, 2013, **4**, 1504.

49. M. Aichler and A. Walch, MALDI Imaging mass spectrometry: current frontiers and perspectives in pathology research and practice, *Lab. Invest.*, 2015, **95**(4), 422–431.

50. M. J. Pittet and R. Weissleder, Intravital imaging, *Cell*, 2011, **147**(5), 983–991.

51. A. M. Laughney, E. Kim, M. M. Sprachman, M. A. Miller, R. H. Kohler, K. S. Yang, *et al.*, Single-cell pharmacokinetic imaging reveals a therapeutic strategy to overcome drug resistance to the microtubule inhibitor eribulin, *Sci. Transl. Med.*, 2014, **6**(261), 261ra152.

52. H. T. Hornig-Do, G. Gunther, M. Bust, P. Lehnartz, A. Bosio and R. J. Wiesner, Isolation of functional pure mitochondria by superparamagnetic microbeads, *Anal. Biochem.*, 2009, **389**(1), 1–5.

53. Y. Wang, T. H. Taylor and E. A. Arriaga, Analysis of the bioactivity of magnetically immunoisolated peroxisomes, *Anal. Bioanal. Chem.*, 2012, **402**(1), 41–49.

54. M. C. Yarema and C. E. Becker, Key concepts in postmortem drug redistribution, *Clin. Toxicol.*, 2005, **43**(4), 235–241.

55. M. Danhof, J. de Jongh, E. C. De Lange, O. Della Pasqua, B. A. Ploeger and R. A. Voskuyl, Mechanism-based pharmacokinetic-pharmacodynamic modeling: biophase distribution, receptor theory, and dynamical systems analysis, *Annu. Rev. Pharmacol. Toxicol.*, 2007, **47**, 357–400.

56. A. E. Waite, *Hermetic and Alchemical Writings of Paracelsus, Part 1*, Kessinger Publishing, 2002.

57. M. Reese, M. Sakatis, J. Ambroso, A. Harrell, E. Yang, L. Chen, *et al.*, An integrated reactive metabolite evaluation approach to assess and reduce safety risk during drug discovery and development, *Chem.-Biol. Interact.*, 2011, **192**(1–2), 60–64.

58. D. A. Smith and E. F. Schmid, Drug withdrawals and the lessons within, *Curr. Opin. Drug Discovery Dev.*, 2006, **9**(1), 38–46.

59. J. L. Walgren, M. D. Mitchell and D. C. Thompson, Role of metabolism in drug-induced idiosyncratic hepatotoxicity, *Crit. Rev. Toxicol.*, 2005, **35**(4), 325–361.

60. J. Uetrecht, Prediction of a new drug's potential to cause idiosyncratic reactions, *Curr. Opin. Drug Discovery Dev.*, 2001, **4**(1), 55–59.

61. M. Chen, J. Borlak and W. Tong, High lipophilicity and high daily dose of oral medications are associated with significant risk for drug-induced liver injury, *Hepatology*, 2013, **58**(1), 388–396.

62. C. Lammert, S. Einarsson, C. Saha, A. Niklasson, E. Bjornsson and N. Chalasani, Relationship between daily dose of oral medications and idiosyncratic drug-induced liver injury: search for signals, *Hepatology*, 2008, **47**(6), 2003–2009.

63. P. Y. Muller and M. N. Milton, The determination and interpretation of the therapeutic index in drug development, *Nat. Rev. Drug Discovery*, 2012, **11**(10), 751–761.

64. J. M. Butler and J. B. Dressman, The developability classification system: application of biopharmaceutics concepts to formulation development, *J. Pharm. Sci.*, 2010, **99**(12), 4940–4954.

65. M. M. Hann and G. M. Keseru, Finding the sweet spot: the role of nature and nurture in medicinal chemistry, *Nat. Rev. Drug Discovery*, 2012, **11**(5), 355–365.

66. J. C. Kalvass, T. S. Maurer and G. M. Pollack, Use of plasma and brain unbound fractions to assess the extent of brain distribution of 34 drugs: comparison of unbound concentration ratios to *in vivo* p-glycoprotein efflux ratios, *Drug Metab. Dispos.*, 2007, **35**(4), 660–666.

67. P. Jeffrey and S. G. Summerfield, Challenges for blood-brain barrier (BBB) screening, *Xenobiotica*, 2007, **37**(10–11), 1135–1151.

68. M. V. Varma, S. J. Steyn, C. Allerton and A. F. El-Kattan, Predicting Clearance Mechanism in Drug Discovery: Extended Clearance Classification System (ECCS), *Pharm. Res.*, 2015, **32**(12), 3785–3802.

69. M. C. Wenlock, Profiling the estimated plasma concentrations of 215 marketed oral drugs, *MedChemComm.*, 2016, **7**(4), 706–719.

70. K. Beaumont and D. A. Smith, Does human pharmacokinetic prediction add significant value to compound selection in drug discovery research?, *Curr. Opin. Drug Discovery Dev.*, 2009, **12**(1), 61–71.

71. N. A. Hosea, W. T. Collard, S. Cole, T. S. Maurer, R. X. Fang, H. Jones, *et al.*, Prediction of human pharmacokinetics from preclinical information: comparative accuracy of quantitative prediction approaches, *J. Clin. Pharmacol.*, 2009, **49**(5), 513–533.

72. Q. Huang and J. E. Riviere, The application of allometric scaling principles to predict pharmacokinetic parameters across species, *Expert Opin. Drug Metab. Toxicol.*, 2014, **10**(9), 1241–1253.

73. H. Jones and K. Rowland-Yeo, Basic concepts in physiologically based pharmacokinetic modeling in drug discovery and development, *CPT: Pharmacometrics Syst. Pharm.*, 2013, **2**, e63.

74. P. Poulin, H. M. Jones, R. D. Jones, J. W. Yates, C. R. Gibson, J. Y. Chien, *et al.*, PhRMA CPCDC initiative on predictive models of human pharmacokinetics, part 1: goals, properties of the PhRMA dataset, and comparison with literature datasets, *J. Pharm. Sci.*, 2011, **100**(10), 4050–4073.

75. K. Beaumont, I. Gardner, K. Chapman, M. Hall and M. Rowland, Toward an integrated human clearance prediction strategy that minimizes animal use, *J. Pharm. Sci.*, 2011, **100**(10), 4518–4535.

76. P. Poulin, R. D. Jones, H. M. Jones, C. R. Gibson, M. Rowland, J. Y. Chien, *et al.*, PHRMA CPCDC initiative on predictive models of human pharmacokinetics, part 5: prediction of plasma concentration-time profiles in human by using the physiologically-based pharmacokinetic modeling approach, *J. Pharm. Sci.*, 2011, **100**(10), 4127–4157.

77. J. E. Sager, J. Yu, I. Ragueneau-Majlessi and N. Isoherranen, Physiologically Based Pharmacokinetic (PBPK) Modeling and Simulation Approaches: A Systematic Review of Published Models, Applications, and Model Verification, *Drug Metab. Dispos.*, 2015, **43**(11), 1823–1837.

78. T. Bueters, B. A. Ploeger and S. A. Visser, The virtue of translational PKPD modeling in drug discovery: selecting the right clinical candidate while sparing animal lives, *Drug Discovery Today*, 2013, **18**(17–18), 853–862.

79. J. Gabrielsson, O. Fjellstrom, J. Ulander, M. Rowley and P. H. Van Der Graaf, Pharmacodynamic-pharmacokinetic integration as a guide to medicinal chemistry, *Curr. Top. Med. Chem.*, 2011, **11**(4), 404–418.

80. S. A. Visser, M. Aurell, R. D. Jones, V. J. Schuck, A. C. Egnell, S. A. Peters, *et al.*, Model-based drug discovery: implementation and impact, *Drug Discovery Today*, 2013, **18**(15–16), 764–775.

81. D. F. McGinnity, J. Collington, R. P. Austin and R. J. Riley, Evaluation of human pharmacokinetics, therapeutic dose and exposure predictions using marketed oral drugs, *Curr. Drug Metab.*, 2007, **8**(5), 463–479.

82. P. Ballard, P. Brassil, K. H. Bui, H. Dolgos, C. Petersson, A. Tunek, *et al.*, The right compound in the right assay at the right time: an integrated discovery DMPK strategy, *Drug Metab. Rev.*, 2012, **44**(3), 224–252.

83. P. H. van der Graaf and N. Benson, Systems pharmacology: bridging systems biology and pharmacokinetics-pharmacodynamics (PKPD) in drug discovery and development, *Pharm. Res.*, 2011, **28**(7), 1460–1464.

84. W. Zheng, N. Thorne and J. C. McKew, Phenotypic screens as a renewed approach for drug discovery, *Drug Discovery Today*, 2013, **18**(21–22), 1067–1073.

85. D. J. Powell, R. P. Hertzberg and R. Macarrón, Design and Implementation of High-Throughput Screening Assays, *Methods Mol. Biol.*, 2016, **1439**, 1–32.

86. H. Hosseinkhani, 3D *in vitro* technology for drug discovery, *Curr. Drug Saf.*, 2012, **7**(1), 37–43.

87. J. Bucher, S. Riedmaier, A. Schnabel, K. Marcus, G. Vacun, T. S. Weiss, *et al.*, A systems biology approach to dynamic modeling and inter-subject variability of statin pharmacokinetics in human hepatocytes, *BMC Syst. Biol.*, 2011, **5**, 66.

88. H. El-Masri, N. Kleinstreuer, R. N. Hines, L. Adams, T. Tal, K. Isaacs, *et al.*, Integration of Life-Stage Physiologically Based Pharmacokinetic Models with Adverse Outcome Pathways and Environmental Exposure Models to Screen for Environmental Hazards, *Toxicol. Sci.*, 2016, **152**(1), 230–243.

89. M. Baker, 1,500 scientists lift the lid on reproducibility, *Nature*, 2016, **533**(7604), 452–454.

90. A. T. Chadwick and M. D. Segall, Overcoming psychological barriers to good discovery decisions, *Drug Discovery Today*, 2010, **15**(13–14), 561–569.

91. P. Chao, A. S. Uss and K. C. Cheng, Use of intrinsic clearance for prediction of human hepatic clearance, *Expert Opin. Drug Metab. Toxicol.*, 2010, **6**(2), 189–198.

92. D. A. Volpe, Drug-permeability and transporter assays in Caco-2 and MDCK cell lines, *Future Med. Chem.*, 2011, **3**(16), 2063–2077.

93. A. M. Palmer and M. S. Alavijeh, Overview of experimental models of the blood-brain barrier in CNS drug discovery, *Curr. Protoc. Pharm.*, 2013, 62. Unit 7 15.

94. K. M. Mahar Doan, J. E. Humphreys, L. O. Webster, S. A. Wring, L. J. Shampine, C. J. Serabjit-Singh, *et al.*, Passive permeability and P-glycoprotein-mediated efflux differentiate central nervous system (CNS) and non-CNS marketed drugs, *J. Pharm. Exp. Ther.*, 2002, **303**(3), 1029–1037.

95. H. Liu, N. Yu, S. Lu, S. Ito, X. Zhang, B. Prasad, *et al.*, Solute Carrier Family of the Organic Anion-Transporting Polypeptides 1A2- Madin-Darby Canine Kidney II: A Promising In Vitro System to Understand the Role of Organic Anion-Transporting Polypeptide 1A2 in Blood-Brain Barrier Drug Penetration, *Drug Metab. Dispos.*, 2015, **43**(7), 1008–1018.

96. D. F. McGinnity, N. J. Waters, J. Tucker and R. J. Riley, Integrated *in vitro* analysis for the *in vivo* prediction of cytochrome P450-mediated drug-drug interactions, *Drug Metab. Dispos.*, 2008, **36**(6), 1126–1134.

97. M. Sinz, G. Wallace and J. Sahi, Current industrial practices in assessing CYP450 enzyme induction: preclinical and clinical, *AAPS J.*, 2008, **10**(2), 391–400.

98. M. P. Grillo, Detecting reactive drug metabolites for reducing the potential for drug toxicity, *Expert Opin. Drug Metab. Toxicol.*, 2015, **11**(8), 1281–1302.

99. A. S. Kalgutkar and D. Dalvie, Predicting toxicities of reactive metabolite-positive drug candidates, *Annu. Rev. Pharm. Toxicol.*, 2015, **55**, 35–54.

100. D. M. Stresser, J. Mao, J. R. Kenny, B. C. Jones and K. Grime, Exploring concepts of *in vitro* time-dependent CYP inhibition assays, *Expert Opin. Drug Metab. Toxicol.*, 2014, **10**(2), 157–174.
101. K. H. Grime, J. Bird, D. Ferguson and R. J. Riley, Mechanism-based inhibition of cytochrome P450 enzymes: an evaluation of early decision making *in vitro* approaches and drug-drug interaction prediction methods, *Eur. J. Pharm. Sci.*, 2009, **36**(2–3), 175–191.
102. C. Prakash, C. L. Shaffer and A. Nedderman, Analytical strategies for identifying drug metabolites, *Mass Spectrom. Rev.*, 2007, **26**(3), 340–369.

9 Predictive Strategies for ADRs – Biomarkers and *In Vitro* Models

Dan Antoine, Neil French and Munir Pirmohamed*

MRC Centre for Drug Safety Science, Department of Molecular and Clinical Pharmacology, Institute of Translational Medicine, University of Liverpool, L69 3GL, UK
*Email: munirp@liverpool.ac.uk

9.1 Introduction

It has been widely reported that currently used biomarkers of drug toxicity lack sensitivity, specificity and a fundamental mechanistic basis. The lack of qualified mechanistic biomarkers has resulted in a significant challenge to investigate the true extent and diagnosis of ADRs.[1] The development and qualification of sensitive and specific drug-safety biomarkers that hold translation between preclinical and clinical studies is urgently required to accelerate the pace of drug development. Improved safety biomarkers may additionally enable patient and/or specific treatment stratification for marketed therapeutics through their predictive characteristics (especially in the case of genomic biomarkers). Irrespective of the source of the biomarkers, the potential for it (them) to provide enhanced understanding of the fundamental mechanisms that result in clinical ADRs is becoming increasingly recognised.

The US National Institute of Health (NIH) defined a biomarker (for example a genetic variant, protein or metabolite) in 2001 as a characteristic that is objectively measured and evaluated as an indicator of

Pharmacology for Chemists: Drug Discovery in Context
Edited by Raymond Hill, Terry Kenakin and Tom Blackburn
© The Royal Society of Chemistry 2018
Published by the Royal Society of Chemistry, www.rsc.org

a normal biological process, a pathogenic process or a pharmaco-logical response to a therapeutic intervention.[2] Biomarkers are clas-sified by the US Food and Drug Administration (FDA) as exploratory, probable valid or known valid. A valid biomarker is further defined as a biomarker that is measured in an analytical test system with well-established performance characteristics and for which there is an established scientific framework or body of evidence that elucidates the physiological, toxicological, pharmacological or clinical signifi-cance of the test result.[3] Since then, recommendations have been set to avoid confusion that the term "validation" should refer to the technical characterisation and documentation of methodological performances, and the term "qualification" refer to the evidentiary process of linking a biomarker to a clinical endpoint or biological process.[4]

Despite the need for new safety biomarkers, the development and clinical integration of such markers over the past 60 years has re-vealed only a limited number of candidates.[5] This concept is perhaps not so surprising given the fact that less focus has been placed on the science of drug safety and the rigorous guidelines we impose on the validation and biological qualification of a potential safety bio-markers[4,6] compared to drug efficacy.[7] Furthermore, the delayed qualification and ultimate scientific acceptance of a potential safety biomarker has been hindered by what has been previously thought of as the competing interest between the various stakeholders. Safety assessment within drug development has traditionally focused on reliable clinical-preclinical concordance. Low baseline variability, specificity and rapid analysis are sought after by clinicians and the ability to provide enhanced mechanistic understanding about tox-icological processes is required by the academic community. Public–private consortia were developed to meet this goal and these consist of leading academic groups, large pharmaceutical companies, small–medium enterprises and clinical units of excellence such as the Pre-dictive Safety Testing Consortium (PSTC) (http://www.c-path.org/pstc.cfm) and the Safer and Faster Evidence-based Translation (SAFE-T) consortium (www.imi-safe-t.eu). These consortia efforts, coupled with feedback and representation from regulatory authorities as external members and advisors, provide an opportunity for collaborative efforts to aid the identification, validation and qualification of novel transla-tional safety biomarkers (TSBM).[4] Two current organ systems that are frequent toxicology targets for marketed drugs or those in development include the liver and kidney. Furthermore, the currently clinical available biochemical tests used for both of these organ systems lack sensitivity and specificity. There is also increasing interest in genomic biomarkers in drug safety, with most of the advances being seen with HLA gene polymorphisms and the risk of immune-mediated diseases affecting different organ systems, most prominently the skin and liver. We will focus on all these areas in this chapter.

9.2 Drug-induced Kidney Injury (DIKI)

The kidneys receive 25% of the resting cardiac output. This, combined with their multifunctional roles of filtration and homeostasis, means that the kidneys are exposed to significant concentrations of medications.[8] There are numerous medications with the potential to cause renal toxicity *via* a variety of mechanisms (see Table 9.1 for some examples).

Drug-induced kidney injury (DIKI) is thought to be a common clinical adverse event and although the true incidence is difficult to determine, US hospital data suggests that drug-induced kidney injury accounts for 18–27% of all acute kidney injury cases in the US[9] and the numbers can rise to 66% in older adults.[10]

During drug development, drug-induced kidney toxicity is considered an important attrition factor and a common side effect amongst many drugs. Nephrotoxicity accounts for 2% of drug failures in preclinical studies,[11] however, when progressing to clinical trials the incidence can rise to 19%. Such a high fraction of DIKI in human studies generates a big concern and underlines the need to improve nephrotoxicity prediction better through either the development of novel circulating biomarkers or *in vitro* predictive models.

Table 9.1 Examples of drugs causing renal injury.

Type of injury	Examples
Nephrotic syndrome	NSAIDs, Ampicillin, Rifampin, Foscarnet, Lithium
Focal segmental glumerulosclerosis	Lithium, heroin, pamidronate
Membranous nephropathy	NSAIDs, gold therapy, mercury, penicillamine
Membranoproliferative glomerulonephritis	Hydralazine
Acute tubular necrosis	Aminoglycosides, Amphotericin B, Radiocontrast dye, Cisplatin
Interstitial nephritis	penicillins, cephalosporins, sulfonamides, vancomycin, NSAIDs, Proton pump inhibitors, calcineurin inhibitors
Obstructive nephropathy	Aciclovir, Sulfadiazine, Methotrexate, Triamterene, Foscarnet, Indinavir

9.2.1 Current Status of Biomarkers for the Assessment of DIKI

The traditional indicator of AKI is a rise in serum creatinine concentration, but this has a delayed response with levels rising significantly above baseline levels only when 25–50% of renal function has been lost.[12] Furthermore, it is a marker of glomerular filtration and not an indicator of damage at other sites in the nephron. It is an insensitive early marker, and reliance upon it means that AKI is frequently not identified early, and that the degree of damage may be underestimated.[13] Interpretation is made more difficult by the variation in the production of creatinine, which depends on age, sex, and weight (in particular muscle mass). Interpretation is also difficult in special populations such as new-born children, in whom serum creatinine initially reflects maternal values.

9.2.2 Novel Biomarkers

A particular focus for the development of biomarkers has been in the area of proximal tubular injury caused by drugs such as aminoglycosides. Here, early identification of DIKI requires a biomarker that is specific for proximal tubules, and that can be quantified earlier than currently used indicators. This would allow for early treatment adjustment, intervention, and the avoidance of further injury.

The Predictive Safety Testing Consortium (PSTC), a collaboration of academic, industry (both pharmaceutical and biotechnology companies) and regulatory (the US Food and Drug Agency, FDA, and the European Medicines Agency, EMA) partners, was established to expedite the qualification of renal biomarkers of nephrotoxicity (Figure 9.1). Their work has led to the qualification of seven renal biomarkers for preclinical use by the FDA, EMA and Japanese Pharmaceuticals and Medical Devices Agency (PMDA): KIM-1, Albumin, Total Protein, β2-Microglobulin, Cystatin C, Clusterin and Trefoil factor-3.[14]

Amongst those approved by the FDA, EMA and PMDA, only KIM-1 is specific purely to the proximal tubule. Albumin, B2-Microglobulin, and Cystatin C may reflect damage to both the proximal tubule and the glomerulus, whereas clusterin may reflect damage to both the proximal and distal tubule.[15] Total protein reflects glomerular injury,[15] and Trefoil factor-3 is expressed in the collecting duct.[16] The albumin/creatinine ratio is widely utilised clinically as a marker of renal disease and response to treatment. However, its urinary concentration may be

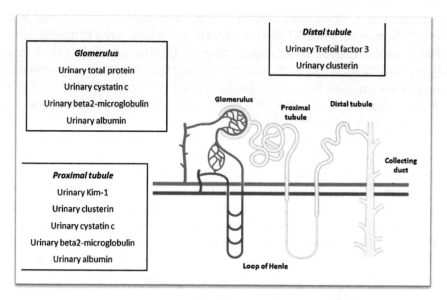

Figure 9.1   Regional specific biomarkers of drug-induced kidney injury.

altered by both changes in glomerular permeability and tubular reabsorption. Furthermore, albumin concentration may be increased by additional factors including fever, exercise, dehydration, diabetes and hypertension, limiting its specificity for AKI.[15] Of those renal biomarkers that have not been qualified through the work of the PSTC, Neutrophil Gelatinase-associated Lipocalin (NGAL) and *N*-acetyl-β-D-glucosaminidase (NAG) have received the most interest as biomarkers of proximal tubular injury, and their potential as biomarkers of proximal tubular injury, using aminoglycosides as an exemplar, will be considered alongside KIM-1 below.

9.2.2.1   Kidney Injury Molecule-1 (KIM-1)

KIM-1 is a cell membrane glycoprotein. Rat models of gentamicin-induced nephrotoxicity show that both the mRNA and the Kim-1 protein (the equivalent protein in rats) are upregulated in proximal tubule epithelial cells.[17,18] In rats, urine and kidney Kim-1 upregulation occurs at lower gentamicin doses than upregulation of urinary NAG levels, serum creatinine or blood urea nitrogen, and parallels the degree of nephropathy seen on histopathology.[19] A multisite validation investigation in animal models treated with a number of nephrotoxins, suggested that Kim-1 (KIM-1 in man) outperformed, with respect to sensitivity and specificity, a number of traditional and novel biomarkers of AKI (serum creatinine, blood urea nitrogen, and NAG), as confirmed by histopathology.[20] In clinical trials KIM-1 has been shown to be an early diagnostic marker of AKI.[21] A systematic review found it to be one of the best performing biomarkers for the differential diagnosis of established AKI and for the prediction of mortality risk following AKI.[22]

In line with preclinical data,[20] KIM-1 outperformed other biomarkers (NGAL, NAG and serum creatinine) in the identification of the potential for aminoglycoside-induced nephrotoxicity in preterm neonates.[23] In children with cystic fibrosis (CF), urinary KIM-1 concentration was significantly correlated with associated with the number of previous courses of aminoglycosides ($r = 0.35, p = 0.012$).[24] Moreover, these observations have been replicated in a UK paediatric CF cohort ($R = 0.70, P < 0.002$).[25] This work also identified acute elevation in KIM-1 concentrations during aminoglycoside exposure,[25] which has also been reported elsewhere.[26]

The assessment of secondary renal injury is a major determinant of poor prognosis in patients with acute liver failure, namely that induced by acetaminophen (APAP) overdose. In patients with APAP overdose, secondary injury to the kidney and specifically the proximal tubule epithelium is a major determinant for mortality. Indeed, biomarkers such as serum creatinine are often incorporated into prognostic algorithms such as the King's College Criteria (KCC).[27] However, the rise in serum creatinine is delayed and data from animal models and humans has repeatedly demonstrated the ability of KIM-1 to become elevated earlier after acute kidney injury.[28,29] In patients with established APAP-induced ALI circulating KIM-1 has been recently demonstrated to be elevated in patients that subsequently died or required a liver transplant compared to spontaneous survivors.[30] The fold change in KIM-1 in this group with the worse prognosis was higher than creatinine and it also outperformed creatinine in a receiver operator characteristic (ROC) curve analysis. Furthermore, circulating KIM-1 was an independent predictor of outcome in a logistic regression model that took into account established markers.[30]

9.2.2.2 *Neutrophil Gelatinase-associated Lipocalin (NGAL)*

NGAL is a 25-kDa protein, part of the lipocalin family, expressed by kidney epithelial cells (and other tissues, as well as neutrophils).[31] Expression of NGAL messenger RNA (mRNA) and protein are upregulated in response to nephrotoxins in mouse models.[32,33] It has been shown to be a highly sensitive and specific early predictor for the development of AKI in adults[22] and children.[34,35]

Urinary NGAL is increased during aminoglycoside exposure in preterm neonates.[23] However, the increase in NGAL was not significant once confounders had been taken into account. In preterm neonates, serum and urinary NGAL correlate with other inflammatory markers,[36] and urinary NGAL has shown promise as a potential biomarker for sepsis.[37,38] If NGAL is already raised due to sepsis, it may not be useful as a biomarker for nephrotoxicity caused by an aminoglycoside prescribed to treat the infection. This may explain the findings in neonates.[23]

9.2.2.3 N-Acetyl-β-D-Glucosaminidase (NAG)

NAG is a 130- to 140-kDa lysosomal enzyme specific to proximal tubule epithelial cells.[39] Its size precludes filtration through the glomerular basement membrane, and therefore its urinary concentration reflects its release from proximal tubule cells. NAG has been utilised experimentally for many years for a number of renal diseases, including nephrotoxicity.[40] It has been used for detecting proximal renal tubular damage including that caused by aminoglycosides.[41,42] However, recent studies have shown that in preclinical settings it is outperformed by KM-1 for the identification of aminoglycoside-induced nephrotoxicity.[19,20] Whilst it is generally regarded as a marker of proximal tubule epithelial cell damage, it may also reflect increased lysosomal turnover in renal tubular cells that occurs secondary to proteinuria.[43]

9.3 Drug-induced Liver Injury (DILI)

DILI represents a significant ADR for both currently used medicines and is a significant impediment to the development of new therapies (Table 9.2). It is a major human health concern as it is a leading cause of patient morbidity and mortality. It has been widely cited that of the 10 000 documented human medicines, more than 1000 are associated with liver injury.[44] The overall incidence of DILI in the general population has been estimated to range from 10–15 cases per 100 000 patient years with the incidence of DILI resulting from an individual drug used in clinical practice ranging from 1 in 10 000 and 1 in 1 000 000 patient years.[45,46] Although DILI accounts for <1% of hospitalised patients presenting to hospital with jaundice,[45,47] it is an

Table 9.2   Examples of some of the drugs withdrawn because of liver injury.

Drug	Year	DILI finding	Impact	Time to onset
Fasiglifam	2013	Liver enzyme elevation	Phase III termination	>1 month
LY2888721	2013	Liver enzyme elevation	Phase II termination	>1 month
Sovaprevir	2013	Liver enzyme elevation	Phase II, clinical hold	>1 month
Sitaxsentan	2009	Liver enzyme elevation, hepatotoxicity	Market withdrawal	>3 months
Lumiracoxib	2007	Liver enzyme elevation, hepatotoxicity	Market withdrawal	>1 month
Ximelagatran	2006	Liver enzyme elevation, hepatotoxicity	Market withdrawal	>1 month
Troglitazone	1997; 2000	Liver enzyme elevations; hepatotoxicity	Market withdrawal	>1 month
Bromofenac	1998	Liver enzyme elevations; hepatotoxicity	Market withdrawal	>1 month
Fialuridine	1993	Lactic acidosis and liver failure	Phase II termination	13 weeks

adverse event that most frequently results in regulatory action leading to black-box warnings or removal of a drug from the market. In the clinic, DILI accounts for more than 50% of acute liver failure cases, and improved detection of DILI before overt liver failure occurs has been the subject of intense investigation.[44]

Drug attrition due to DILI occurs in all phases of the development pipeline, from preclinical testing to clinical trials to the marketplace. In cases where the frequency is high in either animal species or in humans, DILI is considered "intrinsic" in that it is assumed to result from direct hepatocellular damage.[48] However, another concerning manifestation of DILI, termed "idiosyncratic", occurs very rarely in susceptible individuals exposed to therapeutic doses.[45]

Due to the low incidence and multiple contributing factors, currently, a confident diagnosis of DILI in man can only be attained once other possible causes of liver injury have been excluded, which may delay discontinuation of the offending drug. When DILI is suspected in a patient treated with multiple medications, it may be impossible to confidently identify the specific drug responsible and this may force the physician to unnecessarily stop medications, potentially placing the patient at increased risk from the disease(s) being treated.

Furthermore, it is almost impossible to identify which patients are susceptible to DILI before they develop it. Therefore, there is a need to develop new and improved DILI biomarkers that can either confidently establish the diagnosis of DILI, identify the specific drug causing DILI and to predict the course of patients (adapt, survive, develop liver failure).

The presentation of DILI (clinical and histological) can mimic most types of naturally occurring liver diseases. Idiosyncratic hepatocellular liver injury is generally the DILI of greatest concern because it can develop quickly and be life threatening before the development of jaundice. Regardless of type, detection of DILI relies upon a small number of routine laboratory tests. However, there is a lack of specificity and sensitivity in currently utilised markers that leads to poor prediction of toxicity risk for individual patients or patient populations. The assessment of the potential for new chemical entities to elicit clinical hepatotoxicity is heavily dependent upon the histopathological evaluation of hepatotoxicity endpoints in preclinical species coupled with the quantitative assessment of circulating enzymes that are enriched in hepatic tissue.[5,6] However, when clinical trials are performed, current preclinical testing regimens at best successfully correlate to clinical adverse hepatic events in about 50% of cases.[49] In addition, liver biopsies are not routinely taken from clinical trial subjects or patients with overt DILI, leading to incomplete assessment of the mechanisms of injury for a given drug.

The lack of a qualified mechanistic biomarkers has resulted in a significant challenge to investigate the true extent and diagnosis of DILI.[1] The development and qualification of sensitive and specific hepatic biomarkers that hold translation between preclinical and clinical studies is urgently required to accelerate the pace of drug development. Improved DILI biomarkers may additionally enable patient and/or DILI specific treatment stratification for marketed therapeutics. Moreover, the potential for novel biomarkers to provide enhanced understanding of the fundamental mechanisms that result in clinical DILI is becoming increasingly recognised.

9.3.1 Drug-induced Toxicity and the Liver

DILI can be inherent to the drug itself or idiosyncratic (rare, attributed to the individual with multiple potential underlying susceptibility factors and unknown causes) (Figure 9.2). The first-pass exposure to drugs administered orally and the high capacity for xenobiotic metabolism are considered significant reasons why the

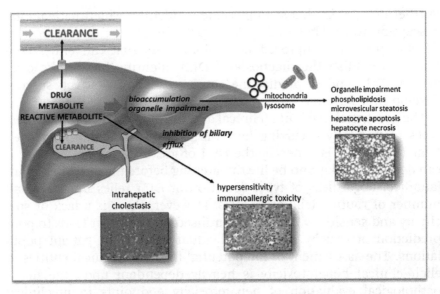

Figure 9.2 A schematic representation of the different mechanisms of drug-induced liver injury.

liver is a target for drug toxicity and this aspect has been widely reviewed.[50] Toxic metabolites can be generated and accumulate within hepatocytes as a result of oxidation–reduction (phase I), conjugation (phase II) and the saturation of transporter (phase III) reactions that normally serve to remove or detoxify xenobiotics.[50,51] The metabolic capacity of the liver coupled to the portal blood supply and the resident immune system contribute to hepatic susceptibility to drug toxicity.

The most frequent cause of DILI in the Western world is as a result of acetaminophen (APAP, paracetamol) overdose which accounted for 38 000 emergency hospital admissions in the financial year 2010–2011 in England alone.[52] Between 1998 and 2007 almost half (46%) of reported cases of acute liver failure had been attributed to APAP with DILI resulting from other drugs accounting for 11%.[53] APAP hepatotoxicity has been widely studied and is reproducible in animal models. Therefore, due to difficulties in studying clinical idiosyncratic DILI in animal models, APAP hepatotoxicity represents an excellent paradigm to identify new biomarkers and to understand mechanisms of clinical-induced liver stress in the exploratory setting that has clear clinical significance.[54,55] The biochemical basis of APAP hepatotoxicity is well defined through cytochrome P450 (CYP2E1, 1A2 and 2D6) mediated reactive metabolite formation (*N*-acetyl-*p*-benzoquinone imine) and hepatic GSH depletion.[56,57] Covalent adduction to cellular proteins then propagates injury through a number of mechanisms, including oxidative stress, which ultimately leads to hepatocyte necrosis and sterile inflammation;[58,59] this has been recently reviewed.[54]

Clinically, antidote treatment with acetylcysteine (AC) is time consuming and results in significant bed occupancy in hospitals (around 47 000 bed days per year in England). Furthermore, adverse reactions to AC are common. The majority of APAP overdose patients present to hospital early, soon after overdose and before acute liver injury (ALI) can be diagnosed, or confidently excluded, using current blood-based biomarkers such as alanine aminotransaminase (ALT).[60,61] Therefore, decisions regarding the initiation of antidote treatment are predominately based on the dose of APAP ingested and a timed blood APAP concentration.[62] This time delay coupled with the early clinical uncertainty of the presence of liver injury prevents treatment being individualised, potentially leading to patients being overtreated with a time-consuming and potentially harmful antidote or undertreated, with increased risk of ALI.

Novel biomarkers that detect ALI at the earliest possible time point are required both from the viewpoint of idiosyncratic hepatotoxicity and for detecting toxicity, and patient stratification of risk resulting from APAP.[63] Moreover, given the multistep and multicellular process of DILI, panels of biomarkers that have the potential to provide insights into the underlying mechanistic basis of ALI are increasingly being recognised as fundamental to efforts in translational research and patient-treatment stratification.

9.3.2 Current Status of Biomarkers for the Assessment of DILI

To date, only a relatively small number of blood-based tests are used to assess DILI in man, while the assessment of DILI in preclinical drug development is heavily dependent upon hepatic histological interpretation.[6] These tests primarily consist of the determination of serum total bilirubin (TBL) concentration and the activity of the enzymes alkaline phosphatase (ALP), aspartate aminotransferase (AST) and alanine aminotransferase (ALT). Elevations in the activity of these enzymes may indicate injury to biliary cells or hepatocytes, whilst elevations of TBL represent either declining hepatic function or obstruction of the bile ducts.[5] Although the assay of ALT has become the primary screening tool to detect DILI, it is not without its problems. Changes in enzyme activities are not specific for DILI and can occur in a number of disease processes, including viral hepatitis, fatty liver disease and liver cancer.[64] Nor are elevations in the aminotransferases unique to liver injury since increases in ALT and AST in circulation can also result from myocardial damage, muscle damage or extreme exercise. Elevations in ALP are not specific for biliary injury as the marker may also be attributed to hyperthyroidism or bone disease. Also, the methods used to quantify ALT activity have not been standardised and a robust definition of normal reference ranges have not been agreed upon; these ranges inevitably depend upon the population group defined as normal and assay measurements will vary between laboratories. Although ALT activity is regarded as generally sensitive for detecting liver injury when it occurs, it is not sensitive with respect to time/kinetics. Furthermore, ALT activity has often been described as having little prognostic value due to the fact that ALT elevation represents probable injury to the liver after it has occurred. From a regulatory point of view, elevations in ALT activity are also worrisome with respect to establishing liver safety during drug treatment. However, frequent and relatively large elevations in ALT activity are associated with treatments that do not pose a clinical liver safety issue, such as heparins and tacrine.[65,66] The challenge is to identify or assess current biomarkers (on their own or in combination) that can distinguish between benign elevations in ALT activity and the potential for a serious DILI outcome.

ALT activity is often combined with the liver-specific assessment of TBL as part of Hy's law. Hy Zimmerman first noted that a patient who presents with jaundice as a result of hepatocellular DILI has at least a 10% chance of developing ALF, regardless of which drug has caused the hepatocellular injury.[67] Hy's law is currently the only accepted regulatory model to assess significant, acute DILI.[7,68,69] Staff at the FDA have developed a liver safety data management tool call eDISH (evaluation of Drug-Induced Serious Liver Injury) which involves data visualisation by plotting the peak serum ALT *versus* the peak serum TBL for each subject in a clinical trial.[69] Although eDISH has revolutionised the standardisation and transparent means of displaying and organising relevant liver safety data from a clinical trial, it remains limited by its reliance on Hy's Law. TBL most likely rises once there has only been a substantial loss of functioning hepatocytes, placing the patient in danger of liver failure. Therefore, serum TBL in this setting is not a biomarker that predicts severe toxicity potential, but instead a confirmation that severe hepatotoxicity has occurred. An ideal biomarker would predict the liver safety of the drug (and the patient) before the injury progresses to the point of TBL elevation.

Therefore, the diagnosis of a Hy's law case is often delayed and death of the patient may still occur despite discontinuation of drug even after relatively subtle liver signals.[7] This highlights the need for improvement and modification to Hy's law criteria that novel biomarkers can fill. Due to the difficulties in defining a true Hy's law case using these biochemical characteristics, an international expert working group of clinicians and scientists reviewed current terminology and diagnostic criteria for DILI based on currently used clinical chemistry parameters to enable uniformed criteria to define a case as DILI and to characterise the spectrum of clinical patterns encompassing it.[70] Currently, a liver biopsy is still the definitive form of diagnosis. Consequently, with the exception of APAP (APAP-adducts, *N*-acetyl-cysteine), there is no specific noninvasive diagnosis, treatment or prevention of DILI except for the early withdrawal of a drug in the case of suspected DILI. Therefore, there has been a considerable effort to identify and develop new biomarkers that can inform the mechanistic basis of DILI and provide potential measures for patient management.

9.3.3 Novel Investigational Circulating Biomarkers for DILI

The identification and eventual qualification of sensitive and specific DILI biomarkers that hold translation between preclinical and clinical studies is urgently required for improved safety screening in drug development and sensitive clinical diagnosis of DILI for patient-treatment stratification. An added benefit of novel biomarkers would be to provide enhanced understanding of the fundamental mechanisms that result in clinical DILI.

Significant resource has recently been directed towards the qualification of new DILI biomarkers and a number of public–private consortia have been developed; namely the Predictive Safety Testing Consortium (PSTC) in the US and the Safer And Faster Evidence based Translation consortium (SAFE-T) in Europe that are currently funding prospective clinical and preclinical DILI biomarker studies.[4] To synergise efforts and minimise overlap, these consortia collaborate closely in what has been described as a highly efficient scientific "meta-consortium" on a global scale.[71]

The strategies for biomarker discovery broadly fall into two categories, unbiased and targeted analysis. Unbiased approaches driven by omic technologies to integrate biological samples are an excellent mechanism to identify novel biomarkers and develop testable theories and have been reviewed with the specific focus on hepatotoxicity.[72,73] Furthermore, efforts have been reported to describe their utility to define proteomic profiles that relate to idiosyncratic hepatotoxicity.[74] However, these strategies are very important but are often scientifically challenging because of the paucity of well-defined clinical samples for a particular drug and because of heterogeneous sample sets and disease manifestations. The alternative is to use model systems to investigate target analytes in biofluids as biomarkers where the chemistry of the drug/molecule reflects the internal biology of the cell/organ and where it is possible to investigate the mechanism of appearance from the cell into the biological matrix and its further clearance. This concept forms the basis of the Innovative Medicines Initiative MIP-DILI (mechanism-based integrated systems for the prediction of drug-induced liver injury) consortium (www.mip-dili.eu). Here, within this review we focus on specific mechanism-based biomarkers that have shown evidence of utility in both preclinical and clinical DILI studies and are of interest to aforementioned biomarker and modelling consortia.

Important and noteworthy progress has been made in the development of biomarkers for renal drug safety evaluation,[75,76] which have been qualified for use by various regulatory authorities and the lessons learnt from these efforts can apply to the liver. In particular, the concept has arisen that new biomarkers are used to complement existing ones and not to replace them to inform the medicinal chemist, clinician, toxicologist, regulator and the public. Novel biomarkers of DILI that have shown clinical utility will be discussed. These novel biomarkers have enhanced sensitivity and specificity for liver injury or offer mechanistic insights into the pathogenic processes that result in DILI.

9.3.3.1 Glutamate Dehydrogenase (GLDH)

GLDH is an enzyme present in matrix-rich mitochondria (liver) and not in cristae-rich mitochondria (cardiac and skeletal muscle). It is important to note that while GLDH is also expressed in the brain and in kidney, its release from these tissues enters the cerebrospinal fluid and tubular lumen, respectively, rather than the blood.[77,78] GLDH is a key enzyme in amino acid oxidation and urea production that is highly conserved across species, making it an attractive biomarker candidate.[79] It is considered relatively liver specific and provides an indicator of leakage of mitochondrial contents into the circulation.[5] GLDH localisation within the liver is regional, with higher concentrations present in the centrilobular area, the region of metabolic activation and site of tissue damage during APAP toxicity. GLDH use as a DILI biomarker is well documented and appears more sensitive and indicative of DILI than other cytosolic enzymes.[80] A recent study in rats subjected to multiple liver injury modalities indicated that GLDH increases were up to 10-fold greater and 3-fold more persistent than ALT elevations.[81] As GLDH is localised to the mitochondrial matrix and due to its relative large size (330 kDa), release of GLDH into the circulation is delayed during hepatocellular necrosis when compared to cytosolic enzymes like the aminotransferases. This property may contribute to increased specificity of GLDH to indicate hepatocellular necrosis. GLDH is additionally elevated in acute liver injury models as it is elevated in blood in both preclinical models and clinical cases of DILI and liver impairment,[80,82,83] highlighting its potential as a translational biomarker. Circulating GLDH has also been shown to rise in healthy volunteers treated with heparins and cholestyramine, treatments that are not associated with clinically important liver injury[66,84] and specific recommendations have been made when drawing conclusions on data regarding sample type and specimen preparation.[85] Measurement of GLDH alone may or may not be useful in distinguishing benign elevations in ALT from those that portent severe DILI potential.

9.3.3.2 Acylcarnitines

Due to the fact that most large enzymes used to investigate mitochondrial dysfunction track changes in ALT activity,[82] a promising approach to the identification of biomarkers of injury that are useful at earlier time points may involve metabolomics. In general, metabolic intermediates are much smaller than proteins and more likely to cross cell membranes and enter blood before the development of injury. In 2009, Chen *et al.*[86] measured increased levels of acylcarnitines in serum from APAP-treated mice. Acylcarnitines are derivatives of long-chain fatty acids that are required for transport of these fatty acids into mitochondria for β-oxidation. First, a coenzyme A (CoA) group is attached in a reaction catalysed by acyl-CoA synthetase. The CoA group is then displaced by carnitine through the action of carnitine palmitoyl transferase I (CPT I), forming an acylcarnitine that can enter the mitochondrial matrix through facilitated diffusion with the help of a carnitine-acylcarnitine translocase (CACT). Because acylcarnitines are broken down within mitochondria by carnitine palmitoyl transferase II (CPT II) and beta-oxidation, mitochondrial dysfunction may result in their accumulation. It has been shown that increases in these fatty acid–carnitine conjugates occur in the serum of mice treated with APAP (mitochondrial-dependent hepatocyte death) but not with mice treated with furosemide (which has been shown to cause liver injury without primarily affecting mitochondrial function).[87] Therefore, circulating acylcarnitines have potential as specific biomarkers of mitochondrial dysfunction. It is important to note that acylcarnitines have been shown to not be elevated in patients with APAP overdose.[87] This is most likely due in part to the standard-of-care treatment *N*-acetylcysteine (AC). However, it might be useful to measure acylcarnitines in other forms of liver injury or APAP patients that present to the "hospital front door" prior to the treatment with AC.[60,61,63]

9.3.3.3 High-mobility Group Box-1 (HMGB1)

HMGB1 is a chromatin binding protein that is passively released by cells undergoing necrosis where it acts as a damage associated molecular pattern (DAMP) molecule by linking cell death to the activation of an immune response by targeting Toll-like receptors and the receptor for advanced glycation end products (RAGE).[88–90] HMGB1 has activity at the intersection between infectious and sterile inflammation. It is also actively secreted as a cytokine by innate immune cells in a hyperacetylated form[91,92] and its biological function is highly dependent upon and is regulated by post-translational redox modifications of three key cysteine residues.[93,94] Furthermore, a recently defined nomenclature has been developed to identify these functional relevant isoforms.[95] Acetylation of lysine residues is also important for the active release of HMGB1 from immune cells and for its release in cell-death mechanisms such as pyroptosis.[91,92,96,97] HMGB1 is an informative and early serum indicator of cell death processes in preclinical models of APAP poisoning[98,99] and in the clinic.[60,61] Circulating levels of total and acetylated HMGB1 displayed different temporal profiles, which in mouse models of APAP toxicity correlate with the onset of necrosis and inflammation, respectively.[98] Serum levels of total HMGB1 correlate strongly with ALT activity and prothrombin time in patients with established acute liver injury following APAP overdose.[100] The prognostic utility of acetylated HMGB1 has also been demonstrated in clinical DILI. In patients with established acute liver injury following APAP overdose, elevations in acetylated HMGB1 associate with a poor prognosis and outcome.[100] Elevations in serum HMGB1 and a secondary rise in acetylated HMGB1 was also observed during treatment of healthy volunteers with heparins.[66] As well as being an important biomarker of APAP toxicity, conditional knock-out animals for HMGB1 and novel therapeutic targeting of these signalling pathways have demonstrated its importance in the pathogenesis of the disease.[101,102] Furthermore, HMGB1 as a mechanistic player and biomarker has also been demonstrated in alcoholic liver disease[103] and preclinical and clinical cholestasis.[104]

9.3.3.4 Keratin-18 (K18)

K18 is a type I intermediate filament protein expressed in epithelial cells and is responsible for cell structure and integrity.[105] Caspase-mediated cleavage of K18 is an early event in cellular structural rearrangement during apoptosis.[106] Caspases 3, 7 and 9 have been implicated in the cleavage of K18 at the C-terminal DALD/S motif. Full-length K18 is released passively during necrotic cell death, whereas fragmented K18 is released with apoptosis.[107] The use of immunoassays directed towards the recognition of caspase-cleaved K18 (apoptosis) and full length K18 (necrosis) have been reported clinically as biomarkers for therapeutic drug monitoring of chemotherapeutic agents and for the quantification of apoptosis during liver disorders such as nonalcoholic steatohepatitis (NASH) and hepatitis C infection,[108,109] and mutations in K18 predispose towards acute liver failure and hepatotoxicity.[110,111] Circulating necrosis K18 and apoptosis K18 have been shown to represent indicators of hepatic necrotic and apoptotic events, respectively, in a mouse model of APAP-induced liver injury[98] and during heparin-induced hepatocellular injury in man.[66] The prognostic utility of K18 has also been demonstrated in clinical DILI and acute liver injury.[100,112] In patients with established acute liver injury following APAP overdose, elevations in absolute levels of necrosis K18 associate with a poor prognosis (KCC) and outcome and a total percentage of K18 attributed to apoptosis correlates with improved survival.[100] Interestingly, in the first blood sample taken at the point of admission following APAP overdose, when currently used markers of liver injury remained within the normal range and prior to antidote treatment, K18 (and also miR-122 and HMGB1) were significantly elevated in the group of patients that subsequently went on to develop liver injury, even in patients that presented less than 8 h postoverdose.[60] Furthermore, the values of K18, HMGB1 and miR-122 at presentation correlated with the peak ALT activity and peak INR recorded during the hospital admission. Interestingly, these data are also supported by a recently published case report highlighting that life-threatening hepatotoxicity following APAP overdose could have potentially been avoided if these biomarkers had been measured.[61] These data demonstrate for the first time in man that these novel biomarkers represent more sensitive biomarkers of DILI in a temporal sense compared to currently used indicators and can be used to aid treatment stratification and identify risk.

9.3.3.5 MicroRNA-122 (miR-122)

MicroRNAs are small noncoding RNAs approximately 22–25 nucleotides in length that predominantly serve to negatively regulate post-transcriptional gene expression. Circulating microRNAs are stable and provide disease state biomarkers spanning diverse therapeutic areas and have been associated with a wide range of tissue-specific toxicities.[113] Many microRNA species show a high degree of organ specificity and cross-species conservation, which makes them attractive candidates as translational safety biomarkers.[114] MicroRNA-122 (miR-122) represents 75% of the total hepatic miRNA content and exhibits exclusive hepatic expression. miR-122 has been shown to be a serum biomarker of APAP-induced ALI in mice, which was more sensitive with respect to dose and time than ALT.[115] The improved tissue specificity of miR-122 *versus* ALT is supported by the observation that clinical ALT elevations associated with muscle injury are not accompanied by concomitant elevations in miR-122.[116] MiR-122 has also been previously shown to serve as a clinical indicator of heparin-induced hepatocellular necrosis.[66] Moreover, as observed in mice, miR-122 is elevated in blood following APAP overdose in man and correlates strongly with ALT activity in patients with established acute liver injury. Furthermore miR-122 has been shown to represent a more sensitive biomarker of APAP hepatotoxicity in humans compared to currently used clinical chemistry parameters.[61,83] In these investigations, elevated miR-122 was observed in patients that present to hospital with normal liver function test values within the normal range but then later develop acute liver injury compared to those that did not develop acute liver injury following APAP overdose. Furthermore, lessons from healthy volunteer studies have also shown that increases in serum livers of miR-122 are associated with individuals that develop liver injury despite taking the therapeutic dose and that miR-122 rise at time points 24 h before ALT activity.[117] In these APAP overdose studies miR-122 correlated strongly with peak ALT levels.[118] Interestingly, serum miR-122 levels in APAP-acute liver injury patients who satisfied King's College Criteria (KCC) for liver transplantation, were higher than those who did not satisfy KCC. However, this did not reach statistical significance, potentially due to small patient numbers.[118] Further prospective and longitudinal biomarker studies in acute liver injury patients will be required to determine whether miR-122 can provide added clinical prognostic value. The translational value of miR-122 as a sensitive circulating biomarker has also been demonstrated in an APAP overdose model in Zebrafish.[119] This represents an important observation for translational research and data interpretation given the increasing utility of this organism for earlier drug-development studies. Despite, the advantages of miR-12, future efforts should be coordinated to developed cross-laboratory validated methods for miRNA isolation and quantification as well as developing a consensus on normalisation standards.[120]

9.3.4 *In Vitro* Models for the Prediction of Human DILI

It is evident from the drug development failures of the past 30 years that only through a better mechanistic understanding of specific drug-induced liver injury will we improve our ability to predict clinical hepatotoxicity. Current *in vitro* models used for the prediction of drug-induced liver injury (DILI) often suffer from uninformative, inappropriate and poorly translatable endpoint measurements, and this may lead to unreliable inferences being drawn. Moreover, there is currently a significant disconnect between *in vitro* DILI endpoints and those used to assess hepatotoxicity in a clinical setting through either traditional endpoints or novel biomarkers.

In vitro models used for early screening for potential hepatotoxic liability are predominantly based on single-cell systems or on human tissue preparations. Liver homogenates, microsomes and slices can all be prepared from human tissue obtained as surgical byproducts, which are more readily available as a resource than in the past. Whilst such models provide valuable insights into the metabolism and covalent binding of new compounds, it is difficult to estimate the propensity for cytotoxicity directly, and consequently most attention is now focused on the use of intact cell models.

The three cell types currently used for *in vitro* hepatotoxicity testing are (i) primary human hepatocytes (PHHs), (ii) transformed liver cell lines and (iii) induced pluripotent stem (iPS) cell-derived hepatocytes. PHHs are considered the gold standard model but the scarcity of suitable tissue and limited life span of the cells in culture is a major issue for their routine use in toxicity screening assays. Furthermore, primary hepatocytes show a marked propensity to dedifferentiate under culture conditions, a feature that is particularly marked with respect to the expression of cytochrome P450 (CYP) enzymes, which decline dramatically almost immediately following isolation. Consequently, immortalised human hepatocyte cell lines, such as HepG2, HuH7 and HepaRG cells, have become popular due to the limitations of PHHs and because of interdonor variability of the primary cells. HepG2 cells exhibit a few of the characteristics of PHHs including expression of some metabolic enzymes and nuclear receptors and are used typically to screen the cytotoxicity potential of new chemical entities at the lead generation stage.[121] However, the low levels of CYPs in HepG2 cells are inappropriate for use in metabolite toxicity testing, which reduces their predictive power considerably, although culturing HepG2 cells in 3D spheroids can improve metabolic competence. To address this limitation, the HepaRG cell line was developed[122] to provide for higher CYP expression and inducibility. Proteomic comparison of these cells with cryopreserved human hepatocytes, however, suggests that the HepaRG cells do not have uniformly high CYP levels, although the CYP3A4 level is comparable to primary cells (University of Liverpool and the MIP-DILI IMI Consortium, unpublished data) and these cells are generally less sensitive to DILI compounds than PHHs, based on cytotoxicity endpoints.

iPS cells and human embryonic stem cells (hESCs) offer exciting opportunities to generate all the cells of the liver, with a genetically identical background of interest. However, the phenotypes of these cells lack many mature hepatocyte features suggesting that the differentiation process and culture systems used require optimisation.[123] It is also unknown whether specific phenotypic and functional characteristics unique to the donor are maintained or lost in culture.

Traditional *in vitro* models based on 2D monolayer cell cultures are considered to be poorly representative of the *in vivo* environment due to their inability to attain normal morphology and do not fully replicate hepatocyte function *in vivo*.[124] Consequently, significant efforts have been undertaken to develop 3D models that improve cell–cell and cell–matrix interactions and hepatocyte polarisation, such as those described below. Such models offer the potential to better reflect the architecture of the liver and to develop more appropriate and relevant endpoints.

Spheroid (3D) culture systems hold promise with regard to improvements in hepatic phenotype. They are traditionally formed using hanging-drop culture and can be both cocultured with other non-parenchymal liver cells, such as stellate cells, and miniaturised relatively easily.[124–128] Systems employing HepaRG cells, HepG2 cells, PHHs and human liver cells have been shown to be stable for up to 28 days[129–131] and suitable for high-throughput screening. However, spheroids lack a perfusion mechanism, which means that, as with standard 2D cultures, the media remains static and this does not accurately reflect the *in vivo* situation.

A recently described multicellular model that may offer promise is the formation of liver buds from pluripotent stem cells cultured with endothelial cells and mesenchymal stem cells.[132] This coculture system self-organises into a 3D bud of cells with a complex vasculature and shows improved hepatic function compared to 2D controls. Despite this, the authors report a lack of biliary cell formation suggesting that this model, whilst being a step forward, requires independent reproduction and further development before fully recapturing the complex architecture of the liver.

Perfusion bioreactors present a highly complex and sophisticated approach for modelling drug-induced hepatotoxicity by offering opportunities to combine a multicellular and 3D hepatic cell system with structural and perfusion capabilities that better mimic the physiological conditions of the liver.[133] Human liver cell cultures in 3D formats placed in hollow-fibre bioreactors have been shown to maintain drug metabolising (CYP) activities and gene expression during prolonged (3–4 week) cultures.[131] These models are capable of producing more reliable data that may lend themselves to measurement of endpoints and markers with greater translational relevance for detecting DILI. For an excellent review on bioreactor technologies in relation to liver cell culture, see ref. 134. There are, however, a number of disadvantages to using this technology. For example, the fibres that attempt to emulate the complex structural components of the liver are also thought to interact with certain classes of drugs, making them ineffective for modelling DILI in some instances.[134,135] The major drawbacks to bioreactors are the large quantities of cells required and the high skill levels and hands-on time needed to use these systems, and these are therefore only truly usable and cost effective when employed with cell lines such as HepaRG.[136–138]

Currently, it is clear that there is no *in vitro* model (and therefore no endpoint or marker) that can recapitulate *in vitro*, the role of the adaptive immune system in DILI.[139] Whilst there has been some progress in modelling the interactions of the innate immune system with hepatocytes through coculture with either monocytes or Kupffer cells, this is still work in progress and needs to be refined based on further elucidation of immune mechanisms underlying human DILI.

In summary, whilst the last 10 years has seen considerable developments and innovations in the use of *in vitro* toxicity models, particularly for predicting acute drug toxicity, there is still no single system that would predict acute hepatotoxicity induced by an agent such as APAP. Prediction of chronic hepatotoxicity is even less advanced, and until it is possible to incorporate components of both the innate and adaptive immune system into the models, it is unlikely that these types of toxicity will be routinely detected preclinically.

9.3.5 Future Perspectives on DILI Biomarkers and *In Vitro* Models

The key questions surrounding the development of new mechanistic biomarkers for DILI in man include:

- Can new biomarkers sensitively identify DILI when it occurs with enhanced specificity?
- Are new biomarkers translational between preclinical models and man?
- Can we use new biomarkers to report clinical mechanisms of DILI?
- Can new biomarkers be used to predict patient prognosis or stratify treatment?
- Do new investigational biomarkers distinguish benign transaminase elevations from serious DILI to build on "Hy's law"?
- What are the hurdles that could prevent clinical adoption?

Significant recent progress has been made and clinical utility has been shown regarding "mechanism-based" biomarkers such as acylcarnitines, HMGB1, K18, GLDH and highly liver specific markers such as miR-122. These biomarkers have been shown to be translational, can report DILI sensitively when it occurs, shed light on mechanistic aspects of clinical DILI and can predict patient prognosis. However, the vast majority of clinical data to date has been obtained from studies of APAP-induced liver injury and have not been assessed in rare cases of idiosyncratic DILI. Moreover, a clear knowledge gap still exists regarding the identification and development of biomarkers that predict serious DILI and reflect hepatic regenerative processes. Therefore, it will become increasingly important to utilise appropriate and well-annotated tissue banks. Important efforts from the US-based DILIN (drug-induced liver injury network)[140] and the Spanish Hepatotoxicity Registry[141] have been collecting serum and urine from all subjects in their registry and where possible has a comprehensive histological evaluation that new biomarkers can be compared against.[142] However, these subjects are enrolled only after the diagnosis of DILI is established; the utility of the biospecimens collected will probably be limited to studies of biomarkers for diagnosis and management of DILI. The qualification of biomarkers that will predict individual subject susceptibility, or that can accurately and safely assess the liver liability of a new drug in development, will require the collection of biospecimens before the start of treatment and during treatment prior to the onset of clinically overt DILI. This effort will require adoption of standardised liver safety databases, standardised protocols for biospecimen collection and storage and the initiation of large prospective clinical trials, involving diverse disease populations and treatment with many different drugs. This should now become a high priority within the pharmaceutical industry.

The lessons learnt from the preclinical qualification of renal safety biomarkers have demonstrated that it is also clear that no single biomarker will be the answer and that a panel approach of novel biomarkers alongside a more intelligent use of currently used biomarkers represents the way forward to inform all stakeholders. Moreover, novel translational biomarkers that reflect the mechanistic basis of DILI are fundamental to efforts in translational research. However, despite significant progress in preclinical renal injury biomarker qualification, to date, clinical biomarker qualification has not been accomplished, although prospective clinical studies are ongoing to achieve this. Defining the context of use for novel biomarkers in man represents an important area of collaborative research interest. Understanding reference ranges for these novel markers in preclinical species and their evaluation in diverse healthy human populations and liver disease cohorts is an important area of investigation and question to address. Further areas of research focus should also be targeted towards the generation of robust cross-species bioanalytical assays that are standardised or point-of-care tests in parallel with a comprehensive understanding of cross-species differences in biomarker expression, mechanisms of release and clearance, distribution and kinetics. It is also important to understand the cost effectiveness of a new biomarker and the added value when moving from an experimental tool to the clinical setting. Defining whether a biomarker is fit for purpose, or what purpose is it fit for in the context of DILI, and understanding what is the right biomarker to assay, and at what particular time for an individual patient or at what stage during drug development is a critical research question to move forward the development and qualification of experimental DILI biomarkers.

The qualification of novel DILI biomarkers will require application to biospecimens obtained from many different patient populations treated with many different drugs, both those that cause clinically important DILI and those that cause elevations in traditional liver chemistries but do not cause clinically important liver injury. It is important that pharmaceutical companies start now to archive samples and link these specimens to the relevant liver safety data. Ideally, liver safety data management tools should be standardised across the industry to facilitate the precompetitive collaborations on biomarker validation and qualification, such as eDISH.[69] As formal biomarker validation and qualification will be a lengthy and challenging process, these proposed biomarkers are likely for some time to retain their exploratory status. A more intelligent use of currently used markers in the meantime can bridge the gap between novel biomarkers being used in the experimental context to formal qualification.

Serum ALT activity is the primary method to assess acute liver injury associated with drugs. Despite not being formally qualified against human histology for DILI, ALT has been widely "qualified" as a DILI biomarker by its extensive use and ease of measurement. Although there are several limitations associated with ALT activity, it can still be an important biomarker of DILI if understood and used properly. This is the topic of a recent review.[68] Here, Senior proposes new interpretations relating to ALT activity and suggests how they might lend themselves to better general use and that suitable revisions in guidance, teaching approaches and clinical practice will need to reflect this. A number of key points were raised and have been built upon to gain a better understanding of serum ALT activity as a biomarker of DILI.[68,143]

Over the past 5 years a paradigm shift towards the thorough elevation of hepatic *in vitro* models has shown that currently available *in vitro* models and, moreover, the endpoints in use with these models lack sufficient sensitivity and specificity to allow meaningful *a priori* risk assessment of the hepatotoxic potential of candidate drugs in human. Considering the multiple molecular initiating events associated with drug-induced liver injury, it is not realistic to suggest a single assay or test will capture all of the potential mechanisms presently described and allow an informed decision on the DILI risk potential of candidate drugs. Single-endpoint cytotoxicity assays have poor concordance with *in vivo* preclinical and clinical readouts, most likely reflecting the measurement of a late event in the pathologic process of liver injury.[121,144] By contrast to cytotoxicity endpoints, panels of molecular cell-based probes permit, by use of high-content cellular imaging, the concomitant time-resolved high content analysis (HCA) of hepatocellular stress response and biochemical endpoints associated with molecular initiating events in liver injury.[145,146] HCA permits a combination of endpoints, including for example, cell loss, nuclear size, mitochondrial potential, DNA damage, apoptosis, GSH depletion and reactive oxygen species (ROS).[146–148] Clearly, the success of HCA depends on the physiological relevance of the cell models that are used, and therefore HCA, whilst serving as a tool to flag candidate drugs for potential DILI risk, is limited by the availability of fully phenotypically and physiologically characterised cell systems. Even with fully functional *in vitro* models there is always a level of uncertainty regarding the "predictivity" of the system in advance of human trials until efficacious doses, exposure and the integration of the whole human response is present.

In summary, DILI is a complex multicellular and multimechanism disease; therefore, it is logical that a battery of complementary biomarkers (both genetic and circulating) and *in vitro* tests with cognate translational read across endpoints that reflect specific cellular processes and predisposition to DILI are required that have been investigated in prospective studies of DILI in the first instance to complement the use of current markers.

9.4 Predictive Strategies for ADRs – Pharmacogenomics

Over the past few years, recent advances in pharmacogenomics have made a significant impact on the way we view the mechanistic basis of ADRs and its prediction. Moreover, such pharmacogenomic studies have also had an impact on our integrated view of maximising efficacy and reducing safety concerns by permitting the appropriate selection of patients for interventional clinical trials.

A number of international consortia exist that recruit idiosyncratic ADR patients into both retrospective and prospective study registries. These efforts represent a valuable resource for the identification and development of new predictive markers. The advantage of DNA-based studies is that the subjects can be enrolled after the diagnosis of an ADR. Here, we will use cases of drug hypersensitivity and liver injury to illustrate the utility of pharmacogenomics in predictive ADR testing.

9.4.1   Drug-induced Hypersensitivity

Genetic analysis of drug hypersensitivity reactions have largely focused on the HLA alleles in the MHC on the short arm of chromosome 6. This is the region of the genome that is highly polymorphic
and has been linked to both autoimmune diseases and infectious
diseases. Since 2001, at least 24 ADRs have been linked to different
HLA alleles and have been reviewed recently.[149] As well as these associations providing important predictive markers of ADR risk, they
also offer a fascinating insight into the complexity of the immune
response, highlighting our incomplete understanding of the pathogenesis of these reactions and there relevance to drug toxicity.

The excitement in the discovery of the association between HLA
markers and drug induced hypersensitivity stems from the identification of HLA-B*57:01 as a risk factor for abacavir hypersensitivity. In
fact, this represents a paradigm for translational research in pharmacogenomics.[149] Work that initially started with a candidate gene
approach in an Australian population[150] was replicated in US[151] and
UK[152] populations, and the utility of genotyping was confirmed in a
randomised controlled trial.[153] In addition, the importance of this
allele has been demonstrated in many different ethnic groups, and
has been shown to be cost effective in most populations. Indeed,
HLA-B*57:01 has been widely adopted into clinical practice, with a
reduction in the incidence of abacavir hypersensitivity.[154]

Carbamazepine is widely used in the treatment of epilepsy, trigeminal neuralgia, and bipolar disorder. Clinical manifestations of
carbamazepine hypersensitivity are more complex and diverse than
those observed with abacavir and include simple maculopapular exanthema, hypersensitivity syndrome, or drug reaction with eosinophilia
and systemic symptoms syndrome and SJS/TEN.[155] Carbamazepine-
induced SJS/TEN has shown a strong (odds ratio >1000) association
with HLA-B*15:02 in the Han Chinese population.[156] This association
has also been replicated in several other Asian populations, but not in
white and Japanese subjects. Apart from being ethnicity specific, the
association with HLA-B*15:02 is phenotype specific in that it is only
valid for SJS/TEN but has not been shown to be important for maculopapular exanthema and hypersensitivity syndrome.[155] The utility of
HLA-B*15:02 testing for preventing carbamazepine-induced SJS/TEN
was also shown in a prospective study from Taiwan.[157] The drug label
for carbamazepine has been changed by many drug regulatory agencies
worldwide, including the EMA and FDA. For other serious cutaneous
ADRs caused by carbamazepine in Chinese patients and for

maculopapular eruptions, drug reaction with eosinophilia and systemic symptoms syndrome, and SJS/TEN in white and Japanese subjects, HLA-A*31:01 has been shown to be a predisposing factor.[158] A recent study has demonstrated that preprescription genotyping for HLA-A*31:01 in the context of the UK NHS would be cost effective.[159] HLA-A*31:01 is mentioned in drug labels in many countries, but testing is not mandatory, and thus it is not routinely used in clinical practice.[155]

9.4.2 Drug-induced Liver Injury

Like hypersensitivity ADRs, the potential to identify predictive genomic markers for DILI has attracted significant attention within the literature. One strategy that has been explored to identify predictive genomic biomarkers is genetic analysis of patients who have experienced DILI. This type of analysis has been greatly facilitated by retrospective evaluation of patient samples from a number of international consortia and registries such as the US Drug-Induced Liver Injury Network (DILIN), Spanish Hepatotoxicity registry and the international Drug-Induced Liver Injury Consortium (iDILIC). To date, studies have implicated variations in genes related to drug metabolism and transport, cellular stress response, and innate and adaptive immune activation in DILI susceptibility.[160] The major class of genes that has been shown to contain DILI risk factors is the major histocompatibility complex (MHC) region. For example, in genome-wide association studies (GWAS), genetic associations have been reported for flucloxacillin-induced DILI and the human leukocyte antigen (HLA)-B*5701 allele.[161] With this predisposition, an 80-fold increase in the risk of developing DILI was attributed (absolute risk of 1:500–1:1000). HLA class II genotypes have been associated with amoxicillin/clavulanate associated DILI.[162] For lumiracoxib, a cyclo-oxygenase-2 (COX-2) selective inhibitor, HLA-DR and HLA-DQ genotypes have been found to be predictive for liver injury.[163] This drug was withdrawn from the market because of drug-related hepatotoxicity. Lapatinib, a tyrosine kinase inhibitor, has been associated with DILI in man in carriers of the HLA allele DQA*02:01.[164] An association between DILI caused by ximelagatran, a direct thrombin inhibitor, and HLA-DRB1*0701 and HLA-DQA1*02 has also been reported.[165] With respect to DILI, even when a genetic association has been observed, the odds ratio of a given haplotype to increase DILI risk has generally been small, unlike the situation with abacavir. The rarity of idiosyncratic DILI has proven to be a hindrance in the detection of DILI susceptibility genes because there is little power to detect a significant association in the few cases that are available for study. However, to date, no patients are genetically tested prospectively before commencing treatment with a drug with liver injury potential, but it is important to note that genetic tests can be used in a variety of ways, and not just predictively.[154]

An alternative approach that has shown some promise is to use genetically diverse mouse populations to model the drug toxicity. Genome-wide association studies conducted in mice generate targeted hypotheses (*i.e.* a small-gene subset) that can be tested in small patient cohorts. This technique has been demonstrated for APAP in which *CD44* was shown to be a risk factor for APAP DILI in mice and in man.[166] Furthermore, a genetically diverse population of mice may improve the prediction of intrinsic toxicity by containing genetically sensitive animals.[167] The technique has yet to be demonstrated for idiosyncratic DILI-causing drugs.

Pharmacogenetic tests, including next-generation sequencing, proteomic or metabolomic approaches for preselecting susceptible patient populations, and tailoring drug therapy to individual patients, have not yet been approved for routine clinical practice. To what extent these new markers will change clinical practice for the prevention of DILI remains to be seen. However, the integration of valuable sample repositories, genetic analysis and data sets from registries represent a significant and valuable opportunity to identify and develop new circulating DILI biomarkers that provide added value to the current armory.

References

1. G. P. Aithal, M. D. Rawlins and C. P. Day, Accuracy of hepatic adverse drug reaction reporting in one English health region, *BMJ*, 1999, **319**(7224), 1541.
2. Biomarkers Definitions Working Group, Biomarkers and surrogate endpoints: preferred definitions and conceptual framework, *Clin. Pharmacol. Ther.*, 2001, **69**(3), 89–95.
3. M. Ratner, FDA pharmacogenomics guidance sends clear message to industry, *Nat. Rev. Drug Discovery*, 2005, **4**(5), 359.
4. K. Matheis, *et al.*, A generic operational strategy to qualify translational safety biomarkers, *Drug Discovery Today*, 2011, **16**(13–14), 600–608.
5. D. J. Antoine, *et al.*, Mechanism-based bioanalysis and biomarkers for hepatic chemical stress, *Xenobiotica*, 2009, **39**(8), 565–577.
6. J. Moggs, *et al.*, Investigative safety science as a competitive advantage for Pharma, *Expert Opin. Drug Metab. Toxicol.*, 2012, **8**(9), 1071–1082.
7. P. B. Watkins, Drug safety sciences and the bottleneck in drug development, *Clin. Pharmacol. Ther.*, 2011, **89**(6), 788–790.
8. D. Choudhury and Z. Ahmed, Drug-associated renal dysfunction and injury, *Nat. Clin. Pract. Nephrol.*, 2006, **2**(2), 80–91.

9. S. S. Taber and D. A. Pasko, The epidemiology of drug-induced disorders: the kidney, *Expert Opin. Drug Saf.*, 2008, 7(6), 679–690.
10. J. Nagai and M. Takano, Molecular-targeted approaches to reduce renal accumulation of nephrotoxic drugs, *Expert Opin. Drug Metab. Toxicol.*, 2010, 6(9), 1125–1138.
11. K. J. Jang, *et al.*, Human kidney proximal tubule-on-a-chip for drug transport and nephrotoxicity assessment, *Integr. Biol.*, 2013, 5(9), 1119–1129.
12. D. J. Askenazi, N. Ambalavanan and S. L. Goldstein, Acute kidney injury in critically ill newborns: what do we know? What do we need to learn? *Pediatr. Nephrol.*, 2009, 24(0931-041; 0931-041; 2), 265–274.
13. S. S. Waikar and J. V. Bonventre, Creatinine kinetics and the definition of acute kidney injury, *J. Am. Soc. Nephrol.*, 2009, 20(3), 672–679.
14. F. Dieterle, *et al.*, Renal biomarker qualification submission: a dialog between the FDA-EMEA and Predictive Safety Testing Consortium, *Nat. Biotechnol.*, 2010, 28(5), 455–462.
15. J. V. Bonventre, *et al.*, Next-generation biomarkers for detecting kidney toxicity, *Nat. Biotechnol.*, 2010, 28(5), 436–440.
16. Y. Yu, *et al.*, Urinary biomarkers trefoil factor 3 and albumin enable early detection of kidney tubular injury, *Nat. Biotechnol.*, 2010, 28(5), 470–477.
17. N. Ozaki, *et al.*, Identification of genes involved in gentamicin-induced nephrotoxicity in rats–a toxicogenomic investigation, *Exp. Toxicol. Pathol.*, 2010, 62(1618-1433; 0940-2993; 5), 555–566.
18. J. Zhang, *et al.*, Immunolocalization of Kim-1, RPA-1, and RPA-2 in kidney of gentamicin-, mercury-, or chromium-treated rats: relationship to renal distributions of iNOS and nitrotyrosine, *Toxicol. Pathol.*, 2008, 36(1533-1601; 0192-6233; 3), 397–409.
19. Y. Zhou, *et al.*, Comparison of kidney injury molecule-1 and other nephrotoxicity biomarkers in urine and kidney following acute exposure to gentamicin, mercury, and chromium, *Toxicol. Sci.*, 2008, 101(1096-6080; 1096-0929; 1), 159–170.
20. V. S. Vaidya, *et al.*, Kidney injury molecule-1 outperforms traditional biomarkers of kidney injury in preclinical biomarker qualification studies, *Nat. Biotechnol.*, 2010, 28(1546-1696; 1087-0156; 5), 478–485.
21. J. V. Bonventre, Diagnosis of acute kidney injury: from classic parameters to new biomarkers, *Contrib. Nephrol.*, 2007, 156(0302-5144; 0302-5144), 213–219.
22. S. G. Coca, *et al.*, Biomarkers for the diagnosis and risk stratification of acute kidney injury: a systematic review, *Kidney Int.*, 2008, 73(1523-1755; 0085-2538; 9), 1008–1016.
23. S. J. McWilliam, *et al.*, Mechanism-based urinary biomarkers to identify the potential for aminoglycoside-induced nephrotoxicity in premature neonates: A proof-of-concept study, *PLoS One*, 2012, 7, 8.
24. T. Lahiri, *et al.*, High-dose ibuprofen is not associated with increased biomarkers of kidney injury in patients with cystic fibrosis, *Pediatr. Pulmonol.*, 2014, 49(2), 148–153.
25. S. J. McWilliam, *et al.*, Association of urinary kidney injury molecule-1 with aminoglycoside exposure in children with cystic fibrosis, *J. Cystic Fibrosis*, 2004, 13, S63.
26. A. Z. Uluer, *et al.*, Urinary biomarkers for early detection of nephrotoxicity in cystic fibrosis, *Pediatr. Pulmonol.*, 2010, 45(S33), A278.
27. N. Chalasani, *et al.*, Model for end-stage liver disease (MELD) for predicting mortality in patients with acute variceal bleeding, *Hepatology*, 2002, 35(5), 1282–1284.
28. S. J. McWilliam, *et al.*, Mechanism-based urinary biomarkers to identify the potential for aminoglycoside-induced nephrotoxicity in premature neonates: a proof-of-concept study, *PLoS One*, 2012, 7(8), e43809.

29. V. S. Sabbisetti, *et al.*, Blood kidney injury molecule-1 is a biomarker of acute and chronic kidney injury and predicts progression to ESRD in type I diabetes, *J. Am. Soc. Nephrol.*, 2014, **25**(10), 2177–2186.
30. D. J. Antoine, *et al.*, Circulating kidney injury molecule 1 predicts prognosis and poor outcome in patients with acetaminophen-induced liver injury, *Hepatology*, 2015, **62**(2), 591–599.
31. L. Kjeldsen, *et al.*, Isolation and primary structure of NGAL, a novel protein associated with human neutrophil gelatinase, *J. Biol. Chem.*, 1993, **268**(14), 10425–10432.
32. J. Mishra, *et al.*, Identification of neutrophil gelatinase-associated lipocalin as a novel early urinary biomarker for ischemic renal injury, *J. Am. Soc. Nephrol.*, 2003, **14**(10), 2534–2543.
33. J. Mishra, *et al.*, Neutrophil gelatinase-associated lipocalin: A novel early urinary biomarker for cisplatin nephrotoxicity, *Am. J. Nephrol.*, 2004, **24**(3), 307–315.
34. D. S. Wheeler *et al.*, Serum neutrophil gelatinase-associated lipocalin (NGAL) as a marker of acute kidney injury in critically ill children with septic shock, *Crit. Care Med.*, 2008, **36**(1530-0293; 0090-3493; 4), 1297–1303.
35. M. Zappitelli, *et al.*, Urine neutrophil gelatinase-associated lipocalin is an early marker of acute kidney injury in critically ill children: a prospective cohort study. *Crit. Care*, 2007, **11**(1466-609; 1364-8535; 4): R84.
36. A. Suchojad, *et al.*, Factors limiting usefulness of serum and urinary NGAL as a marker of acute kidney injury in preterm newborns, *Renal Failure*, 2015, **37**(3), 439–445.
37. J. M. Pynn, *et al.*, Urinary neutrophil gelatinase-associated lipocalin: Potential biomarker for late-onset sepsis, *Pediatr. Res.*, 2015, **78**(1), 76–81.
38. E. Parravicini, *et al.*, Urinary neutrophil gelatinase-associated lipocalin is a promising biomarker for late onset culture-positive sepsis in very low birth weight infants, *Pediatr. Res.*, 2010, **67**(1530-0447; 0031-3998; 6), 636–640.
39. S. Skálová, *The diagnostic role of urinary N-acetyl-beta-D-glucosaminidase (NAG) activity in the detection of renal tubular impairment.* Acta Med. (Hradec Králové)/ Universitas Carolina, *Facultas Medica Hradec Králové*, 2005, **48**(2): 75–80.
40. R. G. Price, The role of NAG (N-acetyl-β-D-glucosaminidase) in the diagnosis of kidney disease including the monitoring of nephrotoxicity, *Clin. Nephrol.*, 1992, **38**(SUPPL. 1), S14–S19.
41. M. Kos, *et al.*, The influence of locally implanted high doses of gentamicin on hearing and renal function of newborns treated for acute hematogenous osteomyelitis, *Int. J. Clin. Pharmacol. Ther.*, 2003, **41**(7), 281–286.
42. E. Ring, *et al.*, Urinary N-acetyl-beta-D-glucosaminidase activity in patients with cystic fibrosis on long-term gentamicin inhalation, *Arch. Dis. Child.*, 1998, **78**(1468-2044; 0003-9888; 6), 540–543.
43. M. P. Bosomworth, S. R. Aparicio and A. W. M. Hay, Urine N-acetyl-β-D-glucosaminidase—A marker of tubular damage? *Nephrol. Dial. Transplant.*, 1999, **14**(3), 620–626.
44. W. M. Lee, Drug-induced hepatotoxicity, *N. Engl. J. Med.*, 2003, **349**(5), 474–485.
45. C. Sgro, *et al.*, Incidence of drug-induced hepatic injuries: a French population-based study, *Hepatology*, 2002, **36**(2), 451–455.
46. Y. Meier, *et al.*, Incidence of drug-induced liver injury in medical inpatients, *Eur. J. Clin. Pharmacol*, 2005, **61**(2), 135–143.
47. R. Vuppalanchi, S. Liangpunsakul and N. Chalasani, Etiology of new-onset jaundice: how often is it caused by idiosyncratic drug-induced liver injury in the United States? *Am. J. Gastroenterol.*, 2007, **102**(3), 558–562, quiz 693.
48. A. Corsini, *et al.*, Current challenges and controversies in drug-induced liver injury, *Drug Saf.*, 2012, **35**(12), 1099–1117.

49. H. Olson, *et al.*, Concordance of the toxicity of pharmaceuticals in humans and in animals, *Regul. Toxicol. Pharmacol.*, 2000, **32**(1), 56–67.
50. B. K. Park, *et al.*, The role of metabolic activation in drug-induced hepatotoxicity, *Annu. Rev. Pharmacol. Toxicol.*, 2005, **45**, 177–202.
51. B. K. Park, *et al.*, Drug bioactivation and protein adduct formation in the pathogenesis of drug-induced toxicity, *Chem. Biol. Interact.*, 2011, **192**(1-2), 30–36.
52. NHS, H.E.S., *http://www.hesonline.nhs.uk/Ease/servlet/ContentServer?siteID=1937.* 2011.
53. W. M. Lee, *et al.*, Acute liver failure: Summary of a workshop, *Hepatology*, 2008, **47**(4), 1401–1415.
54. C. D. Williams and H. Jaeschke, Role of innate and adaptive immunity during drug-induced liver injury, *Toxicol. Res.*, 2012, **1**, 161–170.
55. N. Kaplowitz, Idiosyncratic drug hepatotoxicity, *Nat. Rev. Drug Discovery*, 2005, **4**(6), 489–499.
56. J. R. Mitchell, *et al.*, Acetaminophen-induced hepatic necrosis. I. Role of drug metabolism, *J. Pharmacol. Exp. Ther.*, 1973, **187**(1), 185–194.
57. J. R. Mitchell, *et al.*, Acetaminophen-induced hepatic necrosis. IV. Protective role of glutathione, *J. Pharmacol. Exp. Ther.*, 1973, **187**(1), 211–217.
58. D. J. Jollow, *et al.*, Acetaminophen-induced hepatic necrosis. II. Role of covalent binding in vivo, *J. Pharmacol. Exp. Ther.*, 1973, **187**(1), 195–202.
59. W. Z. Potter, *et al.*, Acetaminophen-induced hepatic necrosis. 3. Cytochrome P-450-mediated covalent binding in vitro, *J. Pharmacol. Exp. Ther.*, 1973, **187**(1), 203–210.
60. D. J. Antoine, *et al.*, Mechanistic biomarkers provide early and sensitive detection of acetaminophen-induced acute liver injury at first presentation to hospital, *Hepatology*, 2013, **58**(2), 777–787.
61. J. W. Dear, *et al.*, Letter to the Editor: Early detection of paracetamol toxicity using circulating liver microRNA and markers of cell necrosis, *Br. J. Clin. Pharmacol.*, 2013, DOI: 10.1111/bcp.12214.
62. R. E. Ferner, J. W. Dear and D. N. Bateman, Management of paracetamol poisoning, *BMJ*, 2011, **342**, d2218.
63. J. W. Dear and D. J. Antoine, Stratification of paracetamol overdose patients using new toxicity biomarkers: current candidates and future challenges, *Expert Rev. Clin. Pharmacol.*, 2014, **7**(2), 181–189.
64. J. Ozer, *et al.*, The current state of serum biomarkers of hepatotoxicity, *Toxicology*, 2008, **245**(3), 194–205.
65. P. B. Watkins, *et al.*, Hepatotoxic effects of tacrine administration in patients with Alzheimer's disease., *JAMA*, 1994, **271**(13), 992–998.
66. A. H. Harrill, *et al.*, The effects of heparins on the liver: application of mechanistic serum biomarkers in a randomized study in healthy volunteers, *Clin. Pharmacol. Ther.*, 2012, **92**(2), 214–220.
67. H. J. Zimmerman, The spectrum of hepatotoxicity, *Perspect. Biol. Med.*, 1968, **12**(1), 135–161.
68. J. R. Senior, *Alanine aminotransferase: a clinical and regulatory tool for detecting liver injury-past, present, and future*, Clin. Pharmacol. Ther., 2012, **92**(3), 332–339.
69. P. B. Watkins, *et al.*, Evaluation of drug-induced serious hepatotoxicity (eDISH): application of this data organization approach to phase III clinical trials of rivaroxaban after total hip or knee replacement surgery, *Drug Saf.*, 2011, **34**(3), 243–252.
70. G. P. Aithal, *et al.*, Case definition and phenotype standardization in drug-induced liver injury, *Clin. Pharmacol. Ther.*, 2011, **89**(6), 806–815.
71. X. Zhang, *et al.*, Involvement of the immune system in idiosyncratic drug reactions, *Drug Metab. Pharmacokinet.*, 2011, **26**(1), 47–59.

72. M. Coen, A metabonomic approach for mechanistic exploration of pre-clinical toxicology, *Toxicology*, 2010, **278**(3), 326–340.

73. A. Van Summeren, *et al.*, Proteomics in the search for mechanisms and biomarkers of drug-induced hepatotoxicity, *Toxicol. In Vitro*, 2012, **26**(3), 373–385.

74. L. N. Bell, *et al.*, Serum proteomic profiling in patients with drug-induced liver injury, *Aliment. Pharmacol. Ther.*, 2012, **35**(5), 600–612.

75. J. V. Bonventre, *et al.*, Next-generation biomarkers for detecting kidney toxicity, *Nat. Biotechnol.*, 2010, **28**(5), 436–440.

76. V. S. Vaidya, *et al.*, Kidney injury molecule-1 outperforms traditional biomarkers of kidney injury in preclinical biomarker qualification studies, *Nat. Biotechnol.*, 2010, **28**(5), 478–485.

77. B. F. Feldman, Cerebrospinal Fluid, in *Clinical Biochemistry of Domestic Animals*, ed. J. Kaneko, Academic Press, San Diego, 1989, pp. 835–865.

78. M. D. Stonard, Assessment of Nephrotoxicity, in *Animal Clinical Chemistry*, ed. G. O. Evans, Taylor & Francis, London, 1996, pp. 87–89.

79. E. S. Schmidt and F. W. Schmidt, Glutamate dehydrogenase: biochemical and clinical aspects of an interesting enzyme, *Clin. Chim. Acta*, 1988, **173**(1), 43–55.

80. S. Schomaker, *et al.*, Assessment of emerging biomarkers of liver injury in human subjects, *Toxicol. Sci.*, 2013, **132**(2), 276–283.

81. P. J. O'Brien, *et al.*, Advantages of glutamate dehydrogenase as a blood biomarker of acute hepatic injury in rats, *Lab. Anim.*, 2002, **36**(3), 313–321.

82. M. R. McGill, *et al.*, The mechanism underlying acetaminophen-induced hepatotoxicity in humans and mice involves mitochondrial damage and nuclear DNA fragmentation, *J. Clin. Invest.*, 2012, **122**(4), 1574–1583.

83. D. J. Antoine, *et al.*, Mechanistic biomarkers provide early and sensitive detection of acetaminophen-induced acute liver injury at first presentation to hospital, *Hepatology*, 2013, **58**(2), 777–787.

84. R. Singhal, *et al.*, Benign elevations in serum aminotransferases and biomarkers of hepatotoxicity in healthy volunteers treated with cholestyramine, *BMC Pharmacol. Toxicol.*, 2014, **15**, 42.

85. H. Jaeschke and M. R. McGill, Serum glutamate dehydrogenase–biomarker for liver cell death or mitochondrial dysfunction?, *Toxicol. Sci.*, 2013, **134**(1), 221–222.

86. C. Chen, *et al.*, Serum metabolomics reveals irreversible inhibition of fatty acid beta-oxidation through the suppression of PPARalpha activation as a contributing mechanism of acetaminophen-induced hepatotoxicity, *Chem. Res. Toxicol.*, 2009, **22**(4), 699–707.

87. M. R. McGill, *et al.*, Circulating acylcarnitines as biomarkers of mitochondrial dysfunction after acetaminophen overdose in mice and humans, *Arch. Toxicol.*, 2014, **88**(2), 391–401.

88. P. Scaffidi, T. Misteli and M. E. Bianchi, Release of chromatin protein HMGB1 by necrotic cells triggers inflammation, *Nature*, 2002, **418**(6894), 191–195.

89. H. Wang, *et al.*, HMG-1 as a late mediator of endotoxin lethality in mice, *Science*, 1999, **285**(5425), 248–251.

90. H. Yang, *et al.*, The many faces of HMGB1: molecular structure-functional activity in inflammation, apoptosis, and chemotaxis, *J. Leukocyte Biol.*, 2013, **93**(6), 865–873.

91. S. Nystrom, *et al.*, TLR activation regulates damage-associated molecular pattern isoforms released during pyroptosis, *EMBO J.*, 2013, **32**(1), 86–99.

92. B. Lu, *et al.*, Novel role of PKR in inflammasome activation and HMGB1 release, *Nature*, 2012, **488**(7413), 670–674.

93. H. Yang, *et al.*, Redox modification of cysteine residues regulates the cytokine activity of high mobility group box-1 (HMGB1), *Mol. Med.*, 2012, **18**(1), 250–259.

94. E. Venereau, *et al.*, Mutually exclusive redox forms of HMGB1 promote cell recruitment or proinflammatory cytokine release, *J. Exp. Med.*, 2012, **209**(9), 1519–1528.

95. D. J. Antoine, *et al.*, A systematic nomenclature for the redox states of High Mobility Group Box (HMGB) proteins, *Mol. Med.*, 2014, **20**(1), 135–137.

96. T. Bonaldi, *et al.*, Monocytic cells hyperacetylate chromatin protein HMGB1 to redirect it towards secretion, *EMBO J.*, 2003, **22**(20), 5551–5560.

97. B. Lu, *et al.*, JAK/STAT1 signaling promotes HMGB1 hyperacetylation and nuclear translocation, *Proc. Natl. Acad. Sci. U. S. A.*, 2014, **111**(8), 3068–3073.

98. D. J. Antoine, *et al.*, High-mobility group box-1 protein and keratin-18, circulating serum proteins informative of acetaminophen-induced necrosis and apoptosis in vivo, *Toxicol. Sci.*, 2009, **112**(2), 521–531.

99. D. J. Antoine, *et al.*, Diet restriction inhibits apoptosis and HMGB1 oxidation and promotes inflammatory cell recruitment during acetaminophen hepatotoxicity, *Mol. Med.*, 2010, **16**(11–12), 479–490.

100. D. J. Antoine, *et al.*, Molecular forms of HMGB1 and keratin-18 as mechanistic biomarkers for mode of cell death and prognosis during clinical acetaminophen hepatotoxicity, *J. Hepatol.*, 2012, **56**(5), 1070–1079.

101. P. Huebener, *et al.*, The HMGB1/RAGE axis triggers neutrophil-mediated injury amplification following necrosis, *J. Clin. Invest.*, 2015, **125**(2), 539–550.

102. H. Yang, *et al.*, MD-2 is required for disulfide HMGB1-dependent TLR4 signaling, *J. Exp. Med.*, 2015, **212**(1), 5–14.

103. X. Ge, *et al.*, High mobility group box-1 (HMGB1) participates in the pathogenesis of alcoholic liver disease (ALD), *J. Biol. Chem.*, 2014, **289**(33), 22672–22691.

104. B. L. Woolbright, *et al.*, Plasma biomarkers of liver injury and inflammation demonstrate a lack of apoptosis during obstructive cholestasis in mice, *Toxicol. Appl. Pharmacol.*, 2013, **273**(3), 524–531.

105. N. O. Ku, *et al.*, Keratins let liver live: Mutations predispose to liver disease and crosslinking generates Mallory-Denk bodies, *Hepatology*, 2007, **46**(5), 1639–1649.

106. C. Caulin, G. S. Salvesen and R. G. Oshima, Caspase cleavage of keratin 18 and reorganization of intermediate filaments during epithelial cell apoptosis, *J. Cell Biol.*, 1997, **138**(6), 1379–1394.

107. B. Schutte, *et al.*, Keratin 8/18 breakdown and reorganization during apoptosis, *Exp. Cell Res.*, 2004, **297**(1), 11–26.

108. A. Wieckowska, *et al.*, *In vivo* assessment of liver cell apoptosis as a novel biomarker of disease severity in nonalcoholic fatty liver disease, *Hepatology*, 2006, **44**(1), 27–33.

109. J. Cummings, *et al.*, Preclinical evaluation of M30 and M65 ELISAs as biomarkers of drug induced tumor cell death and antitumor activity, *Mol. Cancer Ther.*, 2008, **7**(3), 455–463.

110. N. O. Ku, *et al.*, Susceptibility to hepatotoxicity in transgenic mice that express a dominant-negative human keratin 18 mutant, *J. Clin. Invest.*, 1996, **98**(4), 1034–1046.

111. P. Strnad, *et al.*, Keratin variants predispose to acute liver failure and adverse outcome: race and ethnic associations, *Gastroenterology*, 2010, **139**(3), 828–835, 835 e1-3.

112. L. P. Bechmann, *et al.*, Cytokeratin 18-based modification of the MELD score improves prediction of spontaneous survival after acute liver injury, *J. Hepatol.*, 2010, **53**(4), 639–647.

113. P. J. Starkey Lewis, *et al.*, Serum microRNA biomarkers for drug-induced liver injury, *Clin. Pharmacol. Ther.*, 2012, **92**(3), 291–293.

114. K. Zen and C. Y. Zhang, Circulating microRNAs: a novel class of biomarkers to diagnose and monitor human cancers, *Med. Res. Rev.*, 2012, **32**(2), 326–348.

115. K. Wang, *et al.*, Circulating microRNAs, potential biomarkers for drug-induced liver injury, *Proc. Natl. Acad. Sci. U. S. A.*, 2009, **106**(11), 4402–4407.
116. Y. Zhang, *et al.*, Plasma microRNA-122 as a biomarker for viral-, alcohol-, and chemical-related hepatic diseases, *Clin. Chem.*, 2010, **56**(12), 1830–1838.
117. P. Thulin, *et al.*, Keratin-18 and microRNA-122 complement alanine aminotransferase as novel safety biomarkers for drug-induced liver injury in two human cohorts, *Liver Int.*, 2013.
118. P. J. Starkey Lewis, *et al.*, Circulating microRNAs as potential markers of human drug-induced liver injury, *Hepatology*, 2011, **54**(5), 1767–1776.
119. A. D. Vliegenthart, *et al.*, Retro-orbital blood acquisition facilitates circulating microRNA measurement in Zebrafish with Paracetamol hepatotoxicity, *Zebrafish*, 2014, **11**(3), 219–226.
120. J. W. Sharkey, D. J. Antoine and B. K. Park, Validation of the isolation and quantification of kidney enriched miRNAs for use as biomarkers, *Biomarkers*, 2012, **17**(3), 231–239.
121. H. H. Gerets, *et al.*, Characterization of primary human hepatocytes, HepG2 cells, and HepaRG cells at the mRNA level and CYP activity in response to inducers and their predictivity for the detection of human hepatotoxins, *Cell. Biol. Toxicol.*, 2012, **28**(2), 69–87.
122. A. Guillouzo, *et al.*, The human hepatoma HepaRG cells: a highly differentiated model for studies of liver metabolism and toxicity of xenobiotics, *Chem. Biol. Interact.*, 2007, **168**(1), 66–73.
123. R. Kia, *et al.*, Stem cell-derived hepatocytes as a predictive model for drug-induced liver injury: are we there yet? *Br. J. Clin. Pharmacol.*, 2013, **75**(4), 885–896.
124. P. Godoy, *et al.*, Recent advances in 2D and 3D in vitro systems using primary hepatocytes, alternative hepatocyte sources and non-parenchymal liver cells and their use in investigating mechanisms of hepatotoxicity, cell signaling and ADME, *Arch. Toxicol.*, 2013, **87**(8), 1315–1530.
125. L. Riccalton-Banks, *et al.*, Long-term culture of functional liver tissue: three-dimensional coculture of primary hepatocytes and stellate cells, *Tissue Eng.*, 2003, **9**(3), 401–410.
126. R. J. Thomas, *et al.*, The effect of three-dimensional co-culture of hepatocytes and hepatic stellate cells on key hepatocyte functions in vitro, *Cells Tissues Organs*, 2005, **181**(2), 67–79.
127. S. F. Abu-Absi, L. K. Hansen and W. S. Hu, Three-dimensional co-culture of hepatocytes and stellate cells, *Cytotechnology*, 2004, **45**(3), 125–140.
128. M. Inamori, H. Mizumoto and T. Kajiwara, An approach for formation of vascularized liver tissue by endothelial cell-covered hepatocyte spheroid integration, *Tissue Eng. Part A*, 2009, **15**(8), 2029–2037.
129. P. Gunness, *et al.*, 3D organotypic cultures of human HepaRG cells: a tool for in vitro toxicity studies, *Toxicol. Sci.*, 2013, **133**(1), 67–78.
130. S. C. Ramaiahgari, *et al.*, A 3D in vitro model of differentiated HepG2 cell spheroids with improved liver-like properties for repeated dose high-throughput toxicity studies, *Arch. Toxicol.*, 2014, **88**(5), 1083–1095.
131. R. M. Tostões, *et al.*, Human liver cell spheroids in extended perfusion bioreactor culture for repeated-dose drug testing, *Hepatology*, 2012, **55**(4), 1227–1236.
132. T. Takebe, *et al.*, Vascularized and functional human liver from an iPSC-derived organ bud transplant, *Nature*, 2013, **499**(7459), 481–484.
133. K. Zeilinger, *et al.*, Scaling down of a clinical three-dimensional perfusion multicompartment hollow fiber liver bioreactor developed for extracorporeal liver support to an analytical scale device useful for hepatic pharmacological in vitro studies, *Tissue Eng. Part C*, 2011, **17**(5), 549–556.

134. M. R. Ebrahimkhani, *et al.*, Bioreactor technologies to support liver function in vitro, *Adv. Drug Delivery Rev.*, 2014, **69–70**, 132–157.

135. J. J. S. Cadwell, The hollow fiber infection model for antimicrobial pharmaco-dynamics and pharmacokinetics, *Adv. Pharmacoepidemiol. Drug Saf.*, 2012, **S1:007**, DOI: 10.4172/2167-1052.S1-007.

136. V. Cerec, *et al.*, Transdifferentiation of hepatocyte-like cells from the human hepatoma HepaRG cell line through bipotent progenitor, *Hepatology*, 2007, **45**(4), 957–967.

137. M. Darnell, *et al.*, Cytochrome P450-dependent metabolism in HepaRG cells cultured in a dynamic three-dimensional bioreactor, *Drug Metab. Dispos.*, 2011, **39**(7), 1131–1138.

138. M. Darnell, *et al.*, *In vitro* evaluation of major in vivo drug metabolic pathways using primary human hepatocytes and HepaRG cells in suspension and a dynamic three-dimensional bioreactor system, *J. Pharmacol. Exp. Ther.*, 2012, **343**(1), 134–144.

139. J. Uetrecht and D. J. Naisbitt, Idiosyncratic adverse drug reactions: current concepts, *Pharmacol. Rev.*, 2013, **65**(2), 779–808.

140. J. H. Hoofnagle, Drug-induced liver injury network (DILIN), *Hepatology*, 2004, **40**(4), 773.

141. R. J. Andrade, *et al.*, Outcome of acute idiosyncratic drug-induced liver injury: Long-term follow-up in a hepatotoxicity registry, *Hepatology*, 2006, **44**(6), 1581–1588.

142. D. E. Kleiner, *et al.*, Hepatic histological findings in suspected drug-induced liver injury: systematic evaluation and clinical associations, *Hepatology*, 2014, **59**(2), 661–670.

143. D. J. Antoine, *et al.*, Are we closer to finding biomarkers for identifying acute drug-induced liver injury? *Biomarkers Med.*, 2013, **7**(3), 383–386.

144. J. J. Xu, D. Diaz and P. J. O'Brien, Applications of cytotoxicity assays and pre-lethal mechanistic assays for assessment of human hepatotoxicity potential, *Chem. Biol. Interact.*, 2004, **150**(1), 115–128.

145. P. J. O'Brien, *et al.*, High concordance of drug-induced human hepatotoxicity with in vitro cytotoxicity measured in a novel cell-based model using high content screening, *Arch. Toxicol.*, 2006, **80**(9), 580–604.

146. J. J. Xu, *et al.*, Cellular imaging predictions of clinical drug-induced liver injury, *Toxicol. Sci.*, 2008, **105**(1), 97–105.

147. P. Lang, *et al.*, Cellular imaging in drug discovery, *Nat. Rev. Drug Discovery*, 2006, **5**(4), 343–356.

148. S. Wink, *et al.*, Quantitative high content imaging of cellular adaptive stress response pathways in toxicity for chemical safety assessment, *Chem. Res. Toxicol.*, 2014, **27**(3), 338–355.

149. M. Pirmohamed, D. A. Ostrov and B. K. Park, New genetic findings lead the way to a better understanding of fundamental mechanisms of drug hypersensitivity, *J. Allergy Clin. Immunol.*, 2015, **136**(2), 236–244.

150. S. Mallal, *et al.*, Association between presence of HLA-B*5701, HLA-DR7, and HLA-DQ3 and hypersensitivity to HIV-1 reverse-transcriptase inhibitor abacavir, *Lancet*, 2002, **359**(9308), 727–732.

151. S. Hetherington, *et al.*, Genetic variations in HLA-B region and hypersensitivity reactions to abacavir, *Lancet*, 2002, **359**(9312), 1121–1122.

152. D. A. Hughes, *et al.*, Cost-effectiveness analysis of HLA B*5701 genotyping in preventing abacavir hypersensitivity, *Pharmacogenetics*, 2004, **14**(6), 335–342.

153. S. Mallal, *et al.*, HLA-B*5701 screening for hypersensitivity to abacavir, *N. Engl. J. Med.*, 2008, **358**(6), 568–579.

154. A. Alfirevic and M. Pirmohamed, Genomics of adverse drug reactions, *Trends Pharmacol. Sci.*, 2017, **38**(1), 100–109.

155. V. L. Yip and M. Pirmohamed, The HLA-A*31:01 allele: influence on carbamazepine treatment, *Pharmagenomics Pers. Med.*, 2017, **10**, 29–38.
156. W. H. Chung, *et al.*, Medical genetics: a marker for Stevens–Johnson syndrome, *Nature*, 2004, **428**(6982), 486.
157. P. Chen, *et al.*, Carbamazepine-induced toxic effects and HLA-B*1502 screening in Taiwan, *N. Engl. J. Med.*, 2011, **364**(12), 1126–1133.
158. M. McCormack, *et al.*, HLA-A*3101 and carbamazepine-induced hypersensitivity reactions in Europeans, *N. Engl. J. Med.*, 2011, **364**(12), 1134–1143.
159. C. O. Plumpton, *et al.*, Cost-effectiveness of screening for HLA-A*31:01 prior to initiation of carbamazepine in epilepsy, *Epilepsia*, 2015, **56**(4), 556–563.
160. T. J. Urban, *et al.*, Limited contribution of common genetic variants to risk for liver injury due to a variety of drugs, *Pharmacogenet. Genomics*, 2012, **22**(11), 784–795.
161. A. K. Daly, *et al.*, HLA-B*5701 genotype is a major determinant of drug-induced liver injury due to flucloxacillin, *Nat. Genet.*, 2009, **41**(7), 816–819.
162. M. I. Lucena, *et al.*, Susceptibility to amoxicillin-clavulanate-induced liver injury is influenced by multiple HLA class I and II alleles, *Gastroenterology*, 2011, **141**(1), 338–347.
163. J. B. Singer, *et al.*, A genome-wide study identifies HLA alleles associated with lumiracoxib-related liver injury, *Nat. Genet.*, 2010, **42**(8), 711–714.
164. C. F. Spraggs, *et al.*, HLA-DQA1*02:01 is a major risk factor for lapatinib-induced hepatotoxicity in women with advanced breast cancer, *J. Clin. Oncol.*, 2011, **29**(6), 667–673.
165. A. Kindmark, *et al.*, Genome-wide pharmacogenetic investigation of a hepatic adverse event without clinical signs of immunopathology suggests an underlying immune pathogenesis, *Pharmacogenomics J.*, 2008, **8**(3), 186–195.
166. A. H. Harrill, *et al.*, Mouse population-guided resequencing reveals that variants in CD44 contribute to acetaminophen-induced liver injury in humans, *Genome Res.*, 2009, **19**(9), 1507–1515.
167. A. H. Harrill, *et al.*, A mouse diversity panel approach reveals the potential for clinical kidney injury due to DB289 not predicted by classical rodent models, *Toxicol. Sci.*, 2012, **130**(2), 416–426.

10 From Bench to Bedside: The First Studies of a New Molecule in Man

Kate Darwin, Liv Thomsen and Malcolm Boyce*

Hammersmith Medicines Research, Cumberland Avenue, London NW10 7EW, UK
*Email: mboyce@hmrlondon.com

10.1 Introduction

In order to obtain a licence to market a new medicine, the sponsor first has to demonstrate its safety, quality and efficacy through a series of rigorous trials in humans. The start of clinical trials of a potential new medicine – an investigational medicinal product (IMP) – always generates excitement among the chemists and many other scientists in a pharmaceutical or biotechnology company who have nurtured it from discovery through preclinical development. Nowadays, the active molecule is as likely to be a biological product as a chemical substance.

However, before an IMP can be given to humans, the sponsor must first test it thoroughly in animals. The main aims of preclinical studies are to:

- assess the effects of the IMP on body systems (pharmacodynamics [PD]);
- study blood levels of the IMP, and how it is absorbed, distributed, metabolised and eliminated after dosing (pharmacokinetics [PK]);

Pharmacology for Chemists: Drug Discovery in Context
Edited by Raymond Hill, Terry Kenakin and Tom Blackburn
© The Royal Society of Chemistry 2018
Published by the Royal Society of Chemistry, www.rsc.org

- determine if doses of the IMP many times higher than those intended for use in humans are toxic to animals – usually a rodent and a nonrodent – and if so, identify the target organs and the highest dose that causes no harm (the "no-observed-adverse-effect dose level" [NOAEL]), in terms of:
  (a) dose, relative to body weight);[1] and
  (b) IMP exposure – blood levels (toxicokinetics);[2] and
- make a formulation of the IMP suitable for early studies in humans.

Clinical development is traditionally separated into four phases: phases 1 to 3 are done before a licence is granted, and phase 4 is done after licensing. The phases are different in terms of the number and types of subject studied, and the aims (Table 10.1).

The time taken to invent and fully develop a successful IMP varies according to the disease, type and duration of treatment, and the number of patients required. It can take up to 13 years (Figure 10.1). The average time for the clinical part of development in the USA is ~9 years, although FDA expedited and accelerated approval programmes for IMPs for serious or life-threatening conditions have shortened that to ~6 years.[3]

The attrition rate is high; many IMPs are withdrawn from development, mainly because: they are not well tolerated or safe enough in

Table 10.1 The traditional phases of drug development and their aims.

Phase	Number and type of subject	Questions
1	50–200 **healthy subjects** (usually) or **patients** who are not expected to benefit from the IMP	• Is the IMP safe in humans? • What does the body do to the IMP? (*PK*) • What does the IMP do to the body? (*PD*) • Might the IMP work in patients? (*proof of concept*)
2	100–400 **patients with the target disease**	• Is the IMP safe in patients? • Does the IMP seem to work in patients? (*efficacy*)
3	1000–5000 **patients with the target disease**	• Is the IMP really safe in patients? • Does the IMP really work in patients?
4	Many thousands or millions **patients with the target disease**	• Just how safe is the new medicine? (*pharmacovigilance*) • How does the new medicine compare with similar medicines?

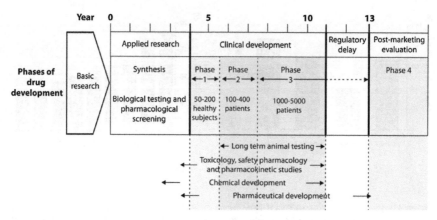

Figure 10.1 Phases in the discovery and development of a "typical" new medicine.

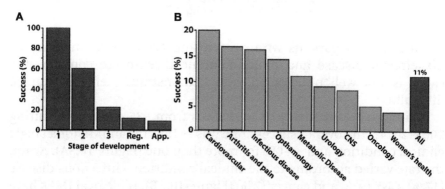

Figure 10.2 Success rates from FIH study to approval. Data from 10 biggest companies in USA and Europe in 1991–2000. Reg = registration; App = approval.
Reprinted by permission from Macmillan Publishers Ltd: *Nat. Rev. Drug Discovery*, (ref. 4), Kola *et al.*, 3, 711–715, Copyright 2004.

humans; their PK or PD profile in humans is disappointing; they do not work or do not work well enough in patients with the target disease; or development is stopped for commercial reasons.

In a review of IMPs from ten of the biggest pharmaceutical companies in the USA and Europe during 1991–2000,[4] only 60% progressed from phase 1 to 2, and a mere 11% became a marketed product (Figure 10.2A). The success rate was highest for IMPs to treat cardiovascular disease (~20%), infections (17%) and arthritis (16%), and lowest for IMPs to treat diseases of the central nervous system (~8%) and cancer (~5%) (Figure 10.2B). Many factors influence clinical success. For example, animal models can be poor predictors

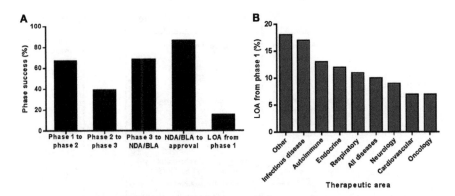

Figure 10.3 (A) Success rates of IMP from one phase to another. NDA = new drug application; BLA = biologic license application. (B) Likelihood of approval (LOA) by the FDA of an IMP according to type.
Reprinted by permission from Macmillan Publishers Ltd: *Nat. Biotechnol.*, (ref. 5) Hay *et al.*, 32, 40–45, Copyright 2014.

of outcomes in patients with cancer, and CNS conditions such as Alzheimer's disease and motor neurone disease are complex, the cause is not well understood, and demonstrating efficacy can be difficult.

In a more recent survey of 4451 IMPs from 835 companies during 2003–2014,[5] the overall success rate was ~15% for an IMP for a single clinical condition and ~10% for more than one (Figure 10.3A). Again, the rate varied according to the clinical condition – infectious disease (17%), CNS (9%) and cancer (7%) (Figure 10.3B). Biological IMPs have an overall success rate (24%) higher than that of small molecules.

The causes of attrition have varied over the years. In the 1990s, adverse PK and bioavailability accounted for 40% of all attrition. By the 2000s, that figure was reduced to 10%, and lack of efficacy (30%) and safety (30%) were the main causes.[4] More recently, lack of efficacy accounted for ~50% of all attrition.[5]

There has been a sea-change in the size and types of company developing IMPs in recent years. The number of companies and the number of new molecules in development have increased 2–3-fold. The majority of IMPs are now coming from smaller companies, especially emerging biotechnology companies, rather than the big pharmaceutical or biotechnology companies (Table 10.2). Most of those smaller companies rely on external investment or partnerships with bigger companies to develop their new molecules. Chemists working in the pharmaceutical or biotechnology industry are as likely, if not more likely, to find themselves working for a relatively new small-to-medium sized company, and perhaps even for a virtual

Table 10.2 Analysis of company size and type. Reprinted by permission from Macmillan Publishers Ltd: *Nat. Biotechnol.*, (ref. 5) Hay *et al.*, 32, 40–45, Copyright 2014.

Company size and type	Company		IMP	
	No.	%	No.	%
Large pharma/biotech	33	4	2075	47
Small to mid-size pharma/biotech	90	11	724	16
Emerging biotech	712	85	1652	37
Total	835	100	4451	100

spin-off company from an academic institution, rather than one of the long-established big companies.

Various factors have led to a drive to reduce development times: the short patent life of an IMP (20 years) and the growth in the use of generic medicines; the increase in the number of companies and IMPs, and the intense competition to get to the market first; the increasing costs of drug development; and the need to satisfy funders of drug research programmes. The traditional development phases 1–3 have become blurred, and there is a trend towards separating development simply into "early" and "late" phases. The aim of the first studies in humans has always been to assess the safety, tolerability, PK and PD of an IMP, and if possible to find the doses suitable for studies in patients with the target disease. Nowadays, the aim is also to obtain evidence of efficacy of an IMP, referred to as "proof-of-concept", as early as possible, perhaps even in the very first study in humans, by use of biomarkers. A biomarker is expected to predict the effect of treatment on a particular disease. In other words, it serves as a substitute for a clinical endpoint. Examples are laboratory measurements, physical signs and imaging tests.[6] It is important to consider potential biomarkers when designing a clinical trial. The overall aim of early studies is to identify IMPs with potential for success as well as exclude failures and thereby prevent unnecessary exposure of an IMP to many more subjects and wastage of money and resources.

Clinical trials are highly regulated. In Europe, they must comply with the Clinical Trials Directive (CTD)[7] (soon to be superseded by the Clinical Trials Regulation). The Directive is based on Good Clinical Practice (GCP)[7,8] and Good Manufacturing Practice (GMP).[9] In the UK, the Medicines and Healthcare products Regulatory Agency (MHRA) inspects sponsors, trial sites and other trial-related activities to check that the CTD is being followed. The aims of the CTD are to simplify and harmonise clinical trials across Europe, give better protection to subjects who take part in them, and enforce by law the principles of GCP and GMP. The CTD covers all commercial and academic clinical trials of IMPs and marketed medicines, apart from trials of marketed medicines prescribed in the usual way. The types of IMP include: chemical entities; biotechnology products; cell therapy products; gene therapy products; plasma-derived products; other extractive products; immunological products, such as vaccines, allergens and immune sera; and radiopharmaceutical products. Also, a placebo, or a marketed product used or assembled in a way different from the approved form, is an IMP when used as a comparator. All nontherapeutic trials of IMP – whether they involve healthy subjects or patients – are now called phase 1 trials.

In the UK, commercial organisations that import, manufacture, assemble or repackage an IMP must have a Manufacturer's Authorisation [MIA (IMP)] from the MHRA and must follow GMP. Sponsors from third countries – countries from outside the European Economic Area (EEA), such as Japan and USA – must manufacture an IMP to at least EU standards of GMP. A clinical trial of an IMP cannot be started without a favourable opinion from a research ethics committee (REC) and a Clinical Trial Authorisation (CTA) from the MHRA. One key component of the application for a CTA is a dossier containing information on the chemistry, manufacture and control of the IMP.

10.2 Preparing for Early Clinical Studies

Before a potential new medicine can be given to humans, it must be well characterised in terms of chemical, physical and biological properties, and its quality must be carefully controlled. Chemists are involved in much of the preparatory work for clinical studies, and ultimately marketing, of a potential new medicine. Research chemists make potential new drug candidates, and development chemists optimise the synthesis of the chosen candidate in an effort to make the method robust and scalable. Formulation chemists transform the candidate into a product that can be administered to humans. Analytical chemists develop and validate methods to test the quality of the product. It is also advantageous for other professionals in related roles, such as regulatory affairs, patent law and quality assurance, to have a background in medicinal chemistry.

Chemistry continues to play an important role in the early clinical phases of drug development: clinical results may lead to the development of a new drug candidate to replace the original, they may guide formulation development, or they may reveal one or more metabolites of the drug that need to be identified and characterised. Hence, although clinical pharmacology is traditionally the realm of physicians, biologists and biochemists, it is important that chemists are involved in, and understand, the clinical development of IMPs.

The goal of drug development is to obtain a licence (a marketing authorisation) for a new medicine. To achieve that, it is necessary to fully define and document the chemistry, pharmacology and formulation of the IMP, for submission to the regulatory authorities as part of a marketing application. Work on a dossier containing that information begins early in drug development. It is continually updated as the IMP undergoes development. Preparation of the dossier is usually the responsibility of technical writers and regulatory affairs specialists, and requires input from a multidisciplinary team, including the chemists responsible for manufacture, analysis and formulation. Since the dossier controls the drug product at all stages of clinical development, close collaboration and clear communication among the team is needed, to ensure that the dossier accurately reflects the product at its current stage of development, and protects clinical trial subjects without placing unnecessary restrictions on the manufacture and control of the product.

This section gives an overview of what is needed to prepare an IMP for clinical studies, and of how early-phase clinical studies might inform further physicochemical development work.

10.2.1 From Candidate Molecule to Potential New Medicine

In the pharmaceutical industry, a candidate drug molecule is referred to as the active pharmaceutical ingredient (API) or the drug substance. The term API can be misleading because it implies pharmacological activity, which is not always the case. Consider aspirin: the API or drug substance is acetylsalicylic acid, but the active moiety, which inhibits the enzyme cyclooxygenase, resulting in pain relief, is salicylic acid. Hence, drug substance may be the more appropriate term, but API is more commonly used. The drug product, or IMP, can be anything from a simple solution of API in water, given by mouth, to a highly formulated tablet or injectable solution.

The pharmaceutical industry is heavily regulated, and a plethora of regulations and guidelines must be followed in the manufacture, development and testing of IMPs. So chemists working in the pharmaceutical industry must work to approved, standard operating procedures and keep scrupulous records, with full traceability to source. Every detail must be documented: it is often said in the pharmaceutical industry "if it isn't written down, it didn't happen".

The European Union (EU) GMP Directive[9] and CTD,[7] and associated guidance documents, set standards for the manufacture of marketed medicines and IMPs in the EU. In addition, the International Council for Harmonisation of Technical Requirements for Pharmaceuticals for Human Use (ICH) has published guidelines[10] for the quality of APIs and IMPs, and development of formulations. The guidelines include limits for impurities or residual solvents, guidance on determination of shelf life, and many other quality aspects. The ICH guidelines apply to marketed products, but it is important to keep them in mind throughout drug development, so a product of consistently high quality can be reliably achieved well before it is marketed. The final, large-scale phase 3 clinical studies needed before marketing a new medicine ideally should be done using API and IMP manufactured in the same way and to the same specifications as the marketed product, or additional data will be needed to confirm the relevance of the phase 3 data. Also, the application for a marketing authorisation must include analytical and stability data on at least three production-scale, GMP-compliant batches made in the same way and to the same specifications as the marketed product. However, during early clinical development, manufacturing methods for both API and IMP are still in development. Hence, good communication between the manufacturer, analytical chemist, regulatory staff and trial site pharmacists is crucial. It is important that everything be tightly controlled, to ensure that the product is safe and effective. However, it is equally important not to set control criteria that are too strict, as that could lead to expensive and unnecessary loss of material, or to delays due to the requirement for regulatory approval of variations to the manufacturing process or product specification for IMPs used in clinical trials.

10.2.1.1 Manufacture of the API

When manufacturing an API, there are many important considerations, such as purity of the product, nature and quantity of individual impurities (including stereoisomers of the API), yield, cost, environmental impact, scalability, and GMP compliance. Some of those factors are not a priority early in the drug development process, but become hugely important as clinical development progresses. For example, only very small amounts of API are needed for initial laboratory studies to test whether the API has the desired pharmacological effect, selectivity and potency, so, at that stage, factors such as yield and cost of reagents are relatively unimportant. However, toxicology studies in animals require huge amounts of API, and late-phase clinical development (and marketing) require even more, so it is crucial to optimise the yield and reduce as much as possible the costs. In addition to scale-up, GMP compliance must be considered during drug development. GMP compliance of the API manufacturing process is not essential early in development, because API for animal toxicology studies and phase 1 clinical trials need not be made to GMP; however, it is required for later-phase studies. Hence, while early experiments using non-GMP API are ongoing, consideration must be given to selecting a future manufacturer able to scale up the process and work in compliance with GMP.

Other factors, such as impurities, are important throughout the drug-development process. API used in animal toxicology studies should have a similar impurity profile to API used in clinical trials, if the animal data are to be used as evidence that the API is safe to give to humans. There are many types of impurities to consider. Some impurities are related to the manufacturing process, for example reagents, catalysts or intermediates; others are a result of breakdown of the API. Some impurities, such as water, are in themselves harmless, but could accelerate the rate at which the API degrades to form new impurities. Impurities may be toxic, so they must be controlled, and tested for, if appropriate. Chemical structures of intermediate products can be assessed for potential carcinogenicity, initially *in silico*, and later by laboratory screening tests. For certain impurities, there are strict defined limits; for others, there is guidance on how to determine what levels of impurities are acceptable. If impurities are present at high levels, they must be qualified – they must be fully characterised and assessed in animals for toxicity.

10.2.1.2 Characterisation of the API

The API must be fully characterised. Analyses typically include elemental analysis, infrared spectrum, mass spectrometry, NMR spectrum, melting point, optical rotation, solubility, appearance, physical form (including polymorphism), stereochemistry, pK_a, hygroscopicity, and partition coefficient. Any reference batch must be fully characterised; other batches may undergo fewer tests because they are compared to the reference. In addition to confirming the structure, purity (including stereochemical purity) must be assessed, and the quantity and nature of impurities investigated.

Stress testing of the API is done to discover how it behaves under extreme conditions, such as high temperature and humidity, exposure to light, oxygen, and extremes of pH. That establishes degradation pathways and identifies likely degradation products. It also helps to establish likely storage conditions.

Based on analysis of preliminary batches, an initial specification is generated, against which future batches will be tested. The specification will be optimised as more data emerge during the drug-development process.

10.2.1.3 Formulation of the API

Bioavailability is a term that describes how much API is absorbed by the body after administration of a dose of IMP. The API must reach the relevant part of the body, in sufficient concentration, for long enough to have the desired effect. The formulation chemist aims to optimise the formulation to maximise bioavailability of the API. Other considerations apply, such as:

- suitability of formulation for the target population – for example, oral formulations for young children, or for patients who have difficulty swallowing, should be in liquid form;
- route of administration (*e.g.* by mouth, applied to the skin, by injection into a vein);
- strength of the IMP;
- stability under appropriate storage conditions; and
- scalability, cost and ease of manufacture.

Formulations in early clinical trials can be very simple (*e.g.* solutions for oral use, API powder hand-filled into capsules). Licensed product formulations are more complex (*e.g.* sterile solutions for injection, capsules containing API and excipients, tablets, dry powder inhalers) and may be highly sophisticated (*e.g.* controlled release forms that release API at a particular rate or in a particular part of the body). For example, some oral products are coated to stop or slow down dissolution in the harsh, acidic environment of the stomach but allow dissolution in the more alkaline small intestine where the API can be absorbed into the bloodstream. Polymers may also be used to control release, by limiting diffusion, increasing retention time in the stomach and small intestine, or through activation in the body (*e.g.* by solvent-triggered swelling, by biodegradation, or in response to other environmental triggers, such as pH or temperature). Some formulations may incorporate multilayers of API plus polymer, polymer alone and other excipients (*e.g.* disintegrants, lubricants) to increase the time for which the API stays in the stomach and small intestine and delivers continuous, controlled release of API from the formulation. However, such sophisticated formulations are not developed before early studies are done in man, and may not even be developed until after a drug has been marketed, as a means to improve product sales and perhaps obtain more patent life.

Solubility is an important factor in formulation – generally, the more water soluble the API, the easier it is for the body to absorb, so the simpler the formulation. For very soluble API, a very simple formulation might be sufficient to achieve blood levels of the API in early clinical trials that are high enough to meet the objectives of the trial. However, an increasing number of drug candidates are sparingly soluble. Hydrophobicity may be an important determinant of an API's interaction with its receptor, and will affect its ability to cross lipid membranes and reach its site of action. In selecting an API for development, solubility/hydrophobicity of candidate molecules will have been considered, and, as far as possible, an appropriate balance reached to achieve the desired characteristics, taking into account receptor binding, ability to cross membranes, ability to cross the blood–brain barrier (which may or may not be desirable, depending on the intended effects), and susceptibility to metabolism. Sparingly soluble APIs pose a formulation challenge. Chemists are required to characterise the physical properties of the API, including crystalline *versus* amorphous state, particle size and uniformity, and glass transition temperature, with a view to increasing solubility. For example, a sparingly soluble crystalline API can be rendered more soluble, and achieve higher blood levels, if it is converted to an amorphous form (for example, by spray drying) or if the particle size can be reduced (for example, by nanomilling). Additionally, if a salt or prodrug of the API can be made, the salt or prodrug may be more soluble than the parent compound.

Compatibility of the API with excipients must be investigated, candidates assessed and the chosen formulation optimised for use in early studies in humans. In parallel with early clinical development, formulation development continues. The formulation may be further optimised, and data must be obtained to show that the chosen formulation is appropriate for use in patients with the condition that the API is intended to treat. The formulation must be fully characterised.

Stability of the formulation is a key factor in IMP development: maximising shelf life reduces waste and cost, and simplifies the supply of IMP for clinical testing. Managing IMP supply across large international phase 3 studies is logistically very challenging, and even more so if shelf life is limited.

Shelf life of API and IMP is assessed by assay, at baseline and at predefined sampling times, of material stored under controlled conditions. Testing follows a stability protocol that specifies storage conditions, sampling time points and the specification against which the material will be tested. To maximise the value of stability data early in development, so called accelerated stability studies may be done, in which the material is stored under conditions of higher temperature and humidity, designed to increase the rate of chemical degradation or physical change. Early in drug development, real-time data may be extrapolated to support a retest period longer than that tested during stability studies, provided that accelerated stability data are acceptable.

10.2.2   Quality of API and IMP

The quality of the API and IMP is, of course, extremely important. Between-batch variation in impurity content or strength of an IMP could result in an ineffective medicine or even harm patients. Hence, the manufacturing process must be reliable, and control measures must ensure a product (API or IMP) that is as pure and consistent as possible.

GMP[9] is a quality standard for manufacture of medicines, to ensure that products are made consistently and to an appropriate specification. The product must be:

- the right medicine, the right strength, and the right purity;
- free from contamination;
- intact – not broken down or "gone off";
- sealed in the right sort of container, and protected from damage, contamination and deterioration; and
- properly labelled.

It would not be appropriate to rely on testing alone to control API and IMP. Testing destroys material, so we can test part of a batch, but not all of it, and we could not be sure that we had tested for everything that could possibly have gone wrong. Hence, GMP applies the principles of quality assurance to manufacture of API and IMP so that quality is built into products. GMP does that by ensuring that a product is made and tested in line with a preapproved process and that records are kept, to demonstrate compliance of the product at every stage of manufacture and testing, and to ensure full batch traceability.

10.2.2.1 Specification

A suitable specification, against which to test the product, is essential. The specification must comply with relevant guidelines (for example, those on residual solvents and impurities), and ensure product safety while allowing for acceptable batch-to-batch variability.

In most cases, the specification for the IMP is still undergoing development during early clinical trials, and so it is subject to change. Good communication between the manufacturer, analytical chemist and regulatory staff is crucial. The specification, and any changes to it, must be justified.

10.2.2.2 Analytical Methods

Analytical methods used to test API and IMP must be demonstrated to give true results every time, and must address all factors relevant to product quality. Release of API and IMP for use depends on the results of analytical testing against the specification: if a batch fails testing, it cannot be used. A key part of analytical testing is to demonstrate that the API is the correct material, of the required purity, and that an IMP contains the correct API at the correct strength. Thus, the quality of the API and IMP cannot be controlled unless there are reliable assays to identify, quantify and characterise the API.

Validation is the process by which an assay is demonstrated to be fit for purpose. Formal validation of an API assay is not required early in the clinical development phase, but it is clearly important to have confidence that the assay produces reliable results, so assays are often validated before the FIH clinical trial starts.

ICH Guidelines[10] describe how drug assays should be validated. The following should be determined:

- specificity for the analyte;
- accuracy: how close the result is to the true value;
- precision: reproducibility of results (within-assay, within-laboratory and between-laboratory precision);
- linearity: the ability (within a given range) to obtain test results that are directly proportional to the concentration of analyte;

- analytical range: the range of concentrations over which the assay has suitable accuracy, precision and linearity;
- detection limit: the lowest amount of analyte that can be detected, but not necessarily quantified;
- quantification limit: the lowest amount of analyte that can be measured with suitable accuracy and precision; and
- robustness: the ability of the assay to remain unaffected by small but deliberate variations in the method (*e.g.* variation in temperature, or duration of incubation).

Changes to the assay or transfer of the method from one laboratory to another should trigger revalidation.

Most FIH trials require measurement of the blood levels of the API in subjects. The assay may need to be adapted for use in clinical trials. Assays used to measure API in blood samples from clinical trials must be validated. Additional factors to consider in addition to those listed above include the effect of: matrix (plasma or serum); and haemolysis (presence of haemoglobin owing to breakdown of red blood cells), which may occur during blood sampling.

Validation must be fully documented. A plan or protocol must be prepared, with predefined acceptance criteria, and the results should be presented and assessed against those criteria in a validation report. Records of raw data, including laboratory notebooks, chromatograms and worksheets, must be retained.

10.2.2.3 Quality by Design

For many years, the quality of APIs and IMPs has been ensured, in part, by tightly controlling the manufacturing process, and making sure that everything is done in exactly the same way every time. Typically, three identical repeats means that the process is valid. That type of process control is easy to implement and follow, but it is not scientifically sound and there is a high risk of wasting material that is suitable simply because it was not produced in the exact same way as previous batches. Hence, the industry is slowly moving towards a more flexible, scientifically founded approach, called quality by design (QbD). Full understanding of the manufacturing process is the key to QbD. The aspects of the formulation and manufacturing process of API and IMP that can influence batch reproducibility, product performance and product quality must be identified and characterised. Experiments are done to examine the effect on specific aspects of product quality (such as impurity levels or dissolution rate) of key factors (such as moisture content, temperature or particle size) during manufacturing processes. The data can then be used to determine a range of operating conditions that reliably produce a product that is within specification. Thus, by understanding the influence of key factors, a degree of flexibility can be incorporated into the manufacturing process. QbD requires more work of everyone involved, and it is essential to ensure that the manufacturing process is presented appropriately in the marketing application to the regulatory authorities, since any deviation from the approved process must have regulatory approval before it may be implemented.

10.2.2.4   Biological Products

Biological products must conform with GMP and be made in line with the principles described above. However, their manufacture, characterisation and control is much more challenging than for small molecules, and is guided by ICH guidelines Q5A–Q5E and Q6B.[10] Proteins and peptides are much larger and more complex than small molecules: they are composed of chains of amino acids that fold into complex secondary and tertiary structures that may be modified post-translationally, for example by glycosylation or phosphorylation. Peptides and proteins are often produced in cell culture, a biological system, so there are additional considerations, such as genetic stability, complexity of culture media, and contamination by infectious agents. Because of the biosynthetic processes involved, there is inherent heterogeneity in structure of peptide and protein products, for example a protein may be differentially glycosylated, and the resulting different glycans might have different biological activities. Minor changes in the manufacturing conditions could affect the quality of the product. Specifications for biologicals focus on critical quality attributes that confirm the quality of the drug substance and drug product, as full characterisation may be challenging. In addition to physicochemical properties, biological activity must usually be considered in the specification. The challenges are even greater if the medicine is an advanced medicinal product, such as cell therapy.

10.2.3   Further Physicochemical Development Work

As noted above, development of the drug product occurs in parallel with its clinical development. Manufacturing methods must be scaled and optimised. Results from clinical studies may show that the bioavailability of the IMP is too low, and could indicate that a salt, alternative salt, or prodrug of the API would be a better candidate medicine, which would lead to additional process development and optimisation. For example, making an API containing a carboxylic acid group less polar, by converting it to an ester, could increase its bioavailability by enhancing its permeability through lipid membranes and, once in the body, esterase enzymes will rapidly cleave the ester bond, releasing the API. Feedback from patients may also drive further development work, for example to improve taste or ease of swallowing. Formulations may be further developed after marketing in an effort to improve a product, for example to speed up absorption, or make a formulation suitable for a paediatric population, and so increase patent life.

As part of the clinical development of an IMP, a study must be done to investigate its absorption by and elimination from the body. One of the objectives is to identify metabolites. A knowledge of the chemistry of the API, and of the body's biochemistry, allows prediction of the likely metabolic pathways of an API and the likely metabolites. Consideration is given during candidate selection to chemical groups in the API structure (toxicophores) that may generate potentially toxic metabolites. However, simply because a toxic metabolite can be generated does not mean that toxicity will occur. The metabolism of paracetamol[11] is a good example (Figure 10.4). The primary metabolic pathways for paracetamol are glucuronidation and sulfation, which yield nontoxic metabolites; however, a small proportion of paracetamol is metabolised by enzymes of the cytochrome P450 family to *N*-acetyl-*p*-benzoquinone imine (NAPQI), which binds covalently to proteins and is extremely toxic to liver tissue. However, NAPQI generated after a therapeutic dose of paracetamol is rendered nontoxic by conjugation with glutathione; it is only after an overdose of paracetamol, when liver glutathione reserves are depleted, that liver failure ensues. During clinical development, biological samples from clinical studies are analysed for the presence of metabolites, and the chemical structures of those metabolites may need to be identified. If metabolites are present in sufficient concentration, they must be synthesised, characterised and tested in animals for toxicity.

10.2.4 The IMP Dossier (IMPD)

During clinical development of an IMP, the IMPD forms part of the application for regulatory approval of each clinical trial. It allows the regulators to assess the quality and the potential toxicity of the IMP,

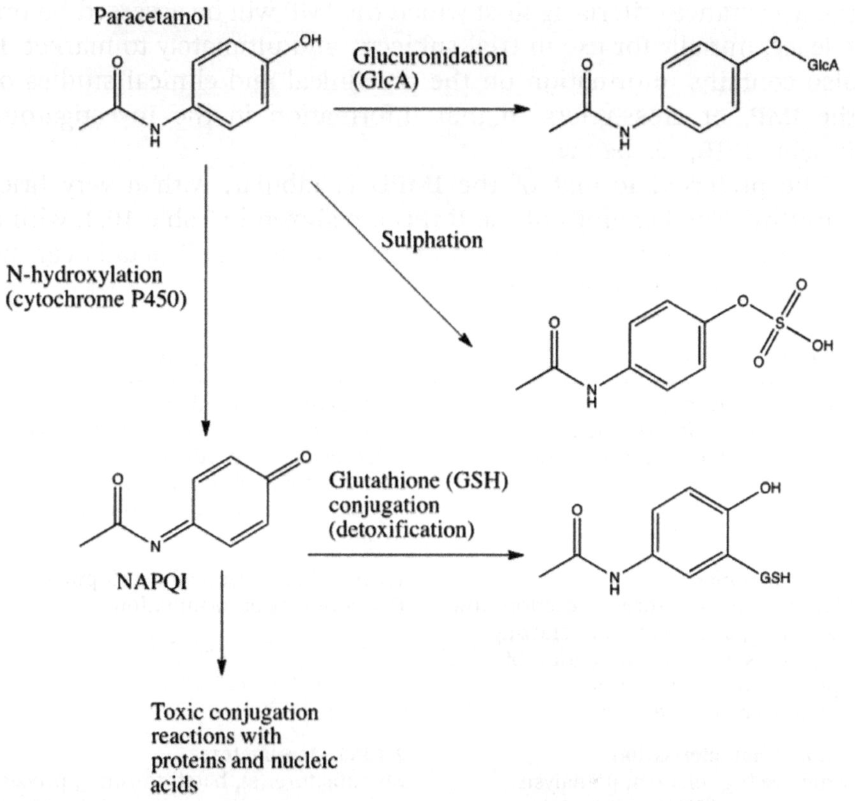

Figure 10.4   Metabolism of paracetamol.

and the safety of its use in the proposed trial. The chemistry, manufacture and control sections of the IMPD mirror those of Module 3 of the Common Technical Document – a document that is needed to support a marketing application for a new medicine.[12] Ultimately, the information in the IMPD will form part of the marketing application for the IMP. However, while the Common Technical Document must demonstrate state-of-the-art quality of a finished product for wide use in patients, the IMPD reflects the current stage of development of the product, and focuses on areas of risk, considering the nature of the product, the population of trial subjects and types and duration of clinical trials.

The IMPD contains detailed physical, chemical and pharmaceutical information on the API and the IMP. It includes details of the manufacturing method, analytical tests and methods, storage conditions, packaging, and shelf life. In particular, it contains the specifications: details of the tests that will be done on each batch, and the acceptance criteria against which the IMP will be assessed before release, initially for use in trial subjects, and ultimately to market. It also contains information on the preclinical and clinical studies of the IMP, or cross-refers to that information in the Investigator's Brochure (IB; see below).

The preferred format of the IMPD is tabular, with a very brief narrative. The headings of the IMPD are shown in Table 10.3, with a very brief guide to the contents of each section. Full details can be found in EU guidance documents.[7]

Table 10.3 Chemistry, manufacture and control sections of the IMPD.

2.1.S Drug substance	**2.1.P Medicinal Product**
2.1.S.1 General information	**2.1.P.1 Description and composition of the medicinal product**
Chemical name, structure, properties	Formulation, nature, quantity and quality of components, role of each component
2.1.S.2 Manufacture	**2.1.P.2 Pharmaceutical development**
Manufacturer(s), process (reaction flow diagram), specification of starting materials, controls, validation of process, development of manufacturing method	Development of formulation
2.1.S.3 Characterisation	**2.1.P.3 Manufacture**
Structure (*e.g.* elemental analysis, infrared and NMR spectra), impurities (theoretical impurities based on manufacturing method, and actual levels)	Manufacturer(s), batch formula, process (flow diagram), controls, validation of manufacturing process
2.1.S.4 Control of drug substance	**2.1.P.4 Control of excipients**
Specification, analytical methods and their validation, batch analysis data, justification of specification	Specifications, analysis, validation, justification of specification, excipients of human/animal origin, any novel excipients
2.1.S.5 Reference standards or materials	**2.1.P.5 Control of medicinal product**
Identity and characterisation of reference material	Specification, analytical methods, validation, batch analysis data, impurities (theoretical and actual levels)
2.1.S.6 Container closure system	**2.1.P.6 Reference standards or materials**
Packaging	Identity and characterisation of reference material
2.1.S.7 Stability	**2.1.P.7 Container closure system**
Storage conditions and retest/expiry date	Packaging
	2.1.P.8 Stability
	Storage conditions and retest/expiry date

The information in the IMPD is derived from the IMP product specification file, which is updated continually during the development of the IMP. Early in clinical development, not all information will be available. For example, analytical assays need not be validated before phase 1 studies can begin, and the formulation is unlikely to have undergone extensive development and characterisation. Thus, the IMPD evolves as more information becomes available and methods are optimised, and Section 2.1.P.2 is updated on an ongoing basis to explain the rationale for any changes to the manufacturing process.

The initial specifications for API and IMP are based on results from preliminary batches. As IMP development continues, and more batches of API and IMP are made, updates may be made to the specification. Example specifications for an API and an IMP are shown in Tables 10.4 and 10.5, respectively. Early in development, suitable criteria might not have been established for all aspects of the specification, so the specification might simply require results to be reported. Also, results from new batches might allow specification criteria to be tightened, as processes are optimised, or might require that certain criteria be relaxed in the light of new information about batch-to-batch variability.

In addition to information about the chemistry, manufacture and control of an IMP, the IMPD is also required to present information about animal and clinical studies. During clinical development, that is usually achieved by cross-referring to another document: the IB. The IB is intended to give the physician responsible for the trial

Table 10.4 Example specification for API.

Test	Acceptance criteria
Appearance (visual inspection)	White to off-white crystals or crystalline powder
Identification (infrared spectrum)	Conforms to reference standard
Identification (HPLC)	Conforms to reference standard
Purity (HPLC area %)	≥99.0%
Single impurities (HPLC area %)	Each impurity ≤0.5%
Optical purity (HPLC area %)	≥99.0%
Optical rotation	specific rotation +115 to +122°
Melting point	230–238 °C
Residual ethanol (gas chromatography)	≤0.5% (5000 ppm)
Water content (semimicrodetermination)	≤0.5%
Heavy metals	<20 ppm

Table 10.5 Example specification for IMP capsules.

Test	Acceptance criteria
Appearance (visual inspection)	White to off-white powder contained in a transparent hard gelatin size 1 capsule
Identification of API by HPLC	Retention time of the principal peak in the sample chromatogram corresponds to that of the principal peak in the analytical standard chromatogram
Assay (% label claim)	90.0–110.0% of the label claim
Related substances (HPLC area %)	Each impurity ≤0.5%; total impurities ≤5%
Average capsule fill weight	±10% of target weight
Dissolution (% API released)	≥80% of the label claim (mean) dissolved by 40 min
Uniformity of content	In sample of 10 capsules, the API content of ≥9 capsules must be within 85–115% of the mean content, with no individual capsule content outside 75–125%.
Uniformity of weight	In a sample of 20 capsules, the weights of ≤2 capsules deviate from the mean by >7.5%, and no individual capsule by >15%

subjects, the Principal Investigator, sufficient information to decide if the proposed trial is justified and safe, and guidance on the expected effects of the IMP on trial subjects.

The IB includes the following information:

- brief summary of the physical, chemical, pharmaceutical properties and formulation of the IMP;
- results of studies in animals, including pharmacology, toxicology, blood levels and metabolism;
- results of any studies in humans, including safety, blood levels and metabolism, and any results of effectiveness in treating patients; and
- summary of data and guidance for the investigator, including details of how to recognise and treat a possible overdose or adverse reaction caused by the IMP.

As more information becomes available, the IB is updated. Anything that might alter the risk/benefit assessment of the IMP, such as a serious finding in an animal or clinical study, must be communicated promptly to the Principal Investigator and the regulatory authority and ethics committee that approved the study.

10.3   First in Human (FIH) Studies

10.3.1   Classical FIH Study

Before an IMP can be tested in man for the first time, there must be:

- sufficient data about its manufacture, chemistry and quality in the IMP dossier;
- evidence that the medicine works and is safe in animals, summarised in the IB; and
- a suitable formulation.

The objectives of the FIH study are primarily to assess the safety and tolerability of the IMP. However, virtually all FIH studies also include an assessment of the PK and PD of the IMP. Wherever possible, they also include an assessment of proof-of-concept.

People who volunteer for phase 1 trials get no therapeutic benefit from the IMP, so the risk of harming the subjects must be minimal. The risk must be fully assessed before each trial, especially during the transition from preclinical studies to the FIH trial, when uncertainty about tolerability and safety of the IMP is usually at its highest. All aspects of the IMP – such as its class, novelty, species specificity, mode of action, potency, dose– and concentration–response relationship for efficacy and toxicity, and route of administration – must be taken into account. Risk must be assessed on a case-by-case basis, and there is no simple formula. The seriousness of possible adverse reactions and the probability of them happening must both be considered.

Consideration of the study population must be included in the risk–benefit analysis of the study. Should the new drug substance be tested in healthy volunteers or in patients? There are many advantages to using healthy volunteers:

- rapid recruitment;
- absence of confounding factors, such as illness or the need for other medication;
- more homogeneous than patients;
- robustness: healthy volunteers can usually tolerate pharmacological insults and an intensive schedule of procedures; and
- can freely give informed consent.

However, healthy volunteers will get no medical benefit from the study (other than the health screen to determine their eligibility), so the risk must be negligible. Thus, some FIH studies, such as those of cytotoxic drugs or involving uncomfortable or painful procedures (such as repeated endoscopies) are done in patients. For some IMPs, a subset of healthy volunteers that resemble the target patient population, such as obese subjects or subjects with high serum cholesterol, may offer some of the advantages of healthy volunteers but also allow more meaningful assessment of the PD of the IMP.

The classical phase 1 study design:

- is placebo controlled – that allows effects (including adverse ones) to be attributed to the drug;
- is double blind: neither investigator nor subject knows whether active or placebo treatment is being administered – that avoids bias in reporting, handling and interpretation of side effects in particular, and increases the credibility of the data;
- tests single, ascending doses: we test a small dose, then review the results before testing higher doses; and
- includes 6–8 subjects per treatment: all subjects may take all treatments (crossover design), different groups of subjects may take different treatments (parallel-group) or subjects may take some of all the treatments tested (incomplete/partial crossover), resulting in treatment comparisons that are within-subject (crossover); between-subject (parallel) or a combination of within- and between-subject (incomplete/partial crossover).

All clinical trials must be scientifically sound and described in a clear, detailed protocol. Compared with the protocol for an IMP at later phases of development, the protocol for FIH phase 1 trial should emphasise:

- the preclinical information about the IMP, including assessment of dose- or concentration–response relations;
- the assessment of risk of harm from the IMP, and trial procedures;
- stopping or withdrawal criteria; and
- the rationale for: the choice of first dose; the maximum allowed dose; rules for increasing the dose; the rate of administration of intravenous doses; the interval between dosing individual subjects; and the number of subjects to be dosed on any one occasion.

The dose escalation is a key feature of the FIH study. We start by testing a small single dose, usually aiming for a blood concentration in man lower than one with minimal pharmacological activity, to allow assessment of any off-target toxicity. If the starting dose is safe and well tolerated, we review the data and agree on a higher dose to test. We continue, reviewing the data after each dose level and choosing an appropriate dose for the next one, ensuring that blood concentrations are not predicted to exceed those that were safe in animals and that the chosen dose is expected to be safe and well tolerated. Predictions on safety will be based on a standard panel of tests, such as laboratory tests of blood and urine, heart rate, blood pressure, ECG, and any additional tests to monitor for side effects seen in animals. At each dose escalation, the results are compared against clearly defined stopping criteria, including blood levels in animals at the NOAEL; if those criteria are met, the dose cannot be increased.

Correct calculation of a safe starting dose is critical. For small molecules, the starting dose is usually calculated by converting the NOAELs from toxicology studies into the human equivalent doses (HEDs), then applying a safety factor.[13] The conversion is based on body surface area. For example, a 1 mg kg^{-1} dose in a rat is divided by 6.2 (the conversion factor for rats) to derive a HED of 0.16 mg kg^{-1}. A safe starting dose is typically derived by applying a safety factor of at least 10-fold to the lowest HED. Higher safety factors may be needed if, for example: the IMP has a novel mechanism of action; the IMP has a step dose–response curve or insufficient information is available about the dose–response relationship; severe or irreversible toxicity was seen in preclinical studies; the IMP caused side effects that are not readily monitored; or the IMP had unfavourable PK in animals (*e.g.* it was poorly absorbed or blood levels were variable). The HED method for calculating starting dose is simple and backed by a wealth of historical evidence.

A complementary method (or alternative when IMP activity is species specific), is to calculate the safe starting dose using the minimal anticipated biological effect level (MABEL) – the lowest dose at which the IMP is expected to have minimal pharmacological activity – and apply a safety factor.[14,15] The calculation of MABEL should use all available information, such as:

- target binding and receptor occupancy studies in target cells from human and the most relevant animal species *in vitro*;
- concentration–response curves in target cells from human and the most relevant animal species *in vitro* and dose/exposure-response in the most relevant animal species *in vivo*; and
- exposures at pharmacological doses in the most relevant animal species.

If the two methods give different results, the lowest dose is used. If the preclinical data are judged to be a poor guide to the effect in humans, the calculated starting dose should be reduced and the dose increased in smaller increments. However, if the IMP is selective and has a pharmacological effect known to cause no safety concerns (such as inhibition of angiotensin-converting enzyme), then the starting dose can be higher than the MABEL.

Phase 1 studies have to be designed flexibly to allow sponsors and investigators to make decisions appropriate to the emerging data. Dose decision meetings are scheduled between treatment groups, for the sponsor and investigator to review the results and choose the next dose to be tested. In the FIH study, the starting dose and each new escalating dose is typically given to one or two "sentinel" subjects first; the remainder of the group receive the same dose 24–48 h later.

Most sponsors need their phase 1 studies to be done to aggressive timelines so that they can get their IMP to market as quickly as possible. Hence, the FIH study must be scheduled as efficiently as possible. Typically, a new dose level will be tested every 1–2 weeks. Plasma samples taken from the subjects at frequent intervals during the first 24 h after dosing are transported to the analytical laboratory for rapid IMP assay.

Accurate, reliable results are essential for the dose-decision process, so the assay must have been validated before the study, and quality control (QC) checks of all results must be done before they are released. As protocols typically do not allow testing of doses that yield blood levels of the IMP exceeding those that were safe in animals, the PK data are critical to control the dose decision and protect the safety of trial subjects. The results are used by a pharmacokineticist to generate concentration–time curves, derive PK parameters and predict blood levels of the IMP at the next planned dose.

The key PK parameters reviewed at dose decision meetings are:

C_{max}	The peak plasma concentration of the IMP.
t_{max}	The time of peak plasma concentration of the IMP.
AUC	The area under the plasma concentration–time curve – this is a measure of overall exposure to the IMP.
$t_{\frac{1}{2}}$	The time taken for plasma concentrations of the IMP to fall by half during the terminal phase of elimination.

If samples are not available from the whole group (*e.g.* insufficient subjects are recruited or a subject withdraws from the study), care must be taken by the pharmacokineticist not to inadvertently unblind the investigator and sponsor. By way of example, consider a study in which subjects are dosed in groups of 8, 2 of whom receive placebo. If only 7 of the planned 8 subjects are recruited, and the PK report lists plasma concentrations and PK parameters for only 5 subjects, then it is clear that the remaining subject (who has not yet been dosed) will receive active treatment.

During dose decision meetings, mean values of C_{max} and AUC in humans are typically compared to those at the NOAEL in animals. Consideration must also be given to the variability in the results: an outlier might have blood levels of the IMP substantially higher than those in the remaining subjects, indicating that blood levels after the next planned dose in some individuals may exceed the predicted mean value. The sponsor and investigator must review the data from previous groups, to determine the relationship between dose and exposure. If blood levels of the drug increase disproportionately with dose, that must be taken into account in the prediction of PK parameters at the next planned dose level. If the predicted C_{max} or AUC at the next planned dose level exceeds the corresponding value at the NOAEL in animals, that dose cannot be tested and a lower dose must be selected.

The flexibility of FIH protocols, and the need to continually review the data to guide the dose escalation schedule, poses logistical challenges. Each step of the process must be carefully timed. Shipping of plasma samples is done as soon as possible after collection of the 24 h time point for the main group (*i.e.* 48 h after the sentinel subjects were dosed). The laboratory must be poised to receive and log the samples and to execute and QC the analysis, including any repeat analyses that may be needed, typically within a few days. Then, the plasma concentration data must be analysed, and the analysis QC checked, also within a matter of a few days, so that a report can be presented at the dose-decision meeting. After the meeting, there must be sufficient time for the clinical site to prepare for the next dose group. The dose decision must be verified to ensure that the chosen dose is permitted by the protocol and supported by the data. Pharmacy must prepare worksheets and labels for the new dose of IMP, have them checked, then manufacture, check and release the new dose.

The study design described above is the classical FIH design: a study of single, ascending doses, to assess the safety, tolerability, PK and PD of an IMP in young, healthy volunteers. However, increasingly, FIH protocols are written to include additional objectives. These so-called "umbrella" protocols contain several parts, for example, to compare results in women or in elderly subjects with those in young men; to compare results in different ethnic groups, and to test the first repeated doses in man. Many FIH protocols also include an assessment of the effect of food on the PK of the IMP. Absorption of an IMP can be increased or reduced by the presence of food, so the FIH protocol often includes a comparison of the PK of an IMP taken by one group of subjects in the fasted state and after a high-fat breakfast.

FIH studies are dynamic and unpredictable. Despite carefully laid plans to cover the logistics of the single ascending dose study, and beyond in the case of umbrella protocols, sponsors and investigators must be able to react appropriately to the emerging data. Unfavourable results after the very first dose in humans could terminate the study, and unexpected results, good or bad, could lead to amendments to the protocol to add extra subjects and assessments. An amendment is *substantial* if it is likely to have a significant impact on:

- the safety or physical or mental integrity of the trial subjects; or
- the scientific value of the trial.

The sponsor decides whether an amendment is substantial. A substantial amendment requires approval of the regulatory authority and/or ethics committee.

The main product of the FIH study is the clinical study report, which summarises the statistical analysis of safety, tolerability, PK and PD data. PK analyses may include characterisation of metabolites and elucidation of the metabolic pathways by which an IMP is eliminated: the result of separation and chemical characterisation of IMP-related compounds in blood and urine. Ultimately, the results of the FIH study will determine whether the IMP continues down the development path or is cast aside, along with the millions invested in it thus far.

10.3.2 Other FIH Studies

In addition to the classical FIH study of a potential new medicine described above, there are other situations where IMPs are administered to humans for the first time. Here are some examples.

- Studies in which a radiolabelled IMP is administered.
- "ADME" studies involve administration of a radiolabelled IMP to allow assay of IMP and metabolites in blood, urine and faeces,

with the aim of characterising the absorption, distribution, metabolism and excretion of the IMP.

- Microdose studies involve administration of a tiny radiolabelled dose; blood levels are monitored by a very sensitive method called accelerator mass spectrometry. A microdose is defined as less than one hundredth of the predicted pharmacological dose but not exceeding 100 µg. Because the risk of harm from a microdose is much lower than from a pharmacological dose, fewer preclinical studies are required to support a microdose trial. Hence, microdoses can be used to compare candidate IMPs and select the best one to enter standard preclinical studies. However, microdoses are more commonly used to assess absolute bioavailability: the proportion of a dose that is absorbed into the bloodstream. In those studies, a pharmacological dose is given by mouth, then an intravenous microdose is administered, and blood levels of unlabelled and radiolabelled IMP are measured and compared:
- Studies in which a new salt of an IMP is tested (salt switches) as part of formulation development.
- Studies of new formulations or of generic copies of licensed medicines.

10.3.3 Safety

Overall, phase 1 trials have a good safety record. In a meta-analysis of 11 028 subjects who participated in 394 studies during 2004–2011 in three centres in Belgium, Malaysia and USA: 36% had no adverse events (AEs); 64% had 24 643 adverse events, of which 85% were mild, and 0.3% serious; and no one died or had a life-threatening AE. Of 36 serious AEs, 11 were related to the IMP, 7 were related to study procedures, and 16 were unrelated to the IMP, including 4 on placebo. 10.3% of all AEs occurred on placebo. The most common AEs were headache (12%), drowsiness (10%), and diarrhoea (7%).[16]

Those findings are typical of surveys of AEs in phase 1 trials in other countries.[17,18] However, a few healthy subjects have died.[19] In a review in 2006,[18] it was noted that 15 deaths had been published during the previous 30 years in Western countries, although probably 100 000 healthy subjects are dosed every year. Some of those 15 examples are described below.

A man died of cardiac arrest after taking an IMP in a trial in Ireland in 1984. When he was screened for the trial, he did not declare that he had recently been given a depot injection of an antipsychotic medicine.[20] A woman died after receiving a high dose of lidocaine – a widely used local anaesthetic – to prevent discomfort from endoscopy in a trial in the USA in 1996.[21] She was discharged soon after the procedure and died at home. Another woman with mild asthma died of lung damage after inhaling hexamethonium in a trial in the USA in 2001.[22] Hexamethonium is an old medicine – given by injection to treat conditions other than asthma – and is now rarely if ever used. Lung toxicity of hexamethonium was first reported many years ago.[23] All of these deaths were probably preventable.

In France, in January 2016, one of six healthy subjects who received daily doses of 50 mg of BIA 10-2474 by mouth was admitted to hospital after the fifth dose and died. BIA 10-2474 is an inhibitor of fatty acid amide hydrolase (FAAH), an enzyme that plays a key role in the regulation of the endocannabinoid system. Four of the five other subjects allocated to BIA 10-2474 also became seriously ill and were hospitalised after they received their sixth dose. Brain magnetic resonance imaging (MRI) of the five symptomatic participants showed deep-brain haemorrhage (bleeding) and necrosis (tissue death).

FAAH inhibitors had been studied safely in previous clinical trials. A Temporary Specialist Scientific Committee (TSSC) was set up by the Director General of the Agence Nationale de Sécurité du Médicament et des produits de santé (ANSM) to investigate the accident during the Phase 1 clinical trial in Rennes. They concluded that the toxicity was most likely caused by BIA 10-2474, acting either directly or indirectly through a metabolite, and that the mechanism that triggered the accident went beyond simple FAAH inhibition. They noted the low specificity of the BIA 10-2474 molecule for its target, the testing of doses much higher than those needed to cause 100% inhibition of FAAH, the lack of dose proportionality of PK parameters and steep dose–response curve, and probable gradual accumulation in the brain. The fact that this type of toxicity was not observed in animals despite administration of very high doses, remains unexplained so far.[24] An initial report into the tragedy by the Inspection Générale des Affaires Sociales criticised the fact that the five other subjects were given their sixth dose of BIA 10-2474 after their fellow volunteer was admitted to hospital, that they were not asked to give written informed consent to continue in the trial and that the authorities were not informed more promptly of the serious side effect.[25]

Although there have never been any deaths in the UK, in a FIH trial in 2006, six subjects experienced a cytokine storm and became seriously ill with organ failure after receiving a single dose of intravenous TGN1412,[26] a monoclonal antibody and superagonist specific for the CD28 receptor. TGN1412 activated T-cells, in particular regulatory T-cells, and had therapeutic activity in rodent models of autoimmunity, inflammation, transplantation and tissue repair. The starting dose in the FIH study was calculated according to the traditional method,[13] as follows. The NOAEL in cynomolgus monkeys was 50 mg kg^{-1}. After applying a correction factor of 3.1 for that species and a safety factor of 10, the HED and maximum recommended starting dose (MRSD) were 16 and 1.6 mg kg^{-1}, respectively. The sponsor applied an additional safety margin and proposed a starting dose of 0.1 mg kg^{-1}. Subjects were dosed at 10 min intervals, and there was no sentinel group of subjects.

In the report of the Expert Scientific Group enquiry into the TGN1412 study, the MABEL dose was deemed between 0.3 and 1.0 mg kg^{-1}.[14] If, as they suggest, the MABEL is taken as 0.5 mg kg^{-1} and the criteria to support trials of a single microdose are applied,[27] the safe starting dose would have been calculated as 0.005 mg kg^{-1} or 5 µg kg^{-1}. The FIH study has been criticised for the size of the starting dose and the lack of a sentinel group.

We now know that TGN1412 caused serious side effects in humans but not in cynomolgus monkeys because T cells that develop into so-called CD4+ effector memory T cells in cynomolgus monkeys lose their CD28 receptors and are not activated by TGN1412, whereas, in humans, CD4+ effector memory T cells retain CD28 receptors on their surface and are activated by TGN1412 to produce proinflammatory cytokines.[28] Hence, the preclinical studies failed to warn of the impending cytokine storm.

With that knowledge, Russian researchers have retested TGN1412 in healthy subjects, starting with an extremely low dose: 0.1% of the dose used in the FIH study, which is 0.0002% of the maximum dose used in the original cynomolgus monkey study.[29] Calculations showed that, at the new starting dose, only 1% of human CD28 receptors would be bound by TGN1412. The dose was increased by small increments in different cohorts. The results indicated that 5% of the dose used in the FIH trial selectively activated regulatory T cells, but not proinflammatory T cells. There were no serious side effects. A study of TGN1412 in patients with rheumatoid arthritis is in progress.[30]

The extraordinary saga of TGN1412 demonstrates the value of: having relevant preclinical studies; understanding the mechanism of action of an IMP; having biomarkers to predict the effect of an IMP in humans; and having the right information to calculate a safe starting dose for a FIH study in humans.

As a result of the TGN1412 tragedy, the MHRA now requires that FIH trials of certain IMPs be reviewed by an expert advisory group, based on assessment of risk. In considering whether an IMP needs expert review, various factors, including the following, are taken into account.

- Does the IMP have a target with potential for large biological amplification of effects?
- Does the IMP target the immune system with a target or mechanism of action that is novel or not well characterised?
- Does the IMP have high potency compared with the natural ligand?
- Is the target poorly characterised or does it differ between healthy subjects and patients?
- Does the IMP act *via* a species-specific mechanism such that animal studies are likely to be poorly predictive of toxicity in humans?

10.3.4   What we can Learn from Early Clinical Studies

We can always obtain safety, tolerability and PK data in early studies of an IMP, and almost always investigate PD effects. Given the high attrition rate, especially in the late phases of clinical development, and even at the registration stage, nowadays it is also essential to try to obtain "proof-of-concept" of an IMP as early as possible during clinical development. That can usually be achieved by use of biomarkers or experimental medicine procedures in healthy subjects. However, sometimes it is necessary to study patients with the target disease. Below are three examples of our proof-of-concept studies. Two studies in healthy subjects provided good evidence to support efficacy of the IMP in patients with the target disease and to define the therapeutic dose for patient studies. The other case study, in patients with the target disease, led to withdrawal of two IMPs early in development.

Case Study 1

At the time of its discovery, CP-88,059-01 (ziprasidone) was a novel IMP considered to have potential advantages over existing treatments for patients with psychosis, such as schizophrenia. Schizophrenia is a mental illness affecting ~1 in every 100 people and characterised by positive symptoms, such as hallucinations and delusions, and negative symptoms, such as lack of motivation, and becoming withdrawn.

Preclinical studies showed ziprasidone acts at various receptors in the brain, but is mainly a dopamine D_2 receptor antagonist. After the FIH studies to assess safety and tolerability of ziprasidone in healthy subjects, we did two studies in healthy volunteers, to determine the dosing regimen of ziprasidone suitable for clinical trials in patients with schizophrenia. We used positron emission tomography (PET) scans to measure binding of ^{11}C-raclopride to dopamine D_2 receptors in the striatum in the brain, in the presence and absence of ziprasidone.

In the first study,[31] we tested single doses of ziprasidone (2–60 mg) and placebo in eight subjects. Seven subjects each took a different dose of ziprasidone and one took placebo, then, five hours later, they had a PET scan. Ziprasidone dose-dependently inhibited binding of the ligand to striatal dopamine D_2 receptors PET scans (Figure 10.5A). A plot of binding *versus* ziprasidone dose (Figure 10.5B) shows that 20–40 mg resulted in >65% occupancy of D_2 receptors, the level of occupancy associated with therapeutic doses of various antipsychotic medicines.

In the second study,[32] 6 healthy subjects each received a single dose of ziprasidone 40 mg and one subject received placebo. Subjects were scanned at 4–36 h after dosing. Figure 10.5C shows a curvilinear increase in binding potential *versus* time after ziprasidone dosing. Results were compared with that from the placebo subject and those from a control group of 9 unmedicated subjects. At 12 h after dosing, the binding potential in those treated with ziprasidone was more than 2 standard deviations (SD) below the mean for unmedicated controls, which is consistent with extensive D_2 receptor binding by ziprasidone. At 18 h, it was within 2 SD of the mean, and at 27 h it was above the mean.

PD (the effect of ziprasidone on blood levels of prolactin) and PK were also assessed in the second study. Blood concentrations of ziprasidone and prolactin increased after dosing (Figure 10.5D). Prolactin, which is a biomarker of D_2 antagonism at the pituitary level, returned to within the normal range by 12 h after dosing in all but one subject. Binding potential correlated both with blood levels of ziprasidone and prolactin at the time of PET scanning.

On the basis of these two studies in small numbers of healthy subjects, the chosen dose regimen of ziprasidone for clinical trials was 20–40 mg twice daily. That proved to be the recommended dose regimen when ziprasidone (Geodon®, Pfizer) was eventually marketed. To determine the dose regimen directly in trials in patients with schizophrenia would have required many patients, taken considerably longer, and been very costly.

Case Study 2

Asthma occurs in children and adults and is common. During attacks, patients experience wheezing, breathlessness and cough due to narrowing (bronchoconstriction), inflammation and swelling (mucosal oedema) in the airways. The main triggers of an attack are: respiratory viral infections; environmental allergens, such as house dust mite,

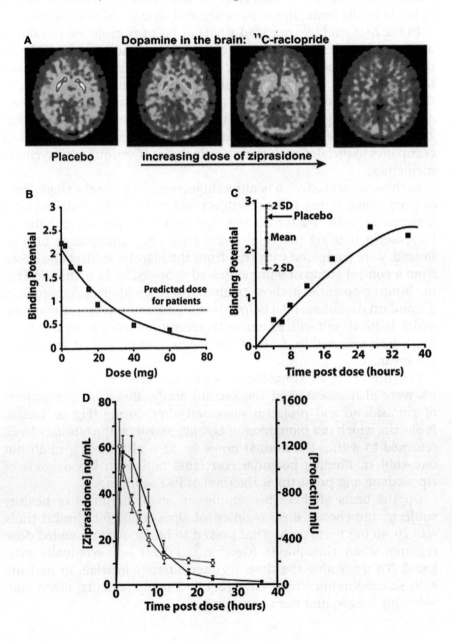

animal dander and pollens; and exercise. For immediate relief of symptoms, patients inhale a short-acting β_2-adrenoceptor agonist (SABA), such as salbutamol, which reduces bronchoconstriction. For long-term control of symptoms, patients inhale a corticosteroid, such as budensonide, with or without a long-acting β_2-adrenoceptor agonist (LABA). Other treatments include a leukotriene antagonist, such as oral montelucast, or, in severe cases, an oral steroid or injections of omalizumab, a monoclonal antibody that neutralises immunoglobulin E.

There are various tests that can be used as biomarkers for efficacy in treatment of asthma. One important biomarker is the bronchial allergen challenge test: lung function, measured as the maximum volume of air that can be exhaled in one second (the forced expiratory volume or FEV_1), is assessed before and after inhalation of an allergen. That test can be used to measure airway sensitivity and thus assess a potential new treatment for asthma. Patients with mild-to-moderate allergic asthma who have not used their SABA inhaler for 8 h, inhale increasing concentrations of a fine mist of an allergen to which the patient is sensitive. We use the data to define, for each patient, the allergen dose that causes an early asthmatic reaction

Figure 10.5 Ziprasidone, a novel antipsychotic agent, in healthy subjects. (A) Example PET scans of brains of healthy subjects. The ligand ^{11}C-raclopride binds selectively to dopamine D_2 receptors in the brain resulting in highlighting of the striatum. Ziprasidone, a dopamine D_2 receptor antagonist, causes dose-dependent inhibition of ^{11}C-raclopride binding and reduction in highlighting. (B) Ziprasidone binds dose dependently to dopamine D_2 receptors. $N=1$ subject per dose. The predicted dose in patients (65% occupancy) was 20–40 mg. (C) There is increased uptake of ^{11}C-raclopride in the striatum with time after ziprasidone 40 mg. $N=1$ subject per time point. The mean and range (2 SD) of binding potential from nine nondosed control subjects is included. The results, together with those from B, indicate that the ziprasidone dose regimen for clinical trials in patients is 20–40 mg twice daily. (D) Mean ($n=7$; $\pm$ sem) plasma concentration–time response curves for ziprasidone (closed squares) and prolactin (open circles). Binding potential correlates with both ziprasidone and prolactin concentrations.
(A) and (B): Reproduced from *Psychopharmacology*, Dose dependent occupancy of central dopamine D2 receptors by the novel neuroleptic CP-88,059-01: a study using positron emission tomography and ^{11}C-raclopride, *Psychopharmacology*, 112, 1993, 308–314, C. J. Bench, A. A. Lammertsma, R. J. Dolan, P. M. Grasby, S. J. Warrington, K. Gunn, M. Cuddigan, D. J. Turton, S. Osman, R. S. J. Frackowiak, copyright Springer-Verlagt 1993, with permission of Springer. C and D: Reproduced from *Psychopharmacology*, The time course of binding to striatal dopamine D2 receptors by the neuroleptic ziprasidone (CP-88,059-01) determined by positron emission tomography, *Psychopharmacology*, 124, 1996, 141–147, C. J. Bench, A. A. Lammertsma, P. M. Grasby, R. J. Dolan, S. J. Warrington, M. Boyce, K. P. Gunn, L. Y. Brannick, R. S. J. Frackowiak, copyright Springer-Verlagt 1996, with permission of Springer.

(EAR): a 20% drop in FEV_1 during 0–2 h postchallenge, reflecting bronchoconstriction. About 40% of patients with an EAR also develop a late asthmatic reaction (LAR) 3–7 h postchallenge as a result of inflammation and mucosal oedema. Typically, a bronchoconstrictor, such as a SABA, will attenuate the EAR, and an anti-inflammatory agent, such as a steroid, will attenuate the LAR.

The ideal study design is a crossover study in which patients are challenged with the same dose of allergen after the test treatment and after placebo, on separate occasions. Figure 10.6 illustrates the results of bronchial allergen challenge tests from three of our studies in patients with mild-to-moderate asthma. Figure 10.6A shows that compared with placebo, a single oral dose of a positive control (montelukast, a marketed leukotriene antagonist) attenuated both the EAR and LAR [unpublished]. Figure 10.6B and C show results of bronchial allergen challenge tests done using two potential new asthma medicines:

- IVL 745, a novel antagonist of $\alpha_4\beta_1$ integrin receptors, which play a major role in the regulation of recruitment of immune cells to sites of inflammation; and
- a novel antagonist of IL-4, a cytokine involved in regulation of inflammation.

Figure 10.6B shows that IVL 745, twice daily for 7 days, affected neither the EAR nor the LAR.[33] Figure 10.6C suggests that, if anything, compared with placebo, the IL-4 antagonist by inhalation for 14 days, increased the EAR and LAR responses to allergen challenge [unpublished].

Development of IVL 745 and the IL-4 antagonist was discontinued as a consequence of those negative bronchial allergen challenge tests. To our knowledge, no treatment that has failed to give a positive result in a bronchial allergen test in patients with asthma has ever made it to the market.

Case Study 3

Alzheimer's disease is a chronic neurodegenerative disease and the main cause of dementia. There is loss of nerve cells (neurons), and connections between them (synapses), in the cortex and subcortical regions of the brain, especially neurons that respond to a chemical messenger (neurotransmitter) called acetylcholine. For that reason, patients with mild-to-moderate cognitive impairment are treated with

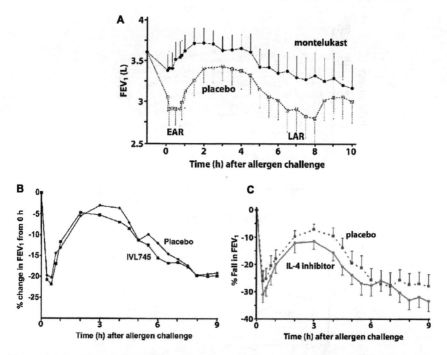

Figure 10.6   Effect of treatment on the early asthmatic response (EAR) and late asthmatic response (LAR) to inhalation of allergen in patients with mild-to-moderate asthma. (A) Oral montelukast ("Singulair"), a leukotriene antagonist, attenuates both EAR and LAR responses (mean ± SD fall in FEV_1; $N = 13$). Unpublished data. (B) Inhalation of IVL745, a novel antagonist of $\alpha 4 \beta 1$ integrin receptors, twice daily for 7 days, affects neither EAR nor LAR response, compared with placebo (mean % changes in FEV1; $n = 16$). Reprinted from *J Allergy Clin Immunol*, 116, Norris V., Choong L., Tran D., Corden Z., Boyce M., Arshad H., Holgate S., O'Connor B., Millet S., Miller B., Rohatagi S., S. Kirkesseli, Effect of IVL745, a VLA-4 antagonist, on allergen-induced bronchoconstriction in patients with asthma, *J. Allergy Clin. Immunol.*, 116, 761–767, Copyright (2005), with permission from Elsevier. (C). Inhalation of a novel IL-4 antagonist, for 14 days, tends to increase the EAR and LAR responses (mean % changes ± SD in FEV1; $n = 20$). Unpublished data.

acetylcholinesterase (ChE) inhibitors, which reduce the rate at which acetylcholine is broken down. They have some benefit for symptoms, but do not delay progression of the disease.

In the FIH studies of the ChE inhibitor galantamine geneserine hydrochloride (CHF 2819), we assessed the safety and side effects of single, rising doses,[34] and then repeated, rising doses for 6 days[35] in healthy subjects. We also measured PD: the effect of the IMP on ChE activity in red blood cells, which is a biomarker of ChE activity in the brain. Single doses of 3–18 mg caused dose-dependent inhibition of erythrocyte ChE (Figure 10.7A). Doses up to 15 mg were well tolerated,

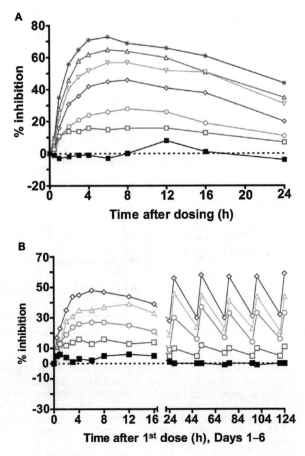

Figure 10.7   Dose-dependent inhibition of erythrocyte cholinesterase activity by galantamine, a novel cholinesterase inhibitor, in healthy men (A) Single, rising doses of galantamine 3 mg (-□-), 6 mg (-○-), 9 mg (-◇-), 12 mg (-▽-), 15 mg (-△-) and 18 mg (-*-) (*n* = 4 per dose) and placebo (-■- *n* = 12). Adapted from Warrington *et al.* 2000.[34] (B) Repeated, rising doses of galantamine 3 mg (-□-), 6 mg (-○-), 7.5 mg(-△-) and 9 mg(-◇-) (*n* = 6 per dose) and placebo (-■- *n* = 8) once daily for 6 days. Adapted from Johnson *et al.* 2000.[35]

whereas 18 mg caused side effects as a result of overstimulation of cholinergic neurons. Repeated, once-daily doses of 3–9 mg caused dose-dependent inhibition of red blood cell ChE, peaking each day at 4 h after dosing, and persisting for 24 h after dosing. On Day 6 of treatment, peak ChE inhibition was 7–58%. The results supported a once-a-day dosing regimen (Figure 10.7B). A daily dose of 9 mg was chosen for clinical trials in patients with Alzheimer's disease. A unit dose of 10 mg was the marketed dose. Thus, our FIH trials helped define the therapeutic dose.

10.4   Conclusions

The first administration of a new IMP to man is an important milestone in the drug-development pathway. Safety of the trial participants is the primary concern, so the quality of the IMP and the design of the study are of paramount importance. However, with careful planning, early studies in man can deliver so much more than a simple assessment of safety, tolerability and PK: proof-of-concept, and information to guide design of studies in patients. However, none of this is possible without the coordinated efforts of a multidisciplinary team of which chemists are an integral part.

Acknowledgements

The authors wish to thank Fatima Yusuf, for preparing Figure 10.4 and for helping to review this chapter, and Lauren Garden for preparing all other figures.

References

1. CPMP/ICH/286/95. Non-clinical studies for clinical trials.
2. CPMP/ICH/384/95. Toxicokinetics and exposure in toxicology studies.
3. J. J. Darrow, J. Avorn and A. S. Kesselheim, New FDA Breakthrough Drug Category – implications for patients, *N. Engl. J. Med.*, 2014, **370**, 1252–1258.
4. I. Kola and J. Landis, Can the pharmaceutical industry reduce attrition rates? *Nat. Rev. Drug Discovery*, 2004, **3**, 711–715.
5. M. Hay, D. W. Thomas, J. L. Craighead, C. Economides and J. Rosenthal, Clinical development success rates for investigational drugs, *Nat. Biotechnol.*, 2014, **32**, 40–51.
6. G. J. Aronson, Biomarkers and surrogate endpoints, *Br. J. Clin. Pharmacol.*, 2005, **59**, 491–494.
7. The Rules Governing Medicinal Products in the European Union, Volume 10 – Clinical Trials, incorporating the EU Clinical Trials Directive (2001/20/EC) and the EU Good Clinical Practice Directive (2006/1928/EC) www.ec.europa.eu/health/documents/eudralex/vol-10/.
8. ICH Guideline for Good Clinical Practice (E6(R1)), 1996.
9. The Rules Governing Medicinal Products in the European Union, Volume 4 – Good Manufacturing Practice, incorporating EU Directive 2003/94/EC (www.ec.europa.eu/health/documents/eudralex/vol-4/).
10. ICH Quality Guidelines (www.ich.org/products/guidelines/quality/article/quality-guidelines).
11. L. L. Mazaleuskaya, K. Sangkuhl, C. F. Thorn, G. A. FitzGerald, R. B. Altman and T. E. Klein, PharmGKB summary: pathways of acetaminophen metabolism at the therapeutic versus toxic doses, *Pharmacogenet. Genomics*, 2015, **25**, 416–426.
12. ICH Multidisciplinary Guidelines M4: The Common Technical Document.

13. FDA Guidance for Industry. Estimating the Maximum Safe Starting Dose in Initial Clinical Trials for Therapeutics in Adult Healthy Volunteers, July 2005.

14. Expert Scientific Group on Phase 1 Clinical Trials. TSO. November 2006.

15. Committee for Medicinal Products for Human use (CHMP) guideline on strategies to identify and mitigate risks for first-in-human clinical trials with investigational medicinal products. EMEA/CHMP/SWP/294648/2007, September 2007.

16. E. J. Emanuel, G. Bedarida, K. Macci, N. B. Gabler, A. Rid and D. Wendler, Quantifying the risks of non-oncology phase I research in healthy volunteers: meta-analysis of phase I studies, *Br. Med. J.*, 2015, **350**, h3271.

17. M. Orme, J. Harry, P. Routledge and S. Hobson, Healthy volunteer studies in Great Britain: the results of a survey into 12 months activity in this field, *Br. J. Clin. Pharmacol.*, 1989, **27**, 125–133.

18. M. Sibille, Y. Donazzolo, F. Lecoz and E. Krupka, After the London tragedy, is it still possible to consider Phase I is safe? *Br. J. Clin. Pharmacol.*, 2006, **62**, 502–503.

19. Guidance for phase 1 clinical trials. Association of the British Pharmaceutical Industry, 2007.

20. A. Darragh, M. Kenny, R. Lambe and I. Brick, Sudden death of a volunteer, *Lancet*, 1985, **1**, 93–94.

21. C. J. Trigg, D. S. Freestone, T. H. Lee and T. G. K. Mant, Death of a healthy student volunteer in a US research study: lessons for bronchoscopic practice, *Int. J. Pharm. Med.*, 1998, **12**, 151–153.

22. F. McLellan, 1966 and all that – when is a literature search done?, *Lancet*, 2011, **358**, 646.

23. R. Steinbrook, Protecting research subjects, *N. Engl. J. Med.*, 2002, **346**, 716–720.

24. Temporary Specialist Scientific Committee (TSSC). Report "FAAH (*Fatty Acid Amide Hydrolase*)", on the causes of the accident during a Phase 1 clinical trial in Rennes in January 2016. ANSM (www.ansm.sante.fr/Dossiers/Essai-Clinique-Bial-Biotrial/Essai-clinique-BIA-102474-101-du-laboratoire-BIAL/(offset)/0). April 2016.

25. Inspection Générale des affaires sociales. Note d'etape: Enquete sur des incidents graves survenus dans le cadre de la realisation d'un essai Clinique. Premieres constatations. (www.social-sante.gouv.fr/actualites/presse/communiques-de-presse/article/suites-de-l-accident-grave-survenu-a-rennes-marisol-touraine-renforce-la). February 2016.

26. G. Suntharalingam, M. R. Perry, S. Ward, S. J. Brett, A. Castello-Cortes, M. D. Brunner and N. Panoskaltsis, Cytokine storm in a phase 1 trial of the anti-CD28 monoclonal antibody TGN1412, *N. Engl. J. Med.*, 2006, **355**, 1018–1028.

27. CPMP/SWP/2599/02. Position Paper on Non-Clinical Safety Studies to Support Clinical Trials with a Single Microdose.

28. D. Eastwood, L. Findlay, S. Poole, C. Bird, M. Wadhwa, M. Moore, C. Burns, R. Thorpe and S. Stebbings, Monoclonal antibody TGN1412 trial failure explained by species differences in CD28 expression on CD4+ effector memory T-cells, *Br. J. Pharmacol.*, 2010, **161**, 512–526.

29. M. J. Kenter and A. F. Cohen, The return of the prodigal son and the extraordinary development route of antibody TGN1412 – lessons for drug development and clinical pharmacology, *Br. J. Clin. Pharmacol.*, 2015, **79**, 545–547.

30. T. Hünig, The rise and fall of the CD28 superagonist TGN1412 and its return as TAB08: a personal account, *FEBS J.*, 2016, DOI: 10.1111/febs.13754.

31. C. J. Bench, A. A. Lammertsma, R. J. Dolan, P. M. Grasby, S. J. Warrington, K. Gunn, M. Cuddigan, D. J. Turton, S. Osman and R. S. J. Frackowiak, Dose dependent occupancy of central dopamine D_2 receptors by the novel neuroleptic CP-88,059-01: a study using positron emission tomography and ^{11}C-raclopride, *Psychopharmacology*, 1993, **112**, 308–314.

32. C. J. Bench, A. A. Lammertsm, P. M. Grasby, R. J. Dolan, S. J. Warrington, M. Boyce, K. P. Gunn, L. Y. Brannick and R. S. J. Frackowiak, The time course of binding to striatal dopamine D_2 receptors by the neuroleptic ziprasidone (CP-88,059-01) determined by positron emission tomography, *Psychopharmacology*, 1996, **124**, 141–147.
33. V. Norris, L. Choong, D. Tran, Z. Corden, M. Boyce, H. Arshad, S. Holgate, B. O'Connor, S. Millet, B. Miller, S. Rohatagi and S. Kirkesseli, Effect of IVL745, a VLA-4 antagonist, on allergen-induced bronchoconstriction in patients with asthma, *J. Allergy Clin. Immunol.*, 2005, **116**, 761–767.
34. S. Warrington, N. Johnson, M. Cattoni and M. Boyce, Tolerability and pharmacodynamic effects of single doses of ganstigmine, a new cholinesterase inhibitor, in healthy men, *Br. J. Clin. Pharmacol.*, 2000, **49**, 503P–504P.
35. N. Johnson, M. Cattoni, S. Warrington and M. Boyce, Tolerability and pharmacodynamics effect of repeated doses of ganstigmine, a new cholinesterase inhibitor, in healthy men, *Br. J. Clin. Pharmacol.*, 2000, **49**, 492P–493P.

11 Nonclinical Safety Assessment in Drug Development

Shayne Gad

Gad Consulting Services, 4008 Barrett Drive, Suite 201, Raleigh, NC 27609, USA
Email: scgad@ix.netcom.com

11.1 Introduction

Requirements for drug safety evaluation as we now know them started in 1937 due to fatalities associated with reformulation of an existing over-the-counter product (elixir of sulfanilamide) and the subsequent passage of the U.S. Food Drug and Cosmetic Act. The requirement for prenotification of the U.S. Food and Drug Administration (FDA) that a drug was going to be tested in humans [an investigational new drug (IND)] with a structured program of nonclinical safety studies, however, did not arise until 1963 following the thalidomide disaster.

Though much improved,[1–3] even with the globalization and harmonization of approval requirements for drugs [under International Conference on Harmonization (ICH)], countries and regions continue to have some unique regulations and guidance as to the safety that drug companies need to meet to gain market approvals.[4,5] The world market for drugs is large, with (as of 2013) about 39% of the pharmaceutical market in the United States, 24% in Europe, 15% in Japan and 22% in emerging markets. The remaining is spread across the rest of the globe in smaller portions. This does not mean, however, that the applicant should ignore the requirements of other countries, such as China, Brazil and India. Approval requirements in

Pharmacology for Chemists: Drug Discovery in Context
Edited by Raymond Hill, Terry Kenakin and Tom Blackburn
© The Royal Society of Chemistry 2018
Published by the Royal Society of Chemistry, www.rsc.org

these countries can, at times, be as rigorous as any other regulatory authority domain.

One factor to consider in the evolving regulatory requirements for early development of new therapeutic entities is the degree of entry barrier that costs may present to the smaller, innovative companies and truly new therapeutic approaches, in particular all requirements must be reviewed from the viewpoint of societal needs for safety versus health.

A second complicating factor in considering the "pharmaceutical" market sector is the diversity of products involved. The most basic expression of this is the division of drugs into "small molecules" (which currently constitute ~ two-thirds of both applications for clinical evaluation of a new drug in humans (INDs and CTAs) and two-thirds of current drug market approvals) and the remainder biotechnology products (which constitute the other third) though biologics/therapeutics (vaccines and cell therapies) are rapidly gaining ground. The challenges in both developing and assessing the safety of these are very different and the regulation of their safety assessment is profoundly different. As will also be seen, if one considers further division into therapeutic claim areas [oncology, anti-infectives, cardiovascular, central nervous system (CNS), ...], the differences become even more marked. Most of what is presented and discussed in this volume speaks of regulatory requirements for nonclinical safety assessment in the general case for either small molecules or protein therapeutics. The reader is advised to bear in mind that the general case model that is first presented almost never applies.

It is also increasingly important that the excipient component of the pharmaceutical product be considered, especially as a seemingly ever high proportion of active drug moieties are poorly soluble and poorly absorbed. Once primarily regulated by nongovernmental organizations, such as *United States Pharmacopoeia* (USP) and International Pharmaceutical Excipients Council, these essential ingredients have come to have regulatory nonclinical safety assessment requirements for new and "novel" inactive ingredients close to those of the active drug components themselves.

The starting place should be the international regulations, which establish how the safety of human pharmaceutical products and medical devices are evaluated. Clearly, the FDA, FMA, MHW, CFDA and their counterparts elsewhere play an enormous role in the development and approval of drugs and devices. Since the mid-1990s, however, the ICH has appeared on the scene in an attempt to gain similarities in guidance and regulations between the United States, Europe, Japan, and other countries. It appears reasonable that additional countries will continue to adopt the ICH guidelines as a means of encouraging consistency in drug approvals.

In a nutshell, the regulations need to require solid scientific evidence of safety and efficacy before a new drug will be permitted in clinical trials, and (at a much higher level of comfort) allowed into the market.

It must be recognized that the guidances present a continuously evolving practice and set of expectations within the regulatory community such that some guidance referenced in this chapter may be out-of-date tomorrow. It is important for all involved in this industry to stay alert to the ever-changing regulatory environment, particularly in those countries where these processes are in their infancy.

11.2 Stages and Components of Nonclinical Safety Assessment

Potential drugs are assessed for safety in nonclinical models in each of four stages, the last three of which are preformed to meet regulatory requirements.

1. Support of compound discovery and selection of development candidate(s).
2. To support initiation of clinical trials in humans (opening of a CTA or IND or equivalent). This stage has the most fixed set of components.
3. To support continued clinical development, through to being able to apply marketing approval.
4. To support marketing approval.

The component sets of studies in these stages evaluate:

1. Systemic toxicity (from single dose to (potentially) chronic repeat-dose studies of six or nine months. These incorporate evaluation of toxicokinetics.
2. Genotoxicity studies.
3. Safety pharmacology studies.
4. Local tissue tolerance studies.
5. Developmental and reproductive toxicity studies.
6. Carcinogenicity studies.

There may be. . .

7. *In vitro* metabolism and ADME studies (generally done as part of the candidate selection phase) and (most recently) *in vitro*

studies of receptor binding to maximize the specificity of therapeutic target receptor binding.

Most of the remainder of this chapter will be devoted to a more detailed evaluation of component studies.

11.3 Strategies for Development

The driving truths behind strategies in developing new drugs are as follows:

1. Most molecules will fail. Although the true success rate is certainly greater than the often quoted one-in-ten-thousand, it is clear that only 3–5% of those new molecules (NMEs) that enter initial clinical evaluation (*i.e.*, for which an IND "opens") become marketed drugs. This rate varies depending on therapeutic class (oncology drugs having a success rate as low as 1–2% and CNS therapeutics being only somewhat higher[6]).

2. The cost of developing drugs is high. While not nearing the currently quoted "average" of $2+ billion (see Goozmer, 2005[7] and Angell, 2005[8] for explanations of how the higher figure is derived), just getting to the point of an IND opening for a new molecular entities (NME) will cost a minimum of $1.4 million (plus the cost of drug synthesis). Biological therapeutics are more expensive yet to get to IND. And the costs of development continue to go up sharply with time/progress—subsequent to a plain vanilla first-in-man (FIM) trial, outlays come to be spoken of first in tens of millions, and (frequently) before a marketing approval filing, in the hundreds of millions. Once the decision is made to develop a molecule into a drug, the process takes years. Again, one can dispute how many (from 5 to 16 about covers the range), and at no point up to the end is success (achieving marketing approval and economically successful therapeutic use) assured.

These truths conspire to produce the principal general goals behind drug development strategy:

(i) Kill the losers as early as possible, before too much money is spent on them; and

(ii) Do all you can to minimize the calendar time spent in developing a drug. Time is more vital than money.

These principles produce a spectrum of strategies in the non-clinical safety assessment of drugs, best illustrated by looking at the two extreme approaches.[9]

11.4   Do Only What You Must

Driven by financial limitations and the plan that, at an optimal point in development (most commonly after either FIM/Phase I trials or a "proof-of-concept" Phase II trial), the candidate therapeutic will be licensed to or partnered with a large company, only the technical and regulatory steps necessary to get a molecule to this point are to be performed. For those pursuing this case, the guidance provided by this book should prove essential (although not generally completely sufficient).

11.5   Minimize the Risk of Subsequent Failure

This is considered the traditional big company model. Studies and technical tasks are not limited to the minimum, but rather are augmented by additional components. Development proceeds through a series of well-defined and carefully considered "go/no go" decision points. Many of the additional components are either limited, or non-GLP forms of studies, which will be required later (such as Ames, acute toxicity, human ether-a-go-go (hERGs) at only one concentration and 7-day to 4-week repeat-dose studies), or studies that are inexpensive and could be done later (CYP inhibition and induction, metabolic stability, and longer than required repeat-dose toxicity studies before proceeding into Phase II). Exactly which "extra" components are included vary from company to company, and frequently reflect past company experiences.

The studies performed to meet regulatory nonclinical safety assessment requirements (which must be considered to include all of the supportive toxicokinetic and metabolism activities and studies) can be thought of as belonging to latter three major stages or categories introduced earlier.

Which specific studies fit into what category (especially beyond those in the first) are somewhat fluid and influenced by what patient population will be served (therapeutic claim), what becomes known about the drug in earlier studies, and the mechanism of action of the drug.

11.6   What You Need to Know to Start

We will start here with explaining the so-called "general case". To understand how the safety assessment of a specific drug needs to be assessed, however, requires understanding the drug itself and its intended use.

Start with knowing what the specific therapeutic claim is intended to be. Knowing this will tell you what the characteristics of the patient population are and any special safety concerns they may have. Also, what is the expected and accepted mode of administration for drugs to treat this disease – that is, whether it is oral, intravenous, dermal, *etc*. Also determined by the claim is the expected treatment regimen, and the length of treatment (for any one individual) required to establish efficacy, and therefore to gain market approval and support label claims. Most of the studies described here are now done by contract research organizations (CROs) on contract, not by the companies actually developing the therapeutic agents.[10]

11.7   Strategies for Development

Start with consideration of the general case approach to nonclinical safety assessment from some fundamental assumptions about the drug under development or to be developed. The first assumption is that the primary intended route of therapeutic administration is oral, as is indeed the case for the vast majority of both existing and (thought decreasingly) new drugs. Most aspects of nonclinical safety assessment do not depend on route, and we will consider in detail the situations where the use of other routes influences what is done for nonclinical safety assessment, and why Table 11.1 provides an overview of the components of the "general" case.

A subset assumption in the general case is that drug administration frequency (or regimen) is once daily, although this assumption is less frequently made (in real life) than the oral route assumption. This regimen assumption has its earliest origin mostly in experimental laboratory practice, and *a priori* assumption of it is frequently a hindrance to developing a therapeutically optimal drug product.

Regulations, costs, and risks acceptance along with adherence to the phased process of clinical drug development cause the task or flow of performances of regulatory nonclinical safety assessment studies to be considered as occurring in three sequential parts.

Table 11.1 General case (oral) drugs.

Test requirement	Species
Initial clinical trial/IND requirements	
(1) Two-phase DRF in rodents	R/M
(2) Two-phase DRF in nonrodents	D/S/P
(3) Genotoxicity: bacterial mutagenicity (Ames)	*In vitro*
(4) Genotoxicity: *in vitro* clastogenicity (HL chrom ab)	*In vitro*
(5) Genotoxicity: *in vivo* (mouse or rat micronucleus)	M/R
(6) Safety pharmacology: CV-hERG	*In vitro*
(7) Safety pharmacology: CV *in vivo*	D/S/P
(8) Multispecies metabolite screen (microsomes or hepatocytes)	*In vitro*
(9) Safety pharmacology: functional observation battery (FOB)/Irwin test	R/M
(10) CYP metabolism screen	*In vitro*
(11) Receptor binding panel screen	*In vitro*
(12) Safety pharmacology: Respiratory- Rodent	R/M
(13) Pivotal/repeat dose in rodents (14/28 days)	R/M
(14) Pivotal/repeat dose in nonrodents (14/28 days)	R/M
To support continued development/approval application	
(15) Immunotoxicity (if indicated by earlier results)	R/M
(16) Pivotal/repeat dose in rodents	R/M
(17) Pivotal/repeat dose in nonrodents	D/S/P
(18) Segment II pilot studies (rodent and rabbit)	R/B
To support marketing approval	
(19) Reproductive/developmental (Segment I & II)	R/Rabbit
(20) Tumorgenicity/carcinogenicity	R/M

11.8   The Stages

FIM (IND/CTA) Enabling (Stage 2)

The nonclinical studies required to initiate initial clinical studies of pharmaceuticals in human beings are variously labeled as "IND enables" or "FIM enabling." The initiation of a program to conduct such studies is a major step in the advancement of a therapeutic candidate into actual development. For many drugs, it comprises the only regulatory nonclinical safety work that will ever be done as they will not be considered for further development after Phase I. All the safety studies that are required to be done should be performed in compliance with GLPs. This means that certain preparatory steps must be performed before the studies listed in Table 11.1 are commenced to achieve such compliance.

  (i) Sufficiently pure drug substances must be produced and characterized.
 (ii) An appropriate GLP compliant analytical method must be developed and validated to verify the purity of the drug substance.

(iii) GLP compliant bioanalytical methods (to measure amounts of drug present in plasma or serum of selected test species – typically a rodent and a nonrodent) must be developed and validated for the concentration range anticipated.

(iv) The stability of either the drug substance or active pharmaceutical ingredient (API) under appropriate storage conditions and in the anticipated animal dosing formulation must be demonstrated.

One or more suitable dosing formulations must be developed. Such formulations (which do not need to be those intended for clinical development) are typically uncomplicated at this point, but tend to channel later efforts to develop more sophisticated formulations.

Once these steps are performed, suitable studies (see Table 11.1) need to be conducted to support the filing of an IND (in the United States), CTA (in the European Union), or equivalent. These studies typically include acute and repeat-dose systemic toxicity studies in a rodent and a nonrodent species, genetic toxicity, and safety pharmacology studies. If a route of administration other than oral is intended, local tissue tolerance studies will also be required.[11]

With these in hand, after a regulatory review period the initial (Phase I and perhaps early Phase II) clinical studies may be conducted.

To Support Continued Clinical Development (Stage 2)

The IND enabling studies typically support repeat dose clinical studies up to a couple (four at most and most commonly) weeks in duration. Drugs more often than not require longer term (more than four weeks of dosing) clinical trials than these to reach the market (ICH, E8), which means that longer-term studies (than the typical 28-day IND enabling repeat-dose toxicity studies) must be conducted in both selected rodent and nonrodent species (if a "small" molecule). Additionally, developmental and (possibly) reproductive toxicity studies (DART) are usually required to allow the inclusion of a broader range of patients in clinical trials. Such longer-term repeat-dose studies are generally conducted in incremental steps so that clinical studies through Phase III can be conducted. This part of the safety assessment process will be addressed later in the text.

To Support Marketing Approval

The last distinct part of the nonclinical safety assessment study package generally consists of studies that are not required until a marketing application (in the United States, an NDA or BLA) is to be submitted. This group is usually limited to carcinogenicity studies (if required) and the final parts of the reproductive toxicity package.

11.9   ICH Requirements: The Global General Case

The ever-changing and growing number of ICH guidelines[†] provides the conceptual starting point for assessing the safety of new medicines, whether they are small molecules or proteins. These two different classes of molecules remain viewed by regulators as vastly different and raising different concerns USFDA, EMA, and Japan are all governed primarily by ICH.[12–16]

Continued revisions of guidances should be taken as an exception, for in no other way will the growth of knowledge of therapeutics and of means and mechanisms of therapeutics producing adverse events be accommodated.

So from these and associated requirements (such as USP), our general case arrives.

11.10   The First Rule

Presented above and throughout this volume is the general case, what usually is expected to be done (and is prescribed in the guidelines). Only rarely in the case of any specific drug dose such a general case apply. This book seeks to point out the exceptions and exclusions that apply, but undoubtedly will have missed some.

So *Caveat emperator* – The reader should always remember that the first rule is that the general case never fully applies.

The challenge of drug discovery and drug development to the pharmaceutical toxicologist is that a drug is supposed to have a biological function and requires a scientific understanding of mechanistically based toxicity. Furthermore, drug development allows the toxicologist to become somewhat flexible in developing the toxicology program that allows the entry into early-stage clinical trials.

EMA, FDA and other regulations require that the "safety and efficacy" of a drug be demonstrated prior to approval and, hence,

---

[†]*Always remember that these are not regulations, and as such allow for flexibility, particularly on the part of regulators.*

commercialization. For the toxicologist, the design and conduct of nonclinical studies to demonstrate the safety of the drug prior to a clinical trial is of prime importance.

The principal responsibility of the toxicologist in developing the IND/CTA is to design, conduct, and interpret appropriate toxicology studies to support the expected and planned initial clinical study and then design the appropriate studies necessary to support each additional phase of clinical investigation. A description of the studies that would be needed to "open" the IND is given later in this chapter.

The GLP Act (21CFR58) in the US specifies standards for study planning, personnel training, data recording, reporting, and so on and was codified in 1978 in response to perceived shoddy practices of the operations of a select few laboratories involved in the conduct of preclinical safety studies.[17] Near global adoption of GLPs revolutionized the standards of practice of toxicology studies and this regulation is, in some circles, considered a yardstick to measure the adequacy of preclinical studies. This is somewhat unfortunate since numerous studies have been conducted over the years as non-GLP but remain studies conducted under good scientific principles. Regardless, with submission of preclinical studies to the ICH/OECD country authorities, there is an expectation that the preclinical toxicology studies specified in ICH M3 (R2) or its equivalent be conducted according to generally accepted GLPs.[18]

Since 1990 or so, the FDA has encouraged study sponsors (those guiding the development of the drug) to meet with the respective FDA division prior to undertaking any of the pharmacology and toxicology studies and before submission of the IND. Such pre-IND meetings are granted upon request by the US FDA without cost. The EMA equivalent, the "scientific consultation", is not free or with actual regulations. Generally, industry has reacted positively to the pre-IND meeting, although these meetings have become an additional burden to the FDA. Nevertheless, these meetings have often been defined for the Sponsor the expectations by the FDA in terms of preclinical testing requirements and development of the clinical protocol. In preparation for the pre-IND meeting, the sponsor needs to submit to the FDA a briefing package that contains a summation of all available pharmacology and toxicology data as well as any chemistry manufacturing and control (CMC) information pertinent to the drug formulation intended for the Phase I trial and a description of the intended first clinical study (typically a synopsis).

11.11   Toxicity Testing: Traditional Pharmaceuticals

Most often, the regulatory development and approval of drugs proceeds in a somewhat fixed and orderly way. As there are always exceptions to the rule, drug development for special case therapies (*e.g.*, AIDS, cancer) do exist where the timeline for safety assessment can be shortened or the "requirements" for testing will vary from the usual small-molecule requirements.

The 1938 FD&C Act required safety assessment studies, but no consistent guidelines were available. Testing guidelines were first proposed in 1949 and published in the *Food, Drug and Cosmetic Law Journal* that year.[19,20] Following several revisions, these guidelines were issued as *The Appraisal Handbook* in 1959. While never formally called a guideline, it set the standard for preclinical toxicity test design for several years. The current basic guidelines for testing required for safety assessment in support of the phases of clinical development of drugs were first outlined by Goldenthal[21] and later incorporated into a 1971 FDA publication entitled *FDA Introduction to Total Drug Quality.*[22]

11.12   Common Mistakes

There is no single correct way to develop a new drug and assess its safety, but there are many wrong ways. Some of the common mistakes to avoid are given in the following sections.

Wrong Test Species

The test species should be selected for (at least) the systemic toxicity studies (a rodent and nonrodent generally) must be pharmacologically responsive to the drug and have at least a similar metabolic profile.

Poor or No Formulation

While the formulation used in nonclinical safety studies does not need (other than for hazard/local tissue tolerance studies) to be the same as for clinical studies, time should be taken to optimize both its tolerance and relative pharmacokinetic properties.

Insufficient Dose

It is essential to demonstrate toxicity (and identify target organs) for toxicity in systemic toxicity studies (at least), or to make every effort to do so. With the general exception of monoclonal antibodies, all too often the high dose in such studies (and even the middle dose) is set too low. This tendency must be avoided.

The FDA settled on and promulgated a procedure for setting a safe starting dose for clinical trials in humans, and put it into a guidance, which is now largely adhered to by the other ICH counties. This procedure starts from the point of "scaling" between species based on body surface area.

Table 11.2 presents the FDA's current values for scaling animal study doses to equivalent human doses.

Use of Nonrepresentative Test Material

Studies (particularly systemic toxicity and cardiovascular safety) should be conducted with API as representative as possible of what is to be evaluated clinically. This strongly suggests the use of material from the first clinical good manufacturing practice (GMP) lot for the 28-day repeat dose and *in vivo* cardiovascular safety studies.

Failure to Verify Adequate Exposure

It is essential to verify systemic or target organ exposure to the drug. In safety studies, toxicokinetics components serve this purpose. The most common failure here is *in vivo* micronucleus studies. A lack of effect in such studies is meaningless if one cannot prove test articles from the bone marrow.

Saving Money While Wasting Time

The most precarious commodity in the drug development process is calendar time. Cost savings by skipping essential or recommended steps or by selecting vendors based purely on cost/price is almost always a mistake.

11.13 Toxicity Testing of Pharmaceuticals – The General Approach

As noted previously, testing of pharmaceuticals usually proceeds in a rather fixed and orderly way. For entry into early clinical trials, testing will be needed to support those trials, and the duration of the repeat-dose studies generally will need to mirror the duration of the clinical trial (Table 11.3).

Table 11.2 Conversion of animal doses to human-equivalent doses based on body surface area.[a]

Species	Reference body weight (kg)	Working weight range[b] (kg)	Body surface area (m²)	To convert doses in mg kg⁻¹ to mg m⁻² multiply by k_m below	To convert animal dose in mg kg⁻¹ to HED[c] in mg kg⁻¹, either	
					Divide animal dose by	Multiply animal dose by
Human[d]	60	—	1.62	37	—	—
Child[d]	20	—	0.80	25	—	—
Mouse	0.020	0.011–0.034	0.007	3	12.3	0.081
Hamster	0.080	0.047–0.157	0.016	5	7.4	0.135
Rat	0.150	0.080–0.270	0.025	6	6.2	0.162
Ferret	0.300	0.160–0.540	0.043	7	5.3	0.189
Guinea pig	0.400	0.208–0.700	0.05	8	4.6	0.216
Rabbit	1.8	0.9–3.0	0.15	12	3.1	0.324
Dog	10	5–17	0.50	20	1.8	0.541
Primates:						
Monkeys[e]	3	1.4–4.9	0.25	12	3.1	0.324
Mamoset	350	0.140–0.720	0.06	6	6.2	0.162
Squirrel monkey	600	0.290–0.970	0.09	7	5.3	0.189
Baboon	12	7–23	0.60	20	1.8	0.541
Micropig	20	10–33	0.74	27	1.4	0.730
Minipig	40	25–64	1.14	35	1.1	0.946

[a]Abbreviation: HED, human-equivalent dose.

[b]For animal weights within the specified ranges, the HED for a 60-kg human is calculated using the standard k_m value and will not vary more than ±20% from the HED calculated using a k_m based on the exact animal weight.

[c]Assumes 60-kg human. For species not listed or for weights outside the standard ranges, human equivalent dose can be calculated from the formula:

$$HED = \text{animal dose in mg/kg} \times 0.33 \left(\frac{\text{animal weight in kg}}{\text{human weight in kg}} \right)$$

[d]The k_m is provided for reference only since the healthy children will rarely be volunteers for Phase I trials.

[e]For example, cynomolgus, rhesus, stumptail.

Table 11.3 General guidelines for duration of repeat dose animal toxicity studies in early development (ICH M3 (R2)).

Route of administration	Duration of clinical trial	Animal study duration	Special studies
Oral or parenteral	Several days to up to 2 weeks Up to 4 weeks Up to 3 months	2 Species; 2 weeks 2 Species; up to 4 weeks 2 Species; up to 3 months	For parentally administered drugs; Compatibility with blood where applicable
Inhalation (general anesthetics)		4 Species; 5 days (3 h day^{-1})	
Dermal	Single application Single or short-term application	2 Species; single 24-h exposure followed by 2-week observation 2 Species; 4 weeks (intact and abraded skin)	Sensitization
Ophthalmic	Single application Multiple application	1 Species; 3 weeks daily applications, as in clinical use 1 Species; duration commensurate with period of drug administration	Eye irritation tests with graded doses
Vaginal or rectal	Single application Multiple application	2 Species; duration and number of applications determined by proposed use	Local and systematic toxicity after vaginal or rectal application in 2 species
Drug combinations		2 Species; up to 3 months	Lethality by appropriate route, compared with components run concurrently in 1 species

As the duration of the clinical study increases, the duration of the enabling systemic preclinical studies also must be longer. It is possible, however, to initiate a single-dose clinical study for some pharmaceuticals by providing to the agency a single-dose toxicity study in two species. These studies must include clinical pathology and histopathology as well as be fully GLP compliant. There needs to be both a toxicokinetic (TK) component to verify systemic exposure levels and an analysis of prepared dosing solutions to verify what was administered to animals. Each of the systemic toxicity studies in these guidelines must be designed and executed in a satisfactory manner. Sufficient animals must be used to have confidence in finding and characterizing any adverse drug actions that may be present. These two features – dosage level and group size – are critical to study designs. Table 11.4 presents general guidance on the least number of animals to be used in systemic toxicity studies. These and other technical considerations for the safety assessment of pharmaceuticals are presented in detail by Gad.[23,24]

The number of animals in each group shown in Table 11.4 is usually the minimal numbers required, and is based on OECD guidelines. Indeed, for four-week studies, 10 rats/gender/group are often used and this number of rats is encouraged for two-week studies. Likewise, the number of nonrodents, for example, dogs, generally has been 3 per gender per group for two- and four-week studies, although the agency has recommended increasing this number for the treatment phase of the study and to allow for some animals not completing the study.

In these studies, the majority endpoints of to be included in the protocol are as outlined in the OECD guidelines. These endpoints would include measurement of body weights, recording of clinical observations, as well as daily observations for mortality and morbidity. Furthermore, hematology and clinical chemical parameters are to be measured in pivotal studies. Although caution should be exercised about including certain measures because of either variability or uncertainty regarding interpretation, for example, creatinine phosphokinase (CPK),

Table 11.4 Numbers of animals per dose group.

Study duration (per gender)	Rodents (per gender)	Nonrodents (per gender)
2–4 weeks	5–10	4
13 weeks	20	6
26 weeks	30	8
39 or 52 weeks (chronic)	N/A	10

isoform of this, or other enzymes, can provide valuable information on changes in cellular function. At necropsy, organ weights are to be measured for major organs, for example, liver, brain, heart, with histopathology of tissues. Also, ophthalmological examinations need to be done in pivotal studies, especially for rabbits and dogs, although including this endpoint in carcinogenicity will be necessary for certain pharmaceutical categories. Systemic exposure must be validated and evaluated by toxicokinetic portions of a study.[25]

The FDA strongly encourages sponsors to include additional animals to examine the reversibility of potential effects observed during the treatment phase of the study. Although this has been generally accepted in studies of four weeks and longer, more recently the agency has urged sponsors to include recovery animals for two-week studies. This practice is currently the subject of some controversy.[26,27] Furthermore, additional rodents are often needed in studies for the collection of blood samples for toxicokinetic determinations.

11.14 Developmental and Reproductive Toxicity Studies

The agency has a special set of concerns with reproductive toxicity, fetal/embryo toxicity, and developmental toxicity. Historically, these studies have been referred to as Segment (Seg) I, II, and III studies. The ICH has issued guidance on these studies (ICH S5(R2)) and are referred to as fertility and early embryonic development (Seg I), pre- and postnatal development, including maternal function (Seg II) and embryofetal development (Seg III).[28] These studies are often conducted during the Phase II clinical trial, but may be required for the IND if the drug represents a possible risk for reproductive toxicity, for example, estrogenic- or androgen-like actions.

The first protocol for DART test is a Segment I study of rats in fertility and general reproductive performance. This Seg I study and the Seg II study are generally completed particularly if women of child-bearing potential are included in the clinical trials. The teratogenicity testing is required in two species – a rodent (rat or mouse) and the rabbit. The use of the rabbit was instituted as a result of the finding that thalidomide was a positive teratogen in the rabbit but not in the rat. On occasion, when a test article is not compatible with the rabbit, teratogenicity data in the mouse may be substituted. There also are some specific classes of therapeutics, for example, quinalone antibiotics, where Seg II studies in primates are effectively required prior to product approval. Both should be completed before entering Phase III clinical trials. The more complicated Seg III DART protocol is generally commenced during Phase III trials and should be part of the marketing application (NDA in the US).

11.15 Genetic Toxicity Assessment

Genetic toxicity testing generally focuses on the potential of a new drug to cause mutations (in single-cell systems) or other forms of genetic damage (disruption of chromosome replication) *in vitro* or *in vivo*. The tests, generally short in duration, often rely on *in vitro* systems and generally have a single endpoint of effect (point mutations, chromosomal damage, and so on). The requirements have been recently (2011) revised to reflect accumulated experience over the years (Table 11.5).[24,29]

The FDA generally expects to see at least some such tests performed and will ask for them if the issue is not addressed. Often, the bacterial reverse mutation assay (Ames) and a study for clastogenicity are needed for the IND. If equivocal results or clear evidence of genotoxicity is observed in any of these *in vitro* studies, the sponsor is expected to perform additional studies prior to submission of the IND to clarify the findings. An *in vivo* (rodent, usually mouse) micronucleus is generally performed under these circumstances, although other *in vitro* or *in vivo* studies may be necessary (*e.g.*, mouse lymphoma tk assay, unscheduled DNA synthesis, SHE transformation assay). Positive results with the bacterial reverse mutation assay or clastogenicity testing do not often impede progress in development,

Table 11.5 ICH safety guidance S2(R1) options for the standard battery.

Option 1	Option 2
i. A test for gene mutation in bacteria	i. A test for gene mutation in bacteria
ii. A cytogenic test for chromosomal damage (the *in vitro* metaphase chromosome aberration test or *in vitro* micronucleus test), or an *in vitro* mouse lymphoma *Tk* gene mutation assay.	ii. An *in vivo* assessment of genotoxicity with two different tissues, usually an assay for micronuclei using rodent hematopoietic cells and a second *in vivo* assay. Typically, this would be a DNA strand breakage assay in liver, unless otherwise justified (see below also Section 4.2 and Note 12).
iii. An *in vivo* test for genotoxicity, generally a text for chromosomal damage using rodent hematopoietic cells, either for micronuclei or for chromosomal aberrations in metaphase cells.	

although the sponsor will be expected to undertake the additional studies to clarify the finding. At the time of writing, the FDA is considering accepting the mouse lymphoma assay in place of the *in vitro* clastogenicity assays. It should be noted that the Japanese regulatory authority, the Ministry of Health, Labor and Welfare, requires the mouse lymphoma tk assay in place of the chromosomal aberration study.[13] Furthermore, there is a consideration of suggesting that the *in vivo* micronucleus be included with repeat-dose toxicity studies, that is, collect peripheral lymphocytes or bone marrow cells at the time of sacrifice in the toxicity study.

There are a wide variety of test protocols that are described under the current OECD guidelines and are accepted by the FDA. These are described below. It should be noted that ICH and FDA have a number of alternative test protocols that have recently been added to the accepted list, namely, the comet assay, the liver unscheduled DNA synthesis test, and a transgenic gene mutation assay performed in mice.

Fifteen common assays described by OECD and accepted by FDA, European medicines agency (EMA), and Ministry of Health and Welfare (MHW) are listed in Table 11.6.

Table 11.6 Assays for gene mutations.

Assays for gene mutations	*In vitro*	*In vivo*
Salmonella typhimurium reverse mutation assay (Ames test, bacteria) [OECD 471]	✓	
Escherichia coli reverse mutation assay (bacteria) [OECD472]	✓	
Gene mutation in mammalian cells in culture [OECD 476]	✓	
Drosophila sex-linked recessive lethal assay (fruit fly) [OECD 477]		✓
Gene mutation in *Saccaromyces cerevisiae* (yeast) [OECD 480]	✓	
Mouse Spot test [OECD 484]		✓
Assays for chromosomal and genomic mutations		
In vitro cytogenetic assay [OECD 473]	✓	
In vivo cytogenetic assay [OICD 475]		✓
Micronucleus test [OECD 474]		✓
Dominant lethal assay [OECD 478]		✓
Heritable translocation assay [OECD 485]		✓
Mammalian germ cell cytogenetic assay [OECD 483]		✓
Assays for DNA effects		
DNA damage and repair: unscheduled DNA synthesis *in vitro* [OECD 482]	✓	
Mitotic recombination in *Saccharomyces cerevisiae* (yeast) [OECD 481]	✓	
In vitro sister chromatid exchange assay [OECD 479]	✓	
Combined assays		
Comet assay (rat dosing)		✓

In certain cases, genotoxicity testing may not be necessary. Clearly such studies have little value if the drug being developed is an oncologic, and where the drug is cytotoxic. For noncytotoxic oncologics, for example, biotechnology products, genotoxicity studies may need to be conducted. Genotoxicity testing of antibiotics, particularly in the bacterial reverse mutation assay may have limited value. However, the FDA has requested these studies at times with the *in vitro* concentrations being below the cytotoxic levels. Finally, for some therapeutics that are not systemically available (*e.g.*, antacids), genotoxicity studies are not usually required.

11.16 Safety Pharmacology Studies

Since 2002, there has been increasing interest on the part of the FDA, as well as other international regulatory authorities, that safety pharmacology studies be included with the IND for a Phase I clinical study.[30] The base set of studies are limited to cardiovascular, respiratory, and CNS function, except in unusual cases there is a specific reason (usually based on a known class effect or the proposed mechanism of action of the drug) to also evaluate another organ system. Other organ systems (renal, gastrointestinal, and liver) are not ignored, but rather are not required to be evaluated until a later point in the development of the drug (generally, concurrently with Phase III). The immune system is not included because it is covered by a separate ICH guidance (S8, promulgated in September 2005).

For the species to be used in these studies, selection of the relevant animal models or other test systems needs to be determined. Selection factors can include the pharmacodynamic response of the animal model to the pharmaceutical, pharmacokinetic profile, species, strain, gender and age of the experimental animals. The time points for the measurements should be based on pharmacodynamic and pharmacokinetic considerations. Most often, the dose levels selected for these studies will be based on the maximum blood concentration of the pharmaceutical determined from other toxicity studies, that is, the dose levels should result in a C_{max} following a single acute administration. In the respiratory and CNS studies, the rat is the preferred species, although other species can be used with scientific justification. For the cardiovascular study, the dog is the preferred species, although the nonhuman primate also has been used.

For the CNS safety pharmacology study, the effects should be assessed by measuring motor activity, behavioral changes, coordination, sensory/motor reflex responses, and body temperature. The functional observation battery (FOB), a modified Irwin's or other appropriate test can be used. In the respiratory safety pharmacology study, respiratory rate, tidal volume, and minute volume should be evaluated. Effects of the test substance on the cardiovascular system should be assessed with heart rate and the electrocardiographic measures. Indeed, prolongation of the QT interval is of paramount importance in this study as a measure of possible *torsades de pointe*.[31]

Although not usually required for Phase I clinical studies, other safety pharmacology studies will be necessary when the mode of action of the pharmaceutical is expected to cause concern on the part of the FDA. Drugs intended for renal or gastrointestinal indications will require these studies. For the renal pharmacology study, specific renal function is examined to include cytology in the urine, glomerular filtration rate, electrolyte concentrations, and so on. Most often these studies are conducted with rats, although dogs, nonhuman primates, and swine also have been used. In the gastrointestinal (GI) pharmacology study, propulsion rate, and alterations in absorption potential are examined with the studies carried out in rats.

Safety pharmacology studies may not be needed for locally applied agents (*e.g.*, dermal agents) or when systemic exposure is demonstrated to be low. Indeed, these studies are often not needed for dermal agents that are intended for topical indications, such as acne, rosacea, and others. However, when systemic exposure is expected because of chronically damaged skin, safety pharmacology studies may be needed. In addition, safety pharmacology studies for cytotoxic agents used in the treatment of end-stage cancer patients may not be necessary. However, for cytotoxic agents with novel mechanisms of action, there may be value in conducting safety pharmacology studies. For biotechnology-derived products that have a novel therapeutic class, an extensive evaluation by safety pharmacology studies should be considered.

11.17 Toxicity Testing: Biotechnology Products

The FDA regulates biologics as described under the 1902 Biologics Control Act, but then uses the rule-making authority granted under the Food and Drug Act to "fill in the gaps." The Bureau of Biologics, now Center for Biologics Evaluation and Research (CBER) was at one time a little-know center within the FDA that was primarily concerned with the regulation of human blood products and vaccines used for mass immunization programs. New technology created in the 1980s and certainly in the 1990s and 2000s, saw a very large increase of new therapies and development of a new industry – the biotech industry. Products, such as recombinant-DNA-produced proteins (*e.g.*, tissue plasminogen activator), biological response modifiers (cytokinins and colony-stimulating factors), monoclonal antibodies, antisense oligo-nucleotides, and self-directed vaccines (raising an immune response to self-proteins, such as gastrin, for therapeutic reasons) began to appear on the market with limited guidance on managing the safety and efficacy of these new products. Table 11.7 presents the basic test requirement matrix for biotechnology-derived therapeutics. Within the scope of this chapter, we are most concerned with the case of small molecules coupled to monoclonal antibodies (Mabs) that serve as selective carriers to target sites.

Therefore, these new products raised a variety of new questions on the appropriateness of traditional methods of evaluating drug toxicity that generated several points-to-consider documents. Some of the safety issues that arose over the years included the following:

- The appropriateness of testing a human-specific peptide hormone in nonhuman species. One must either determine/demonstrate

Table 11.7 Biotechnology-derived drugs.[a,b]

Test requirement	Species[c]
Initial clinical trial/IND requirements	
(1) Acute toxicity in rodents (intended clinical route)	R/M
(2) Acute toxicity in nonrodents (intended clinical route)	D/S/P
(3) 7-Day DRF toxicity in rodents (intended clinical route)	R/M
(4) 7-Day DRF toxicity in nonrodents (intended clinical route)	D/S/P
(5) Genotoxicity only if appropriate (special cases – nonprotein component?)	
(6) Safety pharmacology: CV *in vivo*	D/S/P
(7) Safety pharmacology: respiratory – rodent	R
(8) Pivotal/repeat dose in rodents (14[d]–28 *via* intended clinical route)	R/M
(9) Pivotal/repeat dose in nonrodents (14–28 days *via* intended clinical route)	D/P/S
(10) [e]Five species microsome metabolic panel *In vitro*	*In vitro*
(11) Develop bioanalytical for three species (man/rodent/nonrodent) NA	NA
(12) Antibody-based assay to select appropriate species	
(13) Immunotoxicity	TBD
(14) Pivotal/repeat dose in appropriate species (3/9–12 months *via* intended clinical route). Must also characterize antigenicity	M/D/P/S
To support continued clinical development	
(15) Reproductive toxicity – Seg I	R
(16) Reproductive toxicity – Seg III	R

[a]Species: R, rat; M, mouse; D, dog; S, pig; P, primate; B, rabbit; TBD, to be determined.
[b]*Abbreviations*: CV, cardiovascular; DRF, dose-range finding; IND, investigational new drug.
[c]Less than 14 days clinical use.
[d]May be required.
[e]Recommended.

that the target human pharmacologic response is also present in the relevant animal species to develop such a model species (*i.e.*, a humanized knockout mouse species) or develop an analog compound, which evokes the target receptor response in an animal species and can be evaluated for safety in parallel with the actual therapeutic molecule. Indeed, it is the interpretation of the data and the nuances of the data that led to some rather tragic consequences (*e.g.*, TGN1412).

- The potential that the peptide could break down due to nonspecific metabolism resulting in products that had no therapeutic value or even a toxic fragment.
- The potential sequelae to an immune response (formation of neutralizing antibodies, provoking an autoimmune or a hypersensitivity response), and pathology due to immune precipitation and so on.
- The presence of contamination with oncogenic virus DNA (depending on whether a bacterial or mammalian system was used on the synthesizing agent) or endotoxins.
- The difficulty interpreting the scientific relevance of response to supraphysiological systemic doses of potent biological response modifiers.

Regardless of the type of synthetic pathway, all proteins must be synthesized in compliance with GMPs. Products must be as pure as possible, not only free of rDNA but also free of other types of cell debris (endotoxin). Batch-to-batch consistency with regard to molecular structure must also be demonstrated using appropriate methods. The regulatory thinking and experience has evolved since the late 1980s document, "S6 Preclincial Safety Evaluation of Biotechnology-Derived Pharmaceuticals" prepared by the International Conferences on Harmonization. More recently, however, draft guidance from the FDA on the safety evaluation of biotechnology-derived products was issued[32,33] and then amended in 2012.[33] This new guidance applies to investigational, protein therapeutic, diagnostic, and prophylactic products derived from characterized cells through the use of expression systems, such as bacteria, yeast, insect, plant, and mammalian cells, and produced by cells in culture or by recombinant DNA technology, including transgenic plants and animals. These protein products include monoclonal antibodies, cytokines, growth factors, plasminogen activators, recombinant plasma factors, enzymes, fusion proteins, receptors, hormones, and modified toxins. However, the guidance did not cover antibiotics, allergenic extracts, heparin, vitamins, cellular blood-derived components, conventional bacterial or viral vaccines, DNA vaccines, or cellular and gene therapies. Many of the principles outlined above for small molecules apply to the safety evaluation of biotechnology products. While guidance generally exempts therapeutic proteins for evaluation for safety pharmacology endpoints, recently some FDA divisions have begun to ask for part of the IND package.

It has also been clearly demonstrated in the testing of rDNA protein products that animals will develop antibodies to foreign proteins. The safety testing of any large molecule should include the appropriate assays for determining whether the test system has developed a neutralizing antibody response.[34] Depending on the species, route of administration, intended therapeutic use, and development of neutralizing antibodies (which generally takes about two weeks), it is rare for a toxicity test on an rDNA protein to be longer than four weeks duration. However, if the course of therapy in humans is to be longer than two weeks, formation of neutralizing antibodies must be demonstrated with longer-term testing performed. The second antigen–antibody formation concern is that a hypersensitivity response will be elicited. Traditional preclinical safety assays are generally adequate to guard against this if they are two weeks or longer in duration and the relevant endpoints are evaluated.

For biotechnology products, animal models that mimic the human disease may be used to demonstrate that the product is actually able to bind to the target tissue. Indeed, a number of animal models have become available since the late 1990s for various disease states. However, it is not always clear that the pharmacological activity in a rodent model will behave similarly to that in humans. When the pharmacologic activity of a biopharmaceutical is dependent on specific drug receptor/antigen binding that is not evident in an animal species, a number of different, scientifically rational approaches may be used to obtain these data (*e.g.*, xenograft models, transgenic models). The nonhuman primate often is the animal model of choice in testing of these substances, although the rabbit and dog continue to be used.

11.18 Special Cases

11.18.1 Biologics and Combination Products

It would be easier if we could confine our considerations to just traditional small-molecule drugs, as we knew them but a few decades ago. However, these have been joined by therapeutic proteins and peptides, cell therapies, combination drug device therapeutics,[35] gene therapies, and most recently nanotherapeutics. We should expect this parade to continue through our lifetime, and bring with it new promises and new concerns.

Traditionally, and for all too long, the approach to assessing new concerns has been to retain all existing test paradigms and add more special ones thought to be specific for the newly perceived potential risks. The problem is that not only has such an approach added costs, time in development and increased needs for test animals (especially those more publicly sensitive, such as primates and dogs), but the resulting piecemeal (as opposed to integrative) evaluation has failed to substantially improve the prediction of clinical system/organism-wide effects.

11.18.2 Toxicology Testing: Special Cases

There are a number of special-case situations for preclinical development of pharmaceuticals that do not follow the "normal" pattern of development. These therapies include oral contraceptives, drugs for compassionate use, and orphan drugs. These special situations are described more completely by Gad[23] and elsewhere and are only briefly described here.

Oral Contraceptives

These have recently been modified so that in addition to those preclinical safety tests generally required, the following are also required:[36]

- A three-year carcinogenicity study in beagles (this is a 1987 modification in practice from earlier FDA requirements and the 1974 publication).
- A rat reproductive (Segment II) study including a demonstration of return to fertility.

Life-Threatening Diseases (Compassionate Use)

Drugs for life-threatening diseases are not strictly held to the sequence of testing requirements as described previously[23] because the potential benefit on any effective therapy in these situations is so high. This special case was applied to AIDS-associated diseases and cancer. The development of more effective HIV therapies (protease inhibitors) has now made cancer therapy more the focus of these considerations. Toxicity studies in animals will be required to support initial clinical trials. These studies have multiple goals:

- To determine a starting dose for clinical trials.
- To identify target organ toxicity and assess recovery.
- To assist in the design of clinical dosing regimens.

In general, it can be assumed that most antineoplastic cytotoxic agents will be highly toxic. Studies to support the clinical trial would be of 5–14 days in length, but with longer recovery periods, for example, four weeks. A study in rodents is required that identifies those doses that produce either life-threatening or nonlife-threatening toxicity. Using the information from this first study, a second study in nonrodents (generally the dog) is conducted to determine if the tolerable dose in rodents produces life-threatening toxicity. For antineoplastic agents, dosing would be done on the basis of mg m^{-2} rather than mg kg^{-1} with the staring dose in the initial clinical trial generally one-tenth of that required to produce severe toxicity in rodents or one-tenth the highest dosed in nonrodents that does not cause severe irreversible toxicity. Information on pharmacokinetic information is not usually required but including these data in the study is usually recommended. Special attention is paid to organs with high cell-division rates, such as bone marrow, testes, lymphoid tissue testing, and GI tract. As these agents are almost always given intravenously, special attention needs to be given relatively early in development to intravenous irritation and blood compatibility study, studies that always apply to pharmaceuticals given by the route of administration.

Although not required for the IND, assessment of genotoxicity and developmental toxicity will need to be addressed as the drug progresses in development. As noted above for genotoxicity, it will be important to establish the ratio between cytotoxicity and mutagenicity. *In vivo* models, for example, the mouse micronucleus test, can be particularly important in demonstrating the lack of genotoxicity at otherwise subtoxic doses. For developmental toxicity, ICH stage C–D studies (traditionally known as Segment II studies for teratogenicity in rat and rabbits) will also be necessary.

The emphasis of this discussion has been on purely cytotoxic neoplastic agents. Additional considerations must be given to cytotoxic agents that are administered under special circumstances: those that are photoactivated, delivered as liposomal emulsions, or delivered as antibody conjugates. These types of agents will require additional studies. For example, a liposomal agent will need to be compared with the free agent and a blank liposomal preparation. There are also studies that may be required for a particular class of agents. For example, anthracyclines are known to be cardiotoxic, so comparison of a new anthracycline agent to previously marketed anthracyclines will be expected.

In addition to antineoplastic, cytotoxic agents, there are cancer therapeutic or preventative drugs that are intended to be given on a chronic basis. This includes chemopreventatives, hormonal agents, immunomodulators. The toxicity assessment studies on these will more closely resemble those of more traditional pharmaceutical agents. Chronic toxicity, carcinogenicity, and full developmental toxicity (ICH stage A–B, C–D, E–F) assessments will be required. For a more complete review, the reader is referred to DeGeorge *et al.*[37] Table 11.8 presents a basic test requirement matrix for those drugs (other than oncology) intended for use in the treatment of life-threatening diseases.

Table 11.9 presents a summary of the range of special route-to-approval programs that have been developed to expedite drugs to market approval and therapeutic use for serious medical conditions.[38]

Table 11.8 Life-threatening short-use[a] indications.[b,c]

Test requirement	Species
Initial clinical trial/IND requirements	
(1) Acute toxicity in rodents (intended route)	R/M
(2) Acute toxicity in nonrodents (intended route)	D/S/P
(3) 7-Day DRF toxicity in rodents (intended route)	R/M
(4) 7-day DRF toxicity in nonrodents (intended route)	D/S/P
(5) Genotoxicity: Bacterial mutagenicity (Ames)	*In vitro*
(6) Safety pharmacology: CV/hERG[d,e]	*In vitro*
(7) Safety pharmacology: CV *in vivo*[e]	D/P/S
(8) Pivotal/repeat dose in rodents (14–28 day oral)	R/M
(9) Pivotal/repeat dose in nonrodents (14–28 day oral)	D/P/S
(10) CYP induction/inhibition[d]	*In vitro*
(11) Five species microsome metabolic panel[d]	*In vitro*
(12) Develop bioanalytical for 3 species (man/rodent/nonrodent)	NA
To support continued clinical development	
(13) Pivotal/repeat dose in rodents (3/6 month oral)[d]	R/M
(14) Pivotal/repeat dose in nonrodents (3/9–12 month oral)[f]	DP/S

[a]Less than 14 days clinical use.
[b]Species: R, rat; M, mouse; D, dog; S, pig; P, primate; B, rabbit; TBD, to be determined.
[c]*Abbreviations*: CV/hERG, Cardiovascular human ether-a-go-go; DRF, dose-range finding; IND, investigational new drug.
[d]Recommended.
[e]May be required.
[f]If the drug is not to be used for more than 28 days over a person's life span 13 and 14 are not needed.

Table 11.9 Comparison of FDA's expedited programs for serious conditions (FDA, 2014).[a]

Nature of program	Fast track Designation	Breakthrough therapy Designation	Accelerated approval Approval pathway	Priority review Designation
Reference	• Section 506(b) of the FD&C Act, as added by Section 112 of the Food and Drug Administration Modernization Act of 1997 (FDAMA) and amended by section 901 of the Food and Drug Administration Safety and Innovation Act of 2012 (FDASIA)	• Section 506(a) of the FD&C Act, as added by Section 902 of FDASIA	• 21 CFR part 314, subpart H • 21 CFR part 601, subpart E • Section 506© of the FD&C Act, as amended by Section 901 of FDASIA	• Prescription drug user fee act of 1992
Qualifying criteria	• A drug that is intended to treat a serious condition AND nonclinical or clinical data demonstrate the potential to address unmet medical need OR • A drug that has been designated as a qualified	• A drug that is intended to treat a serious condition AND preliminary clinical evidence indicates that the drug may demonstrate substantial improvement on a clinically significant	• A drug that treats a serious condition AND generally provides a meaningful advantage over available therapies AND demonstrates an effect on a surrogate endpoint that is reasonably likely to predict clinical benefit or on a clinical endpoint that can be measured earlier than irreversible morbidity or mortality (IMM) that is	• An application (original or efficacy supplement) for a drug that treats a serious condition AND, if approved, would provide a significant improvement in safety or effectiveness OR • Any supplement that proposes a labeling change pursuant to a report on a pediatric study under 505A[c] OR

	infectious disease product[b]	endpoint(s) over available therapies	reasonably likely to predict an effect on IMM or other clinical benefit (*i.e.*, an intermediate clinical endpoint)	• An application for a drug that has been designated as a qualified infectious disease product[d] OR • Any application or supplement for a drug submitted with a priority review voucher[e]
When to submit request	• With IND or after • Ideally, no later than the pre-BLA or pre-NDA meeting	• With IND or after • Ideally, no later than the end-of-phase 2 meeting	• The sponsor should ordinarily discuss the possibility of accelerated approval with the review division during development, supporting, for example, the use of the planned endpoint as a basis for approval and discussing the confirmatory trials, which should usually be already underway at the time of approval	• With original BLA, NDA, or efficacy supplement
Timelines for FDA response	• Within 60 calendar days of receipt of the request	• Within 60 calendar days of receipt of the request	• Not specified	• Within 60 calendar days of receipt of original BLA, NDA< or efficacy supplement
Features	• Actions to expedite development and review • Rolling review	• Intensive guidance on efficient drug development • Organizational commitment • Rolling review • Other actions to expedite review	• Approval based on an effect on a surrogate endpoint or an intermediate clinical endpoint that is reasonably likely to predict a drug's clinical benefit	• Shorter clock for review of marketing application (6 months compared with the 10-month standard review)[f]

Table 11.9 (*Continued*)

Nature of program	Fast track Designation	Breakthrough therapy Designation	Accelerated approval Approval pathway	Priority review Designation
Additional considerations	• Designation may be rescinded if it no longer meets the qualifying criteria for fast track[g]	• Designation may be rescinded if it no longer meets the qualifying criteria for breakthrough therapy[h]	• Promotional materials • Confirmatory trials to verify and describe the anticipated effect on IMM or other clinical benefit • Subject to expedited withdrawal	• Designation will be assigned at the time of original BLA, NDA, or efficacy supplement filing

[a]Extracted from FDA (2014b).

[b]Title VIII of FDASIA, Generating Antibiotic Incentives Now (GAIN), provides incentives for the development of antibacterial and antifungal drugs for human use intended to treat serious and life-threatening infections. Under GAIN, a drug may be designated as a qualified infectious disease product (QIDP) if it meets the criteria outlined in the statute. A drug that receives QIDP designation is eligible under the statute for fast-track designation and priority review. However, QIDP designation is beyond the scope of this guidance.

[c]Any supplement to an application under Section 505 of the FD&C Act that proposes a labeling change pursuant to a report on a pediatric study under this section shall be considered a priority review supplement per Section 505A of the FD&C Act as amended by Section 5(b) of the Best Pharmaceuticals for Children Act.

[d]See footnote a above.

[e]Any application or supplement that is submitted with a priority review voucher will be assigned a priority review. Priority review vouchers will be granted to applicants of applications for drugs for the treatment or prevention of certain tropical diseases, as defined in Sections 524(a)(3) and (a)(4) of the FD&C Act and for treatment of rare pediatric diseases as defined in Section 529(a)(3) of the FD&C Act.

[f]As part of its commitments in PDUFA V, FDA has established a review model, the Program. The Program applies to all new molecular entity NDAs and original BLAs, including applications that are resubmitted following a Refuse-to-File action, received from October 1, 2012, through September 30, 2017. For applications filed by FDA under the Program, the PDUFA review clock will begin at the conclusion of the 60 calendar day filing review period that begins on the date of FDA receipt of the original submission.

[g]A sponsor may also withdraw fast track designation if the designation is no longer supported by emerging data or the drug development program is no longer being pursued (see Section A.5. of Appendix 1).

[h]A sponsor may also withdraw breakthrough therapy designation if the designation is no longer supported by emerging data or the drug development program is no longer being pursued (see Section B.5. of Appendix 1).

Antibiotics and Anti-Infectives

Usually, the preclinical development of these agents will follow the traditional small-molecule progression. The exception to this is that for early clinical development, the duration of repeated-dose toxicity studies will need to only be up to 14 days as these agents are not often prescribed for longer than 10 days (it should be noted, however, that conducting 28-day repeat dose systemic toxicity testing may be more efficient in the long term as it will serve to support the full term of clinical evaluation). Genotoxicity studies and developmental toxicity studies will also be necessary, with genotoxicity testing required to open the IND; see above description of the testing needs for these drugs. Also, safety pharmacology studies will be required to open the IND since some of the fluoroquinolone drugs are notorious for inducing cardiovascular effects, for example, *Torsade de Pointes*. Furthermore, additional safety pharmacology studies may be necessary, particularly if there is concern with renal or gastrointestinal toxicity.

Special cases exist even within this subset – because of known class effects, fluroquinalone antibiotics require phototoxicity evaluation and more extensive cardiotoxicity evaluation prior to opening an IND and proceeding to clinical evaluation.

11.19 Imaging Agents (a Case of Direct to FIM Trials)

Imaging agents are regulated as drugs (usually by CDER, although in cases where they are seen as "tagents," which allow tracing of injected cells or tissue therapies, by CBER) and represent a special category as per testing requirements. For these regulated by CDER (traditional imaging agents that do not need to chemically interact with the body to serve their therapeutic agent), initial clinical evaluations can be legally performed without an IND (with the limitation that such a trial be at a single academic center, have Institutional Review Board (IRB) review, and be conducted in a limited number of subjects – generally in 30 or fewer).

Expanding clinical evaluation beyond the single-center structure for single clinical doses requires *in vitro* genotoxicity (mutagenicity and clastrogenicity), expanded acute in a rodent and nonrodent (along the lines outlined in the FDA exploratory IND guidance). Limited multi-dose clinical evaluation (pre-Phase III) adds in a requirement to provide data to address first their safety pharmacology endpoints

Table 11.10   Imaging agents.[a,b]

Test requirement	Species
Initial clinical trial/IND requirements	
(1) Acute toxicity in rodents (IV) – expanded acute as in Phase 0 IND guidance	R/M
(2) Acute toxicity in nonrodents (IV) – expanded acute	D/S/P
(3) Genotoxicity: Bacterial mutagenicity	*In vitro*
(4) Genotoxicity: *in vitro* clastogenicity (mammalian chromosome aberration)	*In vitro*
(5) Genotoxicity: *in vivo* (mouse or rat micronucleus)	R/M
(6) Safety pharmacology: CV *in vivo* (nonrodent)	D/P/S
(7) Safety pharmacology: respiratory – rodent	R
(8) Pivotal/repeat dose in nonrodents (14–28 day IV)	D/P/S
(9) Develop bioanalytical for 2 species (man/nonrodent)	NA
(10) Hemolysis	*In vitro*
To support continued clinical development	
(11) Reproductive toxicity – Seg I	R

[a]Species: R, rat; M, mouse; D, dog; S, pig; P, primate; B, rabbit; TBD, to be determined.
[b]*Abbreviations*: CV, Cardiovascular; IND, investigational new drug; IV, intravenous.

(CNS, respiratory, and cardiovascular, although not necessarily by distinct GLP safety pharmacology studies) and 14-day repeat-dose toxicity studies in a rodent and nonrodent species. Table 11.10 provides the test requirement scheme for imaging agents.

11.20   Excipients

Novel excipients, those for which there is no existing DMF or evidence of use in approved pharmaceutical products (inactive components) that are to be used in formulations administered to humans must now be qualified in accordance with ICH guidance. These guidances set requirements that are identical to what is required for any other new chemical entity (NCE). In practice, for opening an IND these requirements are generally met by (i) supplying baseline genetic toxicity data for the excipient, and (ii) conducting the "pivotal" (14 or 28 days) repeat-dose toxicity studies using the clinical formulation. Table 11.11 presents the ICH guideline requirements for the development and approval of novel excipients.

Pediatric Claims and Juvenile Animal Studies

Relatively few drugs marketed globally (approximately 20%) have pediatric dosing information available. Clinical trials had rarely been

Table 11.11 Excipients (table entries for oral agents—if for use *via* other, that route used for *in vivo* studies).[a,b]

Test requirement	Species[c]
Initial clinical trial/IND requirements	
(1) Acute toxicity in rodents (Oral and IV[d])	R/M
(2) Acute toxicity in nonrodents (Oral)	D/S/P
(3) 7-Day DRF toxicity in rodents (Oral)	R/M
(4) 7-Day DRF toxicity in nonrodents (Oral)	D/S/P
(5) Genotoxicity: Bacterial mutagenicity (Ames)	*In vitro*
(6) Genotoxicity: *in vitro* clastogenicity (CHO chromosome aberration)	*In vitro*
(7) Genotoxicity: *in vivo* (mouse or rat micronucleus)	R/M
(8) [d]Safety pharmacology: CV-hERG	*In vitro*
(9) Safety pharmacology: CV *in vivo*	D/S/P
(10) Safety pharmacology: functional observation battery/Irwin	R/M
(11) Safety pharmacology: Respiratory-Rodent	R
(12) Pivotal/repeat dose in rodents (14–28 day oral)	R/M
(13) Pivotal/repeat dose in nonrodents (14–28 day oral)	D/S/P
(14) [d]CYP induction/inhibition	*In vitro*
(15) [d]Five species microsome metabolic panel	*In vitro*
(16) Develop bioanalytical for 3 species (man/rodent/nonrodent)	NA
To support continued clinical development	
(17) [e]Development Tox (Seg II) – rate and rabbit pilots and rat and rabbit studies	R/B
(18) [e]Immunotoxicity	TBC
(19) [c]Pivotal/repeat dose in rodents (3/6 month oral)	R/M
(20) [c]Pivotal/repeat dose in nonrodents (3/9–12 month oral)	D/S/P
To support marketing approval	
(21) [e]Reproductive Toxicity – Seg I	R
(22) [e]Tumorgenicity/carcinogenicity – rat	R

[a]Species: R, rat; M, mouse; D, dog; S, pig; P, primate; B, rabbit; TBD, to be determined.
[b]*Abbreviations*: CV-hERG, human ether-a-go-go; ING, investigational new drug.
[c]Less than 14 days clinical use.
[d]Recommended.
[e]May be required.

done specifically on pediatric patients. Traditionally, dosing regimens for children have been derived empirically by extrapolating on the basis of body weight or surface area. This approach assumes that the pediatric patient is a young adult, which simply may not be the case. There are many examples (acetaminophen, fluoro-quinolones) of how adults and children differ qualitatively or quantitatively in metabolic and/or pharmacodynamic responses to pharmaceutical agents.[39] In response to the pediatric initiatives, the FDA published policies and guidelines (http//www.fda.gov/cder/pediatric), although the focus of the initiatives had been directed toward clinical trials with limited preclinical toxicology information. In 2006, however, the Agency developed a pediatric guidance that describes the timing for initiation of these studies as well as recommendations on study regimen. Although testing in juvenile animals as a basis for pediatric approvals tends to occur after development of the adult formulation, the sponsor will be required to submit an IND for undertaking a clinical trial in a pediatric population. Pending globally is a new nonclinical safety guidance (ICH S11) addressing pediatric/juvenile toxicity testing.

The FDA-designated levels of postnatal human development and the approximate equivalent ages in various animal models with a comparison to the human for different organ systems provided in the guidance are shown in Table 11.12. The table is not completely accurate, however, because of differences in the stages of development at birth. A rat is born quite underdeveloped when compared with a human being. Therefore, there is no consistent temporal relationship in developing ages of animals compared with humans.[40,41]

In designing the juvenile toxicology study, several considerations are important and include the following:

- the timing of dosing in relation to phases of growth and development in pediatric populations and juvenile animals;

Table 11.12 Cross-species comparative age categories.

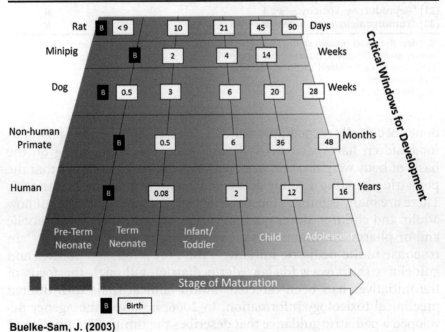

Buelke-Sam, J. (2003)

- the potential differences in pharmacological and toxicological profiles between mature and immature animal models;
- any established temporal developmental differences in animals relative to pediatric populations;
- juvenile animals generally undergo more dynamic development than is seen in the relatively stable adult.

In these studies, juvenile rats and dogs are the preferred species, although pigs, nonhuman primates, and other appropriate species have been used. The duration of testing must consider the length of the treatment period, balancing the developmental age of the animal model and the proposed length of clinical treatment. Where appropriate, changes that will need to be considered for the toxicology study include the developmental landmarks as well as the more standard indicators of target organ toxicity.

11.21 Nonclinical Safety Evaluation Studies Conducted to Support Continued Clinical Development

Once a candidate drug has proceeded through the opening of an INDA or CTA and has been assessed in initial clinical studies, the focus of the nonclinical safety assessment efforts shifts to performing studies to enable longer and more extensive studies (including a wider variety of human subjects).

Much of what must be considered here comes from the ICH M3 Guidance ("Nonclinical Safety Studies for the Conduct of Human Clinical Trials and Marketing Authorization for Pharmaceuticals"), which has recently (August, 2008) been revised (R2).[18]

Repeat-Dose Toxicity

The regulatory expectation is that a new therapeutic must be evaluated for a longer term of exposure than that to which clinical subjects are exposed, up to the limits regulatory expectations as displayed in Table 11.3. Note that requirements are somewhat different for those required to support clinical trials (Table 11.13) and those that, for the same drugs, would be required to support a marketing application (Table 11.14).

These same systematic studies must be conducted in compliance with the other expectation principles: at least three different dose groups must be included in the study, and the high-dose group (at minimum) should serve to identify a target organ toxicity.

Table 11.13 Duration of repeated dose toxicity studies to support the conduct of clinical trials in all regions.

Maximum duration of clinical trial	Maximum duration of repeated dose toxicity studies to support clinical trials	
	Rodents	Nonrodents
Up to 2 weeks	2 weeks	2 weeks
Between 2 weeks and 6 months	Same as clinical trial	Same as clinical trial
>6 months	6 months	9 months[a]

[a]Except perhaps in the United States.

Table 11.14 Duration of repeated dose toxicity studies to support marketing in all regions.

Duration of indicated treatment	Rodent	Nonrodent
Up to 2 weeks	1 month	1 month
>2 weeks to 1 month	3 months	3 months
>1–3 months	6 months	6 months
>3 months	6 months	9 months[a]

[a]Except perhaps in the United States

Currently, these requirements are common across the ICH-conforming community nations. The sole exception is that the US FDA may require 12-month (instead of nine-month) nonrodent studies.

Developmental and Reproductive Toxicology (DART)

There are three separate considerations under DART study timing. The first is for male fertility studies (the classical Segment 1 studies). Such studies (across ICH regions) should be conducted prior to the initiation of Phase III studies. It should be noted, however, that male reproductive organs are evaluated histopathologically in the repeat-dose studies (typically 28-day) conducted prior to human studies and in all subsequent repeat-dose systemic toxicity studies.

The second is for women not of child-bearing potential (*i.e.*, those who have been sterilized). Inclusion of such individuals is governed by the conduct of suitable repeat-dose studies, and there are no explicit DART testing requirements.

Inclusion of women of child-bearing potential is much more restricted (primarily due to developmental toxicity concerns). They may be included in short-term (up to two weeks of treatment) clinical trial with adequate informed consent, pregnancy testing prior to inclusion, and use of a highly effective method of birth control. Note that no DART studies are thus conducted beforehand and that this really applies to women in traditional Phase I clinical studies.

With the completion of adequate developmental toxicity studies (segment II studies, and therefore also the required pilot studies before these in two species), the inclusion of up to 150 women of child-bearing potential may be included in trials, having up to three months of dose administration of the drug. Such segment II studies must be completed prior to the inclusion of women in Phase II studies in the EU and Japan, but not the United States. To all venues, female fertility (segment III) studies must be completed prior to initiation of Phase III clinical studies.

If standard toxicity studies (as described under ICH S8) yield a result identifying the need for an additional study to evaluate immunotoxicity, such studies must be performed prior to the initiation of clinical studies.

11.22 Impurities, Degradants, and Residual Solvents

11.22.1 Impurities

The ICH *Guidance for Industry, Q3A Impurities in New Drug Substances*,[42] published in February 2003, is intended to "provide guidance for registration applications on the content and qualification of impurities in new drug substances produced by chemical syntheses and not previously registered in a region or member state." A new drug substance is not the final marketed product, but the active ingredient used in the marketed product. Impurities in new drug substances are addressed from both a chemistry and safety perspective.

The guidance is not intended to apply to new drug substances during the clinical research stage of development (although such drugs in development must have consideration of meeting these requirements at the time of marketing approval and addresses safety concerns associated with such substances during development) nor does it cover natural product or biological process produced drugs or extraneous contaminants that should not occur in new drug substances and are more appropriately addressed as GMP issues. The guidance further describes the circumstances in which impurities need to be reported, identified, and qualified.

The rationale for the reporting and control, identification, and qualification of impurities is discussed in the guidance. Organic impurities need to be summarized based on the actual and potential

impurities most likely to arise during the synthesis, purification, and
storage of a new drug substance. This discussion can be limited to
those impurities that might reasonably be expected based on knowledge of the chemical reactions and conditions involved.

Studies conducted to characterize the structure of impurities present in a new drug substance at a level greater than the identification
threshold should be described and any impurity from any batch or
degradation product from stability studies should be identified. If
identification of an impurity or degradant is not feasible, a summary
of the laboratory studies demonstrating the unsuccessful effort
should be included in the application. If an impurity is pharmacologically or toxicologically active, identification of the compound
should be conducted even if the impurity level is below the identification threshold. Table 11.15 presents the thresholds for taking such
actions.

The guidance also states that "qualification is the process of
acquiring and evaluating data that establishes the biological safety
of an individual impurity or a given impurity profile at the level(s)
specified. The applicant should provide a rationale for establishing
impurity acceptance criteria that includes safety considerations. The
level of any impurity that is present in a new drug substance that has
been adequately tested in safety and/or clinical studies would be
considered qualified. Impurities that are also significant metabolites present in animal and/or human studies are generally considered qualified. A level of a qualified impurity higher than that
present in a new drug substance can also be justified based on an
analysis of the actual amount of impurity administered in previous
relevant safety studies. If data are unavailable to qualify the proposed acceptance criterion of an impurity, safety studies to obtain
such data can be appropriate when the usual qualification thresholds are exceeded."

Table 11.15 Thresholds for action on impurities in a drug product.

Maximum daily dose[a]	Reporting threshold[b,c]	Identification threshold[c]	Qualification threshold[c]
≤2 g day^{-1}	0.05%	0.10% or 1.0 mg day^{-1} intake (whichever is lower)	0.15% or 1.0 mg day^{-1} intake (whichever is lower)
>2 g day^{-1}	0.03%	0.05%	0.05%

[a]The amount of drug substance administered per day.
[b]Higher reporting thresholds should be scientifically justified.
[c]Lower thresholds can be appropriate if the impurity is unusually toxic.

Q3B(R) describes considerations for the qualification of impurities when thresholds are exceeded.[43] If the level of impurity cannot be decreased to below the threshold, or if adequate data is not available in the scientific literature to justify safety, then additional safety testing should be considered. The studies considered appropriate to qualify an impurity will depend on a number of factors, including the patient population, daily dose, and route and duration of administration. Toxicology studies are discussed briefly later in this chapter and in more detail in other chapters in this volume. Such studies can be conducted on the new drug substance containing the impurities to be controlled, although studies using isolated impurities can sometimes be appropriate.

ICH Q3A states that "safety assessment studies to qualify an impurity should compare the new drug substance containing a representative amount of the new impurity with previously qualified material. Safety assessment studies using a sample of the isolated impurity can also be considered."[42] The latter is especially important to consider for genetic toxicology studies and the importance of testing the isolated impurity is discussed in more detail at the end of this chapter.

Therefore, according to the guidance, if the maximum daily dose of the drug is less than 2 g day^{-1} and the impurity intake is more than 0.15% or 1.0 mg day^{-1}, the qualification threshold has been reached, meaning safety studies will need to be performed. Lower thresholds can be appropriate if the impurity is unusually toxic. In addition, the impurity will need to be reported and identified. These studies include general and genetic toxicology studies, and possibly other specific toxicology endpoints, as appropriate. Discussion of specific toxicity testing with the relevant FDA division is recommended.

If considered desirable, a minimum screen (*e.g.*, genotoxic potential) should be conducted. A study to detect point mutations and one to detect chromosomal aberrations, both *in vitro*, are considered an appropriate minimum screen.

Qualification studies for impurities are essentially bridging studies. If general toxicity studies are desirable, one or more studies should be designed to allow comparison of unqualified to qualified material. The study duration should be based on available relevant information and performed in the species most likely to maximize the potential to detect the toxicity of a degradation product. On a case-by-case basis, single-dose studies can be appropriate, especially for single-dose drugs. In general, a minimum duration of 14 days and a maximum duration of 90 days (always in rodents) would be considered appropriate.

The genetic toxicology studies can include a minimum screen (a study to detect point mutations and one to detect chromosome aberrations, both *in vitro*). The general toxicology studies should include one or more studies designed to allow comparison of unqualified to qualified material. The study duration should be based on available relevant information and performed in the species most likely to maximize the potential to detect the toxicity of an impurity. On a case-by-case basis, single-dose studies can be appropriate, especially for single-dose drugs. In general, a minimum duration of 14 days and a maximum duration of 90 days would be considered appropriate.

Inorganic impurities are normally detected and quantified using pharmacopoeial or other appropriate procedures. The need for inclusion or exclusion of inorganic impurities in a new drug substance specification should be discussed. Acceptance criteria should be based on pharmacopoeia standards or known safety data. The control of residues of the solvents used in the manufacturing process for a new drug substance should be discussed and presented according to ICH Q3C.[44]

The ICH Q3A guidance also states that the specification for a new drug substance should include a list of impurities. Individual impurities with specific acceptance criteria included in the specification for a new drug substance are referred to as specified impurities. Specified impurities can be identified or unidentified. A rationale for the inclusion or exclusion of impurities in a specification should be presented. ICH *Q3B(R) Impurities in New Drug Products*,[43] was published in June 2006 and is intended to provide guidance for registration applications on the content and qualification of impurities in new drug products produced from chemically synthesized new drug substances not previously registered in a region or member state. A new drug product is a finished dosage form, for example, a tablet, capsule, or solution, that contains a drug substance, generally, but not necessarily, in association with one or more other ingredients. The Q3B(R) guidance complements the ICH guidance *Q3A Impurities in New Drug Substances*,[42] which should be consulted for basic principles along with ICH *Q3C Impurities: Residual Solvents*,[44] when appropriate (Table 11.16). There is also now a guidance for levels of elements (metals and boron) – ICH Q3D.

Q3A addresses only those impurities in new drug products classified as degradation products of the drug substance, or reaction products of the drug substance with an excipient and/or immediate container closure system (collectively referred to as degradation products). Generally, impurities present in a new drug substance need

Table 11.16 Threshold for degradation products in new drug products.[a]

Maximum daily dose[b]	Threshold[c,d]
Reporting thresholds	
≤1 g	0.1%
>1 g	0.05%
Identification thresholds	
<1 mg	1.0% or 5 µg TDI, whichever is lower
1–10 mg	0.5% or 20 mg TDI, whichever is lower
>10 mg to 2 g	0.2% or 2 mg TDI, whichever is lower
>2 mg	0.10%
Qualification thresholds	
<10 mg	1.0% or 50 µg TDI, whichever is lower
10–100 mg	0.5% or 200 µg TDI, whichever is lower
>100 mg to 2 g	0.2% or 3 mg TDI, whichever is lower
>2 g	0.15%

[a]*Abbreviation*: TDI, total daily intake.
[b]The amount of drug substance administered per day.
[c]Thresholds for degradation products are expressed either as a percentage of the drug substance or as TDI of the degradation product. Lower thresholds can be appropriate if the degradation product is unusually toxic.
[d]Higher thresholds should be scientifically justified.

not be monitored or specified in new drug product unless they are also degradation products. This guidance does not address impurities arising from excipients present in a new drug product or extracted or leached from the container closure system. This guidance also does not apply to new drug products used during the clinical research stages of development. It also does not cover the same types of products as in 3QA(R): biological/biotechnological, peptides, oligonucleotides, radio-pharmaceuticals, fermentation products and associated semisynthetic products, herbal products, and crude products of animal or plant origin. Also excluded from this guidance are extraneous contaminants that should not occur in new drug products and are more appropriately addressed as GMP issues, polymorphic forms, and enantiomeric impurities. Separate but related issues are leachables and extractables associated with a container or delivery system for the drug.[45–48]

11.22.2   Residual Solvents

ICH Q3C is intended to provide guidance for recommending acceptable amounts for residual solvents in pharmaceuticals for the safety of the patient. The guidance recommends use of less-toxic solvents and describes levels considered to be toxicologically acceptable for some residual solvents. A complete list of the solvents included in this guidance is provided in a companion document entitled ICH *Q3C-Tables and List*,[49] which can be found at the ICH or FDA website. The list is not exhaustive, and other solvents may be used and later added to the list.

Residual solvents in pharmaceuticals are defined here as organic volatile chemicals that are used or produced in the manufacture of drug substances or excipients, or in the preparation of drug products. The solvents are not completely removed by practical manufacturing techniques. Appropriate selection of the solvent for the synthesis of drug substance may enhance the yield, or determine characteristics such as crystal form, purity, and solubility. Therefore, the solvent may sometimes be a critical parameter in the synthetic process. This guidance does not address solvents deliberately used as excipients nor does it address solvates. However, the content of solvents in such products should be evaluated and justified.

As there are no therapeutic benefits from residual solvents, all residual solvents should be removed to the extent possible to meet product specifications, GMPs, or other quality-based requirements. Drug products should contain no higher levels of residual solvents than can be supported by safety data. Some solvents that are known to cause unacceptable toxicities (carcinogens), such as benzene and carbon tetrachloride (Class 1, as per Table 1 in ICH Q3C), should be avoided in the production of drug substances, excipients, or drug products unless their use can be strongly justified in a risk–benefit assessment. Some solvents associated with less severe toxicity (nongenotoxic animal carcinogens or possible causative agents of other irreversible toxicity, such as neurotoxicity or teratogenicity), such as acetonitrile and chlorobenzene (Class 2), should be limited in order to protect patients from potential adverse effects. Ideally, less-toxic solvents, such as acetic acid ad acetone (Class 3), should be used where practical.

This guidance does not specifically apply to potential new drug substances, excipients, or drug products used during the clinical research stages of development, nor does it apply to previously existing marketed drug products.

The guidance applies to all dosage forms and routes of administration. Higher levels of residual solvents may be acceptable in certain cases, such as short-term (30 days or less) or topical application. Justification for these levels should be made on a case-by-case basis and discussed with the appropriate FDA division.

The limits of residual solvents may include a value for the permitted daily exposure (PDE), which is the maximum acceptable intake per day of residual solvent in pharmaceutical products. These limits vary depending on the class.

For solvents where quantities are limited to set values in pharmaceutical products because of their inherent toxicity, the Class 2 list within the ICH Q3H (guidance) should be consulted. PDEs are given to the nearest 0.1 mg day^{-1}, and concentrations are given to the nearest 10 ppm.

For solvents with low toxic potential, solvents in Class 3 may be regarded as less toxic and of lower risk to human health. Class 3 includes no solvent known as a human health hazard at levels normally accepted in pharmaceuticals. However, there are no long-term toxicity or carcinogenicity studies for many of the solvents in Class 3. Available data indicate that they are less toxic in acute or short-term studies and negative in genotoxicity studies. It is considered that amounts of these residual solvents of 50 mg day^{-1} or less (corresponding to 5000 ppm or 0.5% under Option 1) would be acceptable without justification. Higher amounts may also be acceptable provided they are realistic in relation to manufacturing capability and GMPs.

For solvents for which no adequate toxicological data were found, the solvents listed may also be of interest to manufacturers of excipients, drug substances, or drug products. However, no adequate toxicological data on which to base a PDE were previously found for these. Manufacturers should supply justification for residual levels of these solvents in pharmaceutical products.

Acceptable exposure levels in this guidance for Class 2 solvents were established by the calculation of PDE values according to the procedures for setting exposure limits in pharmaceuticals (Pharmacopeial Forum, Nov–Dec 1989), and the method adopted by IPCS for Assessing Human Health Risk of Chemicals (EHC 170, WHO, 1994). These methods are similar to those used by the U.S. EPA (IRIS), the U.S. FDA (Red Book), and others. The method is outlined here to give a better understanding of the origin of the PDE values. It is not necessary to perform these calculations in order to use the PDE values provided in the lists of the ICH Q3C document.

$$PDE = \frac{NOEL \times Weight\ Adjustment}{F1 \times F2 \times F3 \times F4 \times F5} \tag{11.1}$$

PDE is derived from the no effect level (NOEL) or the LOEL in the most relevant animal study as follows:

The PDE is derived preferably from an NOEL. If no NOEL is obtained, the LOEL may be used. Modifying factors proposed here, for relating the data to humans, are the same kind of uncertainty factors used in EHC (EHC 170, WHO, Geneva, 1994), and modifying factors or safety factors in Pharmacopeial Forum. The assumption of 100% systemic exposure is used in all calculations regardless of route of administration.

The modifying factors are as follows:

F1 = A factor to account for extrapolation between species. F1 = 5 for extrapolation from rats to humans.

F1 = 12 for extrapolation from mice to humans. F1 = 2 for extrapolation from dogs to humans.

F1 = 2.5 for extrapolation from rabbits to humans. F1 = 3 for extrapolation from monkeys to humans.

F1 = 10 for extrapolation from other animals to humans.

F1 takes into account the comparative surface area : body weight ratios for the species concerned and for man.

Surface area (S) is calculated as:

$$S = kM0.67$$

in which M = body mass, and the constant k has been taken to be 10. The body weights used in the equation are those shown below in the included table.

F2 = A factor of 10 to account for variability between individuals.

A factor of 10 is generally given for all organic solvents, and 10 is used consistently in this guidance.

F3 = A variable factor to account for toxicity studies of short-term exposure.

F3 = 1 for studies that last at least one half-lifetime (one year for rodents or rabbits; seven years for cats, dogs, and monkeys).

F3 = 1 for reproductive studies in which the whole period of organogenesis is covered.

F3 = 2 for a six-month study in rodents, or a 3.5-year study in nonrodents.

F3 = 5 for a three-month study in rodents, or a two-year study in nonrodents.

F3 = 10 for studies of a shorter duration.

In all cases, the higher factor has been used for study durations between the time points (e.g., a factor of 2 for a nine-month rodent study).

F4 = A factor that may be applied in cases of severe toxicity (*e.g.*, nongenotoxic carcinogenicity, neurotoxicity, or teratogenicity). In studies of reproductive toxicity, the following factors are used:
F4 = 1 for fetal toxicity associated with maternal toxicity. F4 = 5 for fetal toxicity without maternal toxicity.
F4 = 5 for a teratogenic effect with maternal toxicity.
F4 = 10 for a teratogenic effect without maternal toxicity.
F5 = A variable factor that may be applied if the NOEL was not established.

When only an LOEL is available, a factor of up to 10 could be used depending on the severity of the toxicity. The weight adjustment assumes an arbitrary adult human body weight for either gender of 50 kg. This relatively low weight provides an additional safety factor against the standard weights of 60 kg or 70 kg that are often used in this type of calculation. It is recognized that some adult patients weigh less than 50 kg; these patients are considered to be accommodated by the built-in safety factors used to determine a PDE. If the solvent was present in a formulation specifically intended for pediatric use, an adjustment for a lower body weight would be appropriate.

As an example of the application of this equation, consider a toxicity study of acetonitrile in mice that is summarized in *Pharmeuropa*, Vol. 9, No. 1, Supplement, April 1997, page S24. The NOEL is calculated to be 50.7 mg kg^{-1} day^{-1}. The PDE for acetonitrile in this study is calculated as follows:

$$\frac{50.7 \text{ mg kg}^{-1} \text{ day}^{-1} \times 50 \text{ kg}}{}$$
$$PDE = 12 \times 10 \times 5 \times 1 \times 1 = 4.22 \text{ mg day}^{-1}.$$

In this example,

F1 = 12 to account for the extrapolation from mice to humans.
F2 = 10 to account for differences between individual humans.
F3 = 5 because the duration of the study was only 13 weeks.
F4 = 1 because no severe toxicity was encountered. F5 = 1 because the NOEL was determined.

11.23   Residual Metals and Metal Catalysts

If synthetic processes of pharmaceutical substances are known or suspected to lead to the presence of metal residues due to the use of a specific metal catalyst or metal reagent, a concentration limit and validated test for residues of each specific metal should be set. All concentration limits should be realistic in relation to analytical precision, manufacturing capability, and reasonable variation in the manufacturing process. Since the use of metal catalysts or metal reagents during synthesis is restricted to validated and controlled chemical reactions, limitation of their residues in pharmaceutical substances itself will normally be sufficient. A limit for a metal residue in the pharmaceutical substance may, however, be replaced by a limit for that metal residue in the final medicinal product, as described below.

In early 2008, the EMA promulgated a standard for metals as impurities in pharmaceuticals.[50] They organized metals of concern into categories, as presented in Table 11.17. This subsequently formed the basis for the ICH guidance on elements (metals plus boron) ICH Q3D, adopted in December of 2014.

For pharmaceutical products administered *via* the oral, parenteral, or inhalation route of administration, these guidances are presented in Table 11.18. Two options are available when setting a concentration limit for a metal residue.

Option 1: For each metal, the concentration limit in parts per million (ppm) as stated in Table 11.1 can be used. The concentration limits in the guidance (Table 11.15) have been calculated using eqn (11.2) below by assuming a daily dose of 10 g of the drug product.

$$\text{Concentration ppm} = \text{PDE } (\mu\text{g day}^{-1}) \text{ daily dose (g day}^{-1}) \qquad (11.2)$$

If all pharmaceutical substances in a drug product meet option 1 concentration limit for all metals potentially present, then all these substances may be used in any proportion in the drug product as long as the daily dose of the drug product does not exceed 10 g day^{-1}. When the daily dose of the drug product is greater than 10 g day^{-1}, option 2 should be applied.

Option 2a: The PDE in terms of μg day^{-1} can be used together with the actual daily dose of a pharmaceutical substance in the drug product to calculate the concentration of residual metal allowed in that pharmaceutical substance.

Option 2b: Alternatively, it is not considered necessary for each pharmaceutical substance to comply with the limits given in option 1 or the calculated limits using option 2a.

Table 11.17 Class exposure and concentration limits for individual metal catalysts and metal reagents.[a]

Classification	Oral exposure		Parenteral exposure		Inhalation exposure[b]
	PDE (μg day^{-1})	Concentration (ppm)	PDE (μg day^{-1})	Concentration (ppm)	PDE (ng day^{-1})
Class 1A: Pt, Pd	100	10	10	1	Pt: 70[b]
Class 1B: Ir, Rh, Ru, Os	100[c]	10[c]	10[c]	1[c]	
Class 1C: Mo, Ni, Cr, V	250	25	25	2.5	Ni: 100 Cr(NI): 10
Metals of significant safety concern					
Class 2: Cu, Mn	2500	250	250	25	
Metals with low safety concern					
Class 3: Fe, Zn	13 000	1300	1300	130	
Metals with minimal safety concern					

[a]*Abbreviation*: PDE, permitted daily exposure.
[b]Pt as hexachloroplatinic acid.
[c]Subclass limit: the total amount of listed metals should not exceed the directed limit.

Table 11.18   Recommendation for consideration during risk assessment.

Element	Class	If intentionally added (across all routes of administration)	If not intentionally added Oral	Parenteral	Inhalation
As	1	Yes	Yes	Yes	Yes
Cd	1	Yes	Yes	Yes	Yes
Hg	1	Yes	Yes	Yes	Yes
Pb	1	Yes	Yes	Yes	Yes
Co	2A	Yes	Yes	Yes	Yes
Mo	2A	Yes	Yes	Yes	Yes
Se	2A	Yes	Yes	Yes	Yes
V	2A	Yes	Yes	Yes	Yes
Ag	2B	Yes	No	No	No
Au	2B	Yes	No	No	No
Ir	2B	Yes	No	No	No
Os	2B	Yes	No	No	No
Pd	2B	Yes	No	No	No
Pt	2B	Yes	No	No	No
Rh	2B	Yes	No	No	No
Ru	2B	Yes	No	No	No
Tl	2B	Yes	No	No	No
Ba	3	Yes	No	No	Yes
Cr	3	Yes	No	No	Yes
Cu	3	Yes	No	Yes	Yes
Li	3	Yes	No	Yes	Yes
Ni	3	Yes	No	Yes	Yes
Sb	3	Yes	No	Yes	Yes
Sn	3	Yes	No	Yes	Yes

11.24   Safety Assessment of Metabolites

Generally, measurements of circulating concentrations of a parent drug in animals are used as an index of systemic exposure in humans. Quantitative and qualitative differences in metabolite profiles are important when comparing exposure and safety of a drug in a non-clinical species relative to humans during risk assessment. Based on data obtained from *in vitro* and *in vivo* metabolism studies, when the metabolic profile of a parent drug is similar qualitatively and quantitatively across species, we can generally assume that potential clinical risks of the parent drug and its metabolites have been adequately characterized during standard nonclinical safety evaluations. However, metabolic profiles and metabolite concentrations can vary across species, and there are cases when clinically relevant metabolites have not been identified or adequately evaluated during nonclinical safety studies. This may be because the metabolite being formed in humans was absent in the animal test species (unique human metabolite) or because the metabolite was present at much higher levels in humans (major metabolite) than in the species used for drug standard toxicity testing.

The FDA and ICH recommend that, and this guidance encourages, attempts be made to identify as early as possible during the drug development process, the differences in drug metabolism in animals used in nonclinical safety assessments compared with humans. It is especially important to identify metabolites that may be unique to humans. The discovery of unique or major human metabolites late in drug development can cause development delays and could have possible implications for marketing approval. Early identification of unique or major metabolites will allow for timely assessment of potential safety issues.

Generally, I recommend that metabolites identified in human plasma that account for greater than 10% of drug-related material (administered dose or systemic exposure whichever is less) be considered for safety assessment. The rationale for setting the level at greater than 10% for characterization of metabolites reflects consistency with other FDA and EPA regulatory guidance[51,52] and is supported by actual cases, described below, in which it has been determined that the toxicity of a drug could be attributed to one or more metabolites present at greater than 10% of the administered dose. Of the cases that follow, the last two are examples of a situation when a metabolite present at less than 10% caused toxicity. As a result, depending on the situation, some metabolites present at less than 10% should be tested.

The objectives of standard nonclinical safety studies are to evaluate the general toxicity profile of a drug and its metabolites in rodent and nonrodent animal species and to assess the potential for genotoxicity in support of Phase I safety and tolerability studies in humans. Metabolism studies are generally performed through a combination of *in vitro* studies using animal and human tissues and *in vivo* studies in animals. The *in vitro* studies are generally conducted prior to the *in vivo* studies and provide an initial comparative metabolic profile. Results from these studies can assist in the selection of the appropriate animal species for toxicological assessments, should qualitative interspecies differences in metabolism be detected.

Identifying a major metabolite in animals that does not exist in humans can mean that toxicity observed in that animal species may not be relevant to humans. Conversely, identifying a human metabolite during clinical development that did not form at appreciable levels in animals would raise safety concerns because it probably was not evaluated in the nonclinical studies to inadequate exposure.

Additionally, when a potentially clinically relevant toxicity is observed during standard nonclinical studies, it is prudent to determine if metabolites contribute to that finding. In such cases, we recommend that the metabolites be synthesized and directly administered to the appropriate animal species for further pharmacological/toxicological evaluation. When qualitative and/or quantitative species differences in metabolite profiles are discovered, we also recommend investigation of different routes of administration or use of alternative animal species for safety assessments. Discovery of such a metabolite could delay development until the relationship between metabolite exposure and toxicity (if any) is understood.

In vitro studies using liver slices, microsomes, or hepatocytes from animals and humans to identify the drug metabolic profile are generally conducted before initiation of clinical trials. It is important to also try to determine whether the concomitant use of drugs results in the inhibition or the induction of common metabolic pathways. *In vivo* metabolic profiles in nonclinical test species are generally available early in drug development, and their results may reveal significant quantitative and/or qualitative differences in metabolism across species. However, a unique metabolite may only be recognized after completion of *in vivo* metabolic profiling in humans. Therefore, we recommend the *in vivo* metabolic evaluation in humans be performed as early as feasible.

In general, systemic exposure to metabolites varies among species, and it is uncommon for humans to form unique metabolites. Therefore, identification of major human metabolites at levels higher than those measured in the test species used for toxicological assessment is of serious concern. For metabolites detected in humans as well as in nonclinical species (although at lower levels in the latter) adequacy of exposure should be considered on a case-by-case basis. Generally, systemic exposure is assessed by measuring the concentration of the compound in serum or plasma. However, when measurements cannot be made in plasma for any one or a number of reasons, measurements can be made in other biological matrices, such as urine, feces, or bile. Noncirculating metabolites (*i.e.*, excreted in bile or urine) are sometimes identified before clinical trials, but are not usually monitored. It is quite likely that excreted metabolite levels may be more appropriate metrics in many instances. For example, if Phase II conjugation products of a metabolite are present in the excreta, it can be assumed that systemic exposure to the metabolite has occurred. I recommend consulting the ICH Q3A guidance with regard to the development of analytical methods for measuring metabolites in selected matrices. If the systemic exposure in nonclinical species is equivalent to human exposure when measured in plasma and/or excreta, levels may be considered sufficient and alleviate the need for additional toxicity testing. I encourage contacting the Agency early in drug development to discuss these issues.

Early identification of unique human or major metabolites can provide clear justification for nonclinical testing in animals, assist in planning and interpreting clinical studies, and prevent delays in drug development. Sponsors are encouraged to conduct *in vitro* studies to identify and characterize unique human or major metabolites early in drug development. If toxicity studies of human metabolites are warranted, we recommend studies be completed and the study reports be submitted to the Agency before beginning large-scale Phase III trials. In some cases, it may be appropriate for these nonclinical safety studies with unique human metabolites to be conducted before Phase III studies; for example, (i) if the metabolite belongs to a chemical class with known toxicity; (ii) if the metabolite has positive structural alerts for genotoxicity, carcinogenicity, or reproductive toxicity; or (iii) if clinical findings suggest the metabolite or related compounds have indicated special clinical safety concerns, such as QT prolongation.

To optimize and expedite development of drugs for serious and expedite development of drugs for serious or life-threatening diseases that lack an approved effective therapy, the number of nonclinical studies for the unique or major human metabolites may be limited on a case-by-case basis. We recommend sponsors contact the relevant review division to discuss such a situation.

11.25 Local Tissue Tolerance

Local tissue tolerance of the intended clinical formulation to a parenteral route should be evaluated prior to initiation of Phase III studies. Such studies are specified both in relevant pharmacopeia (such as the USP) and in M3 (R2).[18]

Subset: Special or Hazard Studies

All that has been presented so far are appropriate and required for almost all new therapeutics. By the routes other than oral, however, there are additional expected studies that address issues of local tissue response to administered clinical dosage form (or drug product).

What is to be tested is the clinical formulation about to be evaluated in humans. Because the formulation may change several times over the course of clinical development, these tests may need to be repeated several times (for each new formulation). The tests are truly hazard tests – they are generally performed with a strictly defined protocol, with results being evaluated using a set in accordance with subjective preclinical scale, against which it is determined to be pass or fail. The tests include studies, such as hemolysis (for intravenous products), pyrogenicity (for parenteral products), sensitization (for dermal products), route-specific irritation assays (eye, skin, muscle, mucosal, nasal, and so on) and such studies as phototoxicity.[53]

These studies are expected but the expectations are not clearly spelled out in any single guidance. To some degree they are desirable not in a regulatory guidance at all, but rather in the appropriate pharmacopoeia (in the UK the British Pharmacopeia; the US this is USP).

11.26   Supporting Marketing Applications

Attrition of the therapeutic candidates getting to this point is quite high with perhaps only one of every 200 compounds designated as candidates proceeding beyond Phase I. However, there may well be studies, which were desirable before but deferred (CYP inhibition and induction, identification of major metabolites, abuse liability potential, and such).

The test sets immediately required here are more limited than they once were. The two main formats for applications for marketing approvals are the NDA, which can come in four different forms: 505(b)s, which are traditional approvals for NCEs; 505(b2)s for already approved drugs for generic drugs; and BLA used for a biotechnology-derived product); and abbreviated new drug approvals (ANDAs). Most significant nonclinical safety is completed before contemplation of moving to make an application.

11.27   Nontraditional (Other than NDA/505b (1)) Approval Routes and Requirements

Innovative new molecules (NME) are approved for market under the safety assessment process described in this chapter. There are in such processes, a range or variations, which have also been described.

There are, however, two other approaches by which a drug may achieve marketing approval by other forms of NDAs, and for these, the 505(b2) and the ANDA. The nonclinical safety assessment requirements are truncated. For both of these there are two initial requirements: (i) That the active ingredient (or ingredients) be already approved for market in a traditional NDA; and (ii) that all the inactive ingredients present in the drug product be covered by the FDA's inactive ingredient database (and as such, have already been included in an approved marketed drug).

For the ANDA, there is no additional nonclinical safety assessment requirement as such a product must be of the same formulation as the off patent-approved NME. It is only required that the product be bioequivalent, that is, meet guideline requirements as to similarity of pharmacokinetics. It should be noted that this route is restricted to small molecules; there is no approval route in the United States for generic biotherapeutics. Table 11.19 presents the requirements for ANDA nonclinical assessment.

The 505(b2) is, however, different. First, it is not strictly required that there is patent expiration for the innovator NME. Secondly, the route and/or formulation must be different from that of the approved NME; therefore, bioequivalence is not required or relevant. Thirdly, such a product may be a biologic (such as an alternative insulin product).

There is a modest nonclinical safety assessment requirement to support the filing of an IND and therefore before going into humans at least a single repeat dose (typically 30 days in duration, although 90 is sometimes specialized) must be performed to establish safety (Table 11.20). Such a study is typically performed in a nonrodent species, such as the dog, pig, or primate (almost always dog or pig). It is not uncommon that this study, plus adequate information

Table 11.19 Abbreviated new drug application for a generic drug.[a,b]

Test requirement	Species
Initial clinical trial/IND requirements	
(1) Acute toxicity in nonrodents (intended route)[c,d]	D/S/P
(2) Pivotal/repeat dose in nonrodents (28-day bioequivalency by intended clinical route)	D/S/P
(3) Develop bioanalytical for 2 species (man/nonrodent)	NA
To support marketing approval	
(4) In silico and human bioequivalence	

[a]Species: R, rat; M, mouse; D, dog; S, pig; P, primate; B, rabbit; TBD, to be determined.
[b]*Abbreviation*: IND, investigational new drug.
[c]Recommended.
[d]May be required.

Table 11.20   505b(2) registrations.[a,b]

Test requirement	Species
Initial clinical trial/IND requirements	
(1) Acute toxicity in rodents (intended clinical route)[c,d]	R/M
(2) Acute toxicity in nonrodents (intended clinical route)	D/S/P
(3) [c]Safety pharmacology: CV-hERG	*In vitro*
(4) Pivotal/repeat dose in nonrodents (28 days via intended clinical route)	D/S/P
(5) Develop bioanalytical for 3 species (man/rodent/nonrodent)	NA

[a]Species: R, rat; M, mouse; D, dog; S, pig; P, primate; B, rabbit; TBD, to be determined.
[b]*Abbreviation*: CV-hERG, human ether-a-go-go; IND, investigational new drug.
[c]Recommended.
[d]May be required.

obtained from the literature, is sufficient to support marketing approval as long as the formulation (which must be the same as used clinically) and impurity profile remain unchanged.

Carcinogenicity

If treatment is to exceed three months, then an evaluation of the carcinogenicity of the drug must be conducted prior to marketing applications, unless there is a waiver. Drugs that are strictly genotoxic are automatically waived and it is common to set waivers for such testing on drugs, which are proteins (and have not shown any indication of aprenoplastic lesions in repeat-dose studies performed to date) or for compounds for use in a patient population, which is not expected to have more than five years survival.

Pre- and Postnatal Development

These studies, almost always conducted in rodents, are only required (according to ICH M3 (R2))[18] as part of the marketing application package unless there is an identified cause for concern. They may be conducted earlier to meet the needs of other than US jurisdiction or if the subject drug is specifically for use in pregnant women.

Pediatric Population Studies

Such studies are not required unless the drug is specifically intended for use in pediatric populations (that is, humans under the age of 18 years), or where pediatric use labeling is desired or required: ICH changed the requirement to (i) if a consideration the accumulated adult repeat-dose toxicity studies (of appropriate length), core safety pharmacology and the standard package of genotoxicity tests indicator a potential cause for concern, then (ii) a single rodent species pediatric study should be performed. The requirement for a nonrodent pediatric safety study has been dropped.

Special Cases

There are some special studies, the need for which may be triggered by either clinical or nonclinical safety findings during the course of development. These indicate the photocarcinogenicity study and the developmental neurotoxicology studies. The need for these is almost always specified by regulatory authorities prior to the initiation of Phase III studies.

11.28 Special Therapeutic Category and Route of Administration Cases

Within the regulatory systems of the United States, European Union, and Japan, the fixed "general case" approach to nonclinical safety assessment breaks down most frequently (in predictable manner) in the cases of 1: a number of specific therapeutic claims, or 2: of (intended drug) administration by routes other than oral and intravenous.

11.29 Specific Therapeutic Classes

There are a variety of therapeutic claim classes where the nonclinical safety assessment requirements are markedly different from those of the general case as laid out in M3 (R2) and in this text to date.[18] These must each be considered on a therapeutic class basis.

11.30 Oncology Drugs

Both FDA and ICH have promulgated specific guidance for this class (which currently constitutes a third of all new drugs going into development) (ICH S9)[54] presented in the 1998 article by DeGeorge et al.[37]

Most small-molecule anticancer drugs (biologic entities will be considered separately) have some form of selective cytotoxicity as a mechanism. As a starting place, such drugs are not required to have genotoxicity testing prior to first in human testing (or indeed prior to any phase of clinical testing) is not required if clinical evaluation is to be performed in cancer patients only.

Separate safety pharmacology evaluation prior to first in human studies if the clinical evaluation is to be performed in cancer patients and these endpoints are evaluated in the repeat-dose studies.

The regimen (dosing pattern) of animals in repeat-dose studies should mirror what is to be done (or is being done if past initial clinical studies) in patients. That is, it should reflect (usually) the intermittent (*e.g.*, one or two doses a week) nature of clinical dosing. Such drugs fall into three categories: cytotoxic drugs, specific receptor targeting drugs, and protein drugs (monoclonal antibodies, for which testing requirements are presented in a separate section).

Prior to FIM studies, preliminary characterization of the mechanism(s) of action, resistance, and schedule dependencies as well as antitumor activity *in vivo* should have been made. As appropriate, these properties should be further investigated in parallel with Phase II and III studies.

These studies can provide preclinical proof of principle, guide schedules and dose-escalation schemes, provide information for selection of test species, aid in starting dose selection, and in some cases justify pharmaceutical combination where clinical information cannot be obtained.

Secondary pharmacodynamic or off-target effects should be investigated as appropriate.

Careful consideration should be given to whether these pharmaceuticals may also act as tumor promoters, enhance tumor growth, or interfere with effective therapy. An evaluation of the pharmaceuticals' effects on these parameters is thus essential. *In vitro* and *in vivo* data (xenograft, transgenic models, and others) with or without concomitant chemotherapy, may provide valuable insight into the possible adverse consequences of these products, and such information should be provided to support the initial clinical trial. Appropriate *in vitro* and *in vivo* models should be selected based on the target and mechanism of action. Without a clear understanding that the cell line characteristics are related to the pharmacology of the drug it is not required that the same tumor types/models intended for clinical evaluation be studied in these models.

The primary aims of the *in vitro* studies are to obtain mechanistic information about the test substance and characterize the activity profile.

Activity Profile and Mechanism of Action

If a specific target structure is indicated, cell lines expressing different levels of this structure should be studied, if possible. The use of well-characterized cell lines as regards genotype and biochemistry is encouraged.

Mechanism(s) of Resistance

In parallel with the characterization of the mechanism(s) of action, the corresponding profile with respect to possible mechanism(s) of resistance can be obtained. Investigation of the possible indication of resistance by long-term exposure of cell lines to a new drug and further characterization of mechanism(s) of resistance are encouraged.

The primary aims of *in vivo* studies are to obtain further information with respect to antitumor activity, therapeutic index, and schedule dependency.

Studies in animals are usually carried out in rodents, mainly in mice, giving due consideration, when possible, to likely differences to man in pharmacokinetics/dynamics. The selection of a suitable animal model (including species, strain, and tumor type) depends on the properties and proposed therapeutic indications of the anticancer drug and the available information about the response of different tumor cell lines. Suitable criteria for the evaluation of antitumor activity include tumor growth, survival time, and degree of remission or cure.

Pharmacokinetics/Toxicokinetics Including ADME

The evaluation of limited kinetic parameters (*e.g.*, peak plasma levels and area under the curve) in the animal species used for nonclinical studies may facilitate dose escalation during FIM studies. Further information on absorption, distribution, metabolism, and excretion (ADME) in animals should normally be generated in parallel with clinical development.[25]

Systemic Toxicology

The primary objective of FIM clinical trials in patients with cancer is to determine a maximum tolerated dose (MTD) and dose-limiting toxicity. General toxicology studies should be conducted to determine target organ toxicity. If possible, however, the determination of no observed adverse effect level or NOEL is not essential. Toxicology studies should be designed to support the intended clinical schedule. Evaluation of recovery, effects of accumulation and delayed toxicity should also be considered. Nonetheless, for nonrodent studies, dose groups should consist of at least an additional two animals/gender/group for recovery in control and high-dose groups. Both genders should generally be used or justification should be given for specific omissions. To support FIM clinical trial at least one nonclinical study should incorporate a recovery period at the end of the study to assess for reversibility of toxicity findings or the potential that toxicity continues to progress after cessation of drug treatment. For continued clinical development, additional toxicology studies that incorporated the principles discussed in this paragraph should be conducted.

Toxicokinetic evaluation should be conducted as appropriate. Knowledge of relevant physiological, biochemical, and kinetic differences between humans and animal models can help determine the most appropriate species to be used.

Reproductive toxicology studies are not required prior to initial clinical trials. The general expectation is that the reproductive toxicology assessment be available when the marketing application is submitted. In certain patient populations (*e.g.*, adjuvant setting), these studies should be provided prior to submitting Phase III trials.

For small-molecule drugs, embryo-fetal development toxicity studies and peri-/postnatal studies are required with the exception of traditional cytotoxic drugs.

No fertility study is needed but additional endpoints should be included in the repeat-dose toxicity study(ies). However, when a therapy is essentially curative and the study is warranted by the patient population a more complete assessment of fertility should be conducted.

Embryofetal toxicology studies are typically conducted in two species. In cases where embryofetal developmental toxicity study is unambiguously positive for teratogenesis, a confirmatory study in second species is usually not necessary.

Drugs that target rapidly dividing cells (*e.g.*, GI, bone marrow) as assessed generally in toxicology studies, and are positive in genetic toxicology assays are assumed to be developmental toxicants, and therefore developmental toxicity studies do not need to be conducted. However, for this class of compounds, little information exists as to risks to the fetus from treating males. Thus, in the absence of such information, appropriate studies should be provided (mating treated males to untreated dams).

Genotoxicity studies are not necessary to support clinical trials for therapeutics intended to treat patients with late-stage or advanced cancer. If a drug is clearly positive *in vitro* and *in vivo*, study would not be needed. Genotoxicity studies should be performed to support a marketing application.

Carcinogenicity studies are usually not necessary to support marketing for therapeutics intended to treat patients with late-stage or advanced cancer. The need for carcinogenicity assessment for anticancer pharmaceuticals is described in ICH S 1A guidance.

For anticancer pharmaceuticals the design components of the general toxicology studies are considered sufficient to evaluate immunotoxic potential and support marketing. In general, additional studies defined in the ICH S8 guidance are generally not needed for pharmaceuticals intended to treat patients with late-stage or advanced cancers. The concepts outlined in ICH S8 should be considered; however, additional studies are usually not necessary given that the general toxicology evaluation is sufficient to evaluate the immunotoxicity of anticancer agents.

Special Considerations for Biopharmaceutical Studies

Unless otherwise described the principles outlined in ICH S6, and the same considerations above apply to biopharmaceuticals used to treat cancer.

Assessment of pharmacological activity, target distribution, and binding affinity are important in selection of a relevant test species for toxicity testing for biopharmaceuticals. These data should be provided prior to initiation of clinical trials.

Mass balance and excretion studies are not needed for chemically synthesized peptide drugs or biological products used to treat cancer.

In those cases where there is sufficient public information to scientifically justify a class effect, mechanistic studies as outlined in ICH S6 may obviate the need for conducting full reproductive and developmental toxicity evaluation of biopharmaceuticals used to treat cancer.

For biotechnology-derived oncology products, genotoxicity and carcinogenicity studies are not needed.

11.31 Nonclinical Data Evaluation to Support Clinical Trial Design

Start Dose for First Administration in Human

The goal of the start dose is to administer a pharmacologically active dose that is reasonably safe to use. This safety of the dose is determined from toxicology studies in the most sensitive species. The start dose should be scientifically justified and may employ various approaches. In the case of compounds exhibiting low general toxicity it should be considered to set the starting dose in Phase I clinical trials on the basis of expected pharmacologically active dose. For anticancer small molecular weight drugs, the first-in-human maximal starting dose is usually determined from the appropriate general toxicology studies. A common approach is to set a start dose at 1/10 the severely toxic dose (STD) 10 in rodents. If the nonrodent is the most sensitive species then 1/6 the HNSTD is considered an appropriate start dose. The HNSTD is defined as the dose level below that in which observations of lethality, life-threatening toxicities, or irreversible findings were observed. This approach may continue to be followed. Doses that cause excessive lethality are not appropriate to select the safe start dose. For the most systematically administered therapeutics, this conversion should be based on the normalization of doses to body surface area. Although body surface area conversion is the standard way to approximate equivalent exposure if no further information is available, in some cases extrapolating doses based on other parameters may be more appropriate.

In certain circumstances, determined case-by-case, alternative approaches may be acceptable (*e.g.*, cytotoxic drugs). In those cases a repeat-dose toxicity study of appropriate duration in two rodent species may be sufficient.

Dose Escalation and the Highest Dose in a Clinical Trial

In general, nonclinical data do not limit the dose escalation or highest dose investigated in a clinical trial for cancer patients. When a steep dose–responsive curve is observed in nonclinical toxicology studies, or no preceding marker of toxicity is available, a slower escalation should be considered.

Duration and Schedule of Toxicology Studies to Support Initial Clinical Trials

If a more intense schedule (*e.g.*, going from weekly to three times weekly) than those used in the toxicology studies used to support the initial clinical trial is to be used clinically, an appropriate toxicology study in a single species could suffice to support this new schedule and be limited to include clinical signs and clinical chemistry at a minimum.

Duration of Toxicology Studies to Support Continued Development

In order to support continued development of a drug for patients with advanced disease, results from repeat-dose studies of up to three months duration or three to four cycles, as appropriate, should be provided prior to initiation of Phase III studies. For most small molecular weight pharmaceuticals, these studies would be sufficient to support product registration. Longer-term studies may be required in certain circumstances, on a case-by-case basis, to be provided at any phase in development. In Japan, if the indication is for a population without advanced disease, a more extensive evaluation (*e.g.*, six-month studies in two species) should be conducted. In the case of biological therapeutics, studies of six months duration in a relevant animal species are necessary prior to completion of the pivotal registration studies.

Combination of Pharmaceuticals

Pharmaceuticals planned to be used in combination should be well studied individually in separate general toxicology evaluations. Data to support a pharmacological rationale and an assessment for the potential for drug–drug interaction for the combination should be provided prior to starting the clinical study. Based on this information a determination is made whether or not a toxicity study should be conducted. In general, however, toxicology studies investigating the safety of combination of pharmaceuticals intended to treat patients with advanced cancer are not needed.[55]

Studies in Pediatric Populations

The general paradigm that exists for most pharmaceuticals that are investigated in pediatric patients is first to define an MTD in adult populations and to assess some fraction of that dose in initial pediatric studies. Studies in juvenile animals are not usually needed to support inclusion of pediatric populations for treatment of cancer. The requirements outlined elsewhere in this document also apply to this population. Conduct of studies in juvenile animals should be considered when human safety data and previous animal studies are considered insufficient for a safety evaluation in the intended pediatric age group.

Special Considerations for Biologics

The principles described in ICH S6 for dose schedule apply for oncology. However, a dose schedule of weekly five times for products with a long-half administered on an intermittent schedule is usually sufficient to provide support for Phase I clinical trials. Similar to small molecules, doses on a continuous daily basis would be expected to be dosed daily in a nonclinical study.

For nonagonist biologics the starting dose should be based on the same principles as described above for small molecules. For agonist antibodies, however, a minimally biologic active dose should be considered.[56,57]

Conjugated Agents

Conjugated agents are pharmaceuticals covalently bound to carrier molecules (such as Mabs) such as to protein, lipids, or sugars. The safety assessment of the conjugated material is the primary concern. The safety of the unconjugated material including the linker used should have a more limited evaluation. Stability of the conjugate in the test species and human plasma should be provided. A pharmacokinetic evaluation should assess both the conjugated and the unconjugated compound.

Liposomal Formulated Products

The safety assessment should include a complete evaluation of the drug product and a more limited evaluation of the unencapuslated drug and carrier. The special case of nanoparticle formulations is under development, but generally requires specific evaluation of the nanoparticle delivery system materials separately, at least for systemic toxicity. This can generally be done with separate groups added to the systemic (repeat-dose) toxicity studies, with specific attention to potential immune system effects.

Evaluation of Drug Metabolites

In some cases, metabolites have been identified in humans that have not been qualified in safety studies. For these drugs, a separate general toxicology evaluation may not be necessary for patients with late-stage or advanced cancer as the metabolite is not likely to contribute significantly to the overall toxicity profile and the human safety would have to be assessed in Phase I clinical trials. If the parent compound is considered positive in an evaluation for embryofetal and reproductive toxicity, *in vitro* and *in vivo* for genetic toxicity, or in carcinogenicity studies (if necessary), then separate studies for the disproportionate metabolite may not be needed in any cancer indication.

Evaluation of Impurities

It is recognized that impurities are not expected to have any therapeutic benefit, that impurity standards have been based on a negligible risk (*e.g.*, an increase in lifetime risk of cancer of one in 105 or 106 for genotoxic impurities),[58] and that such standards may not be appropriate for antineoplastic drugs intended to treat advanced-stage patients. The limits on impurities in other ICH guidance may be exceeded as justified on a case-by-case basis.

11.32 Routes and Regimens Other than Oral and Daily

The majority of drugs are administered by the oral route, with the second largest category being by intravenous administration. For this reason the general case is presented for an orally administrated drug, especially for one which is orally administered on a daily basis over a chronic (greater than three to six months) period.

However, neither of these situations (oral administration or regular daily administration) are universally the case. There are many other possibilities for both of these. For other routes, specific local tissue tolerance screens are required.

References

1. S. Alder and G. Zbinden, *National and International Drug Safety Guidelines*, M.T.C. Verlog, Zolheron, Switzerland, 1988.
2. J. L. LaMattina, *Drug Truths*, Wiley, Hoboken, NJ, 2008.
3. H. P. Rang, *Drug Discovery and Development*, Churchill Livingstone, Edinburgh, 2006.
4. M. Mathieu, *New Drug Development: A Regulatory Overview*, Parexel, Waltham, 8th edn, 2012.
5. S. C. Gad, *International Regulatory Safety Evaluation of Pharmaceuticals and Medical Devices*, Springer, Berlin, 2010.
6. M. N. Pangalas, L. E. Schechter and O. Hurko, Drug development for CNS disorders: strategies for balancing risk and reducing attrition, *Nat. Rev. Drug Discovery*, 2007, **6**, 521–532.
7. M. Goozmer, *The $800 Million Pill*, University of California Press, Berkeley, CA, 2005.
8. M. Angell, *The Truth About Drug Companies*, Random House, NY, 2005.
9. S. C. Gad, *Development of Therapeutic Agents Handbook*, Wiley-Interscience, Hoboken, NJ, 2011.
10. S. C. Gad, C. B. Spainhour, *Contract Research and Development Organizations: Their Role in Global Product Development*, Springer Verlog, Berlin, 2011.
11. USFDA, *CDER Nonclinical Safety Evaluation of Reformulated Drug Products and Products Intended for Administration by an Alternate Route*, 2008.

12. Japan Pharmaceutical Manufacturer Association (JPMA). Available from: www. jpma.or.jp/english/.
13. Japan (Japanese Ministry of Health, Labour and Welfare), *Guidelines for Toxicity Studies of Drugs. Notification No. 99 of the Pharmaceutical Affairs Bureau of the Ministry of Health & Welfare*, 2009.
14. E. March, Combination products: who's regulating what and how—Part I, *Pharm. Eng.*, 2008, 34–38.
15. USFDA, *CDER Guidance for Industry Safety Testing of Drug Metabolites*, 2005.
16. USFDA, *CDER Guidelines for Industry, Safety Testing of Drug Metabolites*, 2008.
17. P. Swidersky, Quality assurance and good lab practice, *Contract Pharma*, 2007. Available from: www.contractpharma.com.
18. ICH, M3 (R2) *Nonclinical Safety Studies for the Conduct of Human Clinical Trials and Marketing Authorization for Pharmaceuticals*, 2008. Available from: http://www.ich.org/products/guidelines.
19. D. Ball, J. Blanchard, D. Jacobson-Kram, *et al.*, Development of safety qualification thresholds and their use in drug product evaluation, *Toxicol. Sci.*, 2007, **97**, 226–236.
20. J. Burns, Overview of Safety Regulations Governing Food, Drug and Cosmetics, in *The United States in Safety and Evaluation and Regulation of Chemicals 3: Interface between Law and Science*, ed. F. Homberger, Karger, New York, 1983.
21. E. Goldenthal, Current View on Safety Evaluation of Drugs, FDA Papers, 1968, pp. 13–18.
22. USFDA, *FDA Introduction to Total Drug Quality*, US Government Printing Office, Washington DC, 1971.
23. S. C. Gad, *Drug Safety Evaluation*, 3rd edn, Wiley, Hoboken, NY, 2016.
24. ICH, *S2A Guidance on Specific Aspects of Regulatory Geno- toxicity Tests for Pharmaceuticals*, 1995. Available from: http://www.ich.org/products/guidelines.
25. ICH, *S3A Note for Guidance on Toxicokinetics: The Assessment of Systematic Exposure in Toxicity Studies*, 1994. Available from: http://www.ich.org/products/guidelines.
26. K. Pandher, M. W. Leach and L. A. Burns-Naas, Appropriate use of recovery groups in nonclinical toxicity studies: value in a science-driven case-by-case approach, *Vet. Pathol.*, 2012, **49**(2), 357–361.
27. F. Sewell, K. Chapman, P. Baldrick, D. Brewster, A. Broadmeadow, P. Brown, L. A. Burns-Naas, J. Clarke, A. Constan, J. Couch, O. Czupalla, A. Danks, J. DeGeorge, L. deHaan, K. Hettinger, M. Hill, M. Festag, A. Jacobs, D. Jacobson-Kram, S. Kpytek, H. Lorenz, S. G. Moesgaard, E. Moore, M. Pasanen, R. Perry, I. Ragan, S. Robinson, P. M. Schmitt, B. Short, B. S. Lima, D. Smith, S. Sparrow, Y. van Bekkum and D. Jones, Recommendations from a global cross-company data sharing initiative on the incorporation of recovery phase animals in safety assessment studies to support first-in-human clinical trials, *Regul. Toxicol. Pharmacol.*, 2014, **70**(1), 413–429.
28. ICH, *S5 (R2) Detection of Toxicity to Reproduction for Medical Products & Toxicity to Male Fertility*, 2005. Available from: http://www.ich.org/products/guidelines.
29. ICH, *S2 (R1) Guidance on Genotoxicity Testing and Data Interpretation for Pharmaceuticals Intended for Human Use*, 2014. Available from: http://www.ich.org/products/guidelines.
30. ICH, *S7A Safety Pharmacology Studies for Human Pharmaceuticals*, 2001. Available from: http://www.ich.org/products/guidelines.
31. S. C. Gad, *Safety Pharmacology*, 2nd edn, Taylor and Francis, Boca Raton, FL, 2012.
32. ICH, *Guidance for Industry, Q6B Specifications: Test Procedures and Acceptance Criteria for Bio-technological/biological Products*, 1999. Available from: http://www.ich.org/products/guidelines.

33. ICH, *S6 (R2): Addendum to Preclinical Safety Evaluation of Biotechnology Derived Pharmaceuticals, International Conference on Harmonization*, 2012. Available from: http://www.ich.org/products/guidelines.
34. *Preclinical Safety Evaluation of Biopharmaceutical*, ed. J. A.Cavagnaro, Wiley, Hoboken, NJ, 2008.
35. S. Gopalaswamy and V. Gopalaswamy, *Combination Products: Regulatory Challenges and Successful Product Development*, CRC Press, Boca Raton, FL, 2008.
36. V. R. Berliner, U.S. food and drug administration requirements for toxicity testing of contraceptive products, in *Pharmacological Models in Contraceptive Development*, ed. M. H. Briggs and E. Diczbalusy, 1974, vol. 185, Acta Endocrinol. (Copenhagen), pp. 240–253.
37. J. J. DeGeorge, Regulatory considerations for preclinical development of anticancer drugs, *Cancer Chemother. Pharmacol.*, 1988, **41**, 173–185.
38. USFDA, *Guidance for Industry: Expedited Programs for Serious Conditions – Drugs and Biologics*, 2014. Available from: http://www.fda.gov/ucm/groups/fdagov-public/@fdagov-drugs-gen/documents/document/ucm358301.pdf. Accessed 26 Feb 2016.
39. E. Schacter and P. DeSantis, Labeling of drug and biologic products for pediatric use, *Drug Inf. J.*, 1998, **32**, 299–303.
40. R. D. Hood, *Developmental and Reproductive Toxicology: A Practical Approach*, 2nd edn, Taylor & Francis, Philadelphia, 2006.
41. J. Buelke-Sam, Comparative schedules of development in rats and humans: implications for developmental neurotoxicity testing, presented at the 42nd Annual Meeting of the Society of Toxicology, Salt Lake City, UT, March 9–13, 2003.
42. ICH, *Guidance for Industry, Q3A Impurities in New Drug Substances*, 2003. Available from: http://www.ich.org/products/guidelines.
43. ICH, *Guidance for Industry, Q3B(R2) Impurities in New Drug Products*, 2006. Available from: http://www.ich.org/products/guidelines.
44. ICH, *Guidance for Industry, Q3C Impurities: Residual Solvents*, 1997. Available from: http://www.ich.org/products/guidelines.
45. D. L. Norwood, Understanding the challenges of extractables and leachables for the pharmaceutical industry– safety and regulatory environment for pharmaceuticals, *Am. Pharm. Rev.*, 2007, **10**, 32–39.
46. D. L. Norwood, D. Paskiet and M. Ruberto, *et al.*, Best practices for extractables and leachables in orally inhaled and nasal drug products: an overview of the PQRI recommendations, *Pharm. Res.*, 2007, **25**, 727–738.
47. R. E. Osterberg, Extractables and leachables in drug products, *Am. Coll. Toxicol. Newslett.*, 2005, **25**, 1020.
48. R. E. Osterberg, Potential toxicity of extractables and leachables in drug products, *Am. Pharm. Rev.*, 2005, **8**, 64–67.
49. ICH, *Guidance for Industry, Q3C-Tables and List*, 2003. Available from: http://www.ich.org/products/guidelines.
50. EMEA, *Guideline on the Specification Limits for Residues of Metal Catalysts or Metal Reagents*, 2008. Available from: http://www.ema.europa.eu/docs/en_GB/document_library/Scientific_guideline/2009/09/WC500003586.pdf.
51. USFDA, *Guidance for Industry, Metered Dose Inhaler (MDI) and Dry Powder Inhaler (DPI), Drug Products Chemistry, Manufacturing, and Controls Documentation*, Department of Health and Human Services, CDER, 1998. Available from: http://www.fda.gov/cder/guidance/index.htm.
52. USFDA, *Guidance for Industry, NDAs: Impurities in Drug Substances*. US Department of Health and Human Services, CDER, 2000. Available from: http://www.fda.gov/cder/guidance/3622fnl.htm.
53. ICH, *Guidance for industry, QiB photostability testing of new drug substances and products*, 1996. Available from: http://www.ich.org/products/guidelines.

54. ICH, *S9 Nonclinical Evaluation for Anticancer Drugs*, 2009. Available from: http://www.ich.org/products/guidelines.
55. S. A. Segal, *Device and Biologic Combination Products—Understanding the Evolving Regulation, Medical Device and Diagnostic Industry*, 1999, pp. 180–184.
56. K. L. Hastings, Nonclinical safety evaluation of biotechnology-derived pharmaceuticals, *FDA*, 2007. Available from: www.fda.gov.
57. ICH, *S6 Preclinical Safety Evaluation of Biotechnology-Derived Pharmaceuticals*, 1997. Available from: http://www.ich.org/products/guidelines.
58. EMEA, *Guidelines on the Limits of Genotoxic Impurities*, 2006. Available from: http://www.ema.europa.eu/docs/en_GB/document_library/Scientific_guideline/2009/09/WC500002903.pdf.

12 Predicting Dose and Selective Efficacy in Clinical Studies from Preclinical Experiments: Practical Pharmacodynamics

R. G. Hill

Dept of Medicine, Imperial College Medical School, Commonwealth Building, Hammersmith Campus, DuCane Rd, London, UK
Email: raymond.hill@imperial.ac.uk

12.1 Introduction

In vitro experiments allow the prediction of a precise relationship between the concentration of drug applied and the response measured. This is not the case when studies are performed *in vivo* and in a whole animal, or patient, there are many confounding factors that need to be considered. It is necessary to be able to understand the relationship between the tissue concentrations of a drug (obtained from a study of pharmacokinetics and metabolism, see Chapter 8) and how this relates to the observed pharmacology *in vivo*. This relationship is generally given the name ***pharmacodynamics***.[1] *In vivo* pharmacology studies (in addition to toxicology and pharmacokinetic/metabolism studies) are an essential bridge between the *in vitro* evaluation of a new potential drug and its clinical evaluation in human subjects. There

Pharmacology for Chemists: Drug Discovery in Context
Edited by Raymond Hill, Terry Kenakin and Tom Blackburn
© The Royal Society of Chemistry 2018
Published by the Royal Society of Chemistry, www.rsc.org

are a variety of predictive techniques that can be used to estimate the relationship between dose data obtained in animals and what might be expected in human studies.[2-4] This issue is not just one of dose selection, however, and it is also necessary to ensure that the effect being measured in animal experiments truly reflects the therapeutic target and is not attributable to a spurious off-target effect of the drug being studied. The topic of species differences in pharmacology and how experiments in animals relate to those in human subjects, is also important and needs to be taken into consideration. The examples given in this chapter are largely drawn from central nervous system (CNS) drug discovery, as this is the area most familiar to the author. It is also relevant that this is one of the areas where the problem of lack of predictive ability in preclinical experiments is most acute, with only 8% of clinical drug candidates becoming drugs.[5]

12.2 How to Ensure that Robust Data is Generated from Studies Performed *In Vivo*

Small differences in the potency of a drug for its desired target, or for nontarget sites of action, can be discriminated readily *in vitro* so that differences of 10- or a 100-fold between target and off-target activities are sufficient to correctly evaluate the action of the molecule of interest at its desired target. However, when the experiments are *in vivo* these levels of selectivity are often insufficient, especially in the case of agonist molecules for GPCRs where receptor expression numbers may vary between tissues or even show variable expression of members of the same receptor family within a single tissue or part of the brain. For example, the high density of μ opioid receptors, coupled with a low density of κ opioid receptors in the locus coeruleus of the rat brain, meant that only very selective agonists for κ opioid receptors would show κ pharmacology that was distinct from that of μ or unselective ligands in this part of the brain.[6] Imaging techniques, in particular positron emission tomography (PET) can be used to measure receptor abundance and occupancy in particular tissues *in vivo* but this is not always an accessible technique, especially at the early stages of a project when the necessary knowledge and resources to design and make a PET ligand may not be forthcoming or justifiable. If an agonist molecule can be made with 1000× or better selectivity for its target then *in vivo* experiments can be embarked upon with some confidence that a valid positive or negative answer is likely to be obtained and, for antagonist agents, those more than 100× target selective may be useable *in vivo*. Even in this latter case, if the primary activity is largely benign it is possible to give very large doses that then reveal off-target actions (*e.g.* the production of convulsions by GABA-A receptor blockade with naloxone[7]) so *in vivo* data must always be interpreted with caution. If the potential new drug is chiral and it is possible to compare the properties of active and inactive/less active enantiomers this can provide a valuable tool for differentiating between on-target and off-target activity. The pharmacology of the convulsant isoquinoline alkaloid bicuculline illustrates this point. This compound was shown to be a competitive antagonist at the GABA-A receptor and subsequently proved to be a useful tool for mapping where in the brain the amino acid GABA was acting as an inhibitory neurotransmitter.[8,9] However, bicuculline was sometimes difficult to use because of its low aqueous solubility. The solubility could be improved by making a methylhalogen salt such as the methiodide or methylchloride but when applied iontophoretically from multibarrelled micropipettes,[8] these agents were then found to cause a pronounced increase in the background firing rate of CNS neurones as well as blocking the effects of exogenous GABA.[10] This in turn raised the question of whether the increase in firing rate was due to off-target pharmacology or disinhibition by a blockade of the effect of endogenous GABA. Bicuculline has two asymmetric centres and so is a chiral molecule with four possible enantiomers. By comparing the properties of (+1S 9R) bicuculline methochloride, a potent GABA-A antagonist, with those of (−1R 9S) bicuculline methochloride, which is much less active as a GABA-A antagonist (Figure 12.1), it was possible to show that the increase in neuronal firing rate was an off-target activity shared by both these enantiomers but only (+) bicuculline methochloride acted as a GABA-A antagonist[10] (Figure 12.2). It was also possible to show that (+1S 9R) bicuculline was a potent convulsant that blocked presynaptic inhibition whereas (−) bicuculline is at least 400× less potent as a convulsant and had no effect on presynaptic inhibition.[10] Another example of the utility of enantiomeric pairs of compounds to differentiate on- and off-target activity is provided by experiments on compounds that block the action of the peptide, substance P, on NK_1 receptors. First-generation nonpeptide antagonists such as CP-96345 and RP-67580 had a variety of off-target activities including blockade of cation channels[11,12] so when the Merck/MSD Substance P project team wished to explore the range of clinical indications for a NK_1 antagonist

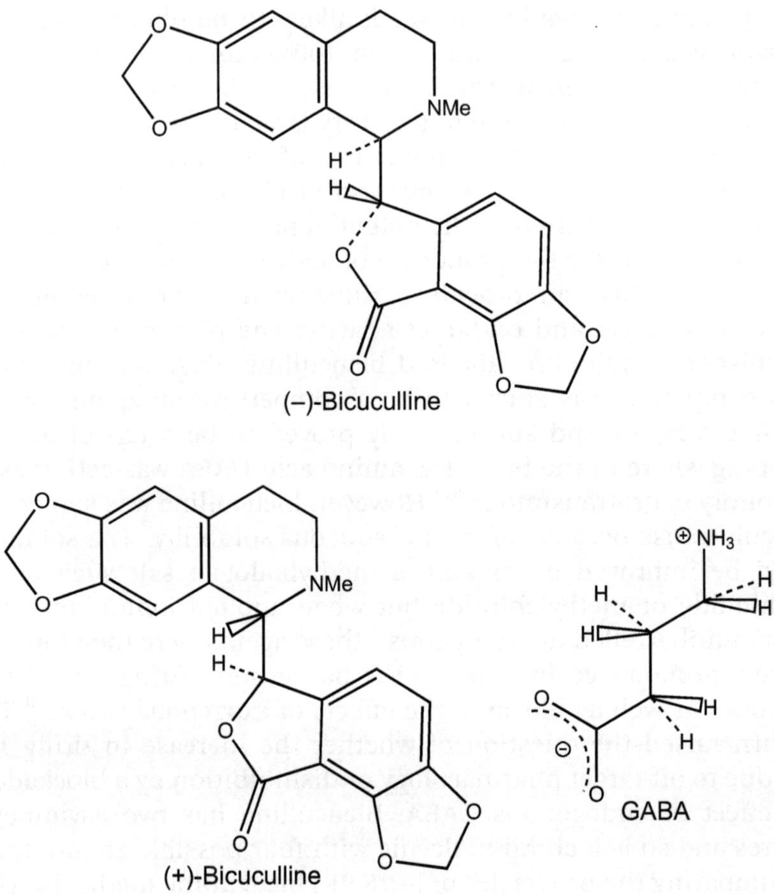

Figure 12.1 Structural formulae showing the configuration of (+) and (−) bicuculline. GABA is shown for comparison in one of its possible conformations. Reprinted by permission from Macmillan Publishers Ltd: *Nature* (ref. 10), copyright 1974.

in a range of *in vivo* animal paradigms it was decided to use the enantiomeric pair of L-733,060 (active; affinity of 0.8 nM at hNK$_1$) and L-733,061 (less active; affinity of 340 nM at hNK$_1$) (Figure 12.3) in order to clearly identify which activities were due to NK$_1$ receptor blockade.[11,12] In particular, it was subsequently shown that the antiemetic action of NK$_1$ antagonists was associated only with the active enantiomer of a number of NK$_1$ antagonists[13,14] confirming that this activity was attributable to NK$_1$ pharmacology. This confirmation of pharmacological specificity was a key step on the path to the introduction of aprepitant to clinical use as the first NK$_1$ antagonist antiemetic.[15,16]

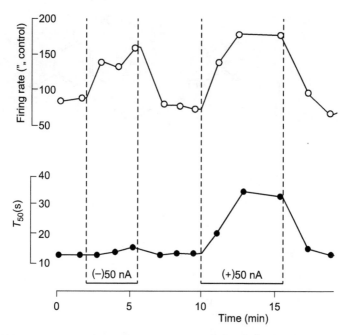

Figure 12.2 Production of an increase in firing rate of a single cortical neuron in the cerebral cortex of an anaesthetised rat (see ref. 10 for experimental details) by both (+) and (−) bicuculline methochloride and GABA antagonism by the (+) isomer. The firing of the neurone was sustained by a continuous microiontophoretic application of d,l homocysteate 5 nA. Upper panel shows firing rate immediately before the micro-iontophoretic application of GABA and the lower panel shows the effects of successive microiontophoretic applications of GABA 40 nA expressed as the time (s) taken to achieve 50% inhibition of firing. The horizontal bars show where first (−) and then secondly (+) bicuculline methochloride was applied for the period indicated with a microiontophoretic current of 50 nA. It can be seen that both isomers of bicuculline methochloride produced an increase in firing rate but only the (+) isomer produces GABA antagonism.
Reprinted by permission from Macmillan Publishers Ltd: *Nature* (ref. 10), copyright 1974.

12.3   Experiments to Target Measures of Efficacy or Surrogate Endpoints

It is not always possible to set up a paradigm where the precise mechanism for clinical efficacy can be studied in an animal model, either because the mechanism is not fully understood or because of human/animal differences. However, if a quantitative surrogate endpoint can be studied in animals then this may allow the relationship of this endpoint to plasma drug levels to be determined, such that a

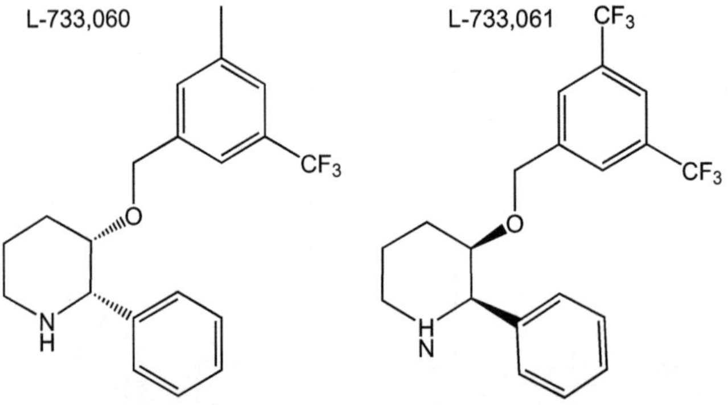

Figure 12.3 The chemical structure and configuration of 2L, 3L L-733060 and its less-active enantiomer 2R, 3R L-733061. The potency of the enantiomers at the NK_1 receptor *in vitro* is given in Table 12.1.
Reprinted from *European Journal of Pharmacology*, Volume 317, G. R. Seabrook, S. L. Shepheard, D. J. Williamson, P. Tyrer, M. Rigby, M. A. Cascieri, T. Harrison, R. J. Hargreaves and R. G. Hill, L-733,060, a novel tachykinin NK1 receptor antagonist; effects in $[Ca^{2+}]$ immobilisation, cardiovascular and dural extravasation assays, 129–135, Copyright 1996, with permission from Elsevier.

clinical dose can be predicted. Where drugs have multiple actions, this relationship is clearly going to be harder to establish but it may still be possible if distinct surrogate endpoints are available for the different actions, especially if some of the measured effects are related to chirality of the molecule being studied and some are not (see above). In the case of the triptans, the gold standard for acute treatment of migraine headache, the plasma concentration needed to produce carotid artery vascular bed constriction in the dog correlates well with the concentration needed to produce headache relief and thus this can be used as a predictive surrogate endpoint. For example, rizatriptan has an ED_{50} for dog carotid bed constriction of 54 $\mu g\,kg^{-1}$, corresponding to a plasma concentration of 16 $ng\,ml^{-1}$[17]. Clinical studies demonstrated antimigraine effects with doses of 5 to 10 mg of rizatriptan, corresponding to plasma levels of 10 to 20 $ng\,ml^{-1}$[17].

One of the most difficult tasks faced by the drug-discovery pharmacologist is to relate endpoints in the CNS to plasma levels of drug and if a measure of free drug in cerebrospinal fluid (CSF) can be obtained this will facilitate obtaining a clear relationship. In the past, agents have progressed as far as phase III clinical trials even though there was a virtual absence of free drug at the CNS target site and this mistake is both expensive and unethical! In animals larger than a mouse, it is usually possible to sample CSF as part of the protocol and alternatively microdialysis allows real-time sampling of some drugs in free CNS interstitial fluid or CSF in animal experiments. This technique is increasingly being used in humans in critical-care situations[18] and thus in due course this raises the possibility of CNS drug-level monitoring in the clinical context (even if only in a neuroscience critical-care situation). If it is not clear whether the drug you are studying acts on the CNS or not then making an analogue that is much less brain penetrant (for example by incorporating a quaternary ammonium group) can help elucidate this. This approach was found to be useful in identifying the locus of action of NK_1 antagonists as antiemetics.[19] It was found that the NK_1 antagonist CP-99,994 would block the emesis produced in ferrets by the cancer chemotherapeutic agent cisplatin and that it also blocked the emesis produced by centrally acting emetogens such as morphine and apomorphine that is resistant to the established $5-HT_3$ receptor antagonist antiemetic drugs.[13,14,19] It was therefore possible that this broader spectrum of action was because the NK_1 antagonists acted from within the CNS. In order to confirm this hypothesis, a pair of NK_1 receptor antagonists, L-741,671 and L-743,310, was selected that were of similar potency as blockers of the NK_1 receptor but differed in their ability to penetrate the blood–brain barrier (Figure 12.4). The brain penetrant agent L-741,671, given intravenously, was found to be effective in reducing the retching and vomiting induced by cisplatin in the ferret whereas the nonbrain penetrant L-743,301 was without effect[19] (Figure 12.5). In a peripheral *in vivo* assay (resinferatoxin-induced vascular leakage in the guinea pig) both of the NK_1 antagonists were equally effective and moreover, when L-743,310 was injected centrally in ferrets *via* a cannula implanted to the brainstem, it was then found to be an effective antiemetic.[19] These experiments thus showed that the antiemetic action of NK_1 antagonists was indeed mediated centrally. Ingenious animal models have also been developed to allow the differentiation of central or peripheral activation or blockade of a particular pharmacological system[20] (an imaginative investigation in the gerbil that showed peripheral activation of NK_1 receptors caused chromodacryorrhorea, whereas central NK_1 receptor activation led to repetitive foot-tapping behaviour). This assay was employed by the Merck/MSD project team to screen in a facile manner for novel NK_1 antagonists with good CNS properties *in vivo*[21,22] (see also Table 12.1) prior to more detailed assessment of their *in vivo* activity.

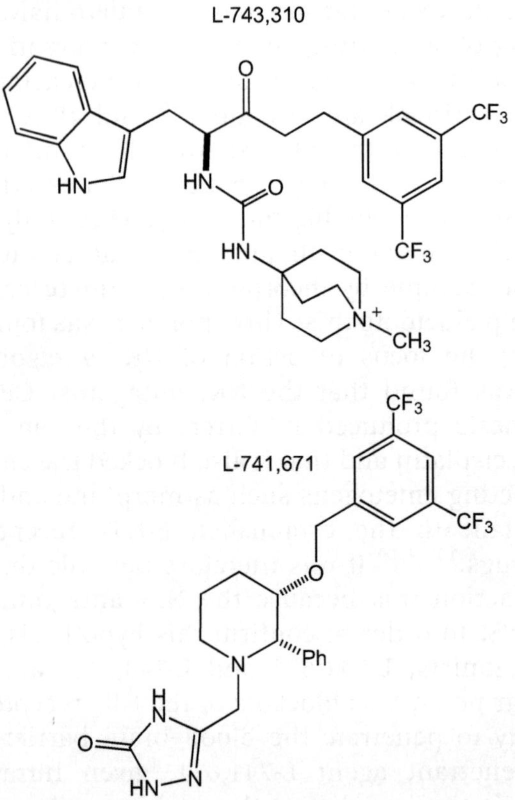

Figure 12.4   The structures of the NK$_1$ antagonists L741671 (Ki hNK$_1$ 0.03 nM; ferret NK$_1$ 0.7 nM) and L743310 (Ki hNK$_1$ 0.06 nM; ferret NK$_1$ 0.1 nM). Note the quaternary group in L743310 that in brain-perfusion experiments in the rat was found to restrict penetration to a value similar to that found with the poorly penetrant reference compound inulin, whereas L741671 was readily brain penetrant (see ref. 19).
Reprinted from *Neuropharmacology*, Volume 35, F. D. Tattersall, W. Rycroft, B. Francis, D. Pearce, K. Merchant, A. M. MacLeod, T. Ladduwahetty, L. Keown, C. Swain, R. Baker, M. Cascieri, E. Ber, J. Metzger, D. E. MacIntyre, R. G. Hill and R. J. Hargreaves, Tachykinin NK1 receptor antagonists act centrally to inhibit emesis induced by the chemotherapeutic agent cisplatin in ferrets, 1121–1129, Copyright 1996, with permission from Elsevier.

12.4   Species Selectivity and the Predictive Value of Studies in Animals

One consequence of the increased availability of cloned and expressed pharmacological targets has been the production of ligands optimised for binding to the human protein. In some cases this is not a major issue, as there is enough resemblance between the binding

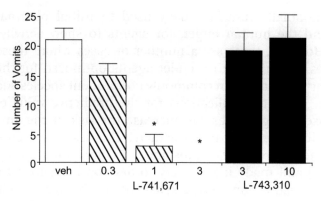

Figure 12.5 Comparison of the effects of L741671 and L743310 on cisplatin-induced vomiting in the male ferret. Full methods are described in ref. 19 but briefly, all drugs were given i.v. by way of a jugular vein cannula with the NK_1 antagonists (at the doses indicated on the histogram in $mg\,kg^{-1}$) or control vehicle being administered three minutes before a dose of $10\ mg\,kg^{-1}$ of cisplatin. Vomiting was measured as number of events for a 4-h period following the injection. The histograms indicate mean values from groups of 4 to 9 animals $+/-$ the SEM. * indicates a value that is significantly different ($p < 0.05$) from the vehicle control (one-way ANOVA followed by Tukey test). The brain-penetrant NK_1 antagonist was effective in reducing the number of vomits, whereas the brain-excluded agent was not (see text).
Reprinted from *Neuropharmacology*, Volume 35, F. D. Tattersall, W. Rycroft, B. Francis, D. Pearce, K. Merchant, A. M. MacLeod, T. Ladduwahetty, L. Keown, C. Swain, R. Baker, M. Cascieri, E. Ber, J. Metzger, D. E. MacIntyre, R. G. Hill and R. J. Hargreaves, Tachykinin NK1 receptor antagonists act centrally to inhibit emesis induced by the chemotherapeutic agent cisplatin in ferrets, 1121–1129, Copyright 1996, with permission from Elsevier.

Table 12.1 Species variants in NK_1 receptor pharmacology (Data is taken from ref. 28, 66). CNS penetration was determined by inhibition of foot-tapping behaviour induced by intracerebroventricular injection of the NK_1 receptor agonist GR73632 in gerbils (as described in ref. 20).[a]

| | IC$_{50}$ for inhibition of [^{3}H]-SP binding (nM) | | | | | CNS penetration |
	Human	Gerbil	Guinea-Pig	Rabbit	Rat	Gerbil ID$_{50}$ mg kg^{-1} i.v.
CP-96,345[b]	28.8	32.3	31.6	41.6	5888	–
MK-869	0.1	0.3	0.31	–	–	0.03
L-733,060	0.87	0.36	0.3	–	550	0.2
L-733,061	350	370	240	–	>1000	–
L-760,735	0.3	0.5	0.34	–	10	0.1
GR205171	0.08	0.06	0.09	–	1.4	0.02
LY 303870	0.15	–	0.22	–	8.7	10
RPR100893	30	–	–	–	1417	>10

[a]Profile compounds in a range of antinociception assays to compare NK_1 receptor antagonists with standard analgesics.
[b]Data taken from Beresford *et al.*, Br. J. Pharmacol., 104, 292.

site in the small animals usually used in initial pharmacological studies and the human target, for agents to show activity in both species. However, there are a number of cases where these species differences are sufficient to render agents optimised for the human target relatively inactive in commonly used rodent species such as rats and mice. This has implications for the predictive value of animal data as use of high doses will increase the risk of the animal data being attributable to off-target activity. This issue was encountered during the Merck/MSD Substance P antagonist project (see Table 12.1) and made it necessary to develop behavioural assays in species that had a human-type sensitivity to the agents being evaluated but that were not routinely used within the pharmaceutical industry. For example, analgesia assays were used in guinea pigs, rabbits and gerbils rather that the more usual rats and mice.[23] In some cases, where it was difficult or impossible to use a paradigm in a species other than the rat, it was possible to use a model compound that had high affinity for rat and human receptors rather than the intended clinical candidate molecule. For example, the compound GR205171 discovered by Glaxo Wellcome (now GSK) has nanomolar affinity for both rat and human NK_1 receptors[24] and has been successfully used as a rat research tool.[24,25]

It is relevant to ask, in a global sense, whether animal tests can always be assumed to be reliably predictive as we try to exploit potential novel drug targets emerging from knowledge of the human genome and those of other animal species. The issue is not one of the simple availability of alternative tests, as many tests are already described in the literature, but whether these tests have cross-species predictive value. In some cases, animal models that ought to be more reliable will become available as they will be based on the use of transgenic or gene knockout/knockdown animals to validate a particular gene product as a target[26] and thus will not depend on the use of a drug as a tool. In other cases, for example where there are observed differences in tissue distribution and the pharmacology of neurotransmitters and their receptors between species, these factors are likely to make animal tests less predictive *per se*. This is especially so where there are differences between the critical binding sites of G-protein coupled receptors (GPCRs) across species such that agents optimised for binding to the human receptor may not be efficacious at the receptors found in small animals commonly used in pharmacological testing and *vice versa*[27] (see reference for discussion in the context of antinociceptive studies). The development of more disease-relevant models[28] and the use of imaging techniques, which allow similar surrogate endpoints to be used in small animals and in human subjects,[29] (see later) is an encouraging step in the right direction. One complication that the experimental pharmacologist faces when scanning the literature for animal models that might be useful in the evaluation of putative drugs, is that a plethora of reasons exist for setting up such models and drug evaluation is not the most common rationale. Much work on peripheral nerve lesions, for example, has had as its driver the desire to produce animals with a phenotype resembling the symptoms of patients experiencing neuropathic pain. Such studies have been remarkably successful and it now possible to reproducibly produce rats that display regional hyperalgesia and allodynia.[30] However, in terms of response to analgesic drugs there is no *a priori* reason why the pharmacological phenotype of such an animal model should resemble its behavioural phenotype. Indeed, it has been postulated that for some purposes separate tests of pharmacological and physiological phenotype might be needed to characterise adequately the actions of a drug in a small laboratory animal.[27] It is also worth considering the hierarchy of quantitative measurements that might be made in order to record the effects of a drug on nociception. The effect of a drug on the nociceptor-evoked firing of a single dorsal horn secondary sensory neuron may be a harder endpoint than a behavioural escape response but it is a reductionist measurement that assumes (perhaps wrongly) that the spinal dorsal horn is the most important site of action for the drug in question. The behavioural measure, although less precise, is a more holistic endpoint and makes no assumptions other than the crucial one of a basic similarity between the physiology and pharmacology of the animal being studied as a model and that of the human. A useful discussion of this dilemma in the context of neuropathic pain is found in the cited references[31-33] which also describe the spectrum of currently available animal antinociception tests and their limitations.

A useful case history is provided by the failure of the clinical evaluation of substance P receptor antagonists to confirm the analgesic potential of these agents suggested by a variety of animal pharmacology tests. This has been described in detail in a number of reviews[22,34-36] and so will only be given a brief treatment here. Substance P has long been considered a prime candidate for a pain messenger, as anatomical and immunochemical studies show substance P is expressed in those small, unmyelinated sensory fibres[37] that transmit noxious information to the spinal cord. Substance P is released into the dorsal horn of the spinal cord following intense noxious stimulation of the periphery[38] and exogenous substance P, when applied onto dorsal horn neurones, produces prolonged excitation that resembled the activation observed following peripheral noxious stimulation.[39] Additionally, it has been shown that, following peripheral noxious stimulation, NK_1 receptors become internalised within dorsal horn neurones and this effect can be blocked by NK_1 receptor antagonists.[21] From this and other evidence (Table 12.2), the reasonable expectation was that centrally acting NK_1 receptor antagonists would be antinociceptive in animals, analgesic in man and constitute a novel class of analgesic drug. One of the initial problems encountered with the evaluation of substance P receptor antagonists as potential analgesics was the marked species difference in NK_1 receptor pharmacology (see above). Compounds that have been optimised for NK_1 receptors expressed in humans, typically have low affinity for rats or mice, the species most commonly used for antinociceptive studies[33] and so the need for higher doses in these species produced data that was confounded by off-target activity.[11] However, there is now good evidence from well-controlled studies in appropriate species with a substance P pharmacology similar to that in human (gerbils, guinea pigs), to demonstrate unequivocally that substance P/NK_1 receptor antagonists do possess antinociceptive effects in animals (see Figure 12.6). Using the gerbil and the well-characterised formalin test it was possible to show that the active member but not the less-active member of a chiral pair of NK_1 antagonists (see earlier) was an effective antinociceptive agent.[40] Overall, there are persuasive preclinical data supportive of an analgesic potential for substance P antagonists, particularly in pain conditions associated with inflammation, nerve injury and visceral pain conditions. Moreover, these data have come from a number of independent research groups working with a variety of different NK_1 antagonists. The analgesic profile of substance P antagonists in animals is similar to that of the nonsteroidal anti-inflammatory drugs

Table 12.2 Evidence for Substance P as a pain neurotransmitter.

- Substance P is expressed in primary sensory afferents.
- Iontophoretic application of SP excites dorsal horn neurones that are activated by noxious stimuli.
- Prolonged or intense noxious stimuli release SP in dorsal horn.
- Substance P injected intrathecally induces hyperalgesia.
- NK_1 receptors are present in dorsal horn.
- Noxious peripheral stimulation induces NK_1 receptor internalisation in dorsal horn.
- Phenotype of NK_1 $-/-$ and PPT $-/-$ partially supportive of a role in nociception.

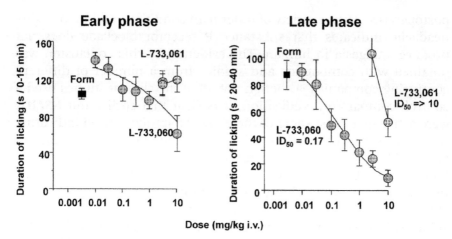

Figure 12.6 Enantioselective inhibition of the paw-licking response to intraplantar injection of formalin in gerbils, treated 3 h previously with either L733060 or L733061 (from ref. 40 with permission). Groups of between 6 and 14 animals were treated and ID_{50} values were computed using nonlinear least squares regression analysis (see ref. 41 for more details). It can be seen that L733060 (hNK_1 IC_{50} 0.87 nM) is clearly more effective in this test than L733061 (hNK_1 IC_{50} 350 nM).
Reprinted with permission from N. M. J. Rupniak, E. Carlson, S. Boyce, J. K. Webb and R. G. Hill, Enantioselective inhibition of the formalin paw late phase by the NK1 receptor antagonist L-733,060 in gerbils, *Pain*, Volume 67, Issue 1, 189–195 (http://www.sciencedirect.com/science/article/pii/0304395996031090).

Table 12.3 The phenotype of homozygous, transgenic mice in which either the gene for encoding preprotachykinin (PPT −/−) has been rendered inactive resulting in failure of synthesis of substance P and neurokinin A or the gene coding for the NK_1 receptor protein has been inactivated (NK_1 −/−) producing animals with no functional NK_1 receptors. wt indicates wild-type animals with unimpaired PPT or NK_1 functionality.

- PPT −/−
  - Two −/− lines made with slight differences in phenotype.
  - No differences from wt in baseline nociception.
  - Reduced response in formalin test.
- NK_1 −/−
  - Higher baseline than wt in reflex withdrawal tests.
  - Windup less easily detected.
  - No difference from wt in baseline nociception.
  - Reduced inflammatory response to capsaicin and carrageenan.
- Only partial attenuation of nociception seen.
- Definitive assessment of analgesic potential needs effective systematically administered receptor antagonist.

(NSAIDs).[28] In addition, the behavioural profile of transgenic animals was broadly supportive of a role for substance P signalling in nociception (Table 12.3). However, the clinical trial data in dental postoperative pain, a variety of other pain conditions and in migraine headache indicates that substance P receptor blockade does not produce analgesia in human. The evidence for this conclusion was obtained with compounds and studies from a number of different competing companies. Although not all of the clinical studies used a control, comparator, medication, it is clear that coxibs and NSAIDs were active in at least some of the clinical paradigms used indicating that these were not failed experiments but good evidence for lack of efficacy.[28] In at least two studies the negative analgesic data was obtained with substance P antagonist drugs in doses that were shown by PET studies to have produced full receptor occupancy in the brain.[41] Other peptide receptor ligands that have been evaluated in human subjects, following evaluation in animal studies, show mixed results. For example, antagonists at CGRP receptors have shown efficacy in migraine headache whereas bradykinin B_1 receptor antagonists have so far proved ineffective in pain studies.[35]

It is important to realise that although animal models of pain states can produce a simple, reproducible and consistent phenotype,[31] the typical human pain patient[42] displays a complex, variable and often progressive phenotype. Put simply, the condition we seek to model is not itself easily characterised. It has been concluded by a clinical expert in the field that a mechanism-based classification of pain is currently not feasible.[43] It is therefore necessary for us to focus on what is achievable and to accept that although clinical pain may not be easily modelled that there may be certain common features between the clinical pain experience and what one can measure in experimental animal tests. These common features may allow successful and predictive drug evaluations to be conducted. The most ideal outcome would be to identify a parameter that could be measured in a small laboratory animal, utilised in human experimental medicine studies in volunteers and, lastly, used as a surrogate endpoint in patients. It is noteworthy that the measures that best illustrate the patient's pain experience (*e.g.* inability to sleep or to walk without assistance or discomfort) are not often the measures chosen by the pain scientist for study in experimental animals. Progress is being made and a number of groups are now trying to study their animal subjects in the same manner as patients using appropriate chronic endpoints. For example,[29] a study in which a subpopulation of a group of rats subjected to sciatic nerve injury displayed social interaction and sleep/waking changes typical of those reported in patients with chronic pain. Kerins *et al.*[44] showed that inflammation of the tempero-mandibular joint in rats produced changes in meal pattern and duration that were normalised following administration of the NSAID analgesic, ibuprofen. Similarly, Nagakura *et al.*[45] were able to dissect out the time course and progression of allodynia and hyperalgesia in rats with complete Freund's adjuvant induced arthritis. There was surprisingly good correlation between the activity of analgesic drugs in these rats and their observed utility in arthritic patients. Morphine, tramadol, indomethacin and diclofenac all effectively relieved joint hyperalgesia.[46] Villaneuva[46] has pointed out that chronic pain in humans is not always associated with tissue damage or injury. This underlines the importance of using novel integrative approaches to bridge the gap between experimental and clinical pain.[47]

12.5 The Importance of Imaging Studies in Drug Discovery

Demonstration of target engagement is especially difficult in the central nervous system (CNS) and the availability of techniques that can visualise the target *in situ* and show displacement with an active ligand are of particular value. PET is perhaps the most useful imaging technique of this type at the present time.[48] The subject of imaging in drug discovery is a very large one that cannot be covered adequately in this chapter but there are a number of useful reviews in the literature.[49]

There is no doubt that broad-based, holistic approaches are needed to evaluate new drugs. However, some techniques that are accessible for animal studies are not easily used in human studies, *e.g.* neuronal ensemble recordings.[46] The advances in the use of imaging techniques, especially in fMRI, provides our best chance of developing a truly bridging technology between animal and human experimental subjects and patient populations.[50] Borsook *et al.*[51] have been able to show somatotopic activation of the human trigeminal ganglion, using fMRI imaging, following either innocuous or noxious stimuli applied to peripheral receptive fields on the head indicating that spatial resolution is no longer a limiting factor. Strigo *et al.*[52] have been able to show differences in the pattern of brain activation following visceral or somatic sensory stimulation. Distinct fMRI activation patterns were recorded, depending on whether the stimuli applied were noxious or innocuous.[52] Studies with analgesic drug regimens are starting to appear and the placebo response is being characterised.[53]

It is therefore appropriate to show a measure of optimism about the prospects for modelling pain states in animals and in man and for the likelihood of being able to obtain information reliably predicting analgesic efficacy.

12.6 Experiments to Explain Unexpected Events Seen in Clinical Studies

In early clinical evaluation of a drug (phase I), unexpected events are sometimes recorded and need to be explained. These may be dose limiting and if they are mechanism specific they will define the safe dose range over which studies of clinical efficacy can be performed. If they are not mechanism related, they may indicate that the clinical candidate should be dropped and replaced by a chemically related compound that is more selective for the target and/or has a larger therapeutic window.

An example of this type of problem is provided by work on the triptans, a class of drug that is now the mainstay for treating acute migraine headache. These agents have their primary effects by way of agonist activity on $5HT1_B$ and $5HT1_D$ receptors. It was observed during clinical evaluation of these drugs that occasional episodes of syncope (fainting) were seen.[54,55] This was especially prevalent where vasovagal output was increased by, for example, painful venepuncture or micturition. This vasovagal activity normally produces bradycardia and vasodilation leading to hypotension that is then compensated for by a sympathetic reflex response. In the absence of such a sympathetic response then syncope may result. It was therefore reasonable to ask if the triptans were capable of blocking sympathetic reflexes and whether there was any difference between the ability of individual triptans to exert this effect. In the early 1990s MSD/Merck was developing rizatriptan (MK-462) as a potential rapidly acting oral treatment for migraine and so experiments were performed to examine its propensity to cause sympatholysis in anaesthetised guinea pigs.[56] It was found that the increase in blood pressure caused by electrical stimulation of spinal sympathetic outflow was reduced in a dose-dependent manner by rizatriptan and by the comparator agents sumatriptan and zolmitriptan. It was therefore concluded that this sympatholysis was a class effect of the triptans and was mechanism related. The rare instance of syncope seen with these agents has subsequently been found not to be a major problem and not to limit the ability to use the triptans to treat migraine.

12.7 Reproducibility of Animal Experiments

This issue has received much recent attention[57,58] but will only be dealt with briefly here as it is a statistical rather than a pharmacological topic.[59,60] The failure of one laboratory to repeat the findings of another has been a stumbling block especially in CNS research. In comparison with clinical trials, many animal experiments are not well designed and are underpowered. We have the paradox that animal models are essential to the translation of data from the laboratory to the clinic, yet the predictive value of such studies is often very poor.[58] As mentioned above (in the discussion of the predictive value of pain models for the evaluation of substance P antagonists as putative analgesics) the face validity of an animal model is often undermined by a lack of understanding of the biology underlying the disease symptoms.[58] The introduction of better experimental design methodology and the use of meta-analysis to combine data from multiple studies performed in different laboratories is probably the way forward. This has been extensively reviewed in particular by Macleod and colleagues.[58,61] It is important to remember that, even if all the legitimate criticism of animal experiment design was addressed by use of sufficient numbers of animals to ensure the assays were adequately powered, there was complete standardisation of protocols and systematic review and meta-analysis of outcome measures were used, many animal models would still fail to be predictive of the human response for the reasons discussed in this chapter.[62] There is no doubt that in future more extensive use of experimental medicine studies in volunteers will help bridge the gap from animals to human. Recent authoritative reviews dealing with similar issues to those raised in this chapter are recommended for further reading.[63-65]

Acknowledgements

Thanks to my many ex-colleagues at Merck/MSD, some of whom are referenced in the bibliography, for their valued collaboration over many years.

References

1. T. Tuntland, Ethell, T. Kosaka, F. Blasco, R. Xu Zang, M. Jain, T. Gould and K. Hoffmaster, Implementation of pharmacokinetic and pharmacodynamics strategies in early research phases of drug discovery and development at Novartis Institute of Biomedical Research, *Front. Pharmacol.*, 2014, 5, 1–16.

2. CDER Guidance for industry. Estimating the maximum safe starting dose in initial clinical trials for therapeutics in adult healthy volunteers. *US DHHS FDA* (2005) July 2005 pp 27.

3. D. E. Mager, S. Woo and W. J. Jusko, Scaling pharmacodynamics from *in vitro* and preclinical animal studies to humans, *Drug Metab. Pharmacokinet.*, 2009, **24**, 16–24.

4. S. Reagan-Shaw, M. Nihal and A. Nihal, Dose translation from animal to human studies revisited, *FASEB J.*, 2014, **22**, 659–661.

5. R. G. Hill, Where will new drugs come from to treat neuropsychiatric diseases? *DANA Alliance – Report on Progress* 2012.

6. I. McFadzean, M. G. Lacey, R. G. Hill and G. Henderson, Kappa opioid activation depresses excitatory synaptic input to rat locus coeruleus neurones *in vitro*, *Neuroscience*, 1987, **20**, 231–239.

7. R. G. Hill, The status of naloxone in the identification of pain control mechanisms operated by endogenous opioids, *Neurosci. Lett.*, 1981, **21**, 217–222.

8. D. R. Curtis, A. W. Duggan, D. Felix and G. A. R. Johnston, GABA, bicuculline and central inhibition, *Nature*, 1970, **226**, 1222–1224.

9. D. R. Curtis, A. W. Duggan, D. Felix and G. A. R. Johnston, Bicuculline, an antagonist of GABA and synaptic inhibition in the spinal cord of the cat, *Brain Res.*, 1971, **32**, 69–96.

10. J. F. Collins and R. G. Hill, (+) and (−) bicuculline methochloride as stereoisomers of a GABA antagonist, *Nature*, 1974, **249**, 845–847.

11. N. M. J. Rupniak, S. Boyce, A. R. Williams, G. Cook, J. Longmore, G. R. Seabrook, M. Caesar, S. D. Iversen and R. G. Hill, Antinociceptive activity of NK1 receptor antagonists: non-specific effects of racemic RP 67580, *Br. J. Pharm.*, 1993, **110**, 1607–1613.

12. G. R. Seabrook, S. L. Shepheard, D. J. Williamson, P. Tyrer, M. Rigby, M. A. Cascieri, T. Harrrison, R. J. Hargreaves and R. G. Hill, L-733,060, a novel tachykinin NK_1 receptor antagonists - effects in $[Ca++]i$ mobilisation, cardiovascular and dural extravasation assays, *Eur. J. Pharm.*, 1996, **317**, 129–135.

13. F. D. Tattersall, W. Rycroft, R. J. Hargreaves and R. G. Hill, The neurokinin 1 receptor antagonist CP-99,994 attenuates cisplatin induced emesis in the ferret, *Eur. J. Pharm.*, 1993, **250**, 5–6.

14. F. D. Tattersall, W. Rycroft, R. G. Hill and R. J. Hargreaves, Enantioselective inhibition of apomorphine-induced emesis in the ferret by the neurokinin-1 receptor antagonist CP-99,994, *Neuropharmacology*, 1994, **33**, 259–260.

15. N. M. Navari, R. R. Reinhardt, R. J. Gralla, M. G. Kris, P. J. Hesketh, A. Khojasten, H. Kindler, T. H. Grote, K. Pendergrass, S. M. Grunberg, A. D. Carides and B. J. Gertz, Reduction of cisplatin-induced emesis by a selective neurokinin-1-receptor antagonist, *N. Engl. J. Med.*, 1999, **340**, 190–195.

16. I. N. Olver, Aprepitant in antiemetic combinations to prevent chemotherapy-induced nausea and vomiting, *J. Clin. Pract.*, 2004, **58**, 201–206.

17. R. J. Hargreaves, J. Longmore, M. Beer, S. Shepheard, D. Cumberbatch, D. Williamson, J. Stanton, Z. Razzaque, B. Sohal, L. Street, G. Seabrook and R. Hill, The pharmacology and mechanisms of action of rizatriptan Monographs, *Clin. Neurosci.*, 2000, **17**, 141–161.

18. P. J. Hutchinson, M. T. O'Connell, P. G. Al-Rawi, L. B. Maskell, R. Kett-White, A. K. Gupta, H. K. Richards, D. B. Hutchinson, P. J. Kirkpatrick and J. D. Pickard, Clinical microdialysis: a methodological study, *J. Neurosurg.*, 2000, **93**, 37–43.

19. F. D. Tattersall, W. Rycroft, B. Francis, D. Pearce, K. Merchant, A. M. MacCleod, T. Ladduwahetty, L. Keown, C. Swain, R. Baker, M. Cascieri, E. Ber, J. Metzger, D. E. McIntyre, R. G. Hill and R. J. Hargreaves, Tachykinin NK1 receptor

antagonists act centrally to inhibit emesis induced by the chemotherapeutic agent cisplatin in ferrets, *Neuropharmacology*, 1996, **35**, 1121–1129.

20. L. J. Bristow and L. Young, Chromodacryorrhoea and repetitive hind paw tapping: models of peripheral and central tachykinin NK1 receptor activation in gerbils, *Eur. J. Pharmacol.*, 1994, **254**, 245–249.

21. P. W. Mantyh, E. DeMaster, A. Malhotra, J. R. Ghilardi, S. D. Rogers, C. R. Mantyh, H. Liu, A. I. Basbaum, S. R. Vigna, J. E. Maggio and D. A. Simone, Receptor endocytosis and dendrite reshaping in spinal neurons after somato-sensory stimulation, *Science*, 1995, **268**, 1629–1632.

22. N. M. J. Rupniak and A. R. Williams, Differential inhibition of foot tapping and chromodacryorhoea in gerbils by CNS penetrant and non-penetrant tachykinin NK1 receptor antagonists, *Eur. J. Pharmacol.*, 1994, **265**, 179–183.

23. N. M. J. Rupniak, F. D. Tattersall, A. R. Williams, W. Rycroft, E. J. Carlson, M. A. Cascieri, S. Sadowski, E. Ber, J. J. Hale, S. G. Mills, M. Maccoss, E. Seward, I. Huscroft, S. Owen, C. J. Swain, R. G. Hill and R. J. Hargreaves, *In vitro* and *in vivo* predictors of the anti-emetic activity of tachykinin NK1 receptor an-tagonists, *Eur. J. Pharm.*, 1997, **326**, 201–209.

24. S. Boyce and R. G. Hill, Discrepant results from preclinical and clinical studies on the potential of substance P-receptor antagonist compounds as analgesics, *Prog. Pain Res. Manage.*, 2000, **17**, 313–324.

25. C. J. Gardner, D. R. Armour, D. T. Beattie, J. D. Gale, A. B. Hawcock, G. J. Kilpatrick, D. J. Twissell and P. Ward, GR205171, a novel antagonist with high affinity for the tachykinin NK1 receptor and potent broad spectrum anti-emetic activity *Reg, Peptides*, 1996, **65**, 45–53.

26. M. J. Cumberbatch, E. Carlson, A. Wyatt, S. Boyce, R. G. Hill and N. Rupniak, Reversal of behavioural and electrophysiological correlates of experimental peripheral neuropathy by the NK1 receptor antagonist GR205171 in rats, *Neuropharmacology*, 1998, **37**, 1535–1543.

27. A. N. Akopian, V. Souslova, S. England, K. Okuse, N. Ogata, J. Ure, A. Smith, B. J. Kerr, S. B. McMahon, S. Boyce, R. Hill, L. C. Stanfa, A. H. Dickenson and J. N. Wood, The tetrodotoxin-resistant sodium channel SNS has a specialized function in pain pathways, *Nat. Neurosci.*, 1999, **2**, 541–548.

28. S. Boyce and R. G. Hill, Substance P (NK1) receptor antagonists – analgesics or not? in *Handbook of Experimental Pharmacology*, ed. P. Holzer and Tachykinins, Springer, Berlin, 2004, vol. 164, pp. 441–457.

29. C. R. Monassi, R. Bandler and K. A. Keay, A sub-population of rats show social and sleep-waking changes typical of chronic neuropathic pain following peripheral nerve injury, *Eur. J. Neurosci.*, 2003, **17**, 1907–1920.

30. R. G. Hill, New targets for analgesic drugs, *Prog. Pain Res. Manage.*, 2003, **24**, 419–436.

31. A. H. Dickenson, E. A. Matthews and R. Suzuki, Central nervous system mechanisms of pain in peripheral neuropathy, *Prog. Pain Res. Manage.*, 2001, **21**, 85–106.

32. P. Hansson, M. Lacerenza and P. Marchettini, Aspects of clinical and experimental neuropathic pain: the clinical perspective, *Prog. Pain Res. Manage.*, 2001, **21**, 1–18.

33. A. Cowan, Animal models of pain. in *Novel Aspects of Pain Management*, ed. J. Sawynok and A. Cowan, Wiley, Liss, New York, 1999, pp. 21–47.

34. A. Tjolsen and K. Hole, Animal models of analgesia, in *The Pharmacology of Pain* ed. A. Dickenson and J.-M. Besson, Handbook of Exp Pharmacol, 1997, vol. 130, pp. 1–20.

35. R. G. Hill and K. R. Oliver, Neuropeptide and kinin antagonists, *Handb. Exp. Pharmacol.*, 2006, **177**, 175–210.

36. R. G. Hill, NK1 (substance P) receptor antagonists – why are they not analgesic in humans?, *Trends Pharmacol.*, 2000, **21**, 244–246.
37. J. I. Nagy, S. P. Hunt, L. L. Iversen and P. Emson, Biochemical and anatomical observations on the degeneration of peptide-containing primary afferent neurons after neonatal capsaicin, *Neuroscience*, 1981, **6**, 1923–1934.
38. A. W. Duggan, C. R. Morton, Z. Q. Zhao and I. A. Hendry, Noxious heating of the skin releases immunoreactive substance P in the substantia gelatinosa of the cat: a study with antibody microprobes, *Brain Res.*, 1987, **403**, 345–349.
39. J. L. Henry, Effects of substance P on functionally identified units in cat spinal cord, *Brain Res.*, 1976, **114**, 439–451.
40. N. M. J. Rupniak, E. Carlson, S. Boyce, J. K. Webb and R. G. Hill, Enantioselective inhibition of the formalin paw late phase by the NK1 receptor antagonist L-733,060 in gerbils, *Pain*, 1996, **67**, 189–195.
41. M. Keller, S. Montgomery, W. Ball, M. Morrison, D. Snavely, G. Liu, R. Hargreaves, J. Hietala, C. Lines, K. Beebe and S. Reines, Lack of efficacy of the substance P (neurokinin 1 receptor) antagonist aprepitant in the treatment of major depressive disorder, *Biol. Psychiatry*, 2005, **59**, 216–223.
42. J. Samanta, J. Kendall and A. Samanta, Polyarthalgia, *Br. Med. J.*, 2003, **326**, 859.
43. P. Hansson, Difficulties in stratifying neuropathic pain by mechanisms, *Eur. J. Pain*, 2003, **7**, 353–357.
44. C. A. Kerins, D. S. Carlson, J. E. McIntosh and L. L. Bellinger, Meal pattern changes associated with temperomandibular joint inflammation/pain in rats: analgesic effects, *Pharmacol. Biochem. Behav.*, 2003, **75**, 181–189.
45. Y. Nagakura, M. Okada, A. Kohara, T. Kiso, T. Takashi, I. Akihiko, W. Fumikazu and T. Yamaguchi, Allodynia and hyperalgesia in adjuvant-induced arthritic rats: time course of progression and efficacy of analgesics, *J. Pharm. Exp. Ther.*, 2003, **306**, 490–497.
46. L. Villaneuva, Is there a gap between preclinical and clinical studies of analgesia?, *Trends Pharmacol. Sci.*, 2000, **21**, 461–462.
47. N. Percie du Sert and A. S. C. Rice, Improving the translation of analgesic drugs to the clinic: animal models of neuropathic pain, *Br. J. Pharmacol.*, 2014, **171**, 2951–2963.
48. P. M. Matthews, Clinical imaging in drug development, in *Drug Discovery and Development*, ed. R. G. Hill and H. P. Rang, Churchill Livingston Elsevier, 2nd edn, 2013, pp. 259–274.
49. T. Krucker and B. S. Sandanaraj, Optical imaging for the new grammar of drug discovery, *Philos. Trans. R Soc., A*, 2011, **369**, 4651–4655.
50. A. K. P. Jones, B. Kulkarni and S. W. G. Derbyshire, Pain mechanisms and their disorders, *Br. Med. Bull.*, 2003, **65**, 83–93.
51. D. Borsook, A. F. M. DaSilva, A. Ploghaus and L. Becarra, Specific and somato-topic functional magnetic resonance imaging activation in the trigeminal ganglion by brush and noxious heat, *J. Neurosci.*, 2003, **23**, 7897–7903.
52. I. A. Strigo, G. H. Duncan, M. Bolvin and C. Bushnell, Differentiation of visceral and cutaneous pain in the human brain, *J. Neurophysiol.*, 2003, **89**, 3294–3303.
53. F. Benedetti, A. Pollo, L. Lopiano, M. Lanotte, S. Vighetti and R. Innocenzo, Conscious expectation and unconscious conditioning in analgesic, motor and hormonal placebo/nocebo responses, *J. Neurosci.*, 2003, **23**, 4315–4323.
54. N. L. Earl, 311C90, a new acute treatment for migraine: an overview of safety, *Cephalalgia*, 1995, **15**(Suppl. 14), Abs 217.
55. J. T. Ottervanger, J. B. Van Witsen, H. A. Valkenburg and B. H. Stricher, Post-marketing study of cardiovascular adverse reactions associated with sumatriptan, *Br. Med. J.*, 1993, **307**, 1185.
56. S. L. Shepheard, D. J. Williamson, D. A. Cook, R. G. Hill and R. J. Hargreaves, Preclinical comparative studies on the sympatholytic effects of 5-HT1D receptor

agonists *in vivo*, in *Headache Treatment: Trial Methodology and New Drugs* ed. J. Olesen and P. Tfelt-Hansen, 1997, Lippincott Raven, Philadelphia, pp. 293–298.

57. B. H. Van der Worp, D. W. Howells, E. S. Sena, M. J. Porritt, S. Rewell, V. O'Collins and M. R. Macleod, Can animal models of disease reliably inform human studies?, *PLoS Med.*, 2010, **7**, 1–7.
58. T. Denayer, T. Stohr and M. Van Roy, Animal models in translational medicine: validation and prediction, *New Horiz Transl. Med.*, 2014, **2**, 5–11.
59. M. F. W. Festing and D. G. Altman, Guidelines for the design and statistical analysis of experiments using laboratory animals, *ILAR J.*, 2002, **43**, 244–258.
60. S. Perrin, Make mouse studies work, *Nature*, 2014, **507**, 423–425.
61. A. Bespalov, T. Steckler, B. Altevogt, E. Koustova, P. Skolnick, D. Deaver, M. J. Millan, J. F. Bastlund, D. Doller, J. Witkin, P. Moser, P. O'Donnell, U. Ebert, M. A. Geyer, Prinnsen, T. Ballard and M. Macleod, Failed trials for central nervous system disorders do not necessarily invalidate preclinical models and drug targets, *Nat. Rev. Drug Discovery*, 2016, **15**, 516–518.
62. R. Greek and A. Menache, Systematic reviews of animal models: methodology versus epistemology, *Int. J. Med. Sci.*, 2013, **10**, 206–221.
63. K. Mullane, R. J. Winquist and M. Williams, Translational paradigms in pharmacology and drug discovery, *Biochem. Pharmacol.*, 2014, **87**, 189–210.
64. P. McGonigle and B. Ruggieri, Animal models of human disease: challenges in enabling translation, *Biochem. Pharmacol.*, 2014, **87**, 162–171.
65. R. A. McArthur, Aligning physiology with psychology: translational neuroscience in neuropsychiatric drug discovery, *Neurosci. Biobehav. Rev.*, 2017, **76**, 4–21.
66. I. J. Beresford, P. J. Birch, R. M. Hagan and S. J. Ireland, Investigation into species variants in tachykinin NK_1 receptors by use of the non-peptide antagonist, CP-96,345, *Br. J. Pharmacol.*, 1991, **104**, 292–293.

Subject Index

Locators in **bold** refer to tables; those in *italic* to figures. Compounds beginning with a number are filed as spelled out.